# Transport Phenomena in Biomedical Engineering

## About the Author

**Kal Renganathan Sharma, Ph.D., P.E.,** has written 9 books, 16 journal articles, and 482 conference papers. Among his books is *Bioinformatics: Sequence Alignment and Markov Models* (McGraw-Hill, 2009). He has earned three degrees in chemical engineering— a B.Tech from the Indian Institute of Technology, Chennai, and an M.S. and a Ph.D. from West Virginia University, Morgantown. Dr. Sharma has held a number of high-level positions at engineering colleges and universities. He currently is an adjunct professor in the Roy G. Perry College of Engineering at Prairie View A&M University in Prairie View, Texas.

# Transport Phenomena in Biomedical Engineering

## Artificial Organ Design and Development and Tissue Engineering

Kal Renganathan Sharma, Ph.D., P.E.
*Adjunct Professor*
*Department of Chemical Engineering*
*Prairie View A&M University*
*Prairie View, Texas*

New York   Chicago   San Francisco
Lisbon   London   Madrid   Mexico City
Milan   New Delhi   San Juan
Seoul   Singapore   Sydney   Toronto

*The McGraw·Hill Companies*

Library of Congress Cataloging-in-Publication Data

Sharma, Kal Renganathan.
    Transport phenomena in biomedical engineering : artificial organ design and
development and tissue engineering / Kal Renganathan Sharma.
        p.    cm.
    Includes bibliographical references and index.
    Summary: "Transport phenomena refer to fluid mechanics within living organisms.
This text offers in-depth coverage of the flow of body fluids, drug delivery systems, and
design of artificial organs"—Provided by publisher.
    ISBN 978-0-07-166397-7 (hardback : alk. paper)
    1. Biological transport.    2. Transport theory.    3. Biomedical engineering.
4. Artificial organs.    5. Tissue engineering.    I. Title.
    [DNLM:    1. Biological Transport—physiology.    2. Artificial Organs.
3. Body Fluids—metabolism.    4. Cell Membrane—metabolism.    5. Drug Delivery
Systems.    6. Tissue Engineering—methods.    QU 120 S531t 2010]
    R857.B52S53        2010
    610.28—dc22                                                              2010001047

1 2 3 4 5 6 7 8 9 0   DOC/DOC   1 6 5 4 3 2 1 0

ISBN 978-0-07-166397-7
MHID 0-07-166397-5

**Sponsoring Editor**
    Taisuke Soda
**Editing Supervisor**
    Stephen M. Smith
**Production Supervisor**
    Richard C. Ruzycka
**Acquisitions Coordinator**
    Michael Mulcahy

**Project Managers**
    Tripti Gupta and
    Vasundhara Sawhney,
    Glyph International
**Copy Editor**
    Lisa McCoy
**Proofreader**
    Kritika Gupta,
    Glyph International

**Indexer**
    Robert Swanson
**Art Director, Cover**
    Jeff Weeks
**Composition**
    Glyph International

Printed and bound by RR Donnelley.

McGraw-Hill books are available at special quantity discounts to use as premiums and sales
promotions, or for use in corporate training programs. To contact a representative, please
e-mail us at bulksales@mcgraw-hill.com.

This book is printed on acid-free paper.

This book is dedicated to my eldest son,
R. Hari Subrahmanyan Sharma (alias Ramkishan),
who turned eight on August 13, 2009.

# Contents

# Preface

This book is a natural outgrowth from the instruction by the author in biofluid dynamics to undergraduate bioengineering students every other semester between 2003 and 2007 at SASTRA University, Thanjavur, India, and to graduate students of nanotechnology for two years. Transport phenomena and biomedical engineering are two vast fields spanning different engineering branches and clinical medicine branches.

Transport phenomena has been the subject of study for the past 67 years by many an engineering student—both in our nation and worldwide. The unified study of heat transfer, mass transfer, and momentum transfer that developed as branches of classical physics many years ago saw the arrival of transport phenomena as a core course in the engineering curriculum. With the plethora of resources available to the student of the next millennium, the emphasis in theory is changing from engineering correlations to mechanistic modeling. Rather than refer to engineering charts in a handbook or wait for experimental data to be measured and published by others, the modern engineer wants to develop mathematical models from first principles, make fewer assumptions, and predict more phenomenological variables more reliably and with an improved understanding of the underlying mechanisms. The advent of personal computers, software for solving ordinary and partial differential equations, and software for flow visualization has shifted the onus to the engineer to make those judicious choices after careful analysis using the resources available and to develop critical thinking skills. Fundamental basis and control volume can be used to develop governing equations for a given problem. The "slice-balance" approach is used to develop mathematical models.

Although a wide range of different applications is possible using this approach, of particular interest is the application of the principles of transport phenomena to bioengineering systems. What can a engineer do in the hospital? He or she can aid the physician with theories and methods to fight and eradicate disease. The goal of eradicating disease by 2050 can be achieved by applying transport phenomena to the human anatomy. The medical significance of this subject is high. This book has been written to reflect physiological significance rather than elaborate mathematics. The equations of continuity, momentum, energy, and mass can be applied to the human anatomical systems. These equations are checked to ensure that they are in accordance with the Clausius inequality, and the solutions are presented showing the term-by-term medical significance. Iterative solutions are used when necessary. Elegant, closed-form analytical solutions to the models are developed using different methods. The range of application of the models is clearly stated. Friction factors are used where appropriate. Flow regimes are delineated, and 50 different flow types are discussed. For the first time, surface tension concepts, viscoplastic fluids, and the finite speed momentum transfer equations are discussed. Worked examples are presented to illustrate the application of the theory developed to various organs in the human anatomy. Boundary and time conditions are selected to provide better insight into the phenomena. Formulation of problems, from the real patient to pencil and paper equations, is emphasized.

Applications that are on the rise include:

- The use of flow visualization by tracer technique to identify the arterial block in the form of an angiogram
- The design of a dialysis machine to cure end-stage renal disease
- Better prediction of how oxygen is transported across the blood capillary wall and into the tissue space
- Better understanding of nitric oxide (NO) transport
- Better prediction of the rheology of blood flow in the capillaries
- Better understanding of the reversible oxygenation of blood
- The development of better drug delivery systems
- Better prediction of drug profiles in the human anatomy using single and multiple pharmacokinetic models
- Better prediction of the work done by the heart
- Applying the Bernoulli law to the flow to the heart
- The design and development of tissue and artificial organs

The student will learn to apply transport theory to complex medical phenomena. The Nobel laureate Krogh's work on oxygen-devoid regions in tissue are derived for Cartesian and cylindrical coordinates.

Literature available in journals and conference proceedings is referred to throughout this book. Patent literature is also cited to ensure that the reader obtains a balanced perspective on the theory and where it is applied. The book is self-contained, with preliminary chapters devoted to fluid mechanics and molecular diffusion. Appendices include a refresher on the Bessel differential equation and a table of Laplace transforms. The utility of this subject is expected to increase as more transport coefficient information is used to scale up into bioartificial organs. As modern patients allow physicians to perform more surgery on them, engineers will find themselves wanted in the hospital.

In order to make this book self-contained, two preliminary chapters review the prerequisite knowledge needed in fluid mechanics and diffusion. In Chap. 3, the three important developments that gave impetus to the emergence of the field of biofluid transport phenomena are discussed in detail: the discovery of osmosis and osmotic pressure, the permeability of a solvent across a membrane and Starling's law, and diffusion of solute across a membrane. Van't Hoff's law to determine osmotic pressure, Darcy's law of permeability, Starling's law for the combined effect of hydrostatic pressure and osmotic pressure, Deen's sieving coefficient, Maxwell's effective diffusion coefficient for suspensions, Kedem-Katchalsky equations, and the Staverman reflection coefficient are elaborated on. The hydraulic conductance of solvent, $L_p$, the permeability of solute, $P_m$, and the Staverman reflection coefficient $\sigma$ are three important parameters in solute transport across membranes. Eight worked examples illustrating the use of theories described are presented. The sieving coefficient and Staverman reflection coefficient are related by $S_e = 1 - \sigma$. Oxygen-depleted regions (identified by the theory of Krogh) are identified by mathematical modeling in cylindrical and Cartesian coordinates. Simultaneous metabolic reactions and diffusion lead to the zone of null transfer after a critical length.

In Chap. 4, blood rheology and transport are discussed. The composition of the blood and the Fahraeus-Lindqvist effect are discussed. The marginal zone theory is elaborated upon. An explicit relation for plasma layer thickness is derived. A list of 46 viscoplastic fluids is given in Table 4.1. The yield stress concept is an idealization and has not been measured directly. The transient velocity profile obtained by the damped wave momentum transfer and relaxation equation is obtained under different geometries. Four regimes of solutions are found. Subcritical damped oscillations in velocity are found for fluids with large relaxation times. The method of relativistic transformation of coordinates, the method of separation of variables, and the method of complex velocity are used to obtain closed-form analytical solutions.

In Chap. 5, the Hill equation is derived. The Bohr shift in the Hill plot is explained. Oxygen-depleted regions of tissue are obtained from mathematical modeling. Michaelis-Menten kinetics are modeled in the asymptotic limits of high and low concentrations. Finite speeds of diffusion are accounted for by the damped wave diffusion and relaxation equation. The Krogh tissue cylinder is modeled under transient conditions, and the kinetics obeyed in the asymptotic limit of high concentration of oxygen (zeroth-order rate) and low concentration of oxygen (first-order rate). For intermediate values, a numerical solution is needed. An infinite Fourier series solution is obtained. NO diffusion in blood and tissue is similar to that of oxygen, but is not the same. NO participates in a set of reactions in parallel. The instantaneous fractional yield of a heme complex during parallel reactions of NO is solved for and shown in Fig. 5.8.

In Chap. 6, pharmacokinetics are discussed. There can be three types of drug concentration as a function of time, as shown in Fig. 6.1: slow absorption, maxima and rapid bolus, and constant-rate delivery. Single-compartment models are developed for first-order absorption with elimination and second-order absorption with elimination. The model solution is obtained by the method of particular integral for a first-order ordinary differential equation (ODE), and given by Eq. (6.39). Single-compartment models also are developed for zeroth-order absorption with elimination, Michaelis-Menten absorption with elimination, and reactions-in-circle absorption with elimination. The conditions when subcritical damped oscillations can be expected are derived. A two-compartment model for absorption with elimination is shown in Fig. 6.17. The concentration that has diffused to the tissue region in the human anatomy is accounted for in addition to the concentration of drug in the blood plasma. The implementation of the pharmacokinetic models on personal computers is discussed.

Tissue design, as discussed in Chap. 7, evolved as a separate discipline from the field of biomaterials during a scientific conclave in 1988. Langer and Vacanti defined tissue engineering as "an interdisciplinary field that applies the principles of engineering and life sciences toward the development of biological substitutes that restore, maintain, or improve tissue function." The 3-D tissue formation is supported by a structure called a scaffold. Scaffolds need to be biodegradable. Lower critical solution temperature (LCST) and upper critical solution temperature (UCST) are also important considerations in the phase separation of polymers. However, they are covalently attached, thus preventing separation at the macroscale. Phase separation is limited to the nanoscale. Biomimetic materials are designed to mimic a natural biological material. Copolymers with block microstructure have been found to self-assemble and organize into periodic nanophases. One property of biomaterials worthy of mimicking is the capability for self-repair.

Chapter 8 is devoted to bioartificial organ design and development. One of the key technical hurdles in the successful transplantation of bioartificial organs is immunoisolation. A bioartificial pancreas can be used to treat diabetes mellitus, and is an improved therapy compared with insulin therapy. Pharmacokinetic models have been developed to describe glucose and insulin metabolism. Much of the research and development in the field of artificial kidney design has been in development of novel dialyzing membranes, autosterilizable membranes, reduction in the 200 to 300 liters of dialyzing fluid required, the development of blood-compatible polymers for the membranes, etc. A hollow-fiber artificial lung used in extracorporeal circulation to remove carbon dioxide ($CO_2$) from the blood and add oxygen to the blood is shown in Fig. 8.5.

Chapter 9 is devoted to bioheat transport. Two important applications of bioheat transport in medicine are thermal therapy and cryopreservation. Nanoscale effects in the time domain are important in a number of applications. The transient temperature profile under damped wave conduction and relaxation is derived for various geometries. Four regimes of solutions are found by the method of relativistic transformation of coordinates. The Taitel paradox is resolved by the use of a final condition in time. For systems with large relaxation times, that is, $\tau_r > (a^2/\pi^2\alpha)$, subcritical damped oscillations can be seen in the temperature. The heat generated within the human anatomy on account of the several metabolic reactions and the heat transfer to the surroundings can be described using the bioheat transfer equation. This was first introduced by Pennes. The issues with regard to body regulation of temperature are discussed. The thermophysical properties of biological properties and other materials are discussed. The bioheat transfer equation may be modified by the damped wave conduction and relaxation equation in order to account for the finite speed of heat propagation.

*Kal Renganathan Sharma, Ph.D., P.E.*

# Transport Phenomena in Biomedical Engineering

# CHAPTER 1

# Fundamentals of Fluid Mechanics

## Learning Objectives

- Review 50 flow types
- Newtonian and non-Newtonian fluids
- "Yield stress" fluids
- Thermodynamic properties of fluids
- Maxwell's relations
- Derive ideal gas law
- System, surroundings, and states
- Viscosity of fluid
- Equation of continuity
- Navier-Stokes equation, Euler equation, Bernoulli equation
- Viscometers–Efflux, rolling ball, Coutte, bubble, cone and plate, falling ball, rotating disk, and torsional

Biomedical engineering is rapidly emerging as a distinct discipline. The fundamentals and basic principles of transport phenomena need to be integrated with biofluid dynamics and quantitative physiology as well as into the biomedical/bioengineering curriculum. The design of hemodialysis devices, oxygen transport to tissues, transport in the kidneys, interstitial transport in solid tumors, drug delivery systems, pharmacokinetic analysis, layered flow of the core and plasma layers of blood, etc., will be discussed in this textbook. In order to render the work self-contained, a preliminary review of fluid mechanics and diffusion is undertaken in the first two chapters.

Sir Isaac Newton published the *Philosophia Naturalis Principia Mathematica* in 1687 [1]. His work started the larger discipline of mechanics. Engineering mechanics (statics and dynamics) is the

1

study of equilibrium and forces on bodies and the kinematic motion of bodies in constant and variable accelerations. Newton devoted his second book to fluid mechanics. Since the days of the sloping wells of the Indus Valley civilization around 2900 B.C., the water systems and aqueducts of Roman civilization, and the lead and clay pipes of the Hellenistic city of Pergamon, Turkey, fluid mechanics has been a fascinating subject of study. The first recognizable contribution came with Archimedes's buoyancy principle in Greece around 250 B.C. Pioneers in the field include L. Vinci, E. Torricelli, B. Pascal, D. Bernoulli, J. Bernoulli, L. Euler, d'Alembert, Lagrange, Laplace, Poisson, Poiseuille, C.L. Navier, G.G. Stokes, L. Prandtl, O. Reynolds, G.I. Taylor, etc.

The application of transport phenomena to human physiology began in recent years.

## 1.1    Fluids

Any form of matter that can flow is considered a *fluid*. It can be a liquid or gas. Thus, a fluid is a gas or a liquid that flows when subjected to sufficient shear stress. *Shear force* is the tangential component of a force field. Divided by the area normal to it, the force serves as the average shear stress over the area. *Shear stress* at a given point is the limiting value of shear force to an area in the limit of the area reduced to a point. In 2001 the Nobel Prize in physics went to work that identified a fourth state of matter: Bose-Einstein condensate. If it flows, it can be considered a fluid.

*Continuum hypotheses* assume that the fluid consists of homogeneous properties, such as uniform density throughout the fluid considered. This is despite the fact that at a molecular level, the mass is concentrated in a small region called the nucleus. The protons and neutrons are where the bulk of the mass lies. The electrons that orbit the protons and neutrons form the volume of the elements. Molecules of gases are separated by vacuum regions. Often, problems of flow are concerned with substances in the larger, macroscopic scale, and the molecular, or microscale, phenomena may be assumed to not make an engineering difference. Thus, it is assumed that the fluid will behave as if it were continuous in structure. Mass and momentum associated with substances within a control volume are regarded as distributed uniformly over that volume instead of being concentrated in a small fraction of it.

*Control volume* refers to a region of volume considered the basis for developing the theory of fluid flow in and out of the region.

## 1.2    56 Fluid Flow Types

Since the pioneering work of Euler, Bernoulli, Navier, and Stokes, for several centuries investigators have been accumulating knowledge in fluid mechanics. Fifty six different fluid flow types can be identified [2, 3]. These are presented in Table 1.1.

| S. No. | Flow Type | S. No. | Flow Type |
|---|---|---|---|
| 1 | Three-dimensional | 29 | Plug |
| 2 | Accelerating | 30 | Poiseuille |
| 3 | Adiabatic | 31 | Prandtl boundary layer |
| 4 | Annular | 32 | Pulsatile |
| 5 | Ballistic | 33 | Raleigh |
| 6 | Buoyant | 34 | Reacting |
| 7 | Bubbly | 35 | Slip |
| 8 | Capillary | 36 | Slug |
| 9 | Choked | 37 | Solenoidal |
| 10 | Compressible | 38 | Sonic |
| 11 | Critical | 39 | Squeeze |
| 12 | Darcy's | 40 | Steady |
| 13 | Electrolytic | 41 | Subcritical |
| 14 | Fanno | 42 | Subsonic |
| 15 | Filtration | 43 | Supercritical |
| 16 | Gravity | 44 | Supersonic |
| 17 | Hele-Shaw | 45 | Tangential |
| 18 | Hypersonic | 46 | Three-phase |
| 19 | Incompressible | 47 | Tranquil |
| 20 | Intraocular | 48 | Transient |
| 21 | Irrotational | 49 | Transition |
| 22 | Jet | 50 | Transonic |
| 23 | Knudsen | 51 | Turbulent |
| 24 | Laminar | 52 | Two-phase |
| 25 | Layered | 53 | Vacuum |
| 26 | Magnetic | 54 | Viscoelastic |
| 27 | Marangoni | 55 | Vortex |
| 28 | Osmotic | 56 | Womersley |

TABLE **1.1** 56 Different Fluid Flow Types

Osborn Reynolds [4] presented his experimental investigation of the circumstances that determine whether the motion of water shall be direct or sinuous and of the laws of resistance in parallel channels to the Royal Society 122 years ago. To this day the dimensionless group $(\rho V d/\mu)$ named after him, called the Reynolds number, is used extensively. It gives the ratio of the inertia forces and viscous forces, and is used to delineate *laminar flow* from *turbulent flow.*

A glass tube was mounted horizontally with one end in a tank and a valve on the opposite end. A smooth bell-mouth entrance was attached to the upstream end with a dye jet arranged so that a fine stream of dye could be injected at any point in front of the bell mouth.

Reynolds took the average velocity, $V$, as the characteristic velocity and the diameter of the tube as the characteristic length. For small flows, the dye stream moved in a straight line through the tube, indicating that the flow was laminar. As the flow rate was increased, Reynolds's number increased, since $d$, $\rho$, and $\mu$ were held constant and $V$ was directly proportional to the rate of flow. With increasing discharge a condition was reached at which the dye stream wavered and then suddenly broke up and was diffused throughout the tube. The nature of the flow had changed to a turbulent one with its violent interchange of momentum that had completely disrupted the orderly movement of laminar flow. By careful manipulation of the variables, Reynolds was able to obtain a value of Re = 12,000 before turbulence set in. Later investigators obtained a value of 40,000 using the same equipment as Reynolds. They let the water stand in the tank for several days before the experiments and took precautions to avoid vibrating the water or equipment. These numbers are referred to as the upper critical Reynolds number. Starting with turbulent flow in a glass tube, Reynolds found that it was always laminar when the velocity is reduced to enable Re < 2000. This is the lower critical Reynolds number. With the usual piping installation, the flow will change from laminar to turbulent in the range of Reynolds numbers from 2000 to 4000. The Reynolds number may be interpreted as the ratio of the bulk transfer of momentum to the momentum by shear stress.

Hele-Shaw [5] refers to two-dimensional laminar flow between closely spaced plates. Laminar flow is defined as flow in which the fluid moves in layers, or laminas, one layer gliding smoothly over an adjacent layer with only a molecular interchange of momentum. Turbulent flow, however, has an erratic motion of fluid particles with a vibrant transverse interchange of momentum. Reynolds number calculations have been popular with many a successful practitioner and have withstood the test of time for more than 12 decades.

In 1904, Prandtl [6] presented the concept of the boundary layer. It provides the important link between ideal fluid flow and real fluid flow. For fluids with small viscosity, the effect of internal friction in a fluid is appreciable only in a narrow region surrounding the fluid boundaries. From this premise, the flow outside the narrow region near the solid boundaries may be considered ideal flow or potential flow. Relations within the boundary layer region can be computed from the general equation for viscous fluid. The momentum equation permits developing an approximate equation for boundary layer growth and drag. When motion is started in a fluid with small viscosity, the flow is initially irrotational. The fluid at the boundaries has zero velocity relative to the boundaries. As a result, there is a steep velocity gradient from the boundary into the flow. The velocity gradient in a real fluid sets up near the boundary shear forces that reduce the flow relative to the boundary. The fluid layer that has had its velocity affected by the boundary shear is called the *boundary layer.*

The velocity in the boundary layer approaches the velocity in the main flow asymptotically. The boundary layer is very thin at the upstream end of a streamlined body at rest in an otherwise uniform flow. As this layer moves along the body, the continual action of shear stress tends to slow down additional fluid particles, causing the thickness of the boundary layer to increase with distance from the upstream point. The fluid in the layer is also subjected to a pressure gradient, determined from the potential flow, that increases the momentum of the layer if the pressure decreases downstream and decreases its momentum if the pressure increases downstream (adverse pressure gradient). The flow outside the boundary layer may also bring momentum into it. For smooth upstream boundaries, the boundary layer starts out as a laminar boundary layer in which the fluid particles move in smooth layers. As the *laminar boundary layer* increases in thickness, it becomes unstable and finally transforms into a turbulent region in which the fluid particles move in zigzag paths, although their velocity has been reduced by the action of viscosity at the boundary. Where the boundary layer has become turbulent, there is still a very thin layer next to the boundary that has laminar motion. It is called the *laminar sublayer.*

*Adiabatic flow* is that flow during which no heat is transferred to or from the fluid. *Isentropic flow* is reversible, adiabatic, and frictionless in nature. *Steady flow* is said to occur when conditions such as velocity and temperature are invariant at a certain point in time. When the conditions of flow do change with time, the flow is said to be unsteady, or *transient.* When all the points in the flow field have the same velocity, the flow is said to be in *plug* or *uniform flow. Vortex flow,* or rotational flow, is said to occur when fluid particles exhibit rotation about any axis. When the fluid within the region has no rotation, the flow is described as *irrotational flow. One-dimensional flow* neglects variations or changes in velocity, pressure, temperature, concentration, etc., transverse to the main flow direction. When there is no change in flow normal to the planes of flow along an identical path, the flow is described as *two-dimensional. Three-dimensional flow,* the generalized description of flow, is described by the $u$, $v$, and $w$ components of the velocity vector as a function of space coordinates $x, y, z$, and $t$.

A *streamline* is the imaginary continuous line drawn through the fluid so that it has the direction of the velocity vector at every point. A stream tube, or stream filament, is a tube with a small or large cross-section of any convenient shape that is entirely bounded by streamlines. A stream tube can be visualized as an imaginary pipe in the mass of flowing fluid through the walls of which no net flow is occurring. A path line is the path followed by a material element of fluid. When flow is steady, the streamline and path line coincide. In transient flow, the path line generally does not coincide with the streamline.

A dye or smoke is frequently injected into a fluid in order to trace its subsequent motion. The resulting dye or smoke trials are called streaklines. For steady fluids, streaklines, path lines, and streamlines are coincidental. In two-dimensional flows, streamlines are contours of the stream function. Streamlines in two-dimensional flows can be obtained by injecting fire-bright particles such as aluminum dust into the fluid, brilliantly lighting one plane and taking a photograph of the streaks made in a short time interval. Tracing on the picture continuous lines that have the direction of the streaks at every point portrays the streamlines for either steady or unsteady flow. Flow patterns may be detected using laser interferometers and Wollaston prism. The tracer particles are illuminated by creating laser sheets, and photographs reveal the streamlines, when a sphere settles in a fluid, for example.

*Incompressible flow* is said to occur when, during study, the density is not changed. *Compressible flow* [7] is when the density changes during flow are more than 5%. The equation of state, in addition to the equation of continuity, equation of mass, equation of momentum, and equation of energy need be considered. The Mach (Ma) number is obtained by taking the ratio of the velocity of fluid to the velocity of sound. When Ma < 1, the flow is said to be *subsonic,* and for Ma > 1, the flow is said to be *supersonic.* When Ma = 1, the flow is said to be *sonic,* or *critical. Isothermal compressible flow* is often encountered in long transport lines where there is sufficient heat transfer to maintain constant temperature. *Annular flow* is found to happen in a cylindrical annulus. *Choked flow* is said to occur at the throat of a convergent divergent nozzle when the fluid reaches the sonic condition. Regardless of how low the exit pressure is, the mass flow remains a constant. The flow properties at the throat and the entire subsonic section of the convergent divergent nozzle are *frozen.* One-dimensional flow with heat addition is called *Raleigh line flow.*

A plot of thermodynamic properties of enthalpy versus entropy is available in the form of a Mollier diagram for such flow. When frictional effects are included, it is referred to as *Fanno-line flow.*

Flow can be classified as *rapid* or *tranquil.* When flow occurs at low velocities so that a small disturbance can travel upstream, it is said to be in tranquil flow conditions. Upstream conditions is affected by downstream conditions, and the flow is controlled by the downstream conditions. The delineating dimensionless group is the Froude number, $F$ $(v/(gl)^{1/2})$ for the tranquil flow $F < 1$. When flow occurs at such high velocities that a small disturbance such as an elementary wave is swept downstream, the flow is described as shooting or rapid ($F > 1$). Small changes in downstream condition do not effect any change in upstream condition. When flow is such that the velocity is just equal to the velocity of an elementary wave, the flow is said to be critical ($F = 1$). *Subcritical* refers to tranquil flow at velocities that are less than critical, and *supercritical* corresponds to rapid flows when velocities are greater than the critical point. Time-dependent flow is a function of the history of fluid.

*Knudsen flow* is said to occur when the mean free path of the molecule is greater than the width of the channel, and the process is described by the pressure and temperature of the system. *Ballistic*, or *relaxational, flow* is said to occur when the accumulation of momentum is higher than an exponential rise; when the width of the channel is small, the velocity of the fluid exhibits *subcritical damped oscillations*. Oscillations exist in *pulsatile flow*—for example, in the inhalation and exhalation of oxygen and carbon dioxide. *Radial flow, or squeeze flow,* is said to happen when the $r$ component of the velocity becomes a salient consideration.

The *Rayleigh–Benard* instabilities arise due to natural convection, and the *Marangoni flow* is said to happen on account of thermocapillary stress. When chemical reactions take place during flow, the condition is described as *reacting flow. Capillary flow* can be said to occur with blood in arteries and veins. Subatmospheric pressure conditions lead to *vacuum flow. Tangential flow* emanates from moving circular objects. *Slip flow* is the transition between molecular and viscous flow. The slip boundary condition permits flow at the wall of the container. *Two-phase flow* refers to the flow of more than one fluid, such as gas-solid, liquid-gas, etc. At certain superficial velocities of gas in liquid during two-phase flows various regimes can be seen, such as:

1. *Bubbly flow.* Gas escapes in the form of bubbles and sometimes there exists a maximum bubble size.

2. *Slug flow.* Slugs are formed. This is when the bubble reaches the size of the apparatus and is called a slug.

*Osmotic flow* was discovered by Dutrochet in the 1800s. The flow of fluid from a region of low solute concentration to a region of higher solute concentration is referred to as osmotic flow. Flow induced by electrolytes or cathode-anode difference is referred to as electrolytic flow. In a similar fashion, *magnetic flow* is said to occur under the influence of magnetic forces. Electrorheological fluids are smart fluids that have been used recently in automatic transmissions of automobiles. They undergo an order-of-magnitude change in viscosity when the electric field is changed externally. *Viscoelastic flow* is said to happen when both elastic and viscous effects can be seen. *Buoyant flow* is said to happen when buoyant forces cause flow. Fluid flow that occurs inside the human eyeball is called intraocular flow.

## 1.3  Thermodynamic Properties of Fluids

Thermodynamics was developed in the 19th century based on the need to describe the operation of steam engines and to set forth the limits of what the steam engines can accomplish. The laws that govern the development of *power* from heat and the applications of heat engines were discussed in this new discipline. The first and second laws of thermodynamics deal with internal energy, $U$ (J/mole); heat, $Q$ (J/mole); work done, $W$ (J/mole); and entropy, $S$ (J/K/mole). These are all *macroscopic* properties. These do not reveal microscopic mechanisms. System and surroundings are defined prior to applying the laws of thermodynamics. The fundamental dimensions that would be used are as follows:

1. Length, $L$ (m)
2. Time, $t$ (s)
3. Mass, $M$ (kg or mole)
4. Temperature, $T$ (K)

The system of units (SI) is preferred in this textbook. A meter is defined as the distance traveled by light in vacuum during 1/299,792,458 of a second. A kg, kilogram, is set as the mass of platinum/iridium cylinder kept at the International Bureau of Weights and Measures at Sevres, France. Kelvin is a unit of temperature and is given as 1/273.16 of the thermodynamic temperature of the triple point of water. The amount of a substance with as many molecules as there are atoms in 0.012 kg of $C_{12}$, carbon, is one gram mole of the substance. One gram mole of any substance consists of Avogadro number of molecules (6.023 E 23 molecules/mole).

The word thermodynamics is coined from the Greek: *therme* means heat and *dynamis* means power. Heat means energy in transit, and power relates to movement. Thus, thermodynamics is a branch of physics where the effects of changes in temperature, pressure, and volume on physical systems are studied at the macroscopic scale by analyzing the collective motion of their particles through the use of statistics. The essence of thermodynamics is the study of the movement of energy and how energy instills movement. The study includes the discussion of the three laws of thermodynamics, the efficiency of engines and refrigerators, entropy, equation of state, thermodynamic potential, internal energy, and system and surroundings. Thermodynamics may be classified as classical thermodynamics and statistical thermodynamics. The term thermodynamics was coined by *James Joule* in 1858 to designate the science of relations between heat and power. The first book on thermodynamics was written in 1859 by William Rankine, originally trained as a physicist. He taught at the University of Glasgow as a civil and mechanical engineering professor.

*Otto von Guericke* designed the world's first vacuum pump in 1650. *Robert Boyle* and *Robert Hooke* built an air pump in 1656. Pressure exerted by a fluid was found to be inversely proportional to volume according to the Boyle's law. *Denis Pipin,* an associate of Boyle, built a bone digester that was used to raise high-pressure steam. The idea of a piston and cylinder emanated from Pipin, although *Tom Savery* built the first engine in 1697.

The father of thermodynamics is *Sadi Carnot.* He wrote *Reflections on the Motive Power of Fire* in 1824. This was a discourse on heat, power, and engine efficiency. The Carnot engine, Carnot cycle, and Carnot equations are named after him. Credit is given to *Rankine, Clausius, Thompson,* and *Kelvin* for the three laws of thermodynamics. Chemical engineering thermodynamics is the study of the interrelation of heat with chemical reactions or with a physical change of state within the laws of thermodynamics. Between 1873 and 1876, *J. W. Gibbs* authored a series of papers on the equilibrium of heterogeneous substances. He developed the criteria whereby a process would occur spontaneously. Graphic analyses and the study of energy, entropy, volume, temperature, and pressure were introduced. The early 20th-century chemists *G. N. Lewis, M. Randall, and E. A. Guggenheim* began to apply the mathematical methods of Gibbs to the analysis of chemical processes. Classical thermodynamics originated in the 1600s. The laws of thermodynamics were developed into the form we use today in the late 1800s. The pre-classical period is the 250 years between 1600 and 1850. Thermometry originated first, and this was followed by the hypotheses of an *adiabatic wall* and led to calorimetry.

The pre-classical period was filled with discussions that were confused and controversial. *Galileo* may be credited with the discovery of thermometry. He attempted to quantitate the subjective experiences of hot and cold. In the Hellenistic era, air was known to expand upon the application of heat. Galileo used this in his bulb and stem device—called a *thermometer*—that is still in use today, although it was once called a barothermoscope. *Torricelli,* a student of Galileo, developed the barometer. He showed that the time taken to drain an open tank using an orifice at the bottom is proportional to the square root of the height of the fluid in the tank. Liquids used in the thermometer evolved from water, to alcohol, to gas, to mercury in the modern era. Thermometry requires two reference temperatures: the freezing point and the boiling point of water at atmospheric pressure. The temperature of a mixture of two liquids at two different temperatures may be obtained by calculating a weighted average of the two. In 1760, *Joe Black* suggested a modification to the mixing rule through the use of specific heat. He pointed out that *heat*, not temperature, was conserved during the mixing process. This discussion formed the subject of *metaphysics*. Twenty years later, Count Rumford showed by experimentation that mechanical work was an infinite source of caloric heat. He called for the revival of a mechanical concept of heat.

Only a century later did Maxwell, Boltzmann, and Gibbs connect the microscale energy to the macroscale calorimetry. In 1824, S. Carnot's ideas led to the replacement of caloric theory by the first and second laws of thermodynamics. The concepts of *heat reservoirs, reversibility,* and requirement of a temperature difference to *generate work from heat* were introduced. The Carnot cycle was analogous to a waterfall in a dam. In 1847, *Helmholtz* came up with the principle of conservation of energy. Joule established the equivalence of mechanical, electrical, and chemical energy to heat. Caloric was later split into energy and entropy. Heat and work were forms of energy and were asymmetric. Entropy is conserved in a reversible process, and energy is conserved during a Carnot cycle. These developments occurred in 1850 when Clausius, Kelvin, Maxwell, Planck, Duhem, Poincare, and Gibbs presented their works.

### 1.3.1   Pressure

Pressure exerted by a fluid is the force per unit area acting on either the external surface of the object or the walls of the enclosed container. Thus:

$$P = \frac{dF}{dA} \tag{1.1}$$

where $F$ is the normal force and $A$ is the area upon which the force is exerted.

Pressure is a scalar quantity. The depth of the oceans is characterized by the hydrostatic pressure, $P = h\rho g$, where $h$ is the depth from the mean sea level, $\rho$ is the density of the fluid, and $g$ is the acceleration due to gravity. The SI units for pressure are Pascal, or $N/m^2$. The standard atmospheric pressure is an established constant, and is 1.01325 E05 $N/m^2$. Other units for pressure include atmosphere (atm), barometric (bar), manometric (mmHg), torr, and imperial units such as pounds per square inch (psi). The absolute pressure is different from gauge pressure. Gauge pressure is given by the amount in excess of atmospheric pressure. Although gauge pressure can take on negative values, especially under vacuum conditions, reports of *negative absolute pressure* are controversial. During the transpiration phase of plants and when the van der Waals interparticle forces become attractive rather than repulsive when they are close to each other, some investigators report a negative absolute pressure. This apparently comes from a negative value for the force.

## 1.3.2 Kinetic Representation of Pressure

Consider a box of gas molecules. Let each of the molecules have a velocity, $v$, with three components: $v_x$, $v_y$, and $v_z$. Let the box be a cube of side $l\,(m)$. When one of the gas molecules collides with one of the walls of the container, assuming an elastic collision, the momentum change during collision may be given by:

Rate of momentum change due to *one* collision

$$= mv_x - (-mv_x) = 2mv_x \tag{1.2}$$

where $m$ is the mass of a molecule.

Assuming a roundtrip of $2l$, the time taken between two collisions of the same molecule with the same wall, the time taken between collisions is

$$= \frac{2l}{v_x} \tag{1.3}$$

Frequency of collisions on account of one molecule:

$$= \frac{v_x}{2l} \tag{1.4}$$

Rate of change of momentum at the wall:

$$= \frac{v_x 2mv_x}{2l} \tag{1.5}$$

Rate of change of momentum at the wall on account of $N$ molecules:

$$= \frac{mv_{x1}^2}{l} + \frac{mv_{x2}^2}{l} + \frac{mv_{x3}^2}{l} + \cdots + \frac{mv_{xN}^2}{l} \tag{1.6}$$

The force exerted by $N$ molecules at the wall is equal to the rate of change of momentum from Newton's second law. The pressure exerted by the fluid from Eq. (1.1) is $F/A$ and hence:

$$P = \frac{mv_{x1}^2}{l^3} + \frac{mv_{x2}^2}{l^3} + \frac{mv_{x3}^2}{l^3} + \cdots + \frac{mv_{xN}^2}{l^3} \tag{1.7}$$

Defining the root-mean-square velocity of the molecule as:

$$N\langle v^2 \rangle = v_1^2 + v_2^2 + v_3^2 + \cdots + v_N^2 \tag{1.8}$$

and accounting for the motion of molecules in three dimensions, combining Eqs. (1.7) and (1.8) gives:

$$P = \frac{mN\langle v^2 \rangle}{3l^3} = \rho\langle v^2 \rangle \tag{1.9}$$

where the density of the fluid, $\rho$, can be seen to be $mN/l^3$. Equation (1.11) gives the kinematic representation of pressure [8].

### 1.3.3 Derivation of Ideal Gas Law

From the Boltzmann equipartition energy theorem, the temperature of the fluid can be written as:

$$\frac{mv^2}{2} = \frac{3k_B T}{2} \tag{1.10}$$

Combining Eqs. (1.9) and (1.10) and multiplying and dividing the numerator and denominator by the Avogadro number, $A_N$:

$$P = \frac{mN3k_B A_N <T>}{A_N 3l^3} = \frac{RT}{V} \tag{1.11}$$

where $V =$ molar volume, m³/mole.

Thus, $PV = RT$ for one mole of the gas can be derived. This is the ideal gas law. The assumptions in the box of molecules were elastic collision and that the gas molecule occupies negligible volume compared to the volume of the container. In Eq. (1.13) it can be seen that $A_N$ is the Avogadro number. $A_N k_B$ yields the universal gas constant $R$ (J/mole/K). Further, $mN/A_N$ gives the number of moles of gas, $N$ present in the box. Also, $A_N l^3/mN$ gives the molar volume of the gas.

### 1.3.4 Maxwell's Relations

Some important parameters of energy will be used in later discussions. Five such parameters are introduced here. These are:

1. Internal energy, $U$ (J/mole)
2. Enthalpy, $H$ (J/mole)
3. Gibbs free energy, $G$ (J/mole)
4. Helmholtz free energy, $A$, (J/mole)
5. Entropy, $S$ (J/K/mole)

These are also called state functions. Some important relationships among the state functions $U$, $H$, $G$, $A$, and $S$ are as follows:

$$H = U + PV \tag{1.12}$$

$$G = H - TS \tag{1.13}$$

$$A = U - TS \tag{1.14}$$

Therefore, $G$ may also be written as $A + PV$ or $U - TS + PV$. $A$ may also be written as $G - PV$.

The free energy, $G$, of a system is the amount of energy that can be converted to work at a constant temperature and pressure. It is named after the thermodynamicist Gibbs. Helmholtz free energy, $A$, of a system is the amount of energy that can be converted to work at a constant temperature. Enthalpy was first introduced by Clapeyron and Clausius in 1827, and represented the useful work done by a system. Entropy of a system, $S$, represents the unavailability of the system energy to do work. It is a measure of randomness of the molecules in the system, and is central to the quantitative description of the second law of thermodynamics. Internal energy, $U$, is the sum of the kinetic energy, potential energy, and vibrational energy of all the molecules in the system.

From the first law of thermodynamics, which shall be formally introduced in the next chapter, it can be seen that:

$$dQ + dW = dU \tag{1.15}$$

where $dQ$ is the heat supplied from the surroundings to the system, $dW$ is the work done on the system, and $dU$ is the internal energy change. When work is done by the system, $dW = -P\,dV$

or

$$dQ - P\,dV = dU \tag{1.16}$$

In Chapter 9, it can be seen that $dQ = T\,dS$. Hence:

$$T\,dS - P\,dV = dU \tag{1.17}$$

It may be deduced from Eq. (1.17) that:

$$\left(\frac{\partial U}{\partial S}\right)_V = T \tag{1.18}$$

$$\left(\frac{\partial U}{\partial V}\right)_S = -P \tag{1.19}$$

The reciprocity relation can be used to obtain the corresponding Maxwell relation. The reciprocity relation states that the order of differentiation does not matter. Thus:

$$\frac{\partial^2 U}{\partial S\,\partial V} = \frac{\partial^2 U}{\partial V\,\partial S} \tag{1.20}$$

Combining Eqs. (1.18) and (1.19) with Eq. (1.20):

$$\left(\frac{\partial T}{\partial V}\right)_S = -\left(\frac{\partial P}{\partial S}\right)_V \tag{1.21}$$

In a similar fashion [9], expressions can be derived from $dH$ as follows:

$$dH = d(U + PV) = dU + P\,dV + V\,dP \tag{1.22}$$

From the first law of thermodynamics, Eq. (1.15):

$$dH = dQ - P\,dV + P\,dV + V\,dP = dQ + V\,dP = T\,dS + V\,dP \qquad (1.23)$$

it may be deduced from Eq. (1.23) that:

$$\left(\frac{\partial H}{\partial S}\right)_P = T \qquad (1.24)$$

$$\left(\frac{\partial H}{\partial P}\right)_S = V \qquad (1.25)$$

The reciprocity relation can be used to obtain the corresponding Maxwell relation. The reciprocity relation states that the order of differentiation does not matter. Thus:

$$\frac{\partial^2 H}{\partial S \partial P} = \frac{\partial^2 H}{\partial P \partial S} \qquad (1.26)$$

Combining Eqs. (1.24) and (1.25) with Eq. (1.26):

$$\left(\frac{\partial T}{\partial P}\right)_S = \left(\frac{\partial V}{\partial S}\right)_P \qquad (1.27)$$

In a similar fashion, the corresponding Maxwell relation can be derived from $dG$:

$$dG = d(H - TS) = dH - T\,dS - S\,dT \qquad (1.28)$$

Combining Eq. (1.28) with the first law of thermodynamics given by Eq. (1.15):

$$dG = T\,dS + V\,dP - T\,dS - S\,dT$$

$$= V\,dP - S\,dT \qquad (1.29)$$

it may be deduced from Eq. (1.29) that:

$$\left(\frac{\partial G}{\partial P}\right)_T = V \qquad (1.30)$$

$$\left(\frac{\partial G}{\partial T}\right)_P = -S \qquad (1.31)$$

The reciprocity relation can be used to obtain the corresponding Maxwell relation. The reciprocity relation states that the order of differentiation does not matter. Thus:

$$\frac{\partial^2 G}{\partial P \partial T} = \frac{\partial^2 G}{\partial T \partial P} \qquad (1.32)$$

Combining Eqs. (1.30) and (1.31) with Eq. (1.32):

$$\left(\frac{\partial V}{\partial T}\right)_P = -\left(\frac{\partial S}{\partial P}\right)_T \tag{1.33}$$

In a similar fashion, the corresponding Maxwell relation can be derived from $dA$:

$$dA = d(U - TS) = dU - T\,dS - S\,dT \tag{1.34}$$

Combining Eq. (1.34) with the first law of thermodynamics given by Eq. (1.15):

$$dA = T\,dS - P\,dV - T\,dS - S\,dT$$
$$= -P\,dV - S\,dT \tag{1.35}$$

it may be deduced from Eq. (1.35) that:

$$\left(\frac{\partial A}{\partial V}\right)_T = -P \tag{1.36}$$

$$\left(\frac{\partial A}{\partial T}\right)_V = -S \tag{1.37}$$

The reciprocity relation can be used to obtain the corresponding Maxwell relation. The reciprocity relation states that the order of differentiation does not matter. Thus:

$$\frac{\partial^2 A}{\partial V\,\partial T} = \frac{\partial^2 A}{\partial T\,\partial V} \tag{1.38}$$

Combining Eqs. (1.36) and (1.37) with Eq. (1.38):

$$\left(\frac{\partial P}{\partial T}\right)_V = \left(\frac{\partial S}{\partial V}\right)_T \tag{1.39}$$

**Example 1.1**   Show for an ideal gas that $C_p - C_v = R$.

$$H = U + PV \tag{1.40}$$

For an ideal gas, $PV = RT$.
   Hence, Eq. (1.40) becomes:

$$H = U + RT \tag{1.41}$$

Differentiating Eq. (1.41) with respect to $T$:

$$\frac{\partial H}{\partial T} = \frac{\partial U}{\partial T} + R \tag{1.42}$$

it can be seen that:

$$\left(\frac{\partial H}{\partial T}\right)_p = C_p \quad \text{and} \quad \left(\frac{\partial U}{\partial T}\right)_v = C_v \tag{1.43}$$

Combining Eqs. (1.42) and (1.43):

$$C_p - C_v = R \text{ for an ideal gas.}$$

### 1.3.5    Work

The work associated with the action of a force from mechanics of particles and rigid bodies may be written as:

$$W = \int F\,ds \cos\theta \tag{1.44}$$

where $\theta$ is the angle made by the line of action of force and the path taken by the particle. In a piston-cylinder arrangement, when the gas in the cylinder expands when heat is supplied to it from the surroundings, the work done by the system can be written as:

$$W = -\int P \cdot A \cdot ds = -\int P\,dV \tag{1.45}$$

The minus sign normalizes the work quantity. When the gas in the cylinder expands, the work is done by the system, $dV$ is positive, the pressure decreases, and the minus sign keeps the work done positive. In the differential form:

$$dW = -P\,dV \tag{1.46}$$

As suggested by Eq. (1.45), the work done by the system consisting of gas is the area under the curve of a $PV$ diagram of the gas.

**Example 1.2**    Ice Cube Sliding Down an Inclined Plane
What happens to the internal energy of an ice cube that slides down an inclined plane with an angle $\theta$ and a length of the incline l (see Fig. 1.1)? Assume that

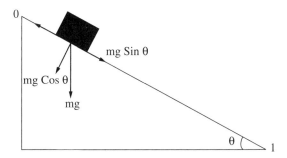

**FIGURE 1.1**    Ice cube on an inclined plane with friction.

the heat gained by the ice during the motion is proportional to the square of its velocity.

$$F = mg \sin\theta - \mu mg \cos\theta \tag{1.47}$$

$$W = \int F \, dl = \int_0^l (mg \sin\theta - \mu mg \cos\theta) \, dl \tag{1.48}$$

$$= mgl (\sin\theta - \mu \cos\theta) \tag{1.49}$$

Change in kinetic energy of the ice cube:

$$v_f^2 = 2gl(\sin\theta - \mu \cos\theta) \tag{1.50}$$

$$Q = c \, 2gl \, (\sin\theta - \mu \cos\theta) \tag{1.51}$$

From the first law of thermodynamics:

$$\Delta U = Q + W = 2cgl \, (\sin\theta - \mu \cos\theta) + mgl \, (\sin\theta - \mu \cos\theta) \tag{1.52}$$

## 1.3.6 Heat

Energy transfers from a hot body to a cold body in a spontaneous manner when they are brought in contact with each other. The degree of hotness or coldness is defined by a quantity called temperature. The units of temperature, $T$, of a system are °C, Celsius, or °F, Fahrenheit. The conversion of Fahrenheit to Celsius can be given by:

$$T(°C) = \frac{5(T(F) - 32)}{9} \tag{1.53}$$

Thermometers are used to measure temperature. They are made of liquid-in-glass constructs. A uniform tube filled with a liquid such as mercury or alcohol is allowed to expand, depending on the degree of hotness or coldness of the system under scrutiny and the length of the column measured. The length of the column is calibrated against standard reference points, such as the freezing point of water at atmospheric pressure at 0°C and the boiling point of water at atmospheric pressure 100°C. These two points are divided into 100 equal spaces called degrees.

The thermodynamic temperature scale is defined by the Kelvin scale. The conversion of °C, degree Celsius, to K, kelvin, can be given by:

$$T(K) = T(°C) + 273.15 \tag{1.54}$$

The lower limit of the Kelvin scale is 0 K or −273.15°C. The International Temperature Scale of 1990 (ITS-90) is used to calibrate thermometers. Fixed points used are the triple point of hydrogen at −259.35°C and the freezing point of silver at 961.78°C. The Rankine temperature scale can be directly related to the Kelvin scale:

$$T(R) = 1.8 \, T(K) \tag{1.55}$$

$Q$ is the amount of heat in joules that is transferred from surroundings into the system. Although the temperature difference is the driving force, the energy transfer is $Q$ in joules of energy. The heat transfer is transient in nature. The study of heat transfer is a separate subject in itself, and will be discussed in detail in later chapters. The modes of heat transfer—conduction, convection, radiation, and of late, microscale mechanisms such as wave heat conduction—shall be discussed later.

A calorie is defined as the quantity of heat when one gram of water was heated or cooled by one unit of temperature. A British thermal unit (Btu) is the quantity of heat that, when transferred, can effect a 1 degree Fahrenheit (F) change in one pound of water. The SI unit of energy is in joules. One joule equals one newton meter (Nm).

The modern notion of heat stemmed from the experiments conducted by James P. Joule in 1850 [10]. He placed known quantities of water, oil, and mercury in an insulated container and agitated the fluid with a rotating stirrer. The amount of work done on the fluid by the stirrer and the temperature changes of the fluid were accurately recorded. He observed that a fixed amount of work was required per unit mass for every degree of temperature raised on account of stirring. A quantitative relationship was established between heat and work. Thus, heat was recognized as a form of energy.

The concepts of *adiabatic wall* and *diathermal wall* are used in discussions about heat engines and heat and work interactions. Consider an object, $A$, at a temperature, $T_A$, immersed in a fluid at a different temperature, $T_B$. The temperature of object $A$ will attain the temperature of fluid $B$ after a certain time. This is the transient response of a step change in temperature at the interfaces of object $A$. Should the temperature of object $A$ remain relatively unchanged after a certain time after the step change in temperature, the wall of object $A$ separating it from fluid $B$ is said to be an *adiabatic wall*. Should the temperature of object $A$ reach the temperature of fluid $B$ instantaneously, the wall separating object $A$ from fluid $B$ is said to be a *diathermal wall*. Depending on the thermal-response characteristics of object $A$, the transient response of temperature $T_A$ to the fluid temperature $T_B$ for all other materials would lie somewhere between the adiabatic wall and the diathermal wall. The adiabatic wall and diathermal wall are idealizations that are used in thermodynamic discussions later on.

### 1.3.7    System, Surroundings, and States of a System

A *closed system* is defined as a set of components under study whose boundaries are impervious to mass flow. Surroundings are the rest of the universe other than the closed system. An open system is defined as a set of components under study whose boundaries permit mass flow across the interfaces. If the closed system is bounded by an adiabatic wall, it is said to be an isolated system. Composite systems consist of two or more systems. Restraints are barriers in a system that do not permit certain changes. In a simple system there are no adiabatic walls, impermeable walls, or external forces. The phase of a system is the state of matter it is in. The phase rule can be written as:

$$F = C - P + 2 \tag{1.56}$$

where $F$ is the degrees of freedom, $C$ is the number of components, and $P$ is the number of phases in the system.

A thermodynamic state is defined as a condition of a system characterized by properties of the system that can be reproduced. States can be at stable equilibrium or unstable or metastable equilibrium. The states can be in nonequilibrium as well. Equilibrium states are those where the macroscale changes are invariant with time. These will figure in the discussions on fugacity and vapor liquid equilibrium later on.

For closed systems with prescribed internal restraints there exist stable equilibrium states that are characterized by two independent variable properties in addition to the masses of the chemical species initially introduced.

A change of state is characterized by a change in at least one property. The path taken refers to the description of changes in the system during a change of state. When the intermediate values during a path are at equilibrium states, the path is said to be quasi-static.

All systems with prescribed internal restraints will change in a fashion so as to approach one and only one stable equilibrium state for each of the subsystems during processes with no net effect on the environment. The entire system is said to be in equilibrium.

Properties of the system may be classified as primitive or derived. Experimental measurements define the primitive property of a system. Properties that can only be defined by changes in the state are derived properties. However, these can be derived from the primitive properties.

## 1.3.8 Reversibility and Equilibrium

When two systems are nearly completely closed by adiabatic walls, except for the one through which they come in contact with each other, the states of the two systems change for some time and cease after a while. This condition is referred to as the state of thermal equilibrium. When two systems are in thermal equilibrium with a third system, they should also be in thermal equilibrium with each other. This shall be stated formally as the zeroth order of thermodynamics as Guggenheim introduced it.

The spontaneous transfer of heat, such as in the example stated previously, is generally *irreversible* in nature. To add to the weightless pulleys and frictionless planes, a reversible process is one where the changes in a series of states are at equilibrium with each other. Change in a continuous succession of equilibrium states is said to be *reversible*. It is quasi-static. In the piston-cylinder assembly discussed in the previous sections, the work done during the reversible process is more than that done during the irreversible one. When the weight in a gauge is removed suddenly, the process is irreversible. A reversible process is more gradual.

Entropy can be defined during a reversible process as follows:

$$\Delta S = Q_{rev}/T$$

$$T\,dS = dQ \tag{1.57}$$

For an irreversible process, entropy can be defined as:

$$T\,dS > dQ \tag{1.58}$$

For a reversible process:

$$T\,dS = dQ \tag{1.59}$$

## 1.4   Viscosity of Fluid

Consider a pair of large, flat, parallel plates, each with a surface area of SA separated by a distance Z. In the space between the plates (Fig. 1.2) is a fluid initially at rest. At time $t = 0$, the upper plate is set in motion at a constant velocity, $V$. As time progresses, momentum is transferred

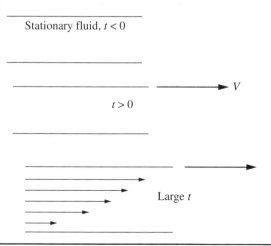

**FIGURE 1.2**   Development of a steady linear velocity profile in a viscous fluid between two plates.

from the top plate to the fluid adjacent to it and then to another layer adjacent to that layer, and so on. At a steady state, a linear velocity profile of the fluid is realized, as shown in Fig. 1.2. This is not chaotic or turbulent, but steady and laminar in character.

A force, $F$, is required to maintain the motion of the upper plate. Such flows can be described by Newton's law of viscosity:

$$\tau_{zx} = \frac{F}{SA} = -\mu \frac{\partial v}{\partial z} \qquad (1.60)$$

where $\tau_{zx}$ = shear stress
$\mu$ = viscosity of the fluid
$v$ = velocity of the fluid at any location $z$
SA = area of the flat plate
$F$ = force required to set the plate in motion

Thus, Newton's law of viscosity states that the shearing force per unit area is proportional to the negative of the velocity gradient. This was derived from empirical observations. A more generalized Newton's law of viscosity that is more applicable for some special types of fluids will be discussed later. The form given in Eq. (1.62), when attempting to derive it from simple kinetic theory of gases, arises as a first term in an expansion, and additional terms can be expected. Viscosity is a property of a fluid that measures the resistance of flow of fluids with molecular weight less than 4,000 to 6,000. Such fluids are called *Newtonian fluids*. Other systems, including polymers with high molecular weight, are classified as *non-Newtonian fluids*.

The flow of viscous fluids can be viewed as momentum-transfer phenomena. In the example considered in Fig. 1.2, momentum transfers from a flat plate to the fluid through contact with layer after layer of fluid. When posed as a problem in momentum transfer, the system becomes analogous to what is encountered in heat-transfer and mass-transfer problems. The equivalent property of the fluid from thermal diffusivity in heat conduction and mass diffusivity in molecular diffusion is *kinematic viscosity* in momentum transfer:

$$v = \frac{\mu}{\rho} \qquad (1.61)$$

Typical viscosity values of industrial systems span a wide range, from $10^{-5}$ kg/m/s for air at ambient temperature to that of glycerol at 1 kg/m/s. Some systems, such as polydimethylsiloxane or silicone oil, are more viscous. Some "smart" fluids, such as electrorheological fluids used in automatic power transmission fluid, undergo an order of magnitude increase in viscosity as the electrical charge applied is doubled. Viscosity of fluids changes with temperature.

A simple expression for viscosity of a fluid can be derived using kinetic theory of gases, as shown in the following paragraphs:

From a molecular view, the viscosity can be derived and the momentum transport mechanism can be illustrated [11]. Consider molecules to be rigid, nonattracting spheres of mass, $m$, and diameter, $d$. The gas is assumed to be at rest, and the molecular motion is considered. The following results of kinetic theory for a rigid sphere dilute gas in which small temperature, pressure, and velocity gradients are used:

$$\text{Mean molecular speed } <u> = \sqrt{\frac{8k_B T}{\pi m}} \qquad (1.62)$$

Wall collision frequency per unit area, $Z = \frac{1}{4} n' <u>$ \qquad (1.63)

$$\text{Mean free path, } \lambda = \frac{1}{\sqrt{2}(\pi d^2 n')} \qquad (1.64)$$

The molecules reaching any plane in the gas have, on average, had their last collision at a distance $a$ from the plane, where:

$$a = 2/3\,\lambda \qquad (1.65)$$

In order to determine the viscosity of a dilute monatomic gas, consider the gas when it flows parallel to the $x$ axis with a velocity gradient $\partial v_x / \partial z$. Assuming the relations for the mean free path of the molecule, wall collision frequency, distance to collision, and mean velocity of the molecule are good during the nonequilibrium conditions, the flux of momentum in the $x$ direction across any plane $z$ is found by summing the $x$ momenta of the molecules that cross in the positive $y$ direction and subtracting the $x$ momenta of those that cross in the opposite direction. Thus:

$$\tau_{zx} = Z\,mv_x\big|_{z-a} - Z\,mv_x\big|_{z+a} \qquad (1.66)$$

It may be assumed that the velocity profile is essentially linear for a distance of several mean free paths. Molecules have a velocity representative of their last collision. Accordingly:

$$v_x\big|_{z-a} = v_x\big|_z - 2/3\,\lambda\,\partial v_x / \partial z$$
$$v_x\big|_{z+a} = v_x\big|_z + 2/3\,\lambda\,\partial v_x / \partial z \qquad (1.67)$$

Substituting Eqs. (1.67) into Eq. (1.66):

$$\tau_{zx} = -1/3\,nm\,<u>\,\lambda\,dv_x / dz \qquad (1.68)$$

Equation (1.68) corresponds to Newton's law of viscosity, with the viscosity given by:

$$\mu = 1/3\,\rho\,<u>\,\lambda \qquad (1.69)$$

This expression for viscosity was obtained by Maxwell in 1860.

Some fluids whose flow does not conform to Newton's law of viscosity but do conform to the following expression are called *non-Newtonian* fluids:

$$\tau_{zx} = -\mu \left( \frac{\partial v}{\partial z} \right)^n$$

where $n$ is the power law exponent. Only when $n = 1$ does the equation revert to the Newtonian law of viscosity. When $n < 1$, the fluid is said to exhibit pseudoplastic behavior; when $n > 1$, the fluid is said to be dilatant.

## 1.5 "Yield Stress" Fluids

For nearly a century, a class of fluids has been referred to as "yield stress" fluids. The shear stress versus shear rate relationship they are expected to follow is shown in Fig. 1.3.

As can be seen in Fig. 1.3, the $y$ intercept is finite and represents a yield stress: a stress below which the fluid behaves like a solid and does not flow. This classification is attributed to Lord Bingham. Examples of such fluids are blood, tomato puree, tomato paste, fermentation broth, suspensions, slurries, etc. Most of the fluids recognized as yield stress fluids are two-component mixtures. The constitutive rheological equations used to describe blood are:

1. Casson model
2. Hershey-Buckley model
3. Bingham model

In a paper, Barnes and Walters [12] posed some questions as to the validity of the yield stress model. Their experimental findings

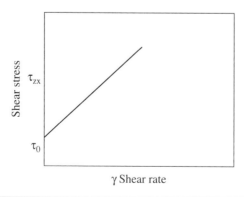

FIGURE 1.3 "Yield stress" fluids.

reveal that as more sophisticated instruments with improved capabilities are used, the yield stress measured for the same fluid becomes lower and lower in value at lower shear rates. Barnes and Walters also pointed out that "yield stress" is often an idealization and not very many investigators report experimental measurements of yield stress.

Barnes and Walters found that "yield stress" is an idealization and when given accurate measurements, no yield stress exists. They used a controlled stress rheometer for commercially available polyvinyl alcohol (PVA) latex with 0.5% aqueous carbopol solution. The shape of the curve of the shear stress/shear rate descended from the linear region in erstwhile yield stress plots to a power law region that can pass through the origin. They used the Cross model to fit the experimental data.

Hartnett and Hu [13] made some experimental measurements spanning several months in an attempt to measure the terminal settling velocity of a nylon ball in carbopol solution. More than six months' movement of a few markings, although infinite for engineering purposes, can be considered as no movement at all. So yield stress is an *engineering* reality.

Yield stress is considered a figment of investigators' extrapolation.

## 1.6    Equation of Conservation of Mass

The equation of conservation of mass for any fluid can be derived as shown in Fig. 1.4.

Consider a stationary volume element $\Delta x \Delta y \Delta z$ through which the fluid is flowing (Figure 1.4):

$$\text{(Rate of mass in)} - \text{(rate of mass out)} \pm \text{(reaction rates)}$$
$$= \text{(rate of mass accumulation)} \tag{1.70}$$

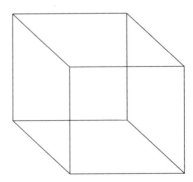

**Figure 1.4**   Region of control volume $\Delta x \Delta y \Delta z$ fixed in space through which fluid is moving.

For the case when there is no chemical reaction:

$$\Delta x \Delta z \left( \rho v_y \big|_y - \rho v_y \big|_{y + \Delta y} \right) + \Delta x \Delta y \left( \rho v_z \big|_z - \rho v_z \big|_{z + \Delta z} \right)$$

$$+ \Delta y \Delta z (\rho \, v_x \big|_x - \rho v_x \big|_{x + \Delta x}) = \Delta x \Delta y \Delta z \, \partial \rho / \partial t \qquad (1.71)$$

Dividing Eq. (1.71) by $\Delta x \Delta y \Delta z$ and taking the limits as the increments in the three directions, $\Delta x$, $\Delta y$, $\Delta z$ goes to zero.

$$\frac{\partial \rho}{\partial t} = -\left( \frac{\partial (\rho v_x)}{\partial x} + \frac{\partial (\rho v_y)}{\partial y} + \frac{\partial (\rho v_z)}{\partial z} \right) \qquad (1.72)$$

Equation (1.71) can be written in terms of the substantial derivative:

$$D\rho/Dt = -\rho(\partial v_x/\partial x + \partial v_y/\partial y + \partial v_z/\partial z) \qquad (1.73)$$

where the total derivative is given by:

$$D\rho/Dt = \partial \rho/\partial t + v_x \partial \rho/\partial x + v_y \partial \rho \, \partial y + v_z \partial \rho/\partial z) \qquad (1.74)$$

at a steady state for a fluid at constant density:

$$0 = -\rho(\partial v_x/\partial x + \partial v_y/\partial y + \partial v_z/\partial z) \qquad (1.75)$$

Equation (1.75) is the differential form of the equation of continuity. An integral form of the equation of continuity can be written as:

$$\partial/\partial t \int_{cv} \rho \, dv + \int_{cs} \rho V \, dA = 0 \qquad (1.76)$$

where $cv$ refers to the control volume and $cs$ to the control surface.

A control volume refers to a region in space, and is useful in analyzing situations where flow occurs into and out of the space. The boundary of a control volume is its control surface. The size and shape of the control volume are entirely arbitrary. They can be made to coincide with solid boundaries in parts. The control volume is also referred to as an open system.

## 1.7    Equation of Motion

The equation of motion can be derived using a momentum balance on the control volume, as shown in Fig. 1.4.

(Rate of momentum in) − (rate of momentum out)

+ (sum of forces acting on system)

= (rate of momentum accumulation) $\qquad (1.77)$

Consider the $x$ component of momentum into and out of the volume element shown in Fig. 1.4. Momentum flows into and out of the volume element by two mechanisms:

1. Convection or bulk fluid flow
2. Molecular transfer (velocity gradients)

$$\Delta y \Delta z \left(\tau_{xx}\big|_x - \tau_{xx}\big|_{x+\Delta x}\right) + \Delta x \Delta y \left(\tau_{zx}\big|_z - \tau_{zx}\big|_{z+\Delta z}\right)$$
$$+ \Delta x \Delta z \left(\tau_{yx}\big|_y - \tau_{yx}\big|_{y+\Delta y}\right) = \text{net transfer of } x \text{ component}$$
$$\text{momentum by molecular transfer}$$

$\tau_{xx}$ is the normal stress on the $x$ face and $\tau_{yx}$ is the tangential stress on the $y$ face from viscous forces. By convection:

$$\Delta y \Delta z \left(\rho\, v_x^2\big|_x - \rho\, v_x^2\big|_{x+\Delta x}\right) + \Delta x \Delta z \left(\rho\, v_y v_x\big|_y - \rho\, v_y v_x\big|_{y+\Delta y}\right)$$
$$+ \Delta x \Delta y \left(\tau\,(\rho\, v_z v_x\big|_z - \rho\, v_z v_x\big|_{z+\Delta z}\right) = \text{net transfer of}$$
$$x \text{ component momentum by convection} \qquad (1.78)$$

The sum of the external forces arises from that of hydrostatic pressure and gravity. The resultant force in $x$ direction is:

$$\Delta y \Delta z \left(p\big|_x - p\big|_{x+\Delta x}\right) + \rho\, g_x\, \Delta x \Delta y \Delta z \qquad (1.79)$$

$$(\Delta x \Delta y \Delta z)\, \partial(\rho v)/\partial t = \text{rate of accumulation of momentum} \qquad (1.80)$$

Substituting Eqs. (1.78) to (1.80) into Eq. (1.77), dividing throughout $\Delta x \Delta y \Delta z$, and obtaining the limits as $\Delta x$, $\Delta y$, $\Delta z$ going to zero, the $x$ component of the equation of motion of the fluid can be obtained:

$$\partial(\rho v_x)/\partial t = - \left[\partial(\rho v_x^2)/\partial x + \partial(\rho v_x v_y)/\partial y + \partial(\rho v_x v_z)/\partial z\right]$$
$$- \left[\partial \tau_{xx}/\partial x + \partial \tau_{yx}/\partial y + \partial \tau_{zx}/\partial z\right]$$
$$- \partial p/\partial x + \rho g_x \qquad (1.81)$$

The equation of momentum in the $x$ can be written in terms of the substantial derivative as:

$$\rho D V_x/Dt = -\nabla p - (\partial \tau_{xx}/\partial x + \partial \tau_{yx}/\partial y + \partial \tau_{zx}/\partial z) + \rho g_x \qquad (1.82)$$

where $\nabla$ is the vector differential operator. Adding the $x$ component, $y$ component, and $z$ components of momenta and using the substantial derivative, the equation of motion, including all three components, can be written as:

$$\rho D V/Dt = -\nabla p + \mu\, \nabla^2 V + \rho g \qquad (1.83)$$

## 1.8    Navier-Stokes, Euler, and Bernoulli Equations

Equation (1.83) is the Navier-Stokes equation [14]. Neglecting the viscous effects, Eq. (1.83) can be reduced to the Euler equation [15]:

$$\rho DV/Dt = -\nabla p + \rho g \qquad (1.84)$$

In one dimension at steady state, the Euler equation can be integrated to yield the Bernoulli equation [16] between two locations of the flowing fluid at 1 and 2:

$$\frac{p_1}{\rho_1} + \frac{v_1^2}{2} + gz_1 = \frac{p_2}{\rho_2} + \frac{v_2^2}{2} + gz_2 \qquad (1.85)$$

The fluid is assumed to be incompressible in the previous equations. For compressible flow, the equations of continuity, momentum, and energy can be derived again.

Between 1730 and 1760 the field of fluid dynamics blossomed. This was largely due to the work of Leonhard Euler and Daniel and Johann Bernoulli. He realized that pressure was a point property and differences in pressure cause an acceleration of fluid elements. The equations of continuity and momentum were developed around this time. The equation of energy came about later, in 1839, due to the work of de Saint Venant.

Euler's legendary fame among 18th-century mathematicians and scientists is due to his work in fluid mechanics. One of Euler's professors was Johann Bernoulli, who tutored Euler in mathematics. Johann Bernoulli, his son Daniel, and Euler were the three men who had a lot to do with the early development of the field of fluid mechanics. Daniel Bernoulli published his book *Hydrodynamica* in 1738. Flow in pipes, manometers, and jet propulsion were some of the topics covered in this work, and the Bernoulli equation is named after him. Johann published the book *Hydraulica*.

Euler succeeded Daniel Bernoulli as a professor of physics. By 1741, Euler had authored 90 papers and the two-volume book *Mechanica*. He prepared at least 380 papers for publication in Berlin. Euler had a major disagreement with Frederick the Great over some financial aspects of Berlin society of Sciences which was transformed into a major academy. First blinded by his insight into fluid dynamics, later in life Euler became physically blind. On September 18, 1783, Euler conducted business as usual, performing some calculations on the motion of balloons and discussing the discovery of the planet Uranus. He developed a brain hemorrhage, and his last words were "I am dying."

Euler was called the "great calculator" of the 18th century. He has made irreversible contributions to mathematical analysis, theory of numbers, mechanics, astronomy, and optics. He is credited with devising calculus of variations, the theory of differential equations, complex variables, and special functions. He also invented the concept of finite differences. The equations developed by Euler are used to this day in modern industrial practice.

## 1.9    Measurement of Viscosity of Fluid

Reliable, accurate, and precise values of viscosity of a fluid at a given temperature and pressure can be obtained using different viscometers. The *accuracy* of the device and *precision* of the device are two different things. A digital readout to the third decimal place may be precise, but not necessarily accurate. Accuracy is the margin of error, or *error bar* or *confidence interval,* surrounding the measured value within which the true value may lie. The error bar denotes the experimental error associated with the instrument and the personnel used to operate the equipment.

### 1.9.1    Efflux Viscometer

The traditional methods of viscosity measurement of a liquid have changed over time. Viscosity used to be measured in terms of seconds needed for the liquid to exit a tube. These viscometers are called *efflux viscometers,* or short tube viscometers. The time needed for a given volume of fluid to discharge under the forces of gravity through a short-tube orifice at the base of the instrument is measured. The viscometers are called Redwood in England, Engler in Germany, and Saybolt in United States. Viscosity is recorded as Redwood or Saybolt seconds. Based on calibration, the Saybolt universal, for example, gives the Stokes viscosity as $0.0226t - 1.95/t$ for fluids with an efflux time between 32 and 100 seconds. It can be seen that for higher-viscosity fluids, the relationships in the calibration are different. Furthermore, viscosity changes with temperature and pressure. The changing relationships of viscosity with temperature and pressure have also been studied by some investigators. For reliable measurements during the test, it is advisable to not let the conditions of pressure and temperature change appreciably.

### 1.9.2    Falling Ball Viscometer

The *falling ball viscometer* is based on Stokes' settling of falling spheres in a fluid attaining its terminal settling velocity. The terminal settling velocity of the falling sphere in a fluid is reached when the forces of gravity are balanced by the forces of buoyancy and drag. Once the terminal settling velocity of a sphere is measured using video photography and the density of the solid and fluid, as well as the diameter of the solid are known, Stokes' law can be used to calculate the viscosity of the fluid. For instance, steel ball bearings are dropped in glycerin to check the viscosity of industrial fluids. Nylon balls were allowed to fall through carbopol solution to measure the type of fluid, whether Newtonian or otherwise. The glass container carrying the sphere must be wide enough so that the wall effects can be neglected. The expression used to calculate the viscosity is the terminal settling velocity of the sphere, and can be written as:

$$V_s = \frac{2r^2 g(\rho_p - \rho_f)}{9\mu}$$

(1.86)

Aspherical particles can also be used to settle. The sphericity of the particle can be used in the calculations. Renganathan, Clark, and Turton [17] developed charts for distance traveled by accelerating spheres in a fluid prior to attaining terminal settling velocity. The change of drag coefficient with Reynolds number is also taken into account. Numerical solutions to the equation of motion were obtained and the results presented in easily usable charts.

### 1.9.3    Cone-and-Plate Viscometer

The liquid whose viscosity needs to be measured is placed between a stationary flat plate and an inverted cone whose apex just contacts the plate (Fig. 1.5). The cone is rotated at a known angular velocity, $\Omega$, in a *cone-and-plate viscometer,* and the torque, $T_y$, required to turn the cone is measured. An expression for viscosity of the liquid in terms of the angular velocity of the rotation of the cone, torque needed, and the angle made by the cone with the plate. This is usually about 1 degree. The expression for the torque required to turn the cone can be shown as:

$$T_y = \frac{2}{3\theta}\pi\mu\Omega R^3$$

(1.87)

where $\theta$ = angle made by the cone with the flat plate
   $R$ = radius of the cone
   $\mu$ = viscosity of the liquid.

### 1.9.4    Coutte Viscometer

A *Coutte viscometer* is a member of a class of rotational rheometers. The torque needed to rotate a solid object in contact with a fluid is measured and the viscosity deduced from the derived expression. A modified Coutte viscometer is called a *Stabiner viscometer.* Here the inner cylinder is hollow and allowed to float, thereby avoiding bearing

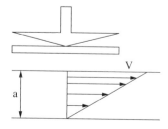

FIGURE 1.5
Side view of cone-
and plate
viscometer and
velocity distribution
in control volume.

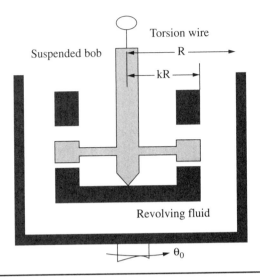

**Figure 1.6**  Coutte viscometer.

friction. Speed and torque measurements are made by remotely rotating a magnetic field.

A schematic of a Coutte viscometer is shown in Fig. 1.6. The cup holds the fluid whose viscosity needs to be measured. It is made to rotate with a constant angular velocity, $\omega_o$. The *revolving* viscous liquid causes the suspended bob to turn. A point is reached when the torque produced on account of momentum transfer in the liquid becomes equal to the product of the torsion constant, $k_t$, and the angular displacement of the bob. The angular displacement, $\theta_o$, is measured using a mirror mounted on the bob by noting the deflection of a light beam. A steady tangential annular flow is maintained between two coaxial cylinders. The end effects due to the bob height, $H$, can be neglected. The equations of continuity and momentum for the liquid in tangential flow can be written as follows [see 11]: The density and viscosity of the liquid are assumed to remain constant and unaffected by the flow.

$$\frac{d}{dr}\left(\frac{1}{r}\frac{d(rv_\theta)}{dr}\right) = 0 \tag{1.88}$$

$$-\frac{\partial p}{\partial z} - \rho g = 0 \tag{1.89}$$

The boundary conditions are:

$$r = R, \qquad v_\theta = 0 \tag{1.90}$$

$$r = kR, \qquad v_\theta = \omega_o \tag{1.91}$$

Integrating Eq. (1.88) twice and solving for the boundary conditions, it can be shown that the tangential velocity of the fluid can be represented at steady state by:

$$v_\theta = \omega_o R \frac{\left( \dfrac{r}{kR} - \dfrac{kR}{r} \right)}{\left( \dfrac{1}{k} - k \right)} \qquad (1.92)$$

Once the velocity distribution is available, the momentum flux can be estimated as:

$$\tau_{r\theta} = -2\mu\omega_o \left( \frac{R}{r} \right)^2 \left( \frac{k^2}{1-k^2} \right) \qquad (1.93)$$

The torque acting on the inner cylinder is then obtained as:

$$T_z = 4\pi\mu\omega_o R^2 H \left( \frac{k^2}{1-k^2} \right) = k_t \theta_o \qquad (1.94)$$

Thus, measuring the angular velocity of the cup and the angular deflection of the bob enable the viscosity of the liquid to be determined. When a critical Reynolds number is reached, Taylor vortices form and turbulent flow ensues upon further increasing the velocity.

### 1.9.5 Parallel Disk Viscometer

In a *parallel disk viscometer*, a liquid whose viscosity needs to be measured is placed in the gap of thickness, $B$, between the two disks of radius, $R$, and held in place by surface tension. The torque needed to turn the upper disk at an angular velocity of $\omega$ is measured and the lower disk is fixed. Assuming creeping, the working equation for obtaining viscosity of the liquid can be shown to be:

$$\mu = \frac{2BT_z}{\pi\omega R^4} \qquad (1.95)$$

In a *parallel disk compression viscometer*, a liquid is allowed to fill completely the region between two circular disks of radius $R$. The bottom disk is fixed, and the upper disk is made to approach the lower one very slowly with a constant speed $v_0$ starting from a initial height $H_0$. The instantaneous height is given by $H(t)$. It can be shown that the force needed to maintain the constant velocity is given by Eq. (1.96), where $V$ is the volume of the liquid sample:

$$F(t) = \frac{3\mu v_0 V^2}{2\pi H(t)^5} \qquad (1.96)$$

The $H(t)$ is measured as a function of time and then the viscosity is determined from:

$$\frac{1}{H(t)^2} = \frac{1}{H_o^2} + \frac{4F_o t}{3\mu\pi R^4} \qquad (1.97)$$

where $F_o$ is a constant applied force.

### 1.9.6 Rolling Ball Viscometer

A *rolling ball viscometer* is designed based upon the results of analyzing laminar flow in a narrow slit. A Newtonian fluid is in laminar flow in a narrow slit formed by two parallel walls a distance $2B$ apart. Edge effects can be omitted as $B \ll W$. Performing a differential momentum balance, the following expressions for the momentum-flux and velocity distributions can be derived:

$$\tau_{xz} = \left(\frac{P_0 - P_L}{L}\right)x \qquad (1.98)$$

$$v_z = \frac{(P_0 - P_L)B^2}{2\mu L}\left(1 - \left(\frac{x}{B}\right)^2\right) \qquad (1.99)$$

where $P = p + \rho g h$.

The rolling ball viscometer is shown in Fig. 1.7. A ball is rolled down the walls of a cylinder held at an incline angle $\beta$ with the horizontal. The sector formed between the cross-sections of the rolling ball and cylinder at any given instant can be shown to be a function of the polar angle, $\theta$ and $z$.

$$s = 2(R - r)\left(\cos^2\left(\frac{\theta}{2}\right) + \frac{R - \sqrt{R^2 - z^2}}{2(R - r)}\right) \qquad (1.100)$$

### 1.9.7 Torsional Oscillatory Viscometer

In a *torsional oscillatory viscometer*, the fluid is sandwiched between a "cup" and a "bob." Sinusoidal oscillations are imposed in the cup in the tangential direction. This causes the suspended bob to oscillate

FIGURE **1.7**
Rolling ball
viscometer.

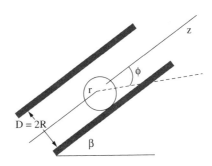

with the same frequency but with a different amplitude and phase. The torsion wire used to suspend the bob also oscillates at the same frequency and amplitude as that of the bob. The ratio of amplitudes between the cup and bob and the phase lag are a function of the viscosity of the liquid. Small oscillations are considered. The problem remains linear for small oscillations. The solution to the governing equations can be obtained by Laplace transform methods.

Newton's second law of motion can be applied to the cylindrical bob for the special case when the annular space between the cup and bob are evacuated. It can be shown that the natural frequency of the system is $\omega_0 = \sqrt{k/I}$, where $I$ is the moment of inertia of the bob and $k$ is the spring constant for the torsion wire. Furthermore, Newton's second law of motion can be applied to the bob:

$$I\frac{d^2\theta_R}{dt^2} = -k\theta_R + (2\pi RL)(R)\left(\mu r \frac{\partial}{\partial r}\left(\frac{v_\theta}{r}\right)\right)_{r=R} \tag{1.101}$$

The initial conditions are:

$$t = 0^{\theta_R = \theta_{R0}} \tag{1.102}$$

$$\frac{d\theta_R}{dt} = 0 \tag{1.103}$$

The governing equation of motion for fluid and the time and space conditions can be written as:

$$\rho\frac{\partial v_\theta}{\partial t} = \mu\frac{\partial}{\partial r}\left(\frac{1}{r}\frac{\partial}{\partial r}(rv_\theta)\right) \tag{1.104}$$

$$t = 0, \qquad v_\theta = 0$$

$$r = R, \qquad v_\theta = R\frac{d\theta_R}{dt} \tag{1.105}$$

$$r = \kappa R, \qquad v_\theta = \kappa R\frac{d\theta_{\kappa R}}{dt} \tag{1.106}$$

$\theta_{\kappa R}(t)$ is a sinusoidal function of time that causes the forced oscillations of the cup and the induced oscillations of the bob. For close clearances between the cup and bob, $\kappa$ is close to 1. The variables are made dimensionless, and the resulting governing equations are solved for by the method of complex velocity. It can be shown that based on these results, the amplitude ratio can be given by:

$$\frac{\theta_R^0}{\theta_{\kappa R}^0} = \frac{AMi\overline{\omega}}{(1-\overline{\omega}^2)\dfrac{\sin h\sqrt{\dfrac{i\overline{\omega}}{M}}}{\sqrt{\dfrac{i\overline{\omega}}{M}}} + AMi\overline{\omega}\cos h\sqrt{\dfrac{i\overline{\omega}}{M}}} \tag{1.107}$$

$$\text{where } M = \frac{v}{(\kappa - 1)^2 R^2} \sqrt{\frac{I}{k}}; \quad \tau = t\sqrt{\frac{k}{I}}; \quad \bar{\varpi} = \frac{\varpi}{\varpi_0} \tag{1.108}$$

$$A = \frac{2\pi R^4 L \rho(\kappa - 1)}{I}; \quad \phi = \frac{2\pi R^3 L \rho(\kappa - 1)}{\sqrt{kI}} v_\theta \tag{1.109}$$

### 1.9.8   Bubble Viscometer

In *bubble viscometers,* the time required for an air bubble to rise in fluid is an inverse measure of the viscosity of the fluid. Calibration is used to standardize the technique in the desired viscosity ranges. Viscometers for non-Newtonian fluids are also called *rheometers* or *plastometers.* A piston-cylinder arrangement is used in *Norcross viscometers.* Material is drawn through the clearance between the piston and cylinder by raising the piston periodically.

## Summary

A preliminary review of fluid mechanics was presented in this chapter. Sir Isaac Newton devoted his entire second book to *fluid mechanics.* Any form of matter that can flow when subjected to a shear stress is considered a fluid. Continuum hypotheses assume that the fluid consists of homogeneous properties, such as uniform density, throughout the fluid, although in atoms the mass is concentrated at the nuclei. Control volume refers to a region of space considered the basis for developing the theory of fluid flow in to and out of the region. Fifty-six different flow types were compared and contrasted with each other (Table 1.1).

The fundamental dimensions are length, time, mass, and temperature. The word thermodynamics comes from the Greek words *therme,* which means heat, and *dynamis,* which means power. Pressure exerted by a fluid is the force per unit area acting on either the external surface of the object or the walls of the enclosed container. Both hydrostatic and kinetic pressures also were discussed. Starting with a box of molecules, an expression for kinetic pressure as a function of the root-mean-square velocity of the molecules was derived. This led to the derivation of the ideal gas law. Maxwell's relations between internal energy, $U$; enthalpy, $H$; Gibbs' free energy, $G$; Helmholtz free energy, $A$; and entropy, $S$, were introduced.

The first law of thermodynamics was written in differential form. Expression for work done as a function of the pressure and volume of the gas in the cylinder of the piston-cylinder arrangement was developed. Joule's experiments and how the modern concept of heat evolved were discussed. The concepts of closed and open systems, surroundings, and states of system were introduced. The phase rule was stated. The concepts of reversibility and equilibrium were introduced.

Viscosity of a fluid and Newton's law of viscosity were reviewed. The myths and realities of yield stress were discussed. Experimental evidence that the yield stress has never been directly measured but inferred by extrapolation was presented. Different methods to measure viscosity of a fluid, such as efflux viscometer, falling ball viscometer, cone-and-plate viscometer, rolling ball viscometer, torsional oscillatory viscometer, bubble viscometer, etc., were discussed.

The equation of conservation of mass was derived in Cartesian coordinates for any fluid in motion. The equation of conservation of mass was presented in cylindrical and spherical coordinates in the differential form. The integral form of the equation of continuity was also presented. The equation of motion for any fluid in motion in Cartesian coordinates was derived. The Navier-Stokes equation was recognized, and the term by term physical significance was discussed. The reduction to the Euler equation in the special case of inviscid flow was shown. The Euler equation was integrated with the Bernoulli equation, which serves as the law of mechanical energy balance.

## References

[1]  I. Newton, *Philosophiæ Naturalis Principia Mathematica*, vols. 1–3, Latin-English translation by R. Bentley, 1687.

[2]  K. R. Sharma, *Fluid Mechanics and Machinery*, Chennai, India: Anuradha Publishers, 2003.

[3]  K. R. Sharma, *Eighteen Different Flow Types and Polymerization Pilot Plant*, AIChE Spring Meeting, New Orleans, LA, April 2004.

[4]  O. Reynolds, "An experimental investigation of the circumstances which determine whether the motion of water shall be direct or sinuous, and of the law of resistance in parallel channels," *Phil. Trans. of the Royal Society*, vol. 174 (1883), 935–982.

[5]  H. J. S. Hele-Shaw, "Investigation of the nature of the surface resistance of water and of streamline motion under certain experimental conditions," *Trans. Inst. Nav. Arch.*,Vol. 40, (1898) 25.

[6]  L. Prandtl, *Essentials of Fluid Dynamics*, New York: Harper, 1952, p. 105.

[7]  J. D. Anderson, *Modern Compressible Flow with Historical Perspective*, New York: McGraw-Hill Professional, 2003.

[8]  D. Halliday, R. Resnick, and J. Walker, *Fundamentals of Physics*, New York: John Wiley, 2004.

[9]  M. Modell and R. C. Reid, *Thermodynamics and Its Applications in Chemical Engineering*, Upper Saddle River, NJ: Prentice Hall, 1974.

[10]  H. J. Steffens, *James Prescott Joule and the Concept of Energy*, New York: Neale Watson Academic Publications, Inc., 1979.

[11]  R. B. Bird, C. F. Curtiss, and W. E. Stewart, "Tangential Flow in Annuli: I. Unsteady State Distribution," *Chem. Eng. Sci.*, 11 (1959), 114–117.

[12]  H. A. Barnes and K. Walters, "The yield stress myth," *Rheologica Acta* (1985), 323–326.

[13]  J. P. Hartnett and R. Y. Z. Hu, "The yield stress: an engineering reality," *Jour Rheology*, vol. 33 (1989), 671.

[14]  C. L. H. M. Navier, *Ann Chimie*, vol. 19 (1821), 244-260; G. G. Stokes, trans. *Camb. Phil. Soc.*, vol. 8 (1845), 287–305.

[15]  L. Euler, *Memoires de Academie de Science*, vol. 11 (1755), 217.

[16]  D. Bernoulli, *Hydrodynamica*, Strousbourg, Germany: Dulsecker, 1738.

[17]  K. Renganathan, R. Turton, and N. C. Clark, "Acceleration motion of geometric and spherical particles in a fluid," *Powder Technology*, vol. 58(4), (1989), 279–284.
[18]  L. D. Landau and E. M. Lifshitz, "Fluid mechanics," *Course of Theoretical Physics*, vol. 6, Elmsford, New York: Pergamon Press, 1987.

# Exercises

## Problems

**1.0**  *Potential flow*. A sphere of diameter, $d$, moves with a velocity, $v_z$, in an incompressible ideal fluid. Determine the potential flow of the fluid past the sphere.

**2.0**  *Reynolds number*. A municipal water distribution system reports a discharge rate of 4 gallons a minute. Determine the Reynolds number of water flowing in a pipe with a diameter of 4 inches at room temperature. Is the fluid laminar or turbulent? The viscosity of water can be taken as 1 cp.

**3.0**  *Mach number*. A supersonic jet is travelling at a speed of 400 m/s. Calculate the Mach number of the fluid flowing past the jet.

**4.0**  *Torricelli's theorem*. Derive the Torricelli's theorem for the efflux velocity of a fluid drained from a cylindrical tank with diameter $D$ and orifice diameter $d$. The height of the fluid in the tank can be taken to be $H$.

**5.0**  Calculate the time taken to drain the tank in Exercise 4.0.

**6.0**  *Friction factor*. What is the length of a ½-inch hose needed to water the kitchen garden over the weekend at a discharge rate of 1 liter/hour?

**7.0**  *Liquid drop*. Calculate the size of a drop formed on account of surface tension as a function of the pressure difference between the atmosphere and within the drop.

**8.0**  *Gravity waves [18]*. The free surface of a liquid is in equilibrium with the gravitational field and exists as a plane. Upon some external perturbation, motion of liquid is propagated by means of what are called gravity waves. They appear mainly on the surface of the liquid and affect the interior of the fluid to a lesser extent. The inertial term in the Euler equation can be neglected (Eq. 1.84) in comparison with the $\partial v/\partial t$ term. During the time interval of the order of the period $\tau$ of the oscillations of the fluid particles in the waves, these entities travel a distance of the order of the amplitude a of the wave. Their velocity is of the order of $a/\tau$ and varies over the time intervals of the period of oscillation and over the distances of wavelength, $\lambda$. The time derivative of velocity is of the order of $v/\tau$ and the space derivatives are of the order of $v/\lambda$. Show that the condition $(v \cdot \nabla)v \ll \partial v/\partial t$ is equivalent to $a \ll \lambda$ i.e., the amplitude of the oscillations in the wave is smaller compared with the wavelength. The inertial term in the Euler equation can be neglected and potential flow can be expected. Show that the following set of equations will govern the motion of the waves:

$$\Delta \phi = 0$$

$$\frac{\partial \phi}{\partial z} + \frac{\partial^2 \phi}{g \partial t^2} = 0$$

**9.0** Show that the following equations can be solutions to the governing equations of gravity waves described in Exercise 8.0:

$$x - x_0 = -A\frac{k}{\omega}e^{kz_0}\cos(kx_0 - \omega t)$$

$$z - z_0 = -A\frac{k}{\omega}e^{kz_0}\sin(kx_0 - \omega t)$$

where $x$, $z$ are the coordinates of the moving fluid particles.

**10.0** The isentropic compressibility can be written as:

$$\kappa_s = -\frac{1}{v}\left(\frac{\partial v}{\partial P}\right)_s$$

Derive the speed of sound to be:

$$a = \sqrt{\frac{v}{\kappa_s}}$$

**11.0** Show that for one mole of an ideal gas:

$$C_p - C_v = R$$

**12.0** Show that for an adiabatic process;

$$PV^\gamma = \text{Const}$$

where, $\gamma = \dfrac{C_p}{C_v}$

**13.0** Describe the significance of each term in the Navier-Stokes equation.

**14.0** How is the Euler equation derived from the Navier-Stokes equation?

**15.0** How is the Bernoulli equation derived from the Euler equation?

**16.0** When the pressure of the ideal gas is reduced by half, what happens to the velocity of the molecules?

**17.0** Two ports are drilled into a pipe with varying cross-sections carrying water. The ports are connected to a manometer, and the pressure differential was found to be 37 cmHg. If the elevation of both ports is the same, what is the velocity of the fluid at the lower pressure port compared with the velocity of the fluid at the larger pressure port? The ratio of the cross-sectional areas between the low pressure and high pressure port can be taken as 4.

**18.0** Calculate the mass flow rate of the air from the fan. The fan spins at 60 RPM. The cross-sectional area of the blades can be taken to be 1/8" × 4". The room is at 37°C. The air can be assumed to obey the ideal gas law. The composition of air can be taken to be 78% nitrogen and 21% oxygen. The radius of the blade is 1.5 ft.

**19.0**   A smoking lounge is to accommodate 20 heavy smokers. According to the American Society of Heating, Refrigerating, and Air-Conditioning Engineers (ASHRAE), the fresh air requirement is 25 lit/s per person. Calculate the minimum duct area and the minimum flow rate required if the maximum velocity of the air is 10 m/s.

**20.0**   Oil is drilled at a refinery from a depth of 200 ft. When 1 lit/min of oil is discharged at the surface, calculate the rating of a pump that operates at 80% efficiency should the diameter of the rig be 2 inches. The density of oil can be assumed to be 1.6 gm/cc.

**21.0**   Amagat of France in the late 19th century used a mercury manometer to perform measurements for the first time in the history of mankind. Estimate the height of the manometer needed for measurements in mine shaft up to 500 bar.

**22.0**   An entourage of engineers visited the moon. They used a spring scale to measure the mass of some ice cubes. At a reading of 21, what is the mass of ice and the weight of the moon? The moon's gravity is 1/6th that of Earth.

**23.0**   What is the difference between hydrostatic pressure and kinetic pressure?

**24.0**   Centuries from now, when there is an acute energy shortage, people will shop for energy in supermarkets. Cylinders of gas may be purchased and connected to any number of Carnot engines or other efficient devices stocked at home. Devise a convenient method to allow a person to comparison shop by providing a unit cost of energy in joules per dollar. The ambient temperature and pressure are 300 K and $1 \times 10^5$ Nm$^{-2}$. The heat capacity at constant volume may be taken as 20.7 J/mol/K. Consider a cylinder 1 m$^3$ in volume initially at $8 \times 10^5$ Nm$^{-2}$ and 400 K that sells for \$0.32.

**25.0**   Show that for any gas:

$$C_p - C_v = \left(\frac{\partial V}{\partial T}\right)_P \left(P + \left(\frac{\partial U}{\partial V}\right)_T\right)$$

Hint: Use the relation in change of variables in differentiation:

$$\left(\frac{\partial f}{\partial x}\right)_z = \left(\frac{\partial f}{\partial x}\right)_y + \left(\frac{\partial f}{\partial y}\right)_x \left(\frac{\partial y}{\partial x}\right)_z$$

**26.0**   Show that for any gas

$$C_p - C_v = -\left(\frac{\partial P}{\partial T}\right)_V \left(\left(\frac{\partial H}{\partial P}\right)_T - V\right)$$

**27.0**   Show that the free energy change during any isothermal and isochoric compression process from $P_1$ to $P_2$ is given by $V \Delta P$.

**28.0**   Show that the entropy change during an isothermal expansion of an ideal gas from $V_i$ to $V_f$ can be given by $R \ln(V_f / V_i)$.

**29.0**   Show that the entropy change during an isobaric expansion of an ideal gas can be written as $C_p \ln(T_f/T_i)$.

**30.0**   Show that the free energy change during an isothermal process in an ideal gas can be written as $RT \ln(P_f/P_i)$.

**31.0**   Show that the solution to Example 29.0 may be written in terms of initial and final volumes as $C_p \ln(V_f/V_i)$.

**32.0**   Derive the Gibbs-Helmholtz equation given in the form:

$$\Delta G = \Delta H + T\left(\frac{\partial \Delta G}{\partial T}\right)_{P_1,P_2,n_1,n_2}$$

**33.0**   Show that:

$$\Delta H = \Delta U + P\left(\frac{\partial \Delta H}{\partial P}\right)_s$$

**34.0**   Show that:

$$\Delta U = \Delta A + S\left(\frac{\partial \Delta U}{\partial S}\right)_y$$

**35.0**   Show that:

$$\Delta A = \Delta U + T\left(\frac{\partial \Delta A}{\partial T}\right)_y$$

**36.0**   Given $\kappa$ is the compressibility factor $-1/V(\partial V/\partial P)$ and $\alpha$ is the coefficient of thermal expansion, show that:

$$\left(\frac{\partial S}{\partial V}\right)_T = \frac{\alpha}{\kappa}$$

**37.0**   Water is in laminar flow in a narrow slit formed by two parallel walls a distance of $2\,w$ apart. Show that with the velocity of water down the walls $v_z$ as a function of $x$, the axial distance between the plates can be shown to be:

$$v_z = \frac{\Delta P w^2}{2\mu L}\left(1-\left(\frac{x}{w}\right)^2\right)$$

**38.0**   A fluid flows down the inclines of a conical surface with the apex at the top. Show that the film thickness as a function of distance, $s$, down the incline is given by:

$$\delta = \left(\frac{3\mu\dot{m}}{\pi s g \rho^2 \sin 2\beta}\right)^{1/3}$$

where $\beta$ is the cone half-angle and $\dot{m}$ the mass flow rate.

**39.0**    *Contactless pick-up device.* Bernoulli's law can be applied to design a device that can be made to pick up objects without coming in contact with the device. Semiconductor wafers can help prevent damage to the device due to mechanical contact. If air is allowed to impinge on the design through an orifice, show that as the air comes in contact with the object to be lifted, the flow changes from azimuthal to radial flow and the velocity of flow increases. This causes a suction pressure, which can be used to pick up the object. How does the suction pressure capable of development vary with the velocity of air, the diameter of the orifice, and other relevant parameters?

**40.0**    *Chimney design.* How tall should a chimney be constructed to create a draft flow of 1 m/s? An exhaust fan is available to generate a flow velocity of 5 m/s across a duct in the roof. What would happen should the chimney be given a taper with a lower diameter on top of the chimney?

# CHAPTER 2

# Principles of Diffusion

## Learning Objectives

- Diffusion phenomena confirmation by experiments in Spacelab
- Fick's laws of diffusion
- Bulk motion and total molar flux
- Damped wave diffusion and relaxation
- Diffusion in gases, liquids, solids, and porous solids
- Steady state and transient diffusion
- Diffusion coefficient as a function of temperature
- Diffusion in polymers
- Transient wave diffusion in semi-infinite medium
- Periodic boundary condition

## 2.1   Diffusion Phenomena

Diffusion is a phenomenon of migration of a species from a region of higher concentration to a region of lower concentration under the driving force of a concentration gradient [1]. There can be other driving forces, such as temperature difference, the large concentration gradient of a second species, osmotic potential, steam sweep, centripetal forces, pressure drop, electromotive forces, surface tension gradient, surface forces, etc., that can cause the transfer of species from one point to another, oftentimes in a secondary manner.

The term *molecular diffusion* refers to the Brownian motion of molecules as observed by Einstein [2] and movement from a region of higher concentration to a region of lower concentration. This is in accordance with the second law of thermodynamics: the Clausius inequality. The movement of a species from a region of lower concentration to a region of higher concentration in a spontaneous manner is infeasible. This is because not all heat can be converted to work without some heat being rejected to the atmosphere. Heat always flows from a hotter temperature to a colder temperature by the phenomenon of molecular conduction. By analogy, the molar species also moves from a region of higher chemical potential to that of a region of lower chemical potential. The direction of transfer equalizes the concentration.

There can be no negative concentration. As the third law of thermodynamics states, the lowest attainable temperature is 0 K. By analogy, the lowest concentration attainable is $0 \text{ mol}/\text{m}^3$. This will be used in later discussions to obtain a plane of zero concentration, a penetration length, etc.

Diffusion is central to separation operations widely used in the chemical and biotechnology industries. It is used to better understand the transport of solutes in the living cells and to design artificial organs, and it plays a pivotal role in the sequence distribution analysis in genome projects. The efficiency of distillation and dispersal of pollutants can be derived from principles of diffusion. As cities around the world face a drought crisis, the desalination of seawater for potable water needs is going to be increasingly relied upon. This is the method of choice in the deserts of the Middle East, where energy is abundant and cheap and drinking water is scarce. The Bhabha Atomic Research Center (BARC), at Trombay in Mumbai, India, has set up the world's largest desalination plant at the atomic power plant at Kalpakkam about 50 km near Chennai. This plant has two sections. One section produces 1.8 million liters of potable water a day from sea water using the reverse-osmosis method, and the other section produces 4.5 million liters a day using the thermal method. Another desalination plant with two units at 50,000 liters per hour was inaugurated at Koodankukulam in Tirunelveli, Tamil Nadu, using reverse-osmosis technology. In order to desalinate sea water, several transfer operations, such as reverse osmosis, electrodialysis, ion exchange, extraction, flash vaporization, molecular sieve filtration, and pervaporation, can be used. The principles of diffusion and mass transfer can help evaluate the technical feasibility of each operation at a large scale at the lowest possible cost without much harm to the environment in a safe manner. The chemical reactions performed on a large scale during the commercial manufacture of products are often conducted in the presence of a catalyst. During the reaction, the critical reactant has to diffuse through the catalyst and approach the active site for reaction and encounter the other reactant prior to reacting and forming the product. Diffusion in the catalyst needs to be understood for better design. The useful product has to be separated from the unreacted reactants and other by-products using mass transfer separation operations where diffusion is a critical governing phenomenon.

Albert Einstein observed that a cube of sugar placed in the bottom of a hot teacup diffuses and a uniform concentration of sugar throughout the entire cup is the result. The term *thermophoresis* refers to processes where the primary driving force of diffusion is the temperature difference. *Diffusophoresis* is when a large drop in concentration of a second species drives the transfer of the first species. *Osmosis* is where osmotic potential drives the flow of solvent from a region of low solute concentration to a region of higher solute concentration. In *reverse osmosis,* the solvent is pumped from a region of higher solute concentration to a region of lower solute concentration. The wilting of lettuce when salted is a good example of an osmosis phenomenon, as the water oozes out of the leafy vegetable and the turgor pressure gives in to a shrunken mass. In *sweep diffusion,* steam sweeps away the solute with it. The centripetal forces during *centrifugation* result in different forces upon different masses, which in turn results in separation. *Pressure diffusion* is characterized by a pressure drop, $\Delta P$, in the direction of transfer. *Electrolysis* refers to the movement of charged particles subject to an electromotive force. *Surface diffusion* is the movement of species of interest on the surface of the solid. Surface tension gradient can be utilized in separation by *foaming.*

## 2.2    Fick's First and Second Laws of Diffusion

In the mid-1800s Fick [3,4] introduced two differential equations that provide a mathematical framework to describe the otherwise random phenomena of molecular diffusion. The flow of mass by diffusion across a plane was proportional to the concentration gradient of the diffusant across the plane. The components in a mixture are transported by a driving force during diffusion. The molecular motion is Brownian. The ability of the diffusant to pass through a body is dependent on the diffusion coefficient, $D$, and the solubility coefficient, $S$. The permeability coefficient, $P$, is given by:

$$P = (DS) \tag{2.1}$$

Fick's laws of diffusion were proposed in the year 1855. Adolf E. Fick, the youngest of five children, was born on September 3, 1829, to a civil engineer. During his secondary schooling, Fick was interested in mathematics and was enamored of the work of Poisson. His brother, a professor of anatomy at the University of Marburg, persuaded him to switch from a career in mathematics to a career in medicine. Carl Ludwig was Fick's tutor at Marburg. Ludwig strongly believed that medicine and life itself have a basis in mathematics, physics, and chemistry. His thesis dealt with the visual errors caused by astigmatism. Ironically, most of Fick's accomplishments do not depend on diffusion studies at all, but on his more general investigations of physiology. He did outstanding work in mechanics in hydrodynamics and hemorheology, and in the visual and thermal functioning of the human body.

In his first diffusion paper [3], Fick interpreted the experiments from Graham with interesting theories, analogies, and quantitative experiments. He showed that diffusion can be described on the same mathematical basis as Fourier's law of heat conduction [5] and Ohm's law of electricity. Fick's first law of diffusion can be written as:

$$J = -AD\frac{\partial C}{\partial x} \tag{2.2}$$

where $J$ is defined as the one-dimensional molar flux. The diffusivity is the proportionality constant that depends on the material under consideration.

Fick's second law of diffusion can be derived by considering a thin shell of thickness, $\Delta x$, with constant cross-sectional area, $A$, across which the diffusion is considered to occur. A mass balance in the incremental volume, considered $A\Delta x$ for an incremental time, $\Delta t$, neglecting any reaction or accumulation of the species, can be written as:

(mass in) − (mass out) ± (mass reacted/generated)

$$= \text{mass accumulated} \tag{2.3}$$

$$\Delta t(J_x - J_{x+\Delta x}) = A\Delta x\, \Delta C \tag{2.4}$$

Dividing Eq. (2.4) throughout by $A\Delta x\Delta t$ and obtaining the limits as $\Delta x$ and $\Delta t$ goes to zero:

$$-\frac{\partial J}{\partial x} = A\frac{\partial C}{\partial t} \tag{2.5}$$

Combining Eq. (2.5) and Eq. (2.2), the governing equation for the diffusing species becomes apparent when the area across which the diffusion occurs is a constant:

$$D\frac{\partial^2 C}{\partial x^2} = \frac{\partial C}{\partial t} \tag{2.6}$$

Equation (2.6) is sometimes referred to as Fick's second law of diffusion [4]. This is a fundamental equation that describes the transient, one-dimensional diffusion of diffusing species. Fick attempted to integrate Eq. (2.6) and was discouraged by the numerical effort needed. He found the second derivative difficult to measure experimentally, and he found that the second difference increases exceptionally with the effect of experimental errors. Finally, he demonstrated in a cylindrical cell the steady-state linear concentration gradient of sodium chloride. He used a glass cylinder containing crystalline sodium chloride in the bottom and a large volume of water in the top. By periodically changing the water in the top volume, he was able to establish a steady-state concentration gradient in the cylindrical cell. He confirmed his equation from this steady-state gradient.

In three dimensions, Fick's first law of diffusion can be written as:

$$J'' = -D \, \nabla C \qquad (2.7)$$

where the differential operator $\nabla$ is given by:

$$\nabla = i \frac{\partial}{\partial x} + j \frac{\partial}{\partial y} + k \frac{\partial}{\partial z} \qquad (2.8)$$

such that:

$$\nabla C = i \frac{\partial C}{\partial x} + j \frac{\partial C}{\partial y} + k \frac{\partial C}{\partial z} \qquad (2.9)$$

In the special case of Eq. (2.9), the one-dimensional case of Eq. (2.7) results. Diffusion may be viewed as a process by which molecules intermingle as a result of their kinetic energy of random motion.

On Earth, density differences within a liquid system often result in convective mixing of the fluid. This gravity-induced convection, coupled with gravity-independent diffusion, contributes to the overall mass transfer within the system. In space, the convective contribution is greatly reduced, and a closer examination of the diffusion contribution can be observed.

## 2.3 Skylab Diffusion Demonstration Experiments

The Skylab science demonstration was the first in a series of investigations designed by Fascimire et al. [6] to study low-gravity diffusive mass transfer. The specific objective of the demonstration was to photographically document the diffusion of tea in water in spacecraft. In preparation for the experiment, Skylab pilot Jack Lousma filled a half-inch-diameter, six-inch-long transparent tube three-fourths of the way full with water. A highly concentrated tea solution was then delivered to the water surface via a 5-cc syringe through a synthetic fiber wad. The tube was then capped. The fiber pad was employed to try to bring the tea and water in contact without any entrapped air. Three attempts to produce the wad were unsuccessful. During the fourth wad/attempt, "a good bubble-free interface" was realized. The next day, Lousma reported that no diffusion of the tea in the liquid had occurred. Thus, the experiment was initiated again.

During this new experimental run, the wad was removed and the tea was delivered on top of the water. After an air bubble between the tea and water was removed via the syringe, a "smooth, continuous interface" was achieved. The tea was allowed to diffuse during the next three days. Post-flight 16-mm photographs of the diffusion were analyzed. In 51.15 hours, the visible diffusion front advanced 1.96 cm. It was noted that the diffusion front became increasingly parabolic during the demonstration and very little diffusion occurred near the container wall.

A similar ground-based experiment was performed for comparison to the space investigation. After 45.5 hours, three different zones were visible: a dark area, an area of medium darkness, and a very light area. The medium-colored area had advanced 1.6 cm in 45.5 hours.

If a few crystals of $K_2CrO_4$, potassium chromate, are placed at the bottom of a tall bottle filled with triple distilled water, the yellow color will slowly spread throughout the bottle. At first, the color will be concentrated in the bottom of the bottle. After a day, it will penetrate upward a few centimeters. After several years, the solution will appear homogeneous. The process responsible for the movement of the colored material is diffusion. Diffusion was studied by Albert Einstein. He noted that as sugar dissolves in water, the viscosity of the solution increases. The Stokes-Einstein equation is used for estimating diffusion coefficients for molecules in liquid phase. Diffusion is a molecular phenomenon. At a microscopic level, molecules do undergo Brownian random motion. However, the driving force sets some direction to the transfer of the species under consideration. In gases, diffusion progresses at a rate of about 10 cm/min; in liquids, the rate is about 0.05 cm/min; in solids, its rate is about 100 nm/min.

Diffusion varies less with temperature, although for polymers, Arrhenius relationships have been reported for changes of diffusion coefficients with temperature. The slow rate of diffusion makes it a rate-limiting step in cases where it occurs sequentially with other phenomena. The rates of distillation are limited by diffusion and that of industrial reactions on porous catalysts. The rate of diffusion limits the speed of absorption of nutrients in the human intestine and the control of microorganisms in the production of penicillin. The rate of corrosion of steel, splat cooling of metallic glasses, dopant diffusion in silicon chip manufacturing, the release of flavor from food, and the delivery of drugs to tumor cells are limited by diffusion.

The equalizing effect of diffusion needs to be distinguished from other methods of producing a uniform mixture, such as bulk convective mixing. Agitation also is used in homogenization. The energy for movement from diffusion comes from the thermal energy of the molecules. The rate of evaporation of water at 25°C into complete vacuum was calculated as 3.3 kg/m²/s. Placing a layer of stagnant air at 1 STP and 100 microns thickness above the water surface reduces the rate of evaporation by a factor of about 600.

## 2.4  Bulk Motion, Molecular Motion, and Total Molar Flux

Consider two containers of $CO_2$ gas and He gas separated by a partition, as shown in Fig. 2.1. The molecules of both gases are in constant motion and make numerous collisions with the partition. If the partition is removed, the gases will mix due to the random velocities of their molecules. In sufficient time, a uniform mixture of $CO_2$ and He molecules will result in the container.

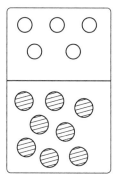

**FIGURE 2.1** Container with a partition separating helium and carbon dioxide.

The helium molecules will move to the bottom and the carbon dioxide molecules will move to the top. If the two species are denoted by $A$ and $B$, and the total fluxes as $N_A$ and $N_B$, then the net flux, $N$, can be written as:

$$N = N_A + N_B \tag{2.10}$$

The movement of $A$ comprises two components, one due to the bulk motion $N$ and the fraction $x_A$ of $N$, which is $A$, and the second component resulting from the diffusion of $A$, $J''_A$:

$$N_A = Nx_A + J''_A \tag{2.11}$$

$$N_B = Nx_B + J''_B \tag{2.12}$$

Adding Eqs. (2.11) and (2.12):

$$N = N + J''_A + J''_B \tag{2.13}$$

or
$$-D_{AB} \frac{\partial C_A}{\partial z} = D_{BA} \frac{\partial C_B}{\partial z} \tag{2.14}$$

If $C_A + C_B$ is constant, then $D_{AB} = D_{BA}$.

**Example 2.1** *Unimolar diffusion.* Consider the diffusion of a liquid, $A$, evaporating into a gas, $B$, in a partially filled tall tube. Assume that the liquid level is maintained at $z = z_1$. At the top of the tube at $z = z_2$ a stream of gas mixture of $A$-$B$ flows steadily past, thereby maintaining the mole fraction of $A$ at $X_{A2}$. At the liquid-gas interface, the gas phase concentration of $A$ expressed as a mole fraction is $X_{A1}$. This is the gas phase concentration of $A$ corresponding to the equilibrium with the liquid at the interface, i.e., $X_{A1}$ is the vapor pressure of $A$ divided by the total pressure, $p_A^{\text{vap}}/P_{\text{tot}}$, provided that $A$ and $B$ form an ideal gas mixture. It is further assumed that the solubility of $B$ in liquid $A$ is negligible. The entire system is presumed to be held at constant temperature and pressure. Gases $A$ and $B$ are assumed to be ideal. When this evaporating surface attains steady state, there is a net motion of $A$ away from the evaporating surface and vapor $B$ is stationary. Obtain the concentration profile at steady state for $A$ in the head space:

$$N_B = 0 \tag{2.15}$$

$$N_A = N_A x_A - D_{AB} \frac{\partial C_A}{\partial z} \tag{2.16}$$

In terms of mole fractions:

$$N_A(1 - x_A) = -C_{tot} D_{AB} \frac{\partial x_A}{\partial z} \tag{2.17}$$

A mass balance over an incremental volume of height $\Delta z$ at a steady state across a constant cross-sectional area, $A$:

$$-\frac{\partial N_A}{\partial z} = 0 \tag{2.18}$$

Combining Eqs. (2.18) and (2.17) and integrating the resulting second-order differential equation with respect to $z$ gives:

$$\frac{1}{(1 - x_A)} \frac{\partial x_A}{\partial z} = c_1 \tag{2.19}$$

A second integration then gives:

$$-\ln(1 - x_A) = c_1 z + c_2 \tag{2.20}$$

The two integration constants can be solved for by using the information given as boundary conditions at locations 1 and 2:

$$\frac{\ln(1 - X_{A1})}{(1 - x_A)} = \frac{(z - z_1)}{(z_2 - z_1)} \frac{\ln(1 - X_{A1})}{(1 - X_{A2})} \tag{2.21}$$

$$\frac{(1 - X_{A1})}{(1 - x_A)} = \left( \frac{1 - X_{A1}}{1 - X_{A2}} \right)^{\left( \frac{z - z_1}{z_2 - z_1} \right)} \tag{2.22}$$

$$N_{Az} = \left( \frac{c D_{AB}}{z_2 - z_1} \right) \ln \left( \frac{X_{B2}}{X_{B1}} \right) \tag{2.23}$$

These expressions are used during the experimental measurement of gas diffusivities.

## 2.5 Diffusivity in Gases

The diffusion coefficients of binary hydrocarbon-hydrocarbon gas systems at low pressures below about 3.5 MPa can be predicted using Gilland's method [7]:

$$D_{12} = (0.1014 T^{1.5}(1/M_1 + 1/M_2)^{0.5}) / P_{tot} (V_1^{1/3} + V_2^{1/3})^2$$

|     | CO$_2$ | H$_2$ | He | Ar | O$_2$ | H$_2$O | N$_2$ | Air | CH$_4$ |
|-----|------|-----|------|------|------|------|------|------|---------|
| CO$_2$ |      | 0.646 | 0.597 | 0.133 | 0.156 | 0.202 | 0.165 | 0.400 | 0.00215 |
| H$_2$ |      |      | 1.706 | 0.902 | 0.891 | 0.915 | 0.779 | 0.71 | 0.726 |
| He  |      |      |      | 0.742 | 0.822 | 0.908 | 0.794 | 0.658 | 0.494 |
| Ar  |      |      |      |      |      | 0.216 |      |      | 0.675 |
| O$_2$ |      |      |      |      |      | 0.282 | 0.181 | 0.176 | 1.1 |
| H$_2$O |      |      |      |      |      |      | 0.293 | 0.260 | 0.212 |
| N$_2$ |      |      |      |      |      |      |      |      | 0.148 |
| Air |      |      |      |      |      |      |      |      | 0.196 |
| CH$_4$ |      |      |      |      |      |      |      |      |      |

TABLE **2.1**    Measured Values of Diffusion Coefficients in Gases at 1 atm Pressure and Data Available at Nearest Temperature to 298 K (cm$^2$/s)

where subscript 1 refers to the solute and subscript 2 refers to the solvent. The units of $T$, $P$, and $V$ are $R_0$, psia, and cm$^3$/gmole, respectively, and diffusivity is given by ft$^2$/hr.

There is no universal theory that predicts a priori all the diffusion coefficients. Experimental measurements have to be relied upon. Sometimes the experimental measurements are difficult to make and the quality of the results is not adequate. The estimates of diffusion coefficients of gases at room temperature are about 0.1 cm$^2$/s, that of liquids 10$^{-5}$ cm$^2$/s, and that of solids 10$^{-10}$ cm$^2$/s. Diffusion coefficients in polymers lie between that of solids and liquids. Binary diffusion coefficients in gases for pairs of gases are given in Table 2.1.

The most widely cited method for theoretical estimation of gaseous diffusion is that developed independently by Chapman and Cowling [8]. This theory is accurate to an average of about 8% yields:

$$D = (1.86 \text{ E-3 } T^{3/2} (1/M_1 + 1/M_2)^{1/2})/p\sigma_{12}^2\Omega \qquad (2.24)$$

where     $D$ = diffusion coefficient in cm$^2$/s
$T$ = absolute temperature in K
$p$ = pressure in atmospheres
$M$ = molecular weight
$\sigma_{12}$ and $\Omega$ = molecular properties
$\sigma_{12}$ = collision diameter; is the arithmetic average of the diameters of the atoms of the species present
$\Omega$ = collision integral

Some other correlations for diffusivities for gases are available in the literature. Reid, Sherwood, and Prausnitz [9] compared predictions from different correlations with 68 experimental values of $D_{AB}$. The amount of error, ease of calculations, range of applicability, empirical parameters, uses, and underlying mechanistic theory from which the correlations were derived are some of the considerations prior to using any of the available correlations.

For nonpolar binary mixtures at low pressure, Wilke and Lee state [10]:

$$D_{AB} = \frac{T^{3/2}\sqrt{M_{AB}}\left(0.0027 - 0.0005\sqrt{M_{AB}}\right)}{P\sigma_{AB}^2\Omega_D} \tag{2.25}$$

For polar binary mixtures and low pressure, Brokaw states [11]:

$$D_{AB} = \frac{0.001858T^{3/2}\sqrt{M_{AB}}}{P\sigma_{AB}^2\Omega_D} \tag{2.26}$$

For self-diffusivity, high pressure, $\rho_r \leq 1.5$, Mathur and Thodos state [12]:

$$D_{AA} = \frac{10.7 * 10^{-5}T_r}{\beta\rho_r} \tag{2.27}$$

For supercritical mixtures, Catchpole and King state [13]:

$$D_{AB} = \frac{5.152RD_cT_r\left(\rho^{-.667} - 0.4510\right)\left(1 + \dfrac{M_A}{M_B}\right)}{\left(1 + \dfrac{V_{CB}}{V_{CA}}\right)^{0.667}} \tag{2.28}$$

Observations show that for many polyatomic gases and mixtures, $D_{AB}P$ is constant up to 1000 K and 700 atm. The characteristic length $\sigma_{AB}$ is in $A$.

## 2.6   Diffusion Coefficients in Liquids

Diffusion coefficients in liquids such as common organic solvents, mercury, and molten metal fall in the order of magnitude of $10^{-5}$ cm$^2$/s. The diffusion can even be slower, sometimes even 100 times slower, for high-molecular-weight solutes like polystyrene and polybutadiene. Diffusion in liquids is slower compared with the gases, and can become the rate-limiting step in simultaneous reaction and diffusion.

A demonstration experiment was performed to illustrate diffusion phenomena in liquids. One hundred milliliters of three solutions were poured into a large filter funnel, with adequate care taken not to mix the layers. The top layer was made up of hydrochloric acid (HCl) dissolved in toluene. The middle layer was a universal indicator in water. The bottom layer was ammonia ($NH_3$) dissolved in chloroform ($CHCl_3$).

The HCl from the top layer diffused into the middle layer, and the color in the second layer changed, indicating an acid. The bottom ammonia layer was also diffusing into the middle layer, and the color changed, indicating a base. Eventually, when the ammonia from the bottom layer and HCl from the top layer diffused into the second layer and mixed, the indicator turned into the color for a neutral solution.

### 2.6.1    Stokes-Einstein Equation for Dilute Solutions

The Stokes-Einstein equation can be used to calculate diffusion coefficients in liquids:

$$D = \frac{k_B T}{f} = \frac{k_B T}{6\pi\mu R_o} \tag{2.29}$$

where $k_B$ is the Boltzmann constant, $f$ is the frictional drag coefficient, $T$ is the temperature, $\mu$ is the viscosity of the surrounding medium, and $R_0$ is the radius of the solute that is diffusing. Equation (2.29) can be derived as follows. A rigid solute sphere is assumed for the molecule diffusing in a common solvent. The frictional drag force acting on the molecule opposing its motion is proportional to the velocity of the sphere:

$$\text{Drag force} = f v_1 \tag{2.30}$$

where $v_1$ is the velocity of the molecule. From Stokes' law [14] for a sphere moving in a fluid, $f = 6\pi\mu R_o$, the driving force was taken by Einstein [2] to be the negative of the chemical potential gradient $(-\nabla\mu_A)$ defined per molecule:

$$-\nabla\mu_A = (6\pi\mu R_o)v_A \tag{2.31}$$

Equation (2.31) is valid when the molecule reaches a steady-state velocity. This is when the net force acting on the molecule is zero. The solution is assumed to be ideal and dilute.

$$\mu_A = \mu_A^0 + k_B T \ln(x_A) = \mu_1^0 + k_B T \ln C_A - k_B T \ln C_B \tag{2.32}$$

For dilute solutions, the concentration of the second species, $C_B$, far exceeds the solute concentration and can be taken as a constant. The gradient at a constant temperature, then, is:

$$\nabla\mu_1 = k_B T \frac{\nabla C_A}{C_A} = -(6\pi\mu R_o)v_A \tag{2.33}$$

$$\frac{-k_B T}{(6\pi\mu R_0)} \nabla C_A = C_A v_A = J/A \tag{2.34}$$

Comparing Eq. (2.34) with Fick's law of molecular diffusion given in Eq. (2.7), the Stokes-Einstein relationship of Eq. (2.29) results.

Equation (2.34) is valid only at a steady state. Oftentimes in transient applications, there is a sudden step-change in concentration, i.e., the driving force is imposed on the system. The molecule will experience an accelerating regime prior to reaching steady state. During the accelerating regime:

$$-\nabla \mu_A - (6\pi\mu R_o)v_A = m\frac{dv_A}{dt} \tag{2.35}$$

where $m$ is the mass of the molecule,

$$mC_A\frac{dv_A}{dt} = -(6\pi\mu R_o)C_A v_A \tag{2.36}$$

or

$$-\frac{k_B T A}{6\pi\mu R_o}\nabla C_A = \frac{m}{(6\pi\mu R_o)}\frac{\partial J}{\partial t} + J \tag{2.37}$$

Equation (2.37) is the generalized Fick's law of diffusion that accounts for the acceleration regime of the molecule as well as the steady-state regime. An expression for the relaxation time for molecular diffusion falls out of the analysis:

$$\tau_r = \frac{m}{(6\pi\mu R_o)} = \frac{mD}{k_B T} \tag{2.38}$$

In terms of $P_{tot}$, the system pressure for ideal gas, the relaxation time can be written as:

$$\tau_r = \frac{MD\rho_m}{P} \tag{2.39}$$

where $\rho_m$ is the molar density of the species migrating. The velocity of mass diffusion is given by:

$$v_m = \sqrt{\frac{D}{\tau_r}} = \sqrt{\frac{k_B T}{m}} \tag{2.40}$$

Equation (2.40) can be rewritten in terms of the molar gas constant and molecular weight as:

$$v_m = \sqrt{\frac{D}{\tau_r}} = \sqrt{\frac{RT}{M}} \tag{2.41}$$

The kinetic representation of pressure can be written after observing that a molecule moving in a one-dimensional cube with a velocity of $v_x$ undergoes a momentum change of $2mv_x$ upon one collision with the wall. The number of collisions on the wall can be estimated by first calculating the time taken by the molecule to make the round trip from the wall after a collision to the opposite wall and return as $2l/v_x$. The number of collisions undergone by a molecule is $v_x/2l$. The rate of transfer of momentum to the surface from the molecular collisions is then $mv_x^2/l$. The total force exerted by all the molecules colliding can be obtained by summing the contribution from each molecule, and the pressure is obtained by dividing the sum by the area of the wall and is given by Resnick and Halliday [15]:

$$P_{tot} = \frac{m}{l^3}\left(v_{x1}^2 + v_{x2}^2 + v_{x3}^2 + \cdots\right) \tag{2.42}$$

Let $N_m$ be the number of molecules in the system and $n$ the number of molecules per unit volume. Then Eq. (2.42) can be rewritten after multiplying the numerator and denominator by $N_m$:

$$P_{tot} = mn\langle v_x^2\rangle = \rho\langle v_x^2\rangle = \frac{1}{3}\rho<v^2> \tag{2.43}$$

As the molecules treated as particles move in random, there is no preferred direction in the box. Hence, $v^2 = v_x^2 + v_y^2 + v_z^2$. The square root of $v^2$ is called the root mean squared speed of the molecule, and is a widely accepted average molecular speed. From the ideal gas law, $P_{tot} = \rho RT/M$. Combining this with Eq. (2.41):

$$\frac{1}{3<v^2>} = \frac{RT}{M} = \frac{A_N k_B T}{M} \tag{2.44}$$

Comparing Eqs. (2.44) and (2.40), it can be seen that the velocity of mass is one-third of the root mean square velocity. This could be due to the fact that only one-dimensional diffusion has been considered. When all three dimensions are considered, these two velocities would be identical, although derived from different first principles.

The governing equation for concentration in Cartesian, cylindrical, and spherical coordinates, taking into account the generalized Fick's law of mass diffusion and relaxation, is given by the following equations:

$$\tau_{mr}\left[\frac{\partial^2 C_A}{\partial t^2} + v_x\frac{\partial^2 C_A}{\partial x\partial t} + v_y\frac{\partial^2 C_A}{\partial y\partial t} + v_z\frac{\partial^2 C_A}{\partial z\partial t}\right] + \frac{\partial C_A}{\partial t} + \frac{\partial C_A}{\partial x}\left[\tau_{mr}\frac{\partial v_x}{\partial t} + v_x\right]$$

$$+ \frac{\partial C_A}{\partial y}\left[\tau_{mr}\frac{\partial v_y}{\partial t} + v_y\right] + \frac{\partial C_A}{\partial z}\left[\tau_{mr}\frac{\partial v_z}{\partial t} + v_z\right] = D\left[\frac{\partial^2 C_A}{\partial x^2} + \frac{\partial^2 C_A}{\partial y^2} + \frac{\partial^2 C_A}{\partial z^2}\right] + R_A$$

$$\tag{2.45}$$

$$\tau_{mr}\left[\frac{\partial^2 C_A}{\partial t^2} + v_r\frac{\partial^2 C_A}{\partial r\partial t} + \frac{v_\theta}{r}\frac{\partial^2 C_A}{\partial\theta\partial t} + v_z\frac{\partial^2 C_A}{\partial z\partial t}\right] + \frac{\partial C_A}{\partial r}\left[\tau_{mr}\frac{\partial v_r}{\partial t} + v_r\right]$$

$$+ \frac{1}{r}\frac{\partial C_A}{\partial\theta}\left[\tau_{mr}\frac{\partial v_\theta}{\partial t} + v_\theta\right] + \frac{\partial C_A}{\partial z}\left[\tau_{mr}\frac{\partial v_z}{\partial t} + v_z\right] + \frac{\partial C_A}{\partial t}$$

$$= D\left[\frac{1}{r}\frac{\partial}{\partial r}\left(r\frac{\partial C_A}{\partial r}\right) + \frac{1}{r^2}\frac{\partial^2 C_A}{\partial\theta^2} + \frac{\partial^2 C_A}{\partial z^2}\right] + R_A \tag{2.46}$$

$$\tau_{mr} \left[ \frac{\partial^2 C_A}{\partial t^2} + v_r \frac{\partial^2 C_A}{\partial r \partial t} + \frac{v_\theta}{r} \frac{\partial^2 C_A}{\partial \theta \partial t} + v_\varphi \frac{1}{r \sin \theta} \frac{\partial^2 C_A}{\partial \varphi \partial t} \right] + \frac{\partial C_A}{\partial r} \left[ \tau_{mr} \frac{\partial v_r}{\partial t} + v_r \right]$$

$$+ \frac{1}{r} \frac{\partial C_A}{\partial \theta} \left[ \tau_{mr} \frac{\partial v_\theta}{\partial t} + v_\theta \right] + \frac{1}{r \sin \theta} \frac{\partial C_A}{\partial \varphi} \left[ \tau_{mr} \frac{\partial v_\varphi}{\partial t} + v_\varphi \right] + \frac{\partial C_A}{\partial t}$$

$$= D \left[ \frac{1}{r^2} \frac{\partial}{\partial r} \left( r^2 \frac{\partial C_A}{\partial r} \right) \right] + \frac{1}{r^2 \sin \theta} \frac{\partial}{\partial \theta} \left[ \sin \theta \frac{\partial C_A}{\partial \theta} \right] + \frac{\partial}{r^2 \sin^2 \theta} \frac{\partial^2 C_A}{\partial \varphi^2} + R_A$$

$$(2.47)$$

Six reasons were listed to seek a generalized Fourier's law of heat conduction (details given in Chap. 9). By analogy, the generalized Fick's law of diffusion needs to be considered:

$$J'' = -D_{AB} \frac{\partial C_A}{\partial x} - \tau_r \frac{\partial J''}{\partial t} \qquad (2.48)$$

The Stokes-Einstein formula for diffusion coefficients is limited to cases in which the solute is larger than the solvent. As a result, other correlations have been derived for cases when the solute and solvent size are similar. Predictions in liquid are not as accurate as in gases. The Wilke and Chang [16] correlation for diffusion in liquids was an empirical correlation, and is given by:

$$D = 7.4\,E - 8\,(\phi M_2)^{1/2} T / \mu V_1^{0.6} \qquad (2.49)$$

**Example 2.2** *Effect of temperature on relaxation time.* Write an expression for the relaxation time during diffusion of the considered species in liquids. Combine this expression with that of the effect of temperature on a diffusion coefficient, and obtain the dependence of relaxation time on temperature.
   Equation (2.38) can be multiplied by the Avogadro number and:

$$\tau_r = DM / RT \qquad (2.50)$$

Combining Eq. (2.50) with Eq. (2.49):

$$\tau_r = 7.4\,E - 8\,(\phi M_2)^{3/2} / R\mu V_1^{0.6} \qquad (2.51)$$

It can be seen that the relaxation time becomes independent of temperature and depends only on the viscosity of the fluid and molecular size parameters.
   Hildebrand adapted a theory of viscosity to self-diffusivity:

$$D_{AA} = B(V - V_{ms}) / V_{ms} \qquad (2.52)$$

where $V$ is the molar volume and $V_{ms}$ is the molar volume when fluidity is zero. The Siddiqi and Lucas correlation [17] for aqueous solutions can be written as:

$$D_{Aw}^0 = 2.98\ E - 7\ V_A^{-0.5473}\mu_w^{-1.026}T \tag{2.53}$$

For hydrocarbon mixtures, the Haydeek-Minhas [18] correlation can be used:

$$D_{AB}^0 = 13.3\ E - 8\ T^{1.47}\mu_B^{(10.2/V_A^{-0.791})}V_A^{-.71} \tag{2.54}$$

When electrolytes are added to a solvent, they dissociate to a certain degree. It would appear that the solution contains at least three components: solvent, anions, and cations. If the solution is to remain neutral in charge at each point, assuming the absence of any applied electric potential field, the anions and cations diffuse effectively as a single component, as with molecular diffusion. The diffusion of the anionic and cationic species in the solvent can thus be treated as a binary mixture. The theory of dilute diffusion of salts is well developed and has been experimentally verified. For dilute solutions of a single salt, the Nernst-Haskell equation is applicable:

$$D_{AB}^0 = RT/F^2\ (|1/n^+|+|1/n^-|)/(1/\lambda_+^0 + 1/\lambda_-^0) \tag{2.55}$$

where $D_{AB}^0$ is diffusivity based on molarity rather than on normality of dilute salt $A$ in solvent $B$ in cm$^2$/s.

## 2.6.2 Diffusion in Concentrated Solutions

The correlations discussed previously pertain to the diffusion in dilute solutions. With increased concentration, some things are different and the considerations will be different. Diffusion coefficients vary with the volume fraction of the solute, oftentimes in a complex manner with a extremamas. Diffusion coefficients are no longer a proportionality constant, but do vary with the concentration and become concentration-dependent. In one approach, the hydrodynamic interaction of the spheres was taken into account and the friction factor $f$ corrected for per Batchelor [19]:

$$f = 6\pi\mu R_o(1 + 1.5\phi_1 + \cdots) \tag{2.56}$$

in which $\phi_1$ is the volume fraction of the solute. Substituting Eq. (2.50) in Eq. (2.34):

$$-\nabla\mu_1 - (6\pi\mu R_o)(1 + 1.5\phi_1 + \cdots)\ v_1 = mdv_1/dt \tag{2.57}$$

$$k_BT\nabla c_1 + mc_1 dv_1/dt = -(6\pi\mu R_o)(1 + 1.5\phi_1 + \cdots]c_1 v_1 \tag{2.58}$$

or $\quad -(k_BT/6\pi\mu R_o)/(1 + 1.5\phi_1 + \cdots]\nabla c_1$

$$= m/6\pi\mu R_o\ /(1 + 1.5\phi_1 + \cdots]\partial J''/\partial t + J'' \tag{2.59}$$

and
$$D = k_B T / 6\pi\mu R_o \, (1 + 1.5\phi_1 + \cdots)$$

$$\tau_r = m / 6\pi\mu R_o \, / (1 + 1.5\phi_1 + \cdots) \tag{2.60}$$

For nonideal solutions, the chemical potential can be written as:

$$\mu_1 = \mu_1^0 + k_B T \ln(c_1\gamma_1) \tag{2.61}$$

where $\gamma_1$ is the activity coefficient.

$$\nabla\mu_1 = k_B T / c_1\gamma_1 (\gamma_1 \nabla c_1 + c_1 \nabla\gamma_1)$$

Substituting Eq. (2.55) in Eq. (2.34):

$$-D\nabla c_1 \, (1 + \partial\ln\gamma_1 / \partial\ln c_1) = J'' + m / (6\pi\mu R_o) \, \partial J'' / \partial t \tag{2.62}$$

The correction for diffusion coefficient given in Eq. (2.59) may be attributed to a cluster of molecules in the solution.

## 2.7    Diffusion in Solids

### 2.7.1    Mechanisms of Diffusion

Atomic diffusion in solids is of increased interest since the phenomenal growth in very large-scale integration (VLSI) of transistors on the silicon chip. Interstitial or substitutional mechanism of diffusion is said to occur when atoms occupy specific sites in a lattice. In an interstitial mechanism of diffusion, an impurity jumps from one interstitial site to the next. In a substitutional mechanism of diffusion, an impurity jumps from one lattice site to the neighboring vacant lattice site. Since the concentration of vacancies is low, substitutional diffusion is much slower than interstitial diffusion. In concentrated diffusion, the atom replaces the lattice atom and moves through the interstices.

The mechanism of diffusion varies greatly, depending upon the crystalline structure and the nature of the solute. For crystals with lattices of cubic symmetry, the diffusivity is isotropic, but not so for noncubic crystals. Interstitial mechanism of diffusion refers to small diffusing solute atoms passing through one interstitial site to the next. The matrix of atoms of the crystal lattice move apart temporarily to provide the necessary space. When there are vacancies and lattice sites are unoccupied, an atom in an adjacent site may jump into such a vacancy. This mechanism is called vacancy mechanism.

The NEC Corporation [20] has developed an interstitial concentration simulation method. Here a mesh is set in a simulation region of a semiconductor device. Under a condition that an area outside of the simulation region is infinite, a provisional interstitial diffusion flux at the boundary of the simulation region is calculated. Then, an interstitial diffusion rate at the boundary of the simulation is calculated by a ratio of the provisional interstitial diffusion flux to the provisional interstitial concentration. Finally, an interstitial diffusion equation is solved for each element of the mesh using the interstitial diffusion rate at the boundary.

Crowd ion mechanism refers to the displacement of an extra atom that can displace several atom positions, thus producing a diffusion flux. The diffusivity in a single crystal is always substantially smaller than that of a multicrystalline sample because the latter has diffusion along the grain boundaries.

For diffusion in metals, Franklin [21] and Stark [22] gave the following expression:

$$D = a_0^2 N_f \omega \tag{2.63}$$

where $a_0$ is the spacing between the atoms, $N_f$ is the fraction of sites vacant in the crystal, and $\omega$ is the jump frequency—that is, the number of jumps per unit time from one position to the next.

**Example 2.3** *Steady diffusion in a hollow cylinder.* Develop the concentration profile in a hollow cylinder when a species is diffusing without any chemical reaction. Consider the concentration of the species to be held constant at the inner and outer surface of the cylinder at $C_{Ai}$ and $C_{Ao}$, respectively.

A mass balance on a thin shell of thickness, $\Delta r$, at radius, $r$, in the cylinder would yield:

$$J_A'' 2\pi r L - J_A'' 2\pi (r + \Delta r) L = 0 \tag{2.64}$$

In the limit when $\Delta r \to 0$:

$$-\frac{\partial (r J_A'')}{\partial r} = 0 \tag{2.65}$$

Upon integration:

$$J_A'' = -\frac{c_1}{r} \tag{2.66}$$

or

$$-D_{AB} \frac{\partial C_A}{\partial r} = -\frac{c_1}{r} \tag{2.67}$$

Upon integration:

$$C_A = \frac{c_1 \ln(r)}{D_{AB}} + c_2 \tag{2.68}$$

From the boundary conditions, $c_1$ and $c_2$ can be solved for:

$$r = R_o, C_A = C_{Ao} \tag{2.69}$$

$$r = R_i, C_A = C_{Ai} \tag{2.70}$$

$$c_1 = \frac{D_{AB}\left(C_{Ao} - C_{Ai}\right)}{\ln\left(\dfrac{R_o}{R_i}\right)}$$ (2.71)

$$c_2 = C_{Ao} - \ln(R_o)\frac{\left(C_{Ao} - C_{Ai}\right)}{\ln\left(\dfrac{R_o}{R_i}\right)}$$

$$c_2 = C_{Ao} - \ln(R_o)\,(C_{Ao} - C_{Ai})/\ln(R_o/R_i)$$ (2.72)

$$\frac{C_A - C_{A0}}{C_{A0} - C_{Ai}} = \frac{\ln(r) - \ln(R_o)}{\ln\left(\dfrac{R_o}{R_i}\right)}$$ (2.73)

Defining a log mean radius:

$$<R_i> = \frac{(R_o - R_i)}{\ln\left(\dfrac{R_o}{R_i}\right)}$$ (2.74)

$$\frac{C_A - C_{Ao}}{C_{Ao} - C_{Ai}} = <R_i> \frac{\ln\left(\dfrac{R_o}{r}\right)}{R_o - R_i}$$ (2.75)

### 2.7.2  Diffusion in Porous Solids

Solute movement by diffusion can be by virtue of concentration difference or by means of pressure difference. Micropores, mesopores, and macropores can be distinguished by means of the pore sizes. Several publications discuss pore diffusion along with gas-solid reactions and catalysis. A Knudsen diffusion may be identified when the mean free path of the molecule is comparable to the pore size. When the pore size to the mean free path of the molecule ratio is about 20, molecular diffusion prevails. When $d/\lambda < 0.2$ rate of diffusion is a function of the collision of the gas molecules and wall, Knudsen diffusion is said to occur:

$$N_A = du_A \Delta p/3RTl$$ (2.76)

where $u_A$ is the molecular velocity of $A$. The Knudsen diffusion coefficient:

$$D_{KA} = d/3\,(8RT/\pi M_A)^{1/2}$$ (2.77)

and the mean free path, $\lambda$, is expressed as:

$$\lambda = 3.2 \, \mu \, (RT/2\pi M) \tag{2.78}$$

In the range of $d/\lambda$ from roughly 0.2 to 20, both molecular diffusion and Knudsen diffusion are important:

$$N_A = (D_{AB,eff} \, p_t / RTz) \, \ln(N_A/N(1 + D_{AB,\,eff}/D_{KA,eff}) - y_{A2})/$$

$$(N_A/N(1 + D_{B,eff}/D_{KA,eff}) - y_{A1}) \tag{2.79}$$

Hydrodynamic flow of gases will occur when there is a difference in absolute pressure across a porous solid. Consider a solid consisting of uniform straight capillary tubes of diameter $d_c$ and length $l$ reaching from the high-pressure side to the low-pressure side. Assuming laminar flow, Hagen-Poiseuille's law for a compressible fluid that obeys the ideal gas law can be written as:

$$N_A = d^2 \, p_{t,av} \, (p_{t,1} - p_{t,2})/32\mu l RT \tag{2.80}$$

The entire pressure difference is assumed to be the result of friction in the pores and ignores entrance and exit losses and kinetic energy effects.

### 2.7.3   Diffusion in Polymers

The diffusion coefficients for polymers lie in between that of solids and liquids. Different systems where diffusion of high-molecular-weight substances become of importance is when the polymer forms a solute of a dilute solution or one component of a polymer-polymer blend. A polymer blend can be miscible, immiscible, compatible, or incompatible. When two polymers are mixed to yield a product with improved property, it is said to be *compatible* blend. When two polymers mix at a molecular level, they are said to be *miscible* blends. A concentrated system where the volume fraction of the polymer solute is large is another category where diffusion has to be treated in a different manner compared to other systems.

A polymer molecule dissolved in a solvent can be envisioned as a necklace comprising spherical beads connected by string [23]. The polymer molecules are separated and only interact through the solvent. The Stokes-Einstein equation for the diffusion coefficient of the polymer can be used for a Flory theta solvent. The root mean square radius of gyration used as a measure of the size of the polymer can be used as the radius of the solute in the Stokes-Einstein formula. These values can be measured by light scattering. For concentrated solutions, the diffusivity is given by:

$$D = D_0(1 + \partial \ln \gamma_1 / \partial \ln \phi_1) \tag{2.81}$$

$D_0$ includes the solute's activation energy. This must be sufficient to overcome any attractive forces that constrain its near neighboring polymer segments. This coefficient can be expected to vary with the free volume of the polymer chains. Only the fraction of the free volume, the hole-free volume, will be accessed by the solute:

$$D_0 = D_0' \exp(-E/RT) \exp(-\omega_1 V_{10} + \omega_2 V_{20}/(\omega_1 K_1 + \omega_2 K_2)) \qquad (2.82)$$

where $E$ is the solute-polymer attractive energy, $\omega_i$ is the mass fractions, $V_{i0}$ is the specific critical free volumes, and $K_i$ is the additional free volume parameters. These parameters are strong functions of the actual temperature minus the glass transition temperatures.

For polymer blends, the Rouse model is suggested:

$$D = k_B T / N_p \zeta \qquad (2.83)$$

where $N_p$ is the degree of polymerization and $\zeta$ is the friction coefficient characteristic of the interaction of a bead with its surroundings.

## 2.8  Transient Diffusion [24–31]

The transient concentration profile due to molecular diffusion can be described using Fick's second law of diffusion and the damped wave diffusion and relaxation equation. The parabolic Fick model and damped wave diffusion and relaxation model for transient mass flux at the surface for the problem of transient diffusion in a semi-infinite medium subject to a step-change in concentration at the surface was found by Sharma [25] to be within 10% of each other for times $t > 2\tau_r$ (Fig. 9.7). This checks out with the Boltzmann transformation—the hyperbolic governing equation reverts to the parabolic at long times. At short times, there is a "blow-up" in the parabolic model. In the hyperbolic model, there is no singularity. This has significant implications in several industrial applications, such as gel acrylamide electrophoresis used in obtaining the sequence distribution of DNA and protein microstructure.

The Fick regime is valid for materials with small relaxation times, long times, and moderate-to-small mass flux rates. The wave regime and the hyperbolic model are valid for short times, high mass flux rates, and materials with large relaxation times. There were some concerns expressed in the literature that the hyperbolic mass diffusion equation violates the second law of thermodynamics. The equation was shown to yield well-bounded solutions in accordance with the second law of thermodynamics by Sharma [24] when final condition in time was used. This condition is a more realistic representation of the transient events in molecular diffusion in practice. The physical significance of the damped wave equation needs to be borne in mind when applying it.

The solution developed by Baumeister and Hamill [32] by the method of Laplace transforms was further integrated into a useful expression. A Chebyshev polynomial approximation was used to approximate the integrand with the modified Bessel composite function of space and time of the first kind and first order. The error involved in Chebyshev economization was $4.1 \times 10^{-5}\, \eta\xi$. The useful expression for transient temperature was shown in Fig. 9.8 for a typical time of $\tau = 5$. The dimensionless temperature as a function of dimensionless distance is shown in Fig. 9.8. The predictions from Baumeister and Hamill and the solution obtained by the method of relativistic transformation are within 12% of each other, on average. Close to the wavefront, the error in the Chebyshev economization is expected to be small and verified accordingly. Close to the surface, the numerical error involved in the Chebyshev economization can be expected to be significant. This can be seen in Fig. 9.8 close to the surface. The method of relativistic transformation yields bounded solutions without any singularities. The transformation variable $\psi$ is symmetric with respect to space and time. It transforms the partial differential equation (PDE) that governs the wave temperature into a Bessel differential equation. The penetration distance beyond which there is no effect of the step change in temperature at the surface for a considered instant in time is shown in Fig. 9.8. The solutions from the relativistic transformation of coordinates is an improvement over the Baumeister and Hamill solution and parabolic Fourier solution in depicting the transient heat events in a semi-infinite medium subject to a step change in boundary temperature. Four regimes in the transient temperature solution for the hyperbolic governing equation using the method of relativistic transformation of coordinates are recognized, and closed-form analytical solutions in each regime are given without any singularities. The transient temperature is also found in accordance with the second law of thermodynamics in all four regimes.

### 2.8.1    Fick Molecular Diffusion—Semi-Infinite Medium

Consider a semi-infinite medium at an initial concentration of a species, $A$, at $C_{A0}$ (Fig. 2.2). For times greater than 0, the surface at $x = 0$ is maintained at a constant surface concentration at $C_A = C_{As}$, $C_{As} > C_{A0}$.

$C_A = C_{As}$

$x = 0$

$x - \infty$

**FIGURE 2.2**    Semi-infinite medium with initial concentration at $C_{A0}$.

The boundary conditions and initial condition are as follows:

$$t = 0, C_A = C_{A0} \tag{2.84}$$

$$x = 0, C_A = C_{As} \tag{2.85}$$

$$x = \infty, C_A = C_{A0} \tag{2.86}$$

The transient concentration in the semi-infinite medium can be obtained by solving the Fick parabolic mass diffusion equations using the Boltzmann transformation $\eta = x/\sqrt{4Dt}$ as follows. The governing equation for molecular diffusion in one dimension using Fick's second law can be written as:

$$\frac{\partial C_A}{\partial t} = D_{AB} \frac{\partial^2 C_A}{\partial x^2} \tag{2.87}$$

Equation (2.87) is a parabolic PDE of the second order in space and time.

Let:

$$u = \frac{(C_A - C_{A0})}{(C_{As} - C_{A0})}; \eta = \frac{x}{\sqrt{4D_{AB}t}} \tag{2.88}$$

Equation (2.87) becomes:

$$-\frac{\eta \partial u}{2t \partial \eta} = \frac{\partial^2 u}{\partial \eta^2} \frac{D_{AB}}{4t D_{AB}} \tag{2.89}$$

or

$$-\frac{2\eta \partial u}{\partial \eta} = \frac{\partial^2 u}{\partial \eta^2} \tag{2.90}$$

The three conditions, one in time and two in space, given by Eqs. (2.84) to (2.86) become:

$$\eta = 0, u = 1 \tag{2.91}$$

$$\eta = \infty, u = 0 \tag{2.92}$$

Thus a PDE of the second order in space and time can be transformed into an ordinary differential equation (ODE) in one variable. The transformation $\eta = x/\sqrt{4D_{AB}t}$ is called the *Boltzmann transformation*. The solution to the ODE in the transformed variable, $\eta$, can be written as:

$$u = c_1 \int_0^\eta e^{-\eta^2} d\eta + c_2 \tag{2.93}$$

The integration constants, $c_1$ and $c_2$, can be solved for using the boundary conditions given by Eqs. (2.91) and (2.92). Thus:

$$u = \frac{(C_A - C_{A0})}{(C_{As} - C_{A0})} = 1 - \mathrm{erf}\left(\frac{x}{\sqrt{4D_{AB}t}}\right) \tag{2.94}$$

The mass flux can be written as:

$$J^* = \frac{J''}{(C_{As} - C_{A0})} \sqrt{\frac{\tau_r}{D_{AB}}} = \frac{\sqrt{\tau_r}}{\sqrt{4\pi t}} \exp\left(-\frac{x^2}{4 D_{AB} t}\right) \qquad (2.95)$$

The dimensionless mass flux at the surface is then given by:

$$q_s^* = \frac{1}{\sqrt{\pi \tau}} \qquad (2.96)$$

## 2.8.2   Damped Wave Diffusion and Relaxation

The semi-infinite medium is considered to study the spatio-temporal patterns that the solution of the non-Fick damped wave diffusion and relaxation equation exhibits. This kind of consideration has been used in the study of Fick mass diffusion. The boundary conditions can be of different kinds, such as the constant wall concentration, the constant wall flux (CWF), pulse injection, convective, impervious, and exponential decay. The similarity or Boltzmann transformation worked out well in the case of the parabolic PDE, where an error function solution can be obtained in the transformed variable. The conditions at infinite width and zero time are the same. The conditions at zero distance from the surface and at infinite time are the same.

Baumeister and Hamill [32] solved the hyperbolic heat conduction equation in a semi-infinite medium subjected to a step change in temperature at one of its ends using the method of Laplace transform. The space-integrated expression for the temperature in the Laplace domain had the inversion readily available within the tables. This expression was differentiated using Leibniz's rule, and the resulting temperature distribution was given for $\tau > X$ as:

$$u = \frac{(C_A - C_0)}{(C_{AS} - C_0)} = \exp\left(\frac{-X}{2}\right) + X\int_X^\tau \exp\left(\frac{-p}{2}\right) \frac{I_1\sqrt{p^2 - X^2}}{\sqrt{p^2 - X^2}} dp \qquad (2.97)$$

The method of relativistic transformation of coordinates is evaluated to obtain the exact solution for the transient temperature. Consider a semi-infinite slab at initial concentration, $C_0$, imposed by a constant wall concentration, $C_s$, for times greater than zero at one of the ends. The transient concentration as a function of time and space in one dimension is obtained. Obtaining the dimensionless variables:

$$u = \frac{(C_A - C_{A0})}{(C_{AS} - C_{A0})} ; \tau = \frac{t}{\tau_{mr}} ; X = \frac{x}{\sqrt{D\tau_{mr}}} ; J^* = \frac{J''}{\sqrt{\frac{D}{\tau_r}}(C_{AS} - C_{A0})} \qquad (2.98)$$

The mass balance on a thin spherical shell at $x$ with thickness $\Delta x$ is written in one dimension as $-\partial J^*/\partial X = \partial u/\partial \tau$. The governing equation can be obtained in terms of the mass flux after eliminating the concentration between the mass balance equation and the non-Fick expression:

$$\frac{\partial J^*}{\partial \tau} + \frac{\partial^2 J^*}{\partial \tau^2} = \frac{\partial^2 J^*}{\partial X^2} \tag{2.99}$$

It can be seen that the governing equation for the dimensionless mass flux is identical in form to that of the dimensionless concentration. The initial condition is:

$$\tau = 0, J^* = 0 \tag{2.100}$$

The boundary conditions are:

$$X = \infty, J^* = 0 \tag{2.101}$$

$$X = 0, C = C_s; u = 1 \tag{2.102}$$

Let us suppose that the solution for $J^*$ is of the form $w \exp(-n\tau)$ for $\tau > 0$ where $W$ is the transient wave flux. Then, when $n = \frac{1}{2}$, Eq. (2.99) becomes:

$$\frac{\partial^2 w}{\partial \tau^2} - \frac{w}{4} = \frac{\partial^2 w}{\partial x^2} \tag{2.103}$$

The solution to Eq. (2.103) can be obtained by the following relativistic transformation of coordinates for $\tau > X$. Let $\eta = (\tau^2 - X^2)$. Then Eq. (2.103) becomes:

$$\frac{\partial^2 w}{\partial \tau^2} = 4\tau^2 \frac{\partial^2 w}{\partial \eta^2} + 2\frac{\partial w}{\partial \eta} \tag{2.104}$$

$$\frac{\partial^2 w}{\partial X^2} = 4X^2 \frac{\partial^2 w}{\partial \eta^2} - 2\frac{\partial w}{\partial \eta} \tag{2.105}$$

Combining Eqs. (2.104) and (2.105) into Eq. (2.103):

$$4(\tau^2 - X^2)\frac{\partial^2 w}{\partial \eta^2} + 4\frac{\partial w}{\partial \eta} - \frac{-w}{4} = 0 \tag{2.106}$$

$$\eta^2 \frac{\partial^2 w}{\partial \eta^2} + \eta\frac{\partial w}{\partial \eta} - \frac{\eta w}{16} = 0 \tag{2.107}$$

Equation (2.107) can be seen to be a special differential equation in one independent variable. The number of variables in the hyperbolic PDE has thus been reduced from two to one. Comparing Eq. (2.107) with the generalized form of Bessel's equation, it can be seen that $a = 1, b = 0, c = 0, s = \frac{1}{2}$, and $d = -\frac{1}{16}$. The order of the solution is calculated as 0 and the general solution is given by:

$$w = c_1 I_0 \left[\frac{\sqrt{\tau^2 - X^2}}{2}\right] + c_2 K_0 \left[\frac{\sqrt{\tau^2 - X^2}}{2}\right] \tag{2.108}$$

The wave flux, $w$, is finite when $\eta = 0$, and hence it can be seen that $c_2$ can be seen to be zero. The $c_1$ can be solved from the boundary condition given in Eq. (2.102). The expression for the dimensionless mass flux for times, $\tau$, greater than $X$ is thus:

$$J^* = c_1 \exp\left(\frac{-\tau}{2}\right) I_0\left[\frac{1}{2}\sqrt{\tau^2 - X^2}\right] \tag{2.109}$$

For large times, the modified Bessel's function can be given as an exponential and reciprocal in square root of time by asymptotic expansion. Consider the surface flux, i.e., when in Eq. (2.109) $X$ is set as zero:

$$J^* = c_1 \exp\left(\frac{-\tau}{2}\right)\frac{\exp\left(\frac{\tau}{2}\right)}{\sqrt{\pi\tau}} = \frac{c_1}{\sqrt{\pi\tau}} \tag{2.110}$$

For times when $\exp(\tau)$ is much greater than the mass flux, it can be seen that the second derivative in time of the dimensionless flux in Eq. (2.99) can be neglected, compared with the first derivative. The resulting expression is the familiar expression for surface flux from the Fourier parabolic governing equation for constant wall concentration in a semi-infinite medium, and is given by:

$$J^* = \frac{1}{\sqrt{\pi\tau}} \tag{2.111}$$

Comparing Eq. (2.111) and Eq. (2.110) it can be seen that $c_1$ is 1. Thus, the dimensionless heat flux is given by:

$$J^* = \exp\left(\frac{-\tau}{2}\right) I_0\left(\frac{\sqrt{\tau^2 - X^2}}{2}\right) \tag{2.112}$$

The solution for $J^*$ needs to be converted to the dimensionless concentration, $u$, and then the boundary conditions applied. From the mass balance:

$$-\frac{\partial J^*}{\partial X} = \frac{\partial u}{\partial \tau} \tag{2.113}$$

Thus, differentiating Eq. (2.112) wrt to $X$ and substituting in Eq. (2.113) and integrating both sides wrt $\tau$. For $\tau > X$:

$$u = \int \exp\left(\frac{-\tau}{2}\right)\left[\frac{I_1\frac{1}{2}\sqrt{\tau^2 - X^2}}{\sqrt{\tau^2 - X^2}}\right] d\tau + c(X) \tag{2.114}$$

It can be left as an indefinite integral and the integration constant can be expected to be a function of space. The $c(X)$ can be solved for by examining what happens at the wavefront. At the wavefront, $\eta = 0$ and time elapsed equals the time taken for a mass disturbance to reach the location $x$ given the wave speed sqrt $\sqrt{D/\tau_{mr}}$. The governing equations for the dimensionless mass flux and dimensionless concentration are identical in form. At the wavefront, Eq. (2.106) reduces to:

$$\frac{\partial w}{\partial \eta} = \frac{w}{16}$$

or

$$w = c' \exp\frac{\eta}{16} = c' \tag{2.115}$$

$$u = c' \exp\frac{-\tau}{2} = c' \exp\frac{-X}{2} \tag{2.116}$$

Thus,

$$c(X) = c' \exp\frac{-X}{2}.$$

Thus,

$$u = \int X \exp\frac{-\tau}{2}\frac{I_1\frac{1}{2}\sqrt{\tau^2 - X^2}}{\sqrt{\tau^2 - X^2}}d\tau + c'\exp\frac{-X}{2} \tag{2.117}$$

From the boundary condition in Eq. (2.102) it can be seen that $c' = 1$. Thus, for $\tau > X$, it can be seen that the boundary conditions are satisfied by Eq. (2.117), and it describes the transient concentration as a function of space and time that is governed by the hyperbolic wave diffusion and relaxation equation. The flux expression is given by Eq. (2.112).

It can be seen that expressions for dimensionless mass flux and dimensionless concentration given by Eq. (2.112) and Eq. (2.117) are valid only in the open interval for $\tau > X$. When $\tau = X$, the wavefront condition results and the dimensionless mass flux and concentration are identical:

$$J^* = u = \exp\frac{-X}{2} = \exp\left(\frac{-\tau}{2}\right) \tag{2.118}$$

When $X > \tau$, the transformation variable can be redefined as $\eta = X^2 - \tau^2$. Equation (2.106) becomes:

$$\eta^2\frac{\partial^2 w}{\partial \eta^2} + \eta\frac{\partial w}{\partial \eta} + \eta\frac{w}{16} = 0 \tag{2.119}$$

The general solution for this Bessel equation is given by:

$$w = c_1 J_0\left[\frac{\sqrt{\eta}}{2}\right] + c_2 Y_0\left[\frac{\sqrt{\eta}}{2}\right] \tag{2.120}$$

The wave temperature, $W$, is finite when $\eta = 0$, and hence it can be seen that $c_2$ can be seen to be zero. The $c_1$ can be solved from the boundary condition given in Eq. (2.56). The expression in the open interval or the dimensionless heat flux for times $\tau$ smaller than $X$ is thus:

$$J^* = c_1 \exp\left(\frac{-\tau}{2}\right) J_0 \left[\frac{\sqrt{X^2 - \tau^2}}{2}\right] \qquad (2.121)$$

On examining the Bessel function in Eq. (2.121), it can be seen that the first zero of the Bessel occurs when the argument becomes 2.4048. Beyond that point the Bessel function will take on negative values, indicating a reversal of heat flux. There is no good reason for the mass flux to reverse in direction at short times. Hence, Eq. (2.121) is valid from the wavefront down to where the first zero of the Bessel function occurs, and the plane of zero transfer explains the initial condition verification from the solution.

By using the expression at the wavefront for the dimensionless mass flux, $c_1$ can be solved for and found to be 1. Equation (2.121) can also be obtained directly from Eq. (2.112) by using $I_0(\eta) = J_0(i\eta)$. The expression for temperature in a similar vein for the open interval $X > \tau$ is thus:

$$u = \int X \exp\left(\frac{-\tau}{2}\right) \frac{J_1\left[\frac{\sqrt{\tau^2 - X^2}}{2}\right]}{\sqrt{\tau^2 - X^2}} d\tau + \exp\left(\frac{-X}{2}\right) \qquad (2.122)$$

Consider a point $X_p$ in the semi-infinite medium. Three regimes can be identified in the mass flux at this point from the surface as a function of time. The series expansion of the modified Bessel composite function of the first kind and zeroth order was used using a Microsoft Excel spreadsheet on a Pentium IV desktop microcomputer. The four regimes and the mass flux at the wavefront are summarized as follows:

1. The first regime is a thermal inertia regime when there is no transfer.

2. The second regime is given by Eq. (2.121) for the mass flux and

$$J^* = \exp\left(\frac{-\tau}{2}\right) J_0 \left[\frac{\sqrt{X^2 - \tau^2}}{2}\right] \qquad (2.123)$$

The first zero of the zeroth-order Bessel function of the first kind occurs at 2.4048. This is when

$$2.4048 = \frac{\sqrt{X^2 - \tau^2}}{2} \quad \text{or} \quad \tau_{lag} = \sqrt{X^2 - 23.132} \qquad (2.124)$$

Thus, $\tau_{lag}$ is the inertial lag that will ensue before the mass flux is realized at an interior point in the semi-infinite medium at a dimensionless distance $X$ from the surface. By way of demonstration, one value of $X$ is used, i.e., 5. Thus, for points closer to the surface the time lag may be zero. Only for dimensionless distances greater than 4.8096 is the time lag finite. For distances *closer than 4.8096 sqrt($\alpha\tau_r$)*, the thermal lag experienced *will be zero*. For distances:

$$x > 4.8096\sqrt{\alpha\tau_{mr}} \qquad (2.125)$$

The time lag experienced is given by Eq. (2.124) and is sqrt$(X^2 - 4\beta_1^2)$ where $\beta_1$ is the first zero of the Bessel function of the first kind and zeroth order, and is 2.4048. In a similar fashion, the penetration distance of the disturbance for a considered instant in time, beyond which the change in initial temperature is zero, can be calculated as:

$$X_{pen} = \sqrt{23.132 + \tau_i^2}$$

3. The third regime starts at the wavefront and is described by Eq. (2.112).

$$J^* = \exp\left(\frac{-\tau}{2}\right) I_0 \left(\frac{\sqrt{\tau^2 - X^2}}{2}\right) \qquad (2.126)$$

4. At the wavefront, $J^* = u = \exp(-X/2) = \exp(-\tau/2)$.

The expressions for transient concentration derived in the previous section need to be integrated prior to use. More easily usable expressions can be developed by making suitable approximations. Realizing that for a PDE, a set of functions instead of constants (as in the case of an ODE) needs to be solved from the boundary conditions, the $c$ in Eq. (2.101) is allowed to vary with time. This results in an expression for transient concentration that is more readily available for direct use. Extensions to three dimensions in space are also straightforward in this method.

In this section, the exact solution for the constant wall concentration problem in a semi-infinite medium in one dimension is revisited since the discussion by the method of Laplace transforms by Baumeister and Hamill. This section will attempt to derive an expression that does not need further integration. Consider a semi-infinite slab at initial concentration, $C_0$, subjected to a sudden change in concentration at one of the ends to $C_s$. The mass propagative velocity is $V_m = $ sqrt $(D_{AB}/\tau_r)$. The initial condition:

$$t = 0, Vx, C = C_0 \qquad (2.127)$$
$$t > 0, x = 0, C = C_s \qquad (2.128)$$
$$t > 0, x = \infty\ C = C_0 \qquad (2.129)$$

Obtaining the dimensionless variables:

$$u = \frac{(C - C_0)}{(C_s - C_0)}; \tau = \frac{t}{\tau_{mr}}; X = \sqrt{D\tau_{mr}} \tag{2.130}$$

The mass balance on a thin spherical shell at $x$ with thickness $\Delta x$ is written. The governing equation can be obtained after eliminating $J''$ between the mass balance equation and the derivative with respect to $x$ of the flux equation and introducing the dimensionless variables:

$$\frac{\partial u}{\partial \tau} + \frac{\partial^2 u}{\partial \tau^2} = \frac{\partial^2 u}{\partial X^2} \tag{2.131}$$

Suppose $u = \exp(-n\tau)\, w\, (X, \tau)$. By choosing $n = \frac{1}{2}$, the damping component of the equation is removed. Thus, for $n = \frac{1}{2}$, the governing equation becomes:

$$\frac{\partial^2 w}{\partial \tau^2} - \frac{w}{4} = \frac{\partial^2 w}{\partial x^2} \tag{2.132}$$

The solution to Eq. (2.132) can be obtained by the following relativistic transformation of coordinates for $\tau > X$. Let $\eta = (\tau^2 - X^2)$. Then Eq. (2.132) becomes:

$$\frac{\partial^2 w}{\partial \tau^2} = 4\tau^2 \frac{\partial^2 w}{\partial \eta^2} + 2\frac{\partial w}{\partial \eta}$$

$$\frac{\partial^2 w}{\partial X^2} = 4X^2 \frac{\partial^2 w}{\partial \eta^2} - 2\frac{\partial w}{\partial \eta} \tag{2.133}$$

Combining Eqs. (2.133) and (2.132):

$$4(\tau^2 - X^2)\frac{\partial^2 w}{\partial \eta^2} + 4\frac{\partial w}{\partial \eta} - \frac{-w}{4} = 0 \tag{2.134}$$

$$\eta^2 \frac{\partial^2 w}{\partial \eta^2} + \eta\frac{\partial w}{\partial \eta} - \frac{\eta w}{16} = 0 \tag{2.135}$$

Equation (2.135) can be seen to be a special differential equation in one independent variable. The number of variables in the hyperbolic PDE has thus been reduced from two to one. Comparing Eq. (2.135) with the generalized form of Bessel's equation, it can be seen that $a = 1$, $b = 0$, $c = 0$, $s = \frac{1}{2}$, and $d = -\frac{1}{16}$. The order of the solution is calculated as 0 and the general solution is given by:

$$w = c_1 I_0\left[\frac{\sqrt{\tau^2 - X^2}}{2}\right] + c_2 K_0\left[\frac{\sqrt{\tau^2 - X^2}}{2}\right] \tag{2.136}$$

The wave temperature, $w$, is finite when $\eta = 0$, and hence, it can be seen that $c_2$ can be seen to be zero. The $c_1$ can be solved from the boundary condition given in Eq. (2.128). For $X = 0$, $u$ is 1. Writing the expression for at $X = 0$:

$$1 = c_1 \exp\left(\frac{-\tau}{2}\right) I_0\left(\frac{\sqrt{\eta}}{2}\right) \tag{2.137}$$

$c_1$ can be eliminated by dividing Eq. (2.136) after setting $c_2 = 0$ by Eq. (2.137) to yield in the open interval of $\tau > X$:

$$u = \frac{I_0\left[\dfrac{\sqrt{\tau^2 - X^2}}{2}\right]}{I_0\left[\dfrac{\tau}{2}\right]} \tag{2.138}$$

In the open interval $X > \tau$:

$$u = \frac{J_0\left[\dfrac{\sqrt{X^2 - \tau^2}}{2}\right]}{I_0\left[\dfrac{\tau}{2}\right]} \tag{2.139}$$

It can be inferred that an expression in time is used for $c_1$. A domain-restricted solution for short and long times may be in order. The dimensionless concentration profile as a function of dimensionless distance for different values of dimensionless times is shown in Fig. 2.3.

### 2.8.3   Periodic Boundary Condition

Consider a semi-infinite slab at initial concentration, $C_0$, imposed by a periodic concentration at one of the ends by $C_0 + C_1 \cos(\omega t)$. The transient concentration as a function of time and space in one dimension is obtained. Obtaining the dimensionless variables:

$$u = \frac{(C - C_0)}{C_1}; \ \tau = \frac{t}{\tau_{mr}}; \ X = \frac{x}{\sqrt{D\tau_{mr}}} \quad u = (C - C_0)/(C_1);$$

$$\tau = t/\tau_r; X = x/\mathrm{sqrt}(D\tau_r) \tag{2.140}$$

The mass balance on a thin shell at $x$ with thickness $\Delta x$ is written. The governing equation is obtained after eliminating $J$ between the mass balance equation and the derivative with respect to $x$ of the flux equation and introducing the dimensionless variables. The initial condition is:

$$t = 0, C = C_0; u = 0 \tag{2.141}$$

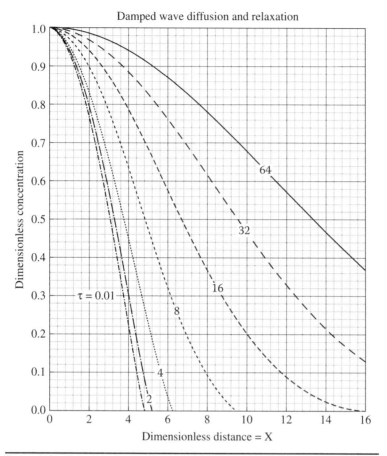

**FIGURE 2.3**    Concentration profile under damped wave diffusion and relaxation in semi-infinite medium.

The boundary conditions are:

$$X = \infty, C = C_0 ; u = 0 \qquad (2.142)$$

$$X = 0, C = C_0 + C_1 \cos(\omega t); u = \cos(\omega^* \tau) \qquad (2.143)$$

Let us suppose that the solution for $u$ is of the form $f(x)\exp(-i\omega^*\tau)$ for $\tau > 0$ where $\omega$ is the frequency of the concentration wave imposed on the surface and $C_1$ is the amplitude of the wave. Then:

$$(-i\omega^*) f \exp(-i\omega^*\tau) + (i^2\omega^{*2})f \exp(-i\omega\tau) = f'' \exp(-i\omega^*\tau) \qquad (2.144)$$

$$i^2 f(\omega^{*2} + i\omega^*) = f''$$

$$f(X) = c \exp(-iX\omega^*\text{sqrt}(\omega^* + i)) \qquad (2.145)$$

$d$ can be seen to be zero as at $X = \infty$, $u = 0$.

$$u = c\exp(-iX\omega^* \sqrt{\omega^* + i})\exp(-i\omega^* \tau) \qquad (2.146)$$

From the boundary condition at $X = 0$:

$$\cos(\omega^*\tau) = \text{real part } (c\exp(-i\omega^*\tau)) \text{ or } c = 1 \qquad (2.147)$$

$$u = \exp(-X\omega^*(A + iB)\exp(-i\omega^*\tau))$$
$$= \exp(-A\omega^* X)\exp(-i(BX\omega^* + \omega^*\tau)) \qquad (2.148)$$

where $\qquad\qquad A + iB = i\,\text{sqrt}(\omega^* + i) \qquad (2.149)$

Squaring both sides:

$$A^2 - B^2 + 2ABi = i^2 (\omega^* + i) = -\omega^* - i \qquad (2.150)$$

$$A^2 - B^2 = -\omega^* \; ; 2AB = -1 \text{ or } B = -1/2A$$

or $\qquad\qquad A^2 - 1/4A^2 = -\omega^* \qquad (2.151)$

$$A^2 = (-\omega^* \pm \text{sqrt } (\omega^{*2} + 1))/2 \; ; B = -1/2A \qquad (2.152)$$

Obtaining the real part:

$$u = \exp(-A\omega^* X)\cos(\omega^*(BX + \tau)) \qquad (2.153)$$

The time lag in the propagation of the periodic disturbance at the surface is captured by the previous relation. Thus, the boundary conditions can be seen to be satisfied by Eq. (2.145). In a similar vein to the supposition of $f(x)\exp(-i\omega^*\tau)$, the mass flux $J''$ can be supposed to be of the form $J^* = g(x)\exp(-i\omega^*\tau)$. Thus:

$$g = \frac{f'}{(1 - i\omega^*)} \qquad (2.154)$$

Combining the $f$ from Eq. (2.145) into Eq. (2.154):

$$J^* = -\omega^*(A + iB) \exp(-X\omega^*(A + iB)\exp(-i\omega^*\tau)) \qquad (2.155)$$

$$= -\omega^*(A + iB) \exp(-A\omega^* X)\exp(-i(BX\omega^* + \omega^*\tau))$$

$$= -\omega^*(A + iB)\exp(-A\omega^* X)(\cos(BX\omega^* + \omega^*\tau) + i\sin(BX\omega^* + \omega^*\tau))$$

Obtaining the real part:

$$J'' = \sqrt{\frac{D}{\tau_{mr}}}\; \omega^*\exp(-A\omega^* X)(B\sin(\omega^*(BX + \tau)) - A\cos(\omega^*(BX + \tau)))$$
$$(2.156)$$

## Summary

Diffusion is a phenomenon whereby a species migrates from a region of higher concentration to a region of lower concentration. The driving force for motion is the concentration gradient. The Skylab demonstration experiments by Fascimire documents the diffusion of tea in water under reduced gravity conditions. The lowest achievable concentration is 0 molm$^{-3}$ by law. Fick's first and second laws can be written as:

$$J = -D_{ij}A\frac{\partial C_i}{\partial x}$$

$$D_{ij}\frac{\partial^2 C_i}{\partial x^2} = \frac{\partial C_i}{\partial t}$$

The $N$ and $J$ fluxes are distinguished from each other. $J$ describes the molecular diffusion and $N$ the migration due to bulk motion. The diffusion coefficient varies with temperature. In gases, the correlations of Chapman and Enskog, Wilke-Lee, Mathur and Thodos, and Catchpole and King are presented. Binary diffusion coefficient values for commonly available gases are given in Table 2.1. For liquids, the Stokes-Einstein relation for diffusion coefficients was derived. During the derivation, when accounting for the acceleration regime of the solute molecule, the generalized Fick's laws of diffusion were derived:

$$J'' + \frac{M_i D_{ij}}{RT}\frac{\partial J''}{\partial t} = -\left(\frac{RT}{6\pi\mu R_o}\right)\frac{\partial C_i}{\partial x}$$

Correlations of Nernst-Haskell for electrolytes were mentioned. The effect of concentration, i.e., dilute versus concentrated solutions, were discussed separately. Correlations of Wilke-Chang, Siddiqi-Lucas, and Haydeek-Minhas were included. The diffusion mechanism in solids was discussed. The different mechanisms of diffusion, such as vacancy mechanism, interstitial mechanism, substitutional mechanism, and crowd ion mechanism were outlined. The Knudsen diffusion when the mean free path of the molecule is greater than the diffusion path, such as in pore diffusion, was discussed. The diffusion in polymers and the Arrhenius dependence of diffusion coefficient with temperature were discussed.

The transient diffusion in a semi-infinite medium was studied under a constant wall concentration boundary condition using Fick's second law of diffusion and the damped wave diffusion and relaxation equation. The latter can account for the finite speed of propagation of mass. A new procedure called the method of relativistic transformation was developed to obtain bounded and physically realistic solutions. These were compared with the solution from Fick's second law of diffusion obtained using Boltzmann transformation and the solution presented in the literature by Baumesiter and Hamill [32]. Four different regimes of the solution were recognized: an inertial regime with zero transfer, a second regime characterized by a Bessel composite function of space and time of the zeroth order and first kind, a third regime characterized by a modified Bessel composite function of space and time of the zeroth order and first kind, and a wavefront regime. The characteristics of the solution to the damped wave diffusion and relaxation equation, subject to the periodic boundary condition by the method of complex concentration, were discussed. The transient concentration profile from the relativistic transformation method was presented in an easy-to-use chart in Fig. 2.3. The profile has a point of inflection and zero curvature at $X = 0$. Mathematical expressions for penetration length and inertial lag time were derived.

# References

[1]   E. L. Cussler, *Diffusion Mass Transfer in Fluid Systems*, Cambridge, UK: Cambridge University Press, 1997.
[2]   A. Einstein, *Annalen der Physik* (1905), 7, 549.
[3]   A. E. Fick, *Poggendorff's Annelen der Physik* (1855a), 94, 59.
[4]   A. E. Fick, *Philosophical Magazine* (1855b), 10, 30.
[5]   J. B. Fourier, *Theorie analytique de la chaleur*, English translation by A. Freeman, New York: Dover Publications, 1955.
[6]   B. Fascimire, NASA Marshall Flight Center, AL, 1973.
[7]   E. R. Gilland, *Ind. Eng. Chem.* (1934) 26, 681.
[8]   S. Chapman, and T. G. Cowling, *The Mathematical Theory of Non-Uniform Gases*, Cambridge, UK: Cambridge University Press, 1970.
[9]   R. C. Reid, T. K. Sherwood, and J. M. Prausnitz, *Properties of Gases and Liquids*, New York: McGraw-Hill, 1977.
[10]  Wilke and Lee, *Ind. Eng. Chem.* (1955), 47, 1253.
[11]  Brokaw, *Ind. Eng. Chem. Process. Des. & Dev.* (1969), 8, 2, 240.
[12]  Mathur and Thodos, *AIChE* (1965), 11, 613.
[13]  Catchpole and King, *Ind. Eng. Chem. Process. Des. & Dev.* (1994), 33, 1828.
[14]  R. H. Stokes, *J. of Amer. Chem. Soc.* (1950), 72, 763, 2243.
[15]  R. Resnick and D. Halliday, *Physics*, Part I, 38th Wiley Eastern Reprint, New Delhi, 1991.
[16]  C. R. Wilke and P. C. Chang, *AIChE J* (1955), 1264.
[17]  Siddiqi and Lucas, *Can. J. of Chem. Eng.* (1986), 64, 839.
[18]  Haydeek and Minhas, *Can. J. of Chem. Eng.* (1982), 60, 195.
[19]  G. K. Batchelor, *J. of Fluid Mechanics* (1972), 52, 245, 71, 1.
[20]  NEC Corporation, U.S. Patent 5,784,300 (1998).
[21]  W. D. Callister, *Materials Science and Engineering*, 7th ed., New York: Wiley, 2007.
[22]  J. P. Stark, *Solid State Diffusion*, New York: John Wiley, 1976.
[23]  J. S. Vrentas and J. L. Duda, *J. of Appl. Polym. Sci.* (1980), 25, 1297.
[24]  K. R. Sharma, *Damped Wave Transport and Relaxation*, Amsterdam: Elsevier, 2005.
[25]  K. R. Sharma, "On the solution of damped wave conduction and relaxation equation in a semi-infinite medium subject to constant wall flux," *International Journal of Heat and Mass Transfer*, vol. 51 (2008), 25–26, 6024–6031.
[26]  K. R. Sharma, "Damped wave conduction and relaxation in a finite sphere and cylinder," *Journal of Thermophysics and Heat Transfer*, vol. 22 (2008), 4, 783–786.
[27]  K. R. Sharma, "Damped wave conduction and relaxation in cylindrical and spherical coordinates," *Journal of Thermophysics and Heat Transfer*, vol. 21 (2007), 4, 688–693.

[28] K. R. Sharma, "Manifestation of acceleration during transient heat conduction," *Journal of Thermophysics and Heat Transfer*, vol. 20 (2006), 4, 799–808.

[29] K. R. Sharma, "A fourth mode of heat transfer called damped wave conduction," 42nd Annual Convention of Chemists, Santiniketan, India, February 2006.

[30] K. R. Sharma, "Solution methods and applications for generalized Fick's law of diffusion," invited lecture, 43rd Annual Convention of Chemists, Aurangabad, December 2006.

[31] K. R. Sharma, "On the second law violation in Fourier conduction," 231st ACS National Meeting, Atlanta, GA, March 2006.

[32] K. J. Baumeister and T. D. Hamill, "Hyperbolic heat conduction equation—a solution for the semi-infinite body problem," *ASME J of Heat Transfer* (1971), 93, 126–128.

# Exercises

## Review Questions

**1.0**  What is the difference between self, binary, and ternary diffusion coefficients?

**2.0**  During Brownian motion, the molecules follow a random zigzag path and sometimes move in the opposite direction, compared with the imposed concentration difference driving the diffusion. Is this a violation of the second law of thermodynamics?

**3.0**  What are the differences between multicomponent diffusion and binary diffusion?

**4.0**  What happens to the formula for total flux during equimolar counterdiffusion, compared with that for molecular diffusion?

**5.0**  Correlations for diffusion in gases, liquids, and solids were discussed. What would be appropriate for liquid diffusing in a solid or gases diffusing in a liquid?

**6.0**  Discuss the units of each term in the equation $P = DS$.

**7.0**  Explain the effect of temperature on the mass propagation velocity. What happens to the diffusion coefficient and relaxation time at high pressure?

**8.0**  Why are insects larger in size in the tropics compared with the insects in the Arctic region?

**9.0**  Are the forces of gravity taken into account in the derivation of the Stokes-Einstein relationship for diffusivity coefficients?

**10.0**  Can you expect a plane of zero concentration or null transfer during drug delivery in the tissue region? How so?

**11.0**  Diffusion coefficient is a proportionality constant in Fick's first law of diffusion, independent of concentration. For concentrated solutions, it is said to vary with concentration. How can this be interpreted?

**12.0**   State the Onsager reciprocal relations. Show that $D_{12} = D_{21}$.

**13.0**   What was Landau's observation of the infinite speed of propagation?

**14.0**   What is penetration length?

**15.0**   What is inertial lag time?

**16.0**   What is the first zero of the Bessel function of the first order? How is this used in the derivation of the penetration length and inertial lag time in a three-dimensional medium?

**17.0**   Examine $I_0(\tau/2)\exp(-\tau/2)$ in terms of extremamas and asymptotic limits. Under what conditions can $I_0(\tau/2)$ be reduced to a simpler expression?

**18.0**   What is the meaning of a negative mass flux? What happens to the ratio of accumulation and diffusion ?

## Problems

**19.0**   *Estimate of diffusion coefficient of argon in hydrogen.* Calculate the diffusion coefficient of argon in hydrogen at 1.5 atm and 310 K. Compare this with the experimental values reported in the literature.

**20.0**   *Parabolic law of oxidation.* During the corrosion of metals, an oxide layer is formed on the metal. Assuming that the oxygen diffuses through the oxide layer, show that the thickness of the oxide layer, $\delta$, can be given by $(C_{bulk}\,D_{AB}t/\rho_m)^{1/2}$ using Fick's law of diffusion. A gentle breeze is blowing at a constant velocity of $U$ over the corroded layer. Is this going to increase the rate of corrosion due to the convection contribution?

**21.0**   *Sacred pond.* Evaporation from ponds is retarded by the introduction of lotus leaves in the sacred ponds in temples. Assume that in a pond of area 9 m × 9 m, 4,130 leaves, each with a diameter of 3", were placed. Calculate the reduction in diffusion rate on account of the reduction in area in the path of evaporation.

**22.0**   *Diffusion of oxygen through spiracles.* Many insects breathe through spiracles. Spiracles are open tubes that extend into the insect's body. Oxygen diffuses from the surrounding air and gas exchange takes place through the walls. For every mole of oxygen diffusing in, there is one mole of $CO_2$ diffusing out. To prevent water loss, the walls of the spiracle are coated with a cuticle of 10 μm thickness. The oxygen concentration outside the cuticle is constant and is 5% of the equilibrium concentration. What is the local oxygen flux in the spiracle to the tissue? Derive an oxygen concentration profile within the tissue. Is the spiracle an efficient method of respiration? (Spiracle radius = 100 μm; spiracle length = 9 mm; $D_{o,cuticle} = 3$ E-5 cm$^2$/s; $D_{o,air} = 0.15$ cm$^2$/s; oxygen solubility in tissue $C_t = 0.2$ mmol/L.)

**23.0**  *Scrubbing of SO$_2$.* During coal combustion, the emission of sulfur dioxide from power plants can be reduced by using CaO scrubbers. In the scrubber:

$$2CaO + 2SO_2 + O_2 \rightarrow 2CaSO_4$$

Consider the diffusion of SO$_2$ into a spherical particle of CaO. Show that a governing equation can be derived from the shell balance as:

$$D_{AB} \frac{\partial}{r^2 \partial r}\left(r^2 \frac{\partial C_A}{\partial r}\right) = k'''C_A$$

Show that the concentration profile of SO$_2$ in the spherical lime particle can be written as:

$$\frac{C_A}{C_{As}} = \frac{I_{1/2}\left(r\sqrt{\dfrac{k'''}{D_{AB}}}\right)}{XI_{1/2}\left(R\sqrt{\dfrac{k'''}{D_{AB}}}\right)}$$

The Thiele modulus is:      $$\phi = R\sqrt{\frac{k'''}{D_{AB}}}$$

**24.0**  *Coextrusion.* In the manufacture of the casings of the solid rocket motor (SRM), the material requirements are bifunctional. They have to have high hoop strength on one side and high ablation resistance on the other. In order to prepare such materials, the technology of coextrusion is utilized. In a twin-screw extruder, both the materials are extruded together. During the residence time of the polymers in the extruder, the interdiffusion of either material into the other occurs. Calculate the interlayer thickness as a function of the extruder residence time and diffusivities of the two materials.

**25.0**  *Diffusion coefficient of milk in the refrigerator.* Estimate the diffusion coefficient of lactic acid in the refrigerator. Compare this with the value at room temperature and that of the milk through the plastic container.

**26.0**  Restriction mapping. *Endonucleases,* or restriction enzymes, cut the unmethylated DNA at several sites and restrict its activity. About 300 restriction enzymes are known, and they act upon 100 distinct restriction sites that are palindromes. Some cut leaves with blunt ends and others leave them sticky. The restriction fragment lengths can be measured by using the technique of gel electrophoresis. The solid matrix is the gel usually agarose or polyacrylamide—which is permeated with a liquid buffer. As DNA is a negatively charged molecule when placed in an electric field, the DNA migrates toward the positive pole. DNA migration is a function of its size. Calibration is used to relate the migration distance as a function of size. Migration distance of DNA under a field for a set time is measured. The DNA molecule is made to fluoresce and made visible under ultraviolet light by stainingthe gel with ethidium bromide. A second method is to tag the DNA with a radio active label and then expose the x-ray film to the gel. Show that the migration under gel electrophoresis can be given by:

$$J_{frag} = -(z_A u_A F)\frac{\partial E}{\partial x} - D_{frag}\frac{\partial C_A}{\partial x}$$

Show that the governing equation can be written in one dimension as:

$$0 = D_{frag}\frac{\partial^2 C_A}{\partial x^2} - (z_A u_A F)\frac{\partial^2 E}{\partial x^2}$$

**27.0**  *Pheromone and insect control.* During insect control, controlled release of pheromones are used. Pheromones are sex attractants released by insects. When mixed with an insecticide and used, it annihilates all of one sex of a particular insect pest. The pheromone sublimation rate in the impermeable holder is given as:

$$S_0 = 9 \text{ E-16 } (1 - 1 \text{ E-6 } C_1)$$

where $C_1$ is the concentration in the vapor. The diffusivity through the polymer is 1.2 E-11 cm$^2$/s. It can be assumed that the pheromone outside the chamber is 0. If the polymeric diffusion barrier is 600 microns thick and has an area of 1.6 cm$^2$, what is the concentration of pheromone in the vapor? How fast is the pheromone released by the device?

**28.0**  *Oxygen transport in the eye.* The cornea is a unique, living tissue and is a transparent window through which light enters the eye to be focused on the retina, thus forming the images of our surroundings and enabling sight. When the eye is open, it receives all of its oxygen requirements from the surrounding air. Other nutrients are likely delivered via the tear duct fluid that bathes the outer surface of the cornea or the aqueous humor, which fills the chamber behind the cornea and in front of the lens. Some oxygen may enter the aqueous humor from a vasculature in the muscle around the periphery of the lens. When the eye is closed, it is cut off from the $O_2$ source in the air. There is a rich microvascular bed (well perfused with high vascular density on the inner surface of the eyelid) that supplies the cornea with oxygen and possibly other nutrients. What is the $pO_2$ at the surface of the cornea when the eye is closed?

| Layer | Thickness (μm) | Diffusion Coefficient (cm$^2$/s) | $VO_2$ (mL $O_2 \cdot$ mL tissue$^{-1}$s$^{-1}$) |
|---|---|---|---|
| Epithelium | 40 | 3.8 E-10 | 2.0 E-4 |
| Stroma | 450 | 3.8 E-10 | 1.0 E-5 |
| Endothelium | 10 | 3.8 E-4 | 2.0 E-4 |

Table of Model Parameters

**29.0**  *Loss from beverage containers.* Soft drink bottles are made out of plastic. The contents diffuse at a slow rate through the walls of the container and out into the air, and result in some losses. It has been suggested to coat the inner wall of the container to reduce the losses. With a coating thickness of 25 μm and a diffusion coefficient in the coating of 1 E-9 m²/s, what would be the benefit to the manufacturer? Assume a thickness of 1.5 mm for the plastic container and a diffusion coefficient of the contents in the plastic container as 1 E-6 m²/s.

**30.0** *Reaction and diffusion in a nuclear fuel rod.* In autocatalytic reactions, such as during nuclear fission, the neutrons can be studied by a first-order reaction. The mass balance in a long cylindrical rod with a first-order autocatalytic reaction can be written at steady state as:

$$\frac{1}{r}\frac{\partial(rJ_r)}{\partial r} + k'''C = 0$$

The long cylindrical rod is at zero initial concentration of autocatalytic reactant, $A$. The surface of the rod is maintained at a constant concentration, $C_s$, for times greater than zero. The boundary conditions are:

$$r = 0, \quad \frac{\partial C}{\partial r} = 0$$

$$r = R, C = C_s$$

Show that the steady-state solution can be obtained as follows after redefining $u^s = C/C_s$:

$$\partial^2 u^s/\partial X^2 + 1/X\, \partial u^s/\partial X + k^* u^s = 0$$

$$X^2 \partial^2 u^s/\partial X^2 + X\, \partial u^s/\partial X + X^2 k^* u^s = 0$$

This equation can be recognized as the Bessel equation. The solution is:

$$u^s = c_1 J_0(X\sqrt{k^*}) + c_2 Y_0(X\sqrt{k^*})$$

It can be seen that $c_2 = 0$ as the concentration is finite at $X = 0$. The boundary condition for surface concentration is used to obtain $c_1$. Thus:

$$c_1 = 1/J_0(R\sqrt{k^*}/D\tau_r)$$

Thus:

$$u^s = J_0(X\sqrt{k^*})/J_0(R\sqrt{k^*}/D\tau_r)$$

**31.0** *Grooming hair with oil.* In order to keep the hair on the human skull from becoming dehydrated, it is oiled or hair cream is applied every day. During the course of the day estimate the loss of the oil from the human hair by diffusion. Show that there are two contributions. One is from the molecular diffusion from the head to the atmosphere in the vertical direction and the other is by convection from a wind blowing in the horizontal direction. Show that the governing equation can be given by:

$$\frac{\partial^2 u}{\partial z^2} = \frac{U d_{hair}}{D}\frac{\partial u}{\partial x}$$

Show that the solution for the concentration profile of the oil in the surrounding region of the human skull at a steady state can be given by:

$$u = 1 - \mathrm{erf}\left( Z\sqrt{\frac{Pe_m}{4X}} \right)$$

Assuming that the diameter of the hair is 2 microns, the velocity of air is 1 m/s, and the diffusivity is 1 E-5 m²/s, estimate the time taken for the layer of cream of 1 micron to be replaced. Make suitable assumptions, such as the cranial area is 2,500 cm² and the length of the hair is 5 cm.

**32.0**   *Dyeing of the wool.* A dye bath at a concentration $C_0$ and a volume $V$ is used to dye wool that is bathed in it. The dye diffuses into the wool. Measuring the concentration of the dye in the wool as a function of time, can you a) estimate the diffusion coefficient of the dye (ff so, how and b) estimate the relaxation time?

**33.0**   *Dopant profile by ion implantation.* Ion implantation is used to introduce dopant atoms into the semiconductor material to alter its electrical conductivity. During ion implantation, a beam of ions containing the dopant is directed at the semiconductor surface. For example, boron atoms are implanted into silicon wafers by Lucent Technologies. Assume that the transfer of boron into the silicon surface is on account of both the convection and diffusion contributions at a steady state. Show that the governing equation for the transfer of boron at the gas-solid interface is given by:

$$-\frac{\partial C_A}{\partial z} = D_{AB}\frac{\partial^2 C_A}{\partial z^2}$$

Given a characteristic length $l$, show that the equation can be reduced to:

$$-Pe_m\frac{\partial u}{\partial Z} = \frac{\partial^2 u}{\partial Z^2}$$

and the solution is:

$$u = 1 - \frac{J_{ss}^*}{Pe_m}e^{-Pe_m Z}$$

**34.0**   *Soot from the steam engine.* The steam engine that powers the train that takes you from Chennai to N. Delhi in 31 hours discharges coal dust at a steady rate of 68 kg-mol/hr. The train moves at a velocity of 90 km/hr. Estimate the thickness of soot that will deposit on a passenger sitting near the window of seat S6 during the entire journey. S6 is about 200 feet from the engine. Assume that the diffusion coefficient of the soot in air is 1 E-6 m²/sec. Repeat the analysis for a wind speed of 10 km/hr. (Hint: Bulk concentration of soot in surrounding air can be calculated by considering a basis of time, such as that taken for the passenger to move 600 feet to the discharge point in fixed space, and in that time, the discharge amount is calculated from the discharge rate and the dispersed region from the penetration length in all three directions.)

**35.0**   *Steady diffusion in a hollow sphere.* Develop the concentration profile in a hollow sphere when a species is diffusing without any chemical reaction.

Consider the concentration of the species to be held constant at the inner and outer surfaces of the cylinder at $C_{Ai}$ and $C_{Ao}$, respectively. Show that:

$$\frac{C_A - C_{Ai}}{C_{Ai} - C_{Ao}} = \frac{\left(1 - \dfrac{R_i}{r}\right)}{\left(1 - \dfrac{R_i}{R_o}\right)}$$

**36.0** *Determination of diffusivity.* Unimolar diffusion can be used to estimate the binary diffusivity of a binary gas pair. Consider the evaporation of $CCl_4$, carbon tetrachloride, into a tube containing oxygen. The distance between the $CCl_4$ level and the top of the tube is 16.5 cm. The total pressure in the system is 760 mmHg and the temperature is $-5°C$. The vapor pressure of $CCl_4$ at that temperature is 29.5 mmHg. The area of the diffusion path in the diffusion tube may be taken as $0.80$ cm$^2$. Determine the binary diffusivity of $O_2$–$CCl_4$ when in an 11-hour period after a steady state, $0.026$ cm$^3$ of $CCl_4$ is evaporated.

**37.0** *Helium separation from natural gas.* McAfee proposed a method to separate helium from natural gas. He noted that Pyrex glass is almost impermeable to all gases but helium. The diffusion coefficient of helium is 25 times the diffusion coefficient of hydrogen. Consider a Pyrex tubing of length, $L$, and inner and outer radii, $R_i$ and $R_o$. Show that the rate at which helium will diffuse through the Pyrex can be given by:

$$J_{He} = \frac{2\pi L D_{He,pyrex}(C_{He,1} - C_{He,2})}{\ln\left(\dfrac{R_o}{R_i}\right)}$$

**38.0** *Solid dissolution into a falling film.* A liquid is flowing in laminar motion down a vertical wall. The wall consists of a species that is slightly soluble in the liquid. Show that the governing equation for species diffusing into the liquid from the wall can be written as:

$$\frac{\partial^2 u}{\partial z^2} = \frac{UL}{D}\frac{\partial u}{\partial x}$$

Show that an error function solution results for this PDE.

**39.0** *Carburizing steel.* Low-carbon steel can be hardened in order to improve the wear resistance by carburizing. Steel is carburized by exposing it to gas, liquid, or solid that provides a high carbon concentration at the surface. Given the percent carbon versus depth graphs for various times at $930°C$, how can the diffusion coefficient be estimated from the graphs?

**40.0** *Electrophoretic term.* For some systems, there is a minus sign in the electrophoretic term, as shown in the following equation. What are the implications of the minus sign in this equation? How will this manifest in applications?

$$-j_A = D\frac{\partial C_A}{\partial z} - \left(\frac{zFm}{RT}\right)C_A + \tau_{mr}\frac{\partial j_A}{\partial t}$$

# CHAPTER 3

# Osmotic Pressure, Solvent Permeability, and Solute Transport

---

## Learning Objectives

- Discuss osmosis, osmotic pressure, and van't Hoff's law
- Learn permeability of a solvent across a membrane, Starling's law
- Familiarize with diffusion mechanisms of a solute across a membrane
- Discuss hindered diffusion of a solute through pores
- Apply the Kedem-Katchalsky equation
- Discuss flow through porous media, Darcy's law
- Derive Starling's law
- Measure a permeability coefficient
- Use Staverman's reflection coefficient and the sieving coefficient
- Estimate effective diffusivity in suspensions
- Design a dialysis system to filter out toxic solutes from the bloodstream
- Characterize body fluids
- Apply the Nernst equation
- Understand electrodialysis and mass exchangers

There were three important developments in the history of biofluid transport phenomena in human anatomy. These are as follows:

1. The discovery of osmosis and osmotic pressure
2. Permeability of a solvent across a membrane and Starling's law
3. Diffusion of a solute across a membrane

## 3.1   Van't Hoff's Law of Osmotic Pressure

The concept of osmotic pressure is illustrated in Fig. 3.1. When a balloon made out of a semipermeable membrane and filled with salt solution is immersed in a bath of pure water, the water will travel from the jar into the balloon and the size of the balloon will increase until equilibrium is reached in terms of the chemical potential on both sides of the membrane. The semipermeable membrane chosen can permit only water and not the solute to a large extent. The flow of water is an example of the concept of osmotic pressure. *Osmosis* is the flow of solvent from a region of low solute concentration to a region of high solute concentration. The pressure difference that causes this flow is called osmotic pressure. This pressure is caused by the presence of solutes. Hence, it is called *colloid osmotic pressure.* For human plasma in the blood, the colloid oncotic pressure is about 28 mmHg. The colloid osmotic pressure is small, compared with the osmotic pressure developed when a human cell is placed in pure water. The total osmotic pressure of the intracellular fluid would be 5450 mmHg at 37°C.

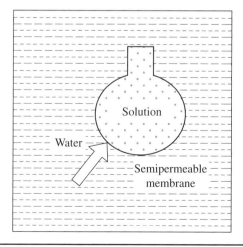

**FIGURE 3.1**   Concept of osmotic pressure.

Osmotic pressure is caused by the presence of solutes such as $K^+$ ions; phosphocreatine; $Mg^{++}$, $Cl^-$, $HCO_3^-$, and $HPO_4^{2-}$ ions; carnosine; amino acids; creatine; lactate; $Na^+$ ions; urea; adenosine triphosphate (ATP), hexose monophosphate; and others. A number of solute molecules contribute to the osmotic pressure. Some of them are dissociating. For example, NaCl dissociates to the $Na^+$ and $Cl^-$ ions. Each ion particle exerts its own osmotic pressure, and the charge of the ion has no bearing on the osmotic pressure. Substances such as glucose do not dissociate, and the osmotic pressure exerted is based upon its concentration. The term *osmole* is introduced to account for the effect of a dissociating solute. Therefore, one osmole is defined as one mole of a nondissociating substance. One mole of dissociating NaCl is equivalent to two osmoles. *Osmolarity* defines the number of osmoles per liter of solution. For physiological solutions, the unit used for convenience is mOs. If a cell is placed within a solution that has a lower concentration of solutes or osmolarity, the cell is in a *hypotonic solution* and establishment of osmotic equilibrium requires the osmotic flow of water into the cell. The influx of water into the cell results in swelling of the cell and a subsequent decrease in its osmolarity. On the other hand, if the cell is placed in a solution with a higher concentration of solutes or osmolarity—that is, a *hypertonic solution*—osmotic equilibrium requires osmotic flow of water out of the cell. An *isotonic solution* is a fluid that has the same osmolarity of the cell. When cells are placed in an isotonic solution, there is neither swelling nor shrinkage of the cell. Examples of isotonic solutions are 0.9 percent by weight NaCl in water solution and 5 percent by weight of glucose solutions with respect to a human cell.

Lettuce leaves in a salad wilt when salt is added. The osmotic pressure exceeds the turgor pressure in the cells of the lettuce, and the water oozes out. The process of wilting is thus accelerated with the addition of common salt. The water droplets on the surface of the leaves come from the interior of the lettuce plant cells. Consequently, the turgor pressure and internal rigidity of the leaves are lowered and they wilt. The process of water transport out of the cells caused by an increase in external salt concentration is an example of osmosis phenomena.

*Dutrochet* discovered the phenomena of osmosis. He made systematic observations of osmotic pressure in the 1800s. He observed that small animal bladders filled with dense solution and then completely closed and plunged in water became turgid and swollen excessively. Water flowed into the bladder so as to dilute the solution inside.

Osmotic phenomena do not violate the second law of thermodynamics. The entropy of the solution is larger than that of the solvent. When brought in contact with each other, the combined system strives to reach a state where the entropy is even higher. This can happen only when the solvent moves from low solute concentration to the solution with higher solute concentration.

*Van't Hoff's law* can be used to determine the osmotic pressure in terms of the concentration of the solution. It is derived from the concept that *fugacity* of different phases needs to be equal at equilibrium. Thus, Fig. 3.1 shows that at equilibrium, the temperature and fugacity of the water and that of the solution must be equal. Fugacity is a measure of the chemical potential of the system. As discussed previously, the difference in the chemical potential between the solution and the solvent causes the osmotic flow from a region of low solute concentration to a region of high solute concentration.

$$f_w(T, P_w) = f_s(T, P_s) \tag{3.1}$$

The fugacity of a solution can be written in terms of the pure-component fugacity using the Poynting factor. The mole fraction of water and the activity coefficient of water are also needed. The Poynting factor corrects for the effect of pressure on the pure-component fugacity where $V_w$ is the molar volume:

$$f_w = \gamma_w x_w f_w \exp\left(-\frac{V_w(P_w - P_s)}{RT}\right) \tag{3.2}$$

The osmotic pressure is given by $(P_s - P_w) = \pi$ and can be solved for from Eq. (3.2) as:

$$\pi = (P_w - P_s) = -\frac{RT}{V_w}\ln(\gamma_w x_w) \tag{3.3}$$

For an ideal solution, the activity coefficient may be taken as 1. For dilute systems using Taylor series expansion, the logarithmic functionality can be approximated as:

$$\ln(x_w) = \ln(1 - x_s) \approx -(x_s) \tag{3.4}$$

Substituting Eq. (3.4) in Eq. (3.3) provides:

$$\pi = \frac{RTx_s}{V_w} = RTC_s \tag{3.5}$$

Equation (3.5) is called van't Hoff's law, and it is used to determine the osmotic pressure. Equation (3.3) may be used when activity coefficient information is available. If the solution contains $N$ ideal solutes, the osmotic pressure can be obtained as a sum of the contributions from each solute:

$$\pi = \frac{RTx_s}{V_w} = RT\sum_{j=1}^{N} C_{sj} \tag{3.6}$$

The number of molecules and not the absolute weight of the solute determines the osmotic pressure of the solution.

**Example 3.1**  *Concentration of orange juice by osmosis.* In the food processing industry, in order to concentrate orange juice, the water needs to be removed. A plastic bag containing orange juice at 1 wt % sucrose concentration is dropped into a brine solution at 35 wt % NaCl by weight. Calculate the osmotic pressure developed that will concentrate the juice.

At equilibrium, the fugacity of water will be equal between the juice phase and brine phase:

$$f^{juice} = f^{brine} \tag{3.7}$$

The fugacity of the solution can be written in terms of the pure-component fugacity using the Poynting factor. The mole fraction of water and the activity coefficient of water are needed. The Poynting factor corrects for the effect of pressure on the pure-component fugacity where $V_w$ is the molar volume. For ideal solutions, the activity coefficient can be taken as 1:

$$1x_{wj}f_w\exp(-V_w\pi_j/RT) = 1x_{wb}f_w\exp(-V_w\pi_b/RT) \tag{3.8}$$

Mole fractions have to be calculated for water in juice and brine solutions:

$$\pi = \pi_b - \pi_j = \ln\pi = \pi_b - \pi_j = -\frac{RT}{V_w}\ln\left(\frac{x_{wb}}{x_{wj}}\right)$$

$$= -\frac{8.314*298}{1.8E-5}\ln\left(\frac{0.86}{0.999}\right) = 204\,\text{atm}.$$

## 3.2  Darcy's Law for Fluid Transport in Porous Media

Oftentimes, the fluid flow and solute transport are across pores in biological transport phenomena. Porous media are solid materials with an internal pore structure. The pores can be *macropores* or *micropores*. The pore size and structure vary from one organ to another and between organisms. Nanostructured materials consist of a regular array of cylindrical pores. Interconnected channels may lead to a sponge or foam structure. Polymer gels form a fiber matrix. Tissues often contain a porous structure. The *extravascular* region can be viewed as a porous medium. The region consists of cells and an interstitial region, and the pores are saturated with interstitial fluid. Pores exist in between cells, much like the spaces between grains in a pile of sand. They also exist in between extracellular fibrous molecules as part of a fiber matrix. A composite material is formed by embedding a fiber-matrix structure in the granular structure. Pores in the interstitial region are either isolated or connected. Tissues are comprised of blood vessels, cells, and interstitial regions. The interstitium is comprised of an extracellular matrix and interstitial fluid.

Synthesized protein and polysaccharide molecules, such as proteoglycans, collagen, elastin, fibronectin, and laminin, form the extracellular matrix. A mechanical scaffold of tissue is provided by the extracellular matrix. It also serves as a substrate for cell adhesion and cell migration. Pores are characterized by their surface area and porosity:

$$s_{sp} = \frac{area(interfacial)}{volume} \tag{3.9}$$

$$\varepsilon = \frac{volume(pores)}{volume(total)} \tag{3.10}$$

where $s_{sp}$ is the specific surface area and $\varepsilon$ is the porosity of the medium. Porous structures are deformable on application of load. The spatial distribution of pores can change on application of mechanical stress.

Pores can be classified depending on their connectivity as follows:

1. Passing penetrable pores

2. Nonpassing penetrable pores

3. Isolated pores

4. Tortuous channels

A pore is considered a passing pore when it connects to at least two subdomains of the outer surface of finite porous media. The passing pores may connect to two boundaries of the material, regardless of its geometry: rectangular, cylindrical, or spherical. A nonpassing pore connects to only one subdomain of the outer surface. Both passing and nonpassing pores are said to be penetrable pores. Pores without any connections to the outer surface of the porous media are considered isolated. Tortuous channels occur when the length of the pore is greater than the thickness of the specimen—that is, tortuosity may be defined as:

$$\tau = \left( \frac{L_{min}}{L} \right)^2 \tag{3.11}$$

where $L$ is the length of the diffusion path through the pore and $L_{min}$ is the shortest distance between the departure and arrival points of the solute in the medium. Accessibility of pores to solutes depends on the molecular structure and property of the solute. Common sense would dictate that a solute with a solute size greater than the pore size would not be penetrable. But there are instances when macromolecules with an initial size greater than the pore size have coiled up and penetrated the pore!

The size of a flexible molecule is defined by its radius of gyration. Oftentimes, not all the pore volume is available for solute transfer. This can be quantitated by using the parameter called the *partition coefficient.* The partition coefficient gives the ratio of the available pore volume for solute diffusion to the porosity of the medium. For example, Ogston [1] developed a statistical model to include the effects of molecular exclusion in an oriented fiber matrix:

$$\phi = \frac{1}{\varepsilon} \exp\left(-\theta\left(1 + \frac{r_{solute}}{r_{fiber}}\right)^2\right) \tag{3.12}$$

where $\theta$ = volume fraction of fibers
$r_{solute}$ = radii of solute
$r_{fiber}$ = radii of fiber
$\phi$ = partition coefficient in the porous media

The partition coefficient is an indicator of solute partitioning at equilibrium, between external solutions, and the void space in the porous media. In human anatomy, for example, the partition coefficient for albumin in the liver was found to be 0.5, 0.61 in the dermis, and 0.9 in the gut. The porosity was 0.163, 0.302, and 0.094 in the liver, dermis, and gut, respectively.

*Fluid flow* through porous media has been studied for more than 150 years. Similar to Ohm's law of electricity, Fourier's law of heat conduction, Fick's law of molecular diffusion, and Newton's law of viscosity, Darcy's law can be written for fluid flow through porous media as follows [2]:

$$Q = -\frac{\kappa A}{\mu}\frac{\partial P}{\partial z} \tag{3.13}$$

where $Q$ = throughput of the fluid
$\kappa$ = permeability of the medium to the fluid
$A$ = the cross-sectional area across which flow occurs
$\mu$ = the viscosity of the fluid
$P$ = pressure
$z$ = distance of the flow field

Darcy observed that water percolates through sand at a flow rate proportional to the pressure gradient and inversely proportional to the viscosity of the fluid. Although derived from empirical observations, attempts have been made to derive Darcy's law after disregarding friction within the fluid. Darcy's law is used to describe fluid flow in interstitium. Extending Eq. (3.13) in three dimensions:

$$v = -K.\nabla P \tag{3.14}$$

where the superficial velocity vector is given as a dot product of permeability and pressure gradient in all three Cartesian coordinates: $x$, $y$, and $z$.

Darcy's law implies the use of a continuum where the material is assumed to be homogeneous throughout. Three length scales are recognized:

1. Average size of pores: $r_{pore}$
2. Length $L$ over which the macroquantities such as fluid velocity and pressure are defined: $L \gg r_{pore}$
3. $r_{pore} \leq a \leq L$ (the volume $a^3$ is a basis volume)

Two phases can be recognized in the basis volume—that is, the void phase and the solid phase. The principle of conservation of mass and the principle of conservation of momentum can be applied to fluid flow in porous media. Equation (1.73) applied to fluid flow in a porous medium will yield:

$$\nabla . v = \theta_{source} - \theta_{sin\,k} \qquad (3.15)$$

Combining Eqs. (3.14) and (3.15):

$$\nabla . v = -\nabla . (K . \nabla P) = \theta_{source} - \theta_{sin\,k} \qquad (3.16)$$

When the source and sink are zero, the Laplace equation results:

$$\nabla^2 P = 0 \qquad (3.17)$$

## 3.3    Starling's Law for Fluid Transport

The combined effect of osmotic pressure and hydrostatic pressure can be seen in *Starling's law* [3], which gives the relation between the flow of fluid across the capillary wall or a porous membrane and the pressure difference across the capillary. The volumetric fluid transfer rate, $J$, across the capillary membrane is given as:

$$\frac{J}{L_p S_c} = \Delta P_h - \Delta \pi = \Delta \bar{P}_h \qquad (3.18)$$

where the effective pressure, $\Delta P_h$, is the result of the hydrodynamic pressure; drop, $\Delta P_h$; and the osmotic pressure difference, $\Delta \pi$. $L_p$ is the hydraulic conductance, and $S_c$ is the effective peripheral surface area through which the fluid flows. The hydraulic conductance is often determined by experiment. It varies from 1 $E-9$ m²s/kg in capillaries in the kidneys' glomeruli to 1 $E-14$ m²s/kg for endothelial cells found in the capillaries of the rabbit brain. One use of Eq. (3.7) is to better understand the flow of plasma across the capillary wall in human anatomy. This can be used in the seawater reverse osmosis (SWRO) systems used to desalinate seawater to drinking potable water. The hydraulic conductance can also be derived from the properties of the system. Thus, should the membrane be viewed as a series of parallel cylindrical pores:

$$L_p = \left( \frac{r^2 A_{pore}}{8 \mu t_w S_c} \right) \tag{3.19}$$

where          $A_{pore}$ = the cross-sectional area of the pore of radius
                        $r$ in the capillary wall
                $S_c$ = the peripheral area
                $t_w$ = the wall thickness.
        The ratio $A_p / S_c$ = the porosity of the capillary wall, $\varepsilon$.

Oftentimes, the solvent moving across the membrane will carry with it the solute molecules. Some molecules will be filtered on account of their large size. Even when the membrane is semipermeable, some solute diffusion will take place. The solute separation on account of size can be accounted for by the introduction of the sieving coefficient, $S_e$. The sieving coefficient is defined as the ratio of the solute concentration in the filtrate, $C_p$, to the solute concentration of the feed solution, $C_f$. Theoretical expressions based on the motion of a spherical solute moving through a cylindrical pore have been developed in order to estimate the value of the sieving coefficient [4]. The expression given by Deen can be written as a seventh-degree polynomial expression for the sieving coefficient in terms of the ratio of solute radius to the capillary pore radius as follows:

$$\frac{C_p}{C_f} = S_e = 1 - 4.67\lambda^2 + 3.837\lambda^3 + 1.67\lambda^4$$

$$- 2.015\lambda^5 + 0.015\lambda^6 + 0.163\lambda^7 \tag{3.20}$$

where $\lambda$ is the ratio of the solute radius, $a$, to the capillary pore radius, $r_{pore}$. The sieving coefficient as a function of the ratio of solute radius to capillary pore radius is shown in Fig. 3.2. With a root mean square (RMS) error of 0.04 percent, the sixth-degree term in Eq. (3.20) can be omitted and Eq. (3.20) written as:

$$\frac{C_p}{C_f} = S_e = 1 - 4.67\lambda^2 + 3.837\lambda^3 + 1.67\lambda^4 - 2.015\lambda^5 + 0.163\lambda^7 \tag{3.21}$$

*Concentration polarization* refers to the formation of a coat of retained solutes on the feed side of the membrane.

At high filtration rates, the formation of a concentration polarization layer has been found, which will change the protein transport.

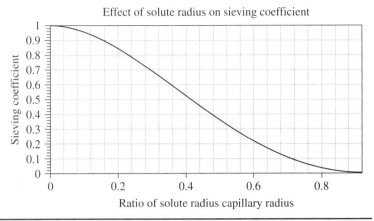

**FIGURE 3.2**    Sieving coefficient as a function of a solute radius.

A sieving coefficient including the polarization effects can be defined as $S_{cp}$ and represents the ratio of the solute concentration in the filtrate $(C_f)$ to that of the solute concentration in the bulk blood, $C_{bulk}$ [5]:

$$S_{cp} = \frac{C_f}{C_{bulk}} = \frac{S_e}{(1-S_e)\exp\left(-\dfrac{j}{k_m}\right)+S_e}$$    (3.22)

where $j$ is the flux of filtration and $k_m$ is the film mass transfer coefficient. For laminar flow, the film mass transfer coefficient can be obtained from the following empirical correlation for Sherwood number $Sh = (k_m D/D_c)$

$$Sh = 3.66 + 0.104 Pe_m \left(\frac{D_c}{L}\right) \frac{1}{\left(1+0.016\left(\dfrac{Pe_m}{L/D_c}\right)^{0.8}\right)}$$    (3.23)

where  $Pe_m$ = the Peclet number (mass) $(VD_c/L)$
        $D_c$ = the diameter of the capillary
        $D$ = the binary diffusivity
        $L$ = the length of the capillary

Equation (3.11) was developed later for cases where velocity and concentration profiles are not yet fully developed. When the flow becomes fully developed, the Sherwood number reaches 3.66, its asymptotic value. For cylindrical channels, the hydraulic diameter $D_H$ can be used in place of the capillary diameter. Hydraulic diameter can be defined as four times the channel cross-sectional area divided by the wetted perimeter.

## 3.4    Solute Diffusion across the Membrane

The solute diffusion across the membrane can be treated with Fick's laws of diffusion at steady state and the generalized Fick's laws of diffusion discussed in Chap. 2 for transient applications. Different diffusivities, such as pore diffusion, diffusion in polymeric systems, and convective effect, can be added together.

Diffusion of solutes through pores can be of different kinds [6]: (1) viscous flow, (2) molecular flow or Knudsen diffusion, (3) surface diffusion, (4) capillary condensation diffusion, (5) molecular sieving diffusion, and (6) diffusion solubility diffusion. When the pore radius is much larger compared with the size of the solute, diffusion happens in a *unfettered* manner, much like water flowing in a circular pipe. When the mean free path of the solute molecule is larger than the pore radius, Knudsen diffusion is said to occur. The classical laws of diffusion can no longer be applied to describe the phenomena. Rather, the kinetic theory of pressure and temperature is used to describe the mechanism. Solute chemistry, or interactions between the solute and the wall, can give rise to surface diffusion. During capillary condensation, there is increased vapor pressure of a liquid inside the pore. It exists when surface tension is a non-negligible factor. The sieving mechanism is found during the transport of linear and branched alkanes using zeolites. The branched alkanes diffuse into the alkanes at one-fifteenth the rate of linear alkanes. The reason is that the linear alkanes are smaller in size and can fit in the pores rather well. Diffusion solubility mechanism involves dissolution of solute in the medium and permeation later.

The size of the solute can be estimated from the Stokes-Einstein relation presented in Chap. 1 [Eq. (1.39)]. The diffusivity of the solute in the liquid needs to be known. If the diffusivity is not known, the solute size can first be estimated from the following equation, assuming that the solute of molecular weight, $M_w$, is a sphere with a density of approximately 1 gm/cm$^3$ and is the same as that of the solute in the solid phase:

$$a = \left( \frac{3M_w}{4\pi\rho A_N} \right)^{1/3} \tag{3.24}$$

Renkin and Curry [7,8] looked at diffusion coefficients for various solutes as a function of the molecular weight for dilute solutions. An empirical equation was developed using a least squares regression fit of the experimental observations at 37°C:

$$D = 1.013 * 10^{-4}(M_w)^{-0.46} \tag{3.25}$$

Biological systems are heterogeneous in nature. The diffusion coefficients of the solute would depend on the medium through which it diffuses. Several diffusion coefficients can be recognized. These are as follows:

1. Solute diffusion coefficient in blood or tissue, $D_{bl}$
2. Solute diffusion coefficient in plasma, $D_{pl}$
3. Solute diffusion within pores of a capillary wall, $D_{pore}$
4. Solute diffusion in the interstitial fluid, $D_{int}$
8. Solute diffusion within cells, $D_{cell}$
9. Solute diffusion in water, $D$

The Stokes-Einstein equation for diffusion coefficients of solutes in dilute solutions may be used to estimate $D_{pl}$, the diffusion coefficient of the solute in plasma. The pore diffusion coefficient can be estimated after considering the available surface area for diffusion. The pore is $A_{pore}$. The path taken by the solute through the pores may be tortuous. Thus, the diffusion distance may be greater than the membrane thickness, $t_{mem}$, in reality.

*Steric exclusion* and *hindered diffusion* can also be accounted for in the expressions for diffusion of the solute, depending on the problem at hand. Steric exclusion refers to the problem that occurs when only the volume in the pores and not that of the solute is available for diffusion. The fraction of the pore volume available to the solute for diffusion is given by a partition coefficient, $K$:

$$K = \left(1 - \frac{a}{r}\right)^2 \tag{3.26}$$

Due to steric exclusion the equilibrium concentration of a solute is less within the pore mouth than in the bulk solution. Attractions between the solute and pores are ignored.

The hydrodynamic drag experienced by the solute is referred to as the hindered diffusion. The Renkin equation [7,8] gives the ratio of the pore diffusivity to that of the bulk diffusivity:

$$\frac{D_{pore}}{D} = 1 - 4.1\lambda + 5.2\lambda^2 - 0.01\lambda^3 - 4.18\lambda^4$$

$$+ 1.14\lambda^5 + 1.9\lambda^6 - 0.95\lambda^7 = K\omega \tag{3.27}$$

where $\lambda$ is the ratio of the solute radius to the pore mouth radius. The partition coefficient, $K$, captures the steric exclusion. The rest of the term accounts for the hydrodynamic drag faced by the diffusing solute through the pore. Gaydos and Brenner [9] give a different expression for pore diffusivity as a function of the ratio of solute radius to the pore radius:

$$\frac{D_{pore}}{D} = 1 + \frac{9}{8}\lambda \ln(\lambda) - 1.54\lambda \tag{3.28}$$

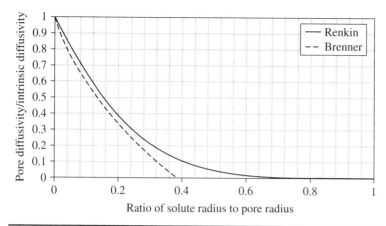

**FIGURE 3.3**   Renkin equation and Brenner equation for pore diffusion coefficient.

The error in writing Eq. (3.28) is $O(\lambda^2)$. This is about 2 percent when the pore diffusion coefficient is about half of the intrinsic diffusion coefficient. Equations (3.27) and (3.28) are shown in Fig. 3.3.

Fick's law of diffusion for a solute in pores may be written as:

$$J_s = -DA_{\text{pore}} K \bar{\omega} \frac{1}{\tau} \frac{\partial C}{\partial x} \tag{3.29}$$

where $K\omega$ = RHS of Eq. (3.27) and $\tau$ is the tortuosity required to take into account the actual path of the solute diffusion through the membrane.

The diffusion of solute through blood and cells can be represented by an effective diffusion coefficient. The transport across suspensions may be applicable here. Maxwell [10] had developed an expression for diffusion in suspensions:

$$\frac{D_{bl}}{D_{\text{int}}} = \frac{2D_{\text{int}} + D_{\text{cell}} - 2\phi(D_{\text{int}} - D_{\text{cell}})}{2D_{\text{int}} + D_{\text{cell}} + \phi(D_{\text{int}} - D_{\text{cell}})} \tag{3.30}$$

Furthermore, $D_{pl}^* D_{bl} = D_{\text{int}}$ and $\phi$ is the volume fraction of the cells in blood. Maxwell had developed his expression for the suspension of spheres. Cells are not spherical. So sphericity for cells may be used to correct for the actual shape of cells. Using Monte Carlo simulations, some investigators [11] have developed a empirical equation for the diffusion coefficient in blood for a wide range of volume fractions, which can be seen to be:

$$\frac{D_{bl}}{D_{\text{int}}} = 1 - \left(1 - \frac{D_{\text{cell}}}{D_{\text{int}}}\right)(1.73\phi - 0.82\phi^2 + 0.09\phi^3) \tag{3.31}$$

The interstitial fluid is a gel of macromolecules. Solute diffusion happens around the random network of macromolecular chains. Reduction in diffusivity due to macromolecules has been accounted for by Brinkman, and is given as a one-parameter equation:

$$\frac{D_{int}}{D} = \frac{1}{1 + \kappa a + \dfrac{\kappa^2 a^2}{3}} \tag{3.32}$$

where $\kappa$ is the one parameter that is a function of the macromolecular structure of the interstitial fluid and can be obtained by fitting experimental data for solute diffusion in gel.

The effective diffusivities of different sizes of solutes through tumor and normal tissue were studied by Jain [12]. They found a greater reduction in solute diffusivity for normal tissue compared with tumor tissue. The interstitial volume in normal tissue was smaller compared with the tumor tissue. They used a fiber-matrix model based on Curry [13], and the diffusion coefficients were given by:

$$\frac{D_{bl}}{D} = \exp\left(-\left(1 + \frac{a}{a_f}\right)\sqrt{v\phi_f}\right) \tag{3.33}$$

The interstitial space is viewed as a matrix of fibers with radius $a_f$, $\phi_f$ is the fiber concentration, and $v$ is the specific volume of the fibers.

The solute diffusion can also be described using a solute permeability similar to the solvent permeability. The solute flux can be written as:

$$J_s = P_m S(C_{sf} - C_{sp}) \tag{3.34}$$

where $P_m$ is the permeability of the solute and $S$ is the membrane surface area. The two concentrations, $C_{sf}$ and $C_{sp}$, are that of the solute in the feed side of the membrane and the permeate side of the membrane, respectively. Solute diffuses from a region of higher concentration to a region of lower concentration. The permeability is given by the product of *effective diffusivity* of the solute in the membrane divided by the thickness of the membrane. Thus:

$$P_m = \frac{D}{t_{mem}} \frac{A_{pore}}{S} \frac{K\bar{\omega}}{\tau} \tag{3.35}$$

Renkin and Curry [8] performed a variety of experiments for solutes with different sizes. They summarized their findings on permeability coefficients in the form of empirical correlations as follows:

$$P_m S = 0.0184 a^{-1.223} \quad a < 1 \text{ nm} \tag{3.36}$$

$$P_m S = 0.0287 a^{-2.92} \quad a > 1 \text{ nm} \tag{3.37}$$

The combined effect of hydraulic pressure and osmotic pressure and the solute flux can be written from the application of irreversible thermodynamics (Kedem and Katchalsky [14]). The cross-coefficients that are from the secondary effects are equal according to the Onsager relations. The relative flow between the solvent and solute is capable of providing a separation of the solute and solvent. The sieving mechanism is called *ultrafiltration,* and when the solute permeability is low, it is referred to as *reverse osmosis:*

$$J = SL_p(\Delta P - \sigma RT\Delta C) \tag{3.38}$$

$$J_s = SL_p\left(-\sigma\Delta P + \frac{L_s}{L_p}RT\Delta C\right) \tag{3.39}$$

where $J$ is the flux of the solvent and $J_s$ is the flux of the solute.

The parameter $\sigma = -L_{sp}/L_p$ is called the *Staverman reflection coefficient* [15]. If the membrane is permeable to solvent and not to the solute, $\sigma = 1$. When $\sigma = 0$, the membrane is equally permeable to both solvent and solute. Equation (3.16) gives the flux of solvent across the semipermeable membrane, and Eq. (3.17) gives the flux of solute.

The total rate of solute transfer through the pores of the capillary wall can be obtained by multiplying the solute concentration by the combined flow rate of the solution due to both applied pressure difference and the concentration difference. Thus:

$$N_s = C_s L_p S\left(\Delta P(1-\sigma) + \left(\frac{L_s}{L_p} - \sigma\right)RT\Delta C\right) \tag{3.40}$$

Substituting Eq. (3.16) in Eq. (3.18):

$$N = C_s J(1-\sigma) + C_s SL_p\left(\frac{L_s}{L_p} - \sigma^2\right)\Delta\pi \tag{3.41}$$

When solute transfers by only diffusion—that is, no flow of solvent:

$$N_{J=0} = C_s L_p SRT\Delta C\left(\frac{L_s}{L_p} - \sigma^2\right) = P_m S\Delta C_{J=0} \tag{3.42}$$

Equation (3.19) can be modified using Eq. (3.20) as shown:

$$N_s = C_s J(1-\sigma) + P_m S\Delta C \tag{3.43}$$

Oftentimes, the problem is in obtaining the three parameters: $L_p$, the hydraulic conductance of the solvent; $P_m$, the permeability of the solute; and $\sigma$, the Staverman reflection coefficient. Anderson and Quinn [9] showed that the sieving coefficient is the same as $1-\sigma$ using a hydrodynamic equation accounting for hindered particle motion in small pores.

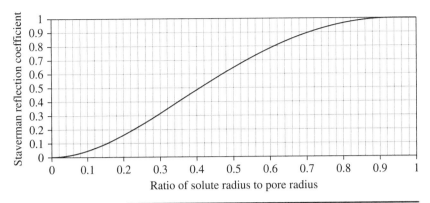

**FIGURE 3.4**    Staverman reflection coefficient as a function of $\lambda$.

Thus, from Eq. (3.9) and the relation between the sieving coefficient and the Staverman reflection coefficient, a polynomial expression for the Staverman reflection coefficient can be written as follows:

$$\sigma = \lambda^2(4.67 - 3.837\lambda - 1.67\lambda^2 + 2.015\lambda^3 - 0.015\lambda^4 - 0.163\lambda^5) \quad (3.44)$$

Equation (3.32) is plotted and shown in Fig. 3.4.

## 3.5    Derivation of Starling's Law

Starling's law, which describes the permeation of solvent across the membrane, can be derived from hydrostatic pressure and chemical potential considerations. The solvent that filters out of the membrane is called the *permeate*, and the solution that is being filtered is called the *feed*. The water flux or solvent flux can be written as:

$$J' = JC_{\text{solvp}} = \frac{D_{\text{solv}} \Delta C_{\text{solv}}}{t_{\text{mem}}} \quad (3.45)$$

The equilibrium chemical and pressure potential in the permeate and feed sides can be written as:

$$\mu_f + V_f P_f + RTC_f = \mu_p + V_p P_p + RTC_p \quad (3.46)$$

Equation (3.46) can be rearranged and a partition coefficient, $K$, introduced:

$$C_p = C_f\left(K \exp\left(\Delta V \frac{\Delta P}{RT}\right)\right) \quad (3.47)$$

where $K$ is the partition coefficient at some average reference pressure, $<P>$.

$$K = \exp\left(\frac{\Delta\mu + \Delta V <P>}{RT}\right) \quad (3.48)$$

Equations (3.47) and (3.48) can be expanded in a Taylor series and:

$$C_p = C_f K \left(1 + \frac{\Delta V \Delta P}{RT}\right) \tag{3.49}$$

and

$$\Delta C = \frac{\Delta P K C_f \Delta V}{RT} \tag{3.50}$$

Substituting Eq. (3.50) in Eq. (3.45):

$$J' = \frac{\Delta P D K \Delta V}{RT t_{mem}} = L_p \Delta P \tag{3.51}$$

where $L_p$ is the hydraulic permeability of membrane to solvent. The net solvent flux, including the osmotic pressure, can be written as:

$$J' = L_p(\Delta P - \sigma \Delta \pi) \tag{3.52}$$

Patlak, Goldstein, and Hoffman [10] account for the diffusion term in Eq. (3.21) as follows:

$$N_s = C_s J(1-\sigma) - D_e S \frac{\partial C_s}{\partial z} \tag{3.53}$$

Assuming constant solute transfer, Eq. (3.22) can be "solved for" as follows:

$$\frac{\partial C_s}{\partial z} = -c + C_s \frac{J(1-\sigma)}{D_e S} \tag{3.54}$$

or

$$\frac{\partial C_s}{C_s Pe - c_1} = \frac{\partial z}{t_{mem}} \tag{3.55}$$

where Peclet number, $Pe$, is defined as $\left(\dfrac{J(1-\sigma)t_{mem}}{D_e S}\right)$ \qquad (3.56)

The physical significance of the Peclet number is that it gives a ratio of the solute transfer by convection divided by the solute transfer by molecular diffusion. When the Peclet number is small and close to zero, the solute transfer is dominated by a molecular diffusion mechanism. When the Peclet number is large and close to infinity, the solute transfer is dominated by bulk convection.

Integrating Eq. (3.24):

$$C_s = c_2 + c_3 \exp\left(\frac{zPe}{t_{mem}}\right) \tag{3.57}$$

The integration constants can be solved for by imposing the following boundary conditions:

$$z = 0, \; C_s = C_f \tag{3.58}$$

$$z = t_{mem}, \; C_s = C_p \tag{3.59}$$

The constants can be seen to be $c_3 = \dfrac{C_f - C_p}{1 - \exp(Pe)}$

$$c_2 = \frac{C_p\left(1 - \exp(Pe)\left(\dfrac{C_f}{C_p} - 1\right)\right)}{(1 - \exp(Pe))} \tag{3.60}$$

This analysis is applicable at steady state for dilute systems where the constant solute transfer assumption yields a rich dividend.

## 3.6    Starling's Law Is Not Universal

The flow of fluid across membranes is governed by Starling's law. When membranes are comprised of uniform macrostructures, the flux of fluid is predicted well using Starling's law. The flow of fluid depends on the pressure differences across the membrane and the hydraulic conductance. Microvessel walls are nonuniform. This can be seen in the glycocalyx, the endothelium, and the basement membrane. Endothelial cells, interendothelial cleft, and junction protein strands are not uniform.

Experimental observations have been made that are inconsistent with the predictions of Starling's law.

Blood in a capillary, for example, with a hydrostatic pressure difference of about 15 mmHg in the arterial end and an osmotic pressure difference of 26 mmHg would be expected to filter water into the interstitial space and out of the artery. The reflection coefficient may be taken as 1. With the osmotic pressure difference remaining the same, the net pressure drop would be $17 - 1(27) \sim -10$ mmHg at the veins. The driving force has changed in direction (Fig. 3.5) and the water can be expected to filter from interstitial space into the veins.

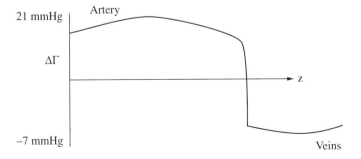

**Figure 3.5**    Filtration pressure drop in arteries and veins.

Experimental observations of Michel and Phillips [16] provide a counterexample to the filtration/reabsorption prediction. Starling's law has been modified by Hu and Weinbaum [10] as:

$$J = L_p S(\Delta P - \sigma \Delta \pi) \qquad (3.61)$$

The model consists of four different regions, such as glycocalyx, a junction cleft between endothelial cells with a junction protein strand in the middle, a semicircular region for albumin mixing at the exit of the cleft, and the extravascular space between the semicircular region and the mid-plane of the microvessel. Three different cases in the model can be identified:

Case (a): $\Delta P > \Delta \pi$, blood pressure is 35 cm water, and $\pi_B$ is 27.2 cm water. Interstitial blood and osmotic pressures are 0. Convective effects are much greater than diffusive effects at break junctions. There is no accumulation. Osmotic pressure drop across the glycocalyx layer is the primary causative factor.

Case (b): Arterial and interstitial blood pressures and osmotic arterial pressure are the same as in Case (a). Osmotic interstitial pressure increases to 27.2 cm water. The osmotic pressure difference is 0. The solvent flux increases compared with Case (a). Experimental observations come close to theoretical predictions.

Case (c): Interstitial blood pressure, osmotic arterial, and interstitial pressures are the same as in Cases (a) and (b). Arterial blood pressure decreases from 35 cm water to 10 cm water. $(\Delta P - \sigma \Delta \pi)$ is lower than in Cases (a) and (b). No fluid reabsorption. Some accumulation due to back-diffusion.

## 3.7 Molecular Probes to Measure Permeability of Transcapillary Pathways (Curry [13])

Single capillary methods were developed at Oxford Laboratory to measure the permeability of solutes across capillary walls. Dyes were developed by the time of World War I. They explored the possibility of using chemically different dyes, large and small, to study the permeability of the capillary wall in greater detail. A micromanipulator was used to cannulate and perfuse the capillaries in frog mesentery with various colored dyes dissolved in frog Ringer's solution. By measuring the time it takes the dye to appear outside the vessels as an index of the permeability of the capillary, the chemical and physical properties of the pathways for solute exchange across the walls of segments of frog microvessels were studied. It was found necessary to measure the capillary pressure. It was realized that the rate at which solutes traverse the capillary wall depends on the permeability of the membrane wall as well as the driving force for solute exchange. The Kedem-Katchalsky equations, with the use of the

Staverman reflection coefficient, can be used to quantitate the exchange of solutes across the capillary membrane. The driving forces are the concentration difference of the solute, $\Delta C$, and the frictional drag force exerted by solvent on diffusing solutes.

The flux of a tracer through a single porous pathway is written using Eq. (3.31). Two mechanisms can be recognized—namely the solute molecular Fickian diffusion and the solvent drag. Whether the process is diffusion-limited or solvent-drag–limited depends on the solute diffusion velocity within the pores relative to the velocity of water convective flow. This can be quantitated using the Peclet number mass as defined by Eq. (3.36). When Peclet number is greater than 3, solvent drag dominates the exchange. The magnitude of solute exchange is determined by the permeability coefficient, $P_m$; the hydraulic conductance, $L_p$; and the Staverman reflection coefficient. The effective osmotic pressure is also captured in the reflection coefficient representation. In the experiments, it is important to ensure that the contribution from other solutes such as plasma proteins is negligible.

Figure 3.6 shows a schematic of a capillary cannulated with two micropipettes at a Y branch. This is used to measure the permeability of a fluorescent solute. Either a control washout solution is used to perfuse the capillary from the pipette on the left or a perfusate containing α-lactalbumin labeled with fluorescent tetramethylrhodamine isothiocyanate from the pipette on the right is used. A rapid

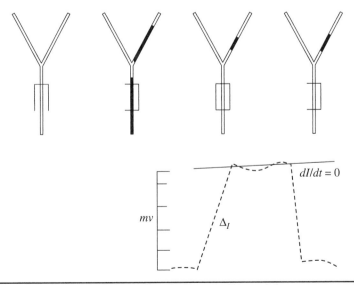

**FIGURE 3.6** Cannulated glass micropipettes at a Y branch for measurement of permeability coefficient.

change in the solution-perfusing capillary lumen can be detected. A photomultiplier tube (PMT) is used as a detector. The output of PMT is plotted as a function of time. Frog Ringer's-albumin solution is used to perfuse the capillary without any fluorescent dye. This establishes the baseline representing no solute transfer. Rapid fill of capillary lumen by the same perfusate is labeled with the fluorescent probe. This is accomplished by a simple switch of perfusion pressures to preset values such that there is no mixing of the perfusates in the capillary lumen. The step-change in fluorescent intensity at fill point is proportional to the number of fluorescently labeled molecules in the capillary lumen. The initial solute transport across the capillary wall was measured from the initial rate of increase in fluorescence intensity $(dI/dt)_0$. When solute was washed out of the capillary lumen, extra capillary solute began to diffuse back into capillary lumen. A 10-sec time interval is provided.

Perfusion of vessel segments with Ringer's perfusate containing tracers with molecular weights greater than 500 gm/mole in the absence of plasma proteins increases the permeability of the microvessel wall. It was realized that not measuring the capillary pressure compromised the interpretation of measurements of the time it took colored tracers to appear in terms of the properties of transcapillary pores. This problem was solved by developing novel methods to measure capillary pressure and transcapillary filtration rate.

Figure 3.6 illustrates a method to measure the permeability of a segment of a microvessel to a fluorescently labeled solute under conditions where both the solute concentration gradient and the hydrostatic and osmotic pressures determining water flow across the wall are measured directly. The true diffusive permeability coefficient is measured only when the net filtration pressure in all pathways is 0—that is, when there is no coupling of solute flux to transcapillary water flows. When these conditions are not adhered to an *apparent permeability coefficient* is measured. This is larger than the true diffusive permeability. Transcapillary solute flux is measured at a series of pressures, and the true permeability is identified by extrapolation to the condition of zero volume flow. The membrane coefficients that capture solvent drag, $L_p(1 - \sigma)$, are identified from the increase in transcapillary flux as pressure increases. This method has been successfully applied by several investigators to frog microvessels and in mammalian vessels.

**Example 3.2**  *Effective pore size of gel.* Transplantation of insulin-secreting cells in a pancreas using nanotechnology can be a way to cure type 1 diabetes. Nanoporous biocapsules are bulk and surface micromachined to make available uniform and controlled pore sizes as small as 7 nm, tailored surface chemistries, and precise microarchitectures. This provides immunoisolating microenvironments for cells. Such a design may overcome some prior limitations associated with conventional encapsulation and delivery technologies, including chemical instabilities, material degradation, and fracture and broad membrane pore sizes.

For immunoprotection of pancreatic cells, the immunoprotection membrane ought to allow permeability of glucose, insulin, oxygen, and other metabolic products to ensure islet functionality and therapeutic effectiveness. The nanopore microfabricated membranes were tested (Desai [14]) for diffusion of biomolecules such as glucose with a molecular weight of 180 kDa, human albumin with a molecular weight of 67 kDa, human IgG with a molecular weight of 150 kDa, vitamin $B_{12}$ with a molecular weight of 1200 kDa, myoglobin with a molecular weight of 17,000 kDa, and bovine serum albumin (BSA) with a molecular weight of 69,000 kDa.

Tests were conducted at 37°C over 4 hours in a diffusion chamber with two compartments, $A$ and $B$, with fixed volumes of 2 mL separated by the desired membrane and sealed with $O$ rings and screwed together. The measured diffusion coefficients are seen to be in million cm²/s for glucose as 4.5, human albumin and human IgG as 0.13, vitamin $B_{12}$ as 1.7, myoglobin as 0.4, and BSA as 0.1. What is the effective pore size of the membrane? Use a suitable equation described in the text.

| | Pore Radius | 252 | nm | | | | | | | | |
|---|---|---|---|---|---|---|---|---|---|---|---|
| | | kDa | cm^2/s | Nm | | | | | cm^2/s | | Min. |
| # | Solute | MW | Dbl | Solute Radius | a/r | K | w | Kw | D | Dbl/D | Error |
| | | | | | 0.0000 | | | 1.00 | | | |
| 1 | Human albumin | 67 | 9.00E–06 | 30.4 | 0.1205 | 0.77 | 0.75 | 0.58 | 1.46E–05 | 0.615 | 0.001 |
| 2 | Glucose | 180 | 4.50E–06 | 42.2 | 0.1674 | 0.69 | 0.66 | 0.46 | 9.29E–06 | 0.484 | 0.001 |
| 3 | Human IgG | 150 | 1.30E–07 | 39.7 | 0.1575 | 0.71 | 0.68 | 0.48 | 1.01E–05 | 0.013 | 0.219 |
| 4 | Vitamin $B_{12}$ | 1,200 | 1.70E–06 | 79.3 | 0.3149 | 0.47 | 0.40 | 0.19 | 3.88E–06 | 0.438 | 0.062 |
| 5 | Myoglobin | 17,000 | 4.00E–07 | 191.8 | 0.7612 | 0.06 | 0.08 | 0.00 | 1.15E–06 | 0.349 | 0.118 |
| 6 | BSA | 69,000 | 1.00E–07 | 305.8 | 1.2137 | 0.05 | –0.31 | –0.01 | 6.02E–07 | 0.166 | 0.033 |
| | | | | | | | | | | | 0.433 |

Equation (3.12) was used to calculate the solute radius given the molecular weight, Eq. (3.13) was used to estimate the diffusion coefficient of the solute in water, and Eq. (3.15) was used to calculate the diffusivity ratio of solute in blood and in water. The density of solute was taken to be 1 gm/cm³. The Renkin equation prediction is shown as a solid curve. The pore radius was iterated until the least squared error was minimized using a Microsoft Excel spreadsheet. At 252 nm, the error was minimized. The measured diffusivity ratios and predicted ratios are shown in Fig. 3.7.

**Example 3.3** *Effective diffusivity through spherical suspensions.* Islets of Langerhans are spheroidal aggregates of cells that are located in the pancreas and secrete hormones that are involved in glucose metabolism [17]. Type 1 diabetes can be cured by transplanting isolated islets. Islets removed from the pancreas lose their internal vascularization and are dependent on the diffusion of oxygen from the external environment and through the oxygen-consuming islet tissue to satisfy the metabolic requirements of the cells. Islets can be viewed as a suspension of tissue spheres. The diffusivity of oxygen was

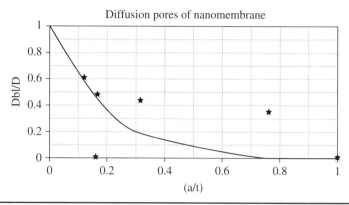

**FIGURE 3.7**   Measured diffusivity ratio and prediction from Renkin equation for pore diffusion.

measured. The islets were isolated from male rats using a modified digestion and purification technique under a dissecting microscope and cultured for a day in nonattacking polystyrene Petri dishes containing 5.6 mm of glucose, 50 U/mL of penicillin, 50 µg/mL streptomycin, and 10 percent newborn calf serum. The material was placed in an incubator at a temperature of oxygen uptake measurements. A known number of islets were placed in a tube that contained 45 mL of culture medium and 5 mL of air. The tubes were intermittently rotated to prevent settling and aggregation of the islets and to enhance oxygen transfer. The oxygen uptake chamber was equipped for measuring the oxygen-dependent lifetime of Pd-coproporphyrin phosphorescence to provide rapid and accurate measurements of oxygen concentration down to values as low as 0.05 µM. The chamber was a glass cuvette that contained a small Teflon-coated magnetic stirring bar rotated at a speed of 1,200 rpm. A sample of 1,500 islets was loaded in the cuvette, which was filled with phosphate-buffered saline (pH 7.4) containing 0.35 gm/lit HEPES buffer, 0.5 gm/lit bovine serum albumin, and 300 mg/L glucose supplemented with 0.01 µM palladium coproporphyrin and 1–5 U/mL catalase. The cuvette was capped with a ground-glass stopper to eliminate the gas phase. The measured effective diffusivity of oxygen through the islets was found to be 1.31 E – 5 cm²/s. Take the diffusivity of oxygen in interstitial fluid to be 2.1 E – 5 cm²/s. Should the diffusivity in the cells be 1.72 E – 5 cm²/s, calculate the volume fraction of the islets in the suspension.

From Eq. (3.18):

$$\frac{D_{bl}}{D_{int}} = 1 - \left(1 - \frac{D_{cell}}{D_{int}}\right)(1.73\phi - 0.82\phi^2 + 0.09\phi^3) \qquad (3.62)$$

$$D_{bl} = 1.31 \text{ E} - 5 \text{ cm}^2/\text{s}$$

$$D_{int} = 2.1 \text{ E} - 5 \text{ cm}^2/\text{s}$$

$$D_{cell} = 1.72 \text{ E} - 5 \text{ cm}^2/\text{s}$$

$$0.762 = 1.73\phi - 0.82\phi^2 + 0.09\phi^3$$

This equation required a numerical solution. Using a Microsoft Excel spreadsheet, the volume fraction of the islets was found to be 0.6.

**Example 3.4**  *Plasmapheresis membranes made of polycarbonate.* Plasmapheresis is a blood separation procedure used to isolate blood cells from plasma. In hemofiltration, the "cut-off" for the passage of molecules through the membrane is $10^3 - 5*10^4$ Dalton molecular weight and the cut-off in molecular weight of species in plasmapheresis is $3*10^6$ Dalton. A German manufacturer developed a polycarbonate membrane with an average pore diameter of 0.4 μm and a porosity of 0.65. The membrane thickness was about 100 μm. A polycarbonate polymer solution was cast onto a smooth surface and contacted with a gel medium, followed by precipitation of membrane and gelled layer to form the membrane. Calculate the hydraulic conductance of capillary flow. From Eq. 3.8:

$$L_p = \frac{0.65(0.2E-6)^2}{8*100E-6*1E-3} = 3.25E-6 \, m^2 s/kg \tag{3.63}$$

The viscosity of plasma fluid was assumed to be that of water at room temperature.

**Example 3.5**  *Saline water injection.* What ought to be the pump pressure to inject 3.6 mL/hr of saline water at a 10 wt % NaCl into the human bloodstream across a membrane of thickness 0.1 microns? The membrane has a porous structure with a pore radius of 500 nm.

The pump pressure head has to overcome the osmotic pressure and filter through the pores in accordance with Starling's law.

Hydraulic conductance:

$$L_p S = \frac{A_p r^2}{8 \mu t_{mem}} = \frac{3.14*(500E-9)^4}{8*0.001*1E-9} = 2.45E-16 \, m^4 s/kg \tag{3.64}$$

From Starling's law:

$$\Delta \bar{P} = \frac{J}{L_p S} = \frac{1E-9}{2.45E-16} = 40.2 \, atm$$

Now, the pressure head at the pump is given by:

$$\Delta P = \Delta \bar{P} - RT \ln(x_w)\frac{1}{V_w} = 40.2 + 46 = 86.2 \, atm \tag{3.65}$$

**Example 3.6**  *Blood-purifying hematocatharsis unit.* The human urinary system is made up of kidneys, the bladder, two ureters, and a single urethra. The kidneys are a pair of organs resembling kidney beans measuring around four to five inches in length and two to three inches in width. They are situated against the rear wall of the abdomen in the middle of the back, with on located on either side of the spine, beneath the liver on the right and the spleen on the left. Healthy kidneys in the average adult person process about 125 mL/min, or 180 liters of blood per day, and filter out about 2 liters of waste product and extra water in the urine. The kidneys remove excess minerals and wastes, and regulate the composition of such inorganic ions as sodium, phosphorous, and chloride in the blood plasma at a nearly constant level. Potassium is controlled by the kidneys for proper functioning of the nerves and muscles, particularly those of the heart.

Blood urea nitrogen (BUN), a waste product produced in the liver as the end product of protein metabolism, is removed from the blood by the kidneys in the Bowman's capsule along with creatinine, a waste product of creatinine phosphate—an energy-storing molecule produced largely as a result of muscle breakdown. Most kidney diseases, such as diabetes and high blood pressure, are caused by an attack on nephrons, causing them to lose their filtering capacity. The damaged nephrons cannot filter out the poison as they should. If the problem worsens and renal function drops below 10 to 15 percent that person has end-stage renal disease. When a person's kidneys fail, harmful wastes build up in the body, their blood pressure elevates, and the blood retains fluid. The person will soon die unless his or her life is temporarily prolonged by a kidney transplant [18]. In order to prevent the immune system from attacking the foreign kidney, the patient will take immunosuppressant medications for the rest of his or her life.

When the kidneys are functioning properly and the concentration of an ion in the blood exceeds its kidney threshold value, the excess ions and proteins in the filtrate are not reabsorbed but are released in the urine, thus maintaining near-constant levels. Maintaining constant levels is achieved by the mechanism of reverse osmosis, osmosis, and ion-exchange filtration.

Dialysis machines are the most widely used temporary lifesaving invention for patients with end-stage renal disease (Fig. 3.8). Hemodialysis machines are described as large, stationary, hydromechanical devices. In order to function, they require the following accessories:

1. Arterial line
2. Blood pump
3. Heparin infusion pump
4. Dialyzer filter
5. Venous line
6. Blood flow and pressure monitors
7. Air/foam detectors
8. Motors
9. Regulators and piping to carry 500–800 mL/min of dialysis solution
10. Aqueous solutions of Ca, Mg, Na, K, and other minerals from large mixing-holding vats to the patient's dialyzer and from there to the drain

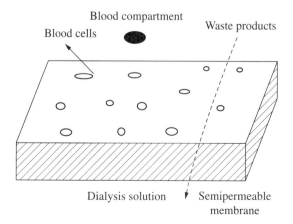

**FIGURE 3.8**   Blood purifying hemocatharsis unit.

With hemodialysis, the patient must be dialyzed three times a week. Each treatment lasts approximately three to four hours. Although the dialyzers are removing poisons, there are side effects, caused primarily by the dialyzers themselves.

Dialyzer filters are made of cellulose acetate, polysulfone, or similar materials and are sterilized with a solution of ethylene oxide, bleach, or formaldehyde. Dialyzer filters have just one membrane pore size, with a cut-off point larger than creatinine at 113.1 amu. Removed with creatinine is urea at 60.1 amu, water, and essential electrolytes, such as Na, K, Ca, Mg; however, these are not replaced during dialysis. Phosphorous molecules at 123.9 amu are not removed by dialysis, and large amounts are deadly to the patient. Find the pore size of the filter.

| Creatine | | |
|---|---|---|
| MW | 113 | gm/mole |
| a | 3.45671E–08 | 34.56711 |
| density | 1.09 | gm/cc |
| urea | 60 | |
| density | 1.3 | gm/cc |
| a | 2.63953E–08 | 26.39531 |
| Se | | 0.763596 |

Molecular Formula for Creatinine: $C_4H_7N_3O$

Creatinine comes from the Greek word *kreas*, which means flesh. It is a breakdown product of creatinine phosphate in the muscle. It is produced at a constant rate by the human anatomy. Creatinine is actively filtered out by the kidneys. Some of it is secreted by the kidneys into the urine. Creatine levels in blood and urine may be used to calculate the all-important glomerular filtration rate. It is clinically important in the evaluation of renal function.

From Fig. 3.9, the ratio of solute radius to pore radius can be read as 0.25. Therefore, the pore radius is 105.8 nm. Equation (3.9) has been plotted in Fig. 3.4.

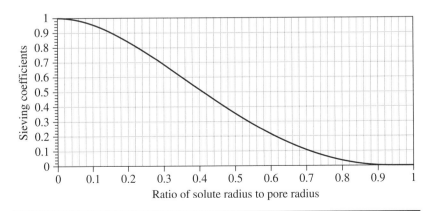

FIGURE 3.9    Sieving coefficient versus ratio of solute radius to pore radius.

Equation (3.9) is also shown in table form after computing the values using a Microsoft Excel spreadsheet.

| λ | Se |
|------|----------|
| 0 | 1 |
| 0.05 | 0.988814 |
| 0.1 | 0.957284 |
| 0.15 | 0.908568 |
| 0.2 | 0.845926 |
| 0.25 | 0.772647 |
| 0.3 | 0.691976 |
| 0.35 | 0.607046 |
| 0.4 | 0.520815 |
| 0.45 | 0.436003 |
| 0.5 | 0.355039 |
| 0.55 | 0.280006 |
| 0.6 | 0.2126 |
| 0.65 | 0.15409 |
| 0.7 | 0.105285 |
| 0.75 | 0.066516 |
| 0.8 | 0.037617 |
| 0.85 | 0.017918 |
| 0.9 | 0.006257 |
| 0.95 | 0.000985 |
| 1 | 0 |

**Example 3.7** *Glucose transport using the Kedem-Katchalsky equation.* The transport of glucose across the capillary wall is $3.0 * 10^{-5}$ μmole/hr. Glucose is a water-soluble and lipid-insoluble solute. The mean pressure of blood in the capillary is 17.3 mmHg; the interstitial pressure of blood is –3 mmHg. The colloid osmotic pressure inside the capillary and the interstitial fluid are 28 and 8 mmHg, respectively. The capillary length is 1 mm, and the inside diameter is 10 μm. It can be assumed that all the glucose transported to the extracapillary space is consumed rapidly by the cells. All plasma protein is retained by the capillary wall. The average concentration of glucose in plasma is 7 μmole/mL. The filtration rate is 5.75 E – 6 μL/hr. Calculate the Staverman reflection coefficient and the pore radius through which the solute transfers.

Glucose comes from the Greek words *glukus,* meaning sweet, and *ose,* meaning sugar. An important carbohydrate in human physiology, it is a monosaccharide and is a source of energy and metabolic intermediate compound.

Molecular formula for glucose: $C_6H_{12}O_6$
Molecular weight: 180.16 gm/mole
Density: 1.54 gm/cc
From Eq. (3.12), the molecular radius of glucose can be calculated as = 36 nm.

From Eq. (3.24), the permeability coefficient can be calculated as:

$$P_m S = 0.0287(36)^{-2.92} = 8.21 * 10^{-7} \, \text{nm}^3/\text{s} \qquad (3.65)$$

$$P_m = \frac{8.21 * 10^{-7}}{2 * \pi * 10^6 * 10^4} = 1.31E - 8 \, \text{m/s} \qquad (3.66)$$

$J = 5.75$ E–6 μL/hr
$N = 3.0$ E–5 μmole/hr
From Eq. (3.31), the Staverman reflection coefficient can be calculated as 0.255.
From Fig. 3.2, $\lambda = 0.26$.
Hence, the pore radius = 0.26*36 = 9.36 nm.

**Example 3.8**   *Thermofiltration of plasma*. Diseases usually have undesirable elevated levels of plasma solutes, such as toxins, excessive antibodies, and other metabolites. Plasma filtration has been used to separate undesirable solutes from blood plasma. Successful treatment of such diseases involves removal of undesirable plasma solutes from the blood plasma using membrane filtration.

Cholesterol has been determined to be an important component of arterial plaque formation in atherosclerosis as well as in hypercholesterolemia. Cholesterol circulates in the blood and is linked to large protein molecules. One form of cholesterol-carrying protein, called low-density lipoprotein (LDL), is known to promote atherosclerosis. About two-thirds or more of the total blood cholesterol is transported in LDL. Another form, called high-density lipoprotein (HDL), is known to be protective against the disease process. Therefore, the selective removal of LDL and maintenance of HDL is important in the treatment of atherosclerosis and the therapeutic control of hypercholesterolemia.

The observed sieving coefficients for different solutes are given in the following table. What is the pore diameter used in filtration?

| Solute | Sieving Coefficient at 25°C |
|---|---|
| Albumin | 0.71 |
| Fibrinogen | 0.05 |
| LDL cholesterol | 0.03 |
| HDL cholesterol | 0.71 |

Based on Fig. 3.6, the corresponding $\lambda$ is read from the charts for albumin, fibrinogen, LDL cholesterol, and HDL cholesterol as 0.3, 0.8, 0.88, and 0.3, respectively. Given that the molecular weight of cholesterol is 386 ($C_{27}H_{46}O$), the solute radius can be estimated from Eq. (3.12) as (3*386/4/Pi/6.023E23/1.5)^.333 = 0.47 nm.

The membrane pore radius is then 0.47*0.3 = 2.4 A.

Equation (3.12) is not reliable in calculating the solute radius given the molecular weight, as these solutes are macromolecules. However, Eq. (3.12) is valid should the solute be approximated to a spherical solid.

## 3.8   Body Fluids

The human anatomy contains three types of fluids: extracellular fluids, intracellular fluids, and transcellular fluids. Sixty percent of human anatomy is comprised of fluids. Thus, a 100-kg male would

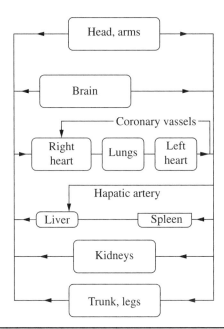

**FIGURE 3.10**  Adult circulatory system for blood.

contain 60 liters of fluid volume at room temperature. Interstitial fluid circulates within the spaces between cells. About 36 wt % of the body mass consists of extracellular fluids, about 21 wt % of body mass is comprised of interstitial fluid, and 4 wt % of body mass is made up of plasma. Figure 3.10 shows the main organs where the fluids flow in to and out of.

The bulk of the body mass is water. The rest consists of fat, proteins, and carbohydrates. Sodium, potassium, calcium, magnesium, chlorine, phosphorous, sulfur, iron, and iodine are also present in trace amounts. Food, air, and water enter the human anatomy every day, and air, sweat, urine, and feces are excreted every day. Metabolic activities consume part of the energy in the food ingested, and water is produced along with the metabolic reactions. Some water is lost through the human dermis.

Blood volume is about one-eighth the total body fluid volume. Sixty percent of the blood volume is comprised of plasma, and the rest is the cells in the blood, such as red blood cells (RBC), white blood cells (WBC), and blood platelets. The cells are filled with intracellular fluid. *Hematocrit* denotes the blood volume occupied by the red blood cells. This can be measured using a centrifuge. Corrections can be allowed for trapped plasma in the cells. After correcting for the trapped quantities, it is called true hematocrit. Transcellular fluids are cerebrospinal, intraocular, pleural, pericardial, synovial, sweat, and digestive fluids. Tracer techniques have been developed to measure the quantities and composition of these fluids. For example, radioactive water is used to measure total body mass consisting of water and radioactive sodium is used to measure extracellular fluid volume. Interstitial fluid volume can be obtained by carefully accounting for the different fluids.

The smallest element of the cardiovascular system is the capillaries. Interstitial fluid is formed here from the plasma as filtrate. Vital substances are exchanged in the capillaries. The capillary wall consists of a porous, semipermeable membrane. The typical dimensions of a capillary are 10 μm in diameter, 1 mm in length, and the representative residence times in the capillary are one to two seconds. There are three types of capillaries: continuous, fenestrated, and discontinuous. Oftentimes, the solute diffusion through the capillary wall can become an important consideration. The muscles, skin, and lungs consist of continuous capillaries. Continuous capillaries can also be seen in fat, nervous system, and connective tissues. A cross-sectional view [7] of a capillary reveals the basement membrane, pinocytotic channels, endothelial cells, pinocytotic vesicles, and an intercellular cleft. The paths of solute movements are several. The pinocytotic channel occupies 1/100th of the total capillary surface area, and typical dimensions are 6 to 7 nm.

Transport of solutes across the wall can be by several different mechanisms. One such mechanism is the intercellular cleft and pinocytotic vesicles and channels. The cleft is a slit-pore of about 6 to 7 nm. The plasma proteins' molecular size is greater than the capillary slit-pore diameter. Thus, their entry into the capillary is blocked. Smaller molecules, such as ions, glucose, and metabolic waste products, will readily pass through the capillary wall.

Oxygen and carbon dioxide are lipid-soluble. They can diffuse directly through the endothelial cells that line the cell wall. No pore diffusion is involved. The rates of diffusion are observed to be higher compared with the water-soluble substances. There are two other mechanisms by which solutes can be transported across the capillary walls other than through the slit-pores. These mechanisms are called pinocytosis and *receptor-mediated transcytosis.* The pinocytosis mechanism is similar to how the microorganism amoeba ingests substances. Stimulated by the presence of the solute, the plasma membrane engulfs the solute, grows in size, and upon migration to the posterior of the membrane, is released. During receptor-mediated transcytosis, the solute/ligand first binds with receptors that complement them, then concentrates the solute regardless of its specificity, followed by complexation of the ligand receptor, and then it is endocytosed. Release of solutes can lead to 80 percent transport of insulin by this process.

Cell membranes are comprised of a lipid bilayer with a head-to-tail configuration. The head of the lipid layer is *hydrophilic* and the tail is *hydrophobic*. The hydrophilic heads face into the aqueous environment inside and outside of the cell, and the hydrophobic tails are sandwiched between the heads of the lipid molecules. In addition to the lipids, proteins are found scattered in the cell membrane. These proteins also participate in the transport of specific molecules across the cell membrane. Others serve as catalysts in reactions, while some serve as enzymes. Proteins can be classified as transmembrane or peripheral. Carrier proteins and channel proteins are membrane transport proteins. The cell membrane is usually impermeable to polar or other water-soluble molecules.

Sometimes, solutes are pumped against their electrochemical gradient by a process called *active transport*. This is an uphill phenomenon that requires cellular energy. Proper functioning of the cells requires that the concentration differences of ions such as sodium and potassium be maintained in order to preserve the resting membrane potential. The energy needed for active transport is made available by adenosine triphosphate (ATP) molecules. ATP is a nucleotide and consists of an adenine base, ribose sugar, and a triphosphate group. ATP is converted to adensoine diphosphate (ADP) by the action of the enzyme ATPase, which is an example of active transport in the sodium-potassium pump. The K-Na pump transports sodium ions out of the cell and at the same instant transfers potassium ions into the cell. The carrier protein protrudes through the two sides of the cell membrane. It has three receptor sites for binding sodium ions and also has ATPase activity. Two receptor sites for the carrier protein are available outside the cell membrane. ATPase is activated upon binding to receptor sites, and a high-energy phosphate bond from ATP is liberated. The energy in the phosphate bond causes a conformational change in the carrier protein that allows for the passage of the sodium and potassium ions. Active transport can also be driven by ion gradients during secondary active transport.

## 3.9  Nernst Equation

The Gibbs free energy change of diffusion and the movement of ions in the presence of an electric field can be estimated as follows. The Gibbs free energy change due to the movement of a solute by diffusion from a region of high concentration to a region of low concentration is given by:

$$\Delta G_D = -RT \ln\left(\frac{C_0}{C_i}\right)$$  (3.67)

where  $C_0$ = the region of high concentration
$\quad\quad C_i$ = the region of low concentration
$\quad\quad \Delta G_D$ = the free energy change due to diffusion

In a similar fashion, the Gibbs free energy change for the movement of an ion across the cell membrane and into a cell at a voltage, $V$, relative to the outside is given by:

$$\Delta G_{\text{elec}} = zFV \qquad (3.68)$$

where $z$ is the charge on the ion and $F$ is the Faraday's constant. The Nernst equation can be derived by balance or equilibrium of the concentration and voltage gradients for an ion. The driving force for the transport of solutes is the combined effect of their concentration gradient and the electrical potential difference that is found across the membrane. The electrochemical gradient denotes the combined effect of charge and solute concentration on the transport of a molecule. Membrane potential is created when the charged molecules flow through channels in a cell membrane. For example, a higher concentration of sodium ions within the cell compared with the surroundings will cause leakage currents out of the cell. The loss of charged ions will make the interior of the cell negative in charge. This creates the membrane potential. As sodium ions are lost, the membrane potential grows. A point is reached where the negative charge created within the cell begins to inhibit the loss of charged ions due to the differences in sodium concentration. Thus, equilibrium membrane potential for the cell is reached.

Equating Eqs. (3.15) and (3.16):

$$V = -\frac{RT}{zF}\ln\left(\frac{C_0}{C_i}\right) \qquad (3.69)$$

Equation (3.17) is the Nernst equation and can be used to calculate the equilibrium membrane potential. $R$ is the universal molar gas constant, $T$ is the absolute temperature in Kelvin, $F$ is the Faraday's constant ($2.3*10^4$ cal/V/gmol), and $z$ is the charge on the ion.

## Electrodialysis

*Dialysis* is a membrane-separation technique used to remove toxic metabolites from blood in patients suffering from kidney failure. The first artificial kidney was developed in 1940 and was based on cellophane. In the 1990s, most artificial kidneys were based on hollow-fiber modules with a membrane area of 1 m². Cellulose fibers were replaced with polycarbonate, polysulfone, and other polymers, which have higher fluxes and are less damaging to the blood. Blood is circulated through the center of the fiber, while isotonic saline, the dialysate urea, creatinine, and other low-molecular-weight metabolites in the blood diffuse across the fiber wall and are removed with the saline solution. The process is slow, requiring several hours to remove the required low-molecular-weight metabolites from the patient, and must be repeated one to two times per week. More than 100,000 patients in hospitals use these devices on a regular basis.

The largest application of membranes is the artificial kidney. Similar hollow-fiber devices are being explored for medical uses, including an artificial pancreas, in which islets of Langerhans supply insulin to patients with diabetes, or an artificial liver in which adsorbent materials remove bilirubin and other toxins. In *carrier facilitated transport,* the membrane used to perform the separation contains a carrier that preferentially reacts with one of the components to be transported across the membrane. Liquids containing a complexing agent are used. Membranes are formed by holding the liquids through capillary action in the pores of a microporous film. The *carrier agent* reacts with the solute on the feed side of the membrane and then diffuses across the membrane to release the solute on the product side of the membrane. The carrier agent is then reformed and diffuses back to the feed side of the membrane. The carrier acts as a shuttle to transport one component of the feed to the other side.

Metal ions can be transported selectively across a membrane driven by the flow of hydrogen or hydroxyl ions in the other direction pumped counter currently around the outside of the fibers (Fig. 3.11). High membrane selectivities can be achieved using facilitated transport. There are no commercial processes yet using this method. This is due to the instability of the membrane and the carrier agent. Dialysis is the earliest molecularly separative membrane process discovered. Fick's law of diffusion and the generalized laws of diffusion are applicable in describing the transport of solute molecules across the membrane to the other side. A multistage dialysis separation procedure can be envisioned for desalinating sea water.

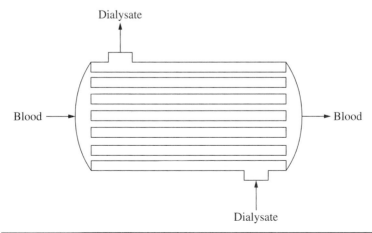

**FIGURE 3.11**   Schematic of a hollow-fiber mass exchanger used as an artificial kidney dialyzer used to remove urea and toxic metabolites from blood.

Depending on the ratio of the pore size of the membrane and the solute radius, the salt concentration on the other side of the first stage can be estimated using the sieving coefficient expressions discussed in the previous sections. Up to one-half the feed concentration of the feed of solute can be obtained on the permeate side of the membrane. Thus, at every stage, a maximum of halving of concentration takes place. Starting with 3.6 wt % NaCl of sea water, a suitable dialysis membrane can, in after $n$ stages, reduce the concentration of the NaCl in both sides of the membrane down to 100 ppm or 0.01 percent. After the first stage, if sufficient time is allowed for equilibrium to be attained by diffusion, both sides of the dialysis membrane will be at 1.8 wt %. Repeated over $n$ stages, this would be 1.8, 0.9, 0.45, 0.225, 0.113, 0.055, 0.028, 0.014, and 0.0007, respectively. For example, after nine stages, the concentration of sea water will reach potable water allowable limits down to less than 100 ppm (Figs. 3.12, 3.13).

The recovery of caustic from hemicellulose in the rayon process was well established in the 1930s and has been used in modern times in the paper pulp industry. *Isobaric dialysis* as a unit operation is emerging and is used to remove alcohol from beverages and in the production of products derived from biotechnology. By the end of

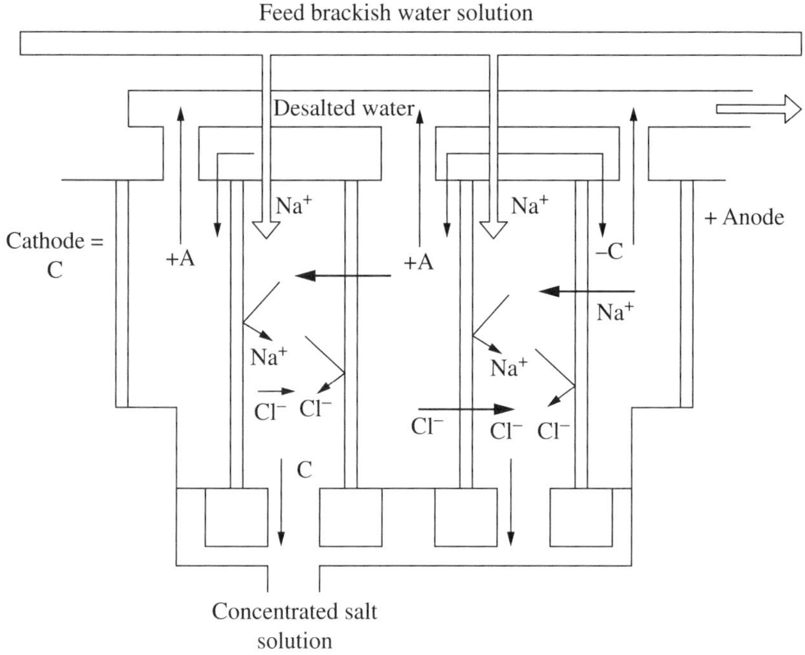

**FIGURE 3.12**    Schematic of electrodialysis apparatus with alternating anode and cathode up to 100 cell pairs.

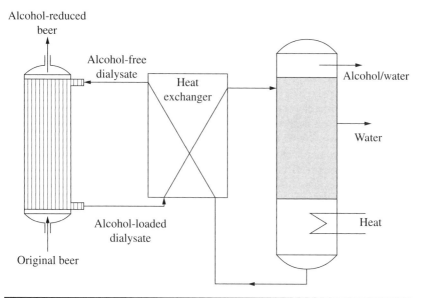

**FIGURE 3.13**    Countercurrent dialysis and distillation to separate alcohol from beverages.

1992, 40 key beer breweries had installed worldwide industrial dialysis plants with an annual capacity of 189 million liters. The schematic of an example industrial dialysis process is shown in Fig. 3.11.

Alcohol is removed from beer by dialysis, and the alcohol is distilled from the dialysate. The raffinate is recycled as a distillate stream. The combination of dialysis and distillation preserves the flavor of the product. Dialysis is an isothermal operation. It is an important parameter in the biotechnology industry. Dialysis facilitates the removal of salts from heat-sensitive or mechanically labile compounds such as vaccines, hormones, enzymes, and other bioactive cell secretions. Dialysis is combined with ultrafiltration to offer *diafiltration* which offers a more efficient process efficiency. The media and extracellular environment in bioreactors can be controlled using diafiltration. Novel bioreactor designs are possible using dialyzers. The extraluminal region of a hollow-fiber dialyzer provides an excellent growth environment for mammalian cells when the lumen is perfused with oxygen and nutrients. In the production of monoclonal antibodies, for example, a bench-top bioreactor can readily equal the antibody production of several thousand mice.

This technology is in the developmental stage. Attempts have been made to separate biological fluids using a dialysis membrane. Removal of a buffer from a protein solution or concentrating polypeptide and hyperosmotic dialysate are examples. Microdialysis is a specialized application of the technique. A U-shaped dialysis capillary is surgically implanted into the tissue of a living animal. Isotonic dialysate is pumped through the tubing at a flow rate low enough to allow equilibration with small solutes in the host's extracellular portion of the tissue. It helps in sampling tissues. Perfusate rate is low, at 1 nL per minute.

Electrodialysis (ED) [20] has been found to be the most economical to desalinate brackish water at a feed NaCl concentration of a little over 1 wt %. The apparatus for ED as shown in Fig. 3.12 is basically an array of anode and cathode membranes terminated by electrodes. The membranes are separated from each other by gaskets, which form fluid compartments. Compartments that have the *A* membrane on the side facing the positively charged anode are *electrolyte depletion compartments*. The remaining compartments are *electrolyte concentrating compartments*. The concentrating and depleting compartments alternate throughout the apparatus. The feed solution is supplied to all compartments. Piping is provided in a fashion so that the concentrated solution is removed from one end and the diluate is removed from the other. In the case of desalination of brackish water, for example, the feed contains a little more than 1 wt % salt solution, the diluate is the potable drinking water at less than 100 ppm NaCl concentration, and the concentrated solution is the brine solution that can be allowed to segregate and removed at the bottom of the apparatus.

Holes in the gaskets and membranes register with each other to provide two pairs of internal hydraulic manifolds to carry fluid into and out of the compartments. One pair communicates with the depletion compartments and the other with the concentrated compartment. Much effort has been spent on the design of the entrance and exit channels from the manifolds to the compartments to prevent unwanted cross-leak of fluid intended for one class of compartment into the other class. As the trend in membrane architecture leads to thinner membranes, the design becomes more difficult. A *cell pair* refers to a contiguous group of two membranes and the associated two fluid compartments. A group of cell pairs and the associated end electrodes is called a *stack or pack*. Generally, *100 to 600 cell pairs* are arranged in a single stack. The choice depends on the capacity of ED, the uniformity of flow distribution achieved among the several compartments of the same class in stack, and the maximum total direct current potential desired. One or more stacks may be arranged in a filter press configuration designed to compress the membranes and gaskets against the force of fluid flowing through the compartments, thereby preventing fluid leaks to the outside and internal cross-leaks between compartments. Hydraulic rams are used for large presses, and rods provide the compression for small presses. Commercial membranes have a thickness of 150 to 500 μ. The compartments between the membranes have a typical thicknesses of 0.5 to 2 mm. The thickness of a cell pair is, therefore, in the range of 1.3 to 5.0 mm. One hundred cell pairs have a combined thickness of 30 cm. The effective area of a cell pair for current conduction is generally on the order of 0.2 to 2 m². Electric current applied to the stack is limited by economic considerations. The power consumption is I²R. In relatively dilute electrolyte, the electric current that can be applied is diffusion-limited. This is the ability of ions to diffuse through the membranes.

The membranes used in the ED apparatus are ion-selective. The stack of membranes is prepared on a plate-and-frame concept. Anion-exchange membranes contain positively charged entities, such as quaternary ammonium groups, fixed to the polymer backbone. These membranes may permit negatively charged ions and exclude positive ions. Cation exchange membranes contain fixed negatively charged groups such as sulfonic acid groups. They permit positively charged ions to move through them.

## 3.11   Oxygen-Depleted Regions by Theory of Krogh in Cylindrical Coordinates

A microscopic view reveals a repetitive arrangement of capillaries surrounded by a cylindrical layer of tissue. An idealized sketch of the capillary bed and the corresponding layer of tissue idealized into a cylinder is shown in Fig. 3.14. Let the radius of the tissue layer be $r_T$. The residence time of the blood in the capillary is in the order of 1 sec. The wave diffusion and relaxation time is comparable in magnitude to the residence time in the blood. Krogh [21] developed this cylindrical capillary tissue model to study the supply of oxygen to muscle. The tissue space surrounding the capillary is considered a continuous phase, although it consists of discrete cells. An effective diffusivity, $D_T$, can be used to represent the diffusion process in the tissue. The driving force for the diffusion is the consumption of the solute by the cells within the tissue space.

The Michaelis-Menten equation can be used to describe the metabolic consumption of the solute in the tissue space. The equation may be written as:

$$R = \frac{V_m C_T}{(K_m + C_T)} \tag{3.70}$$

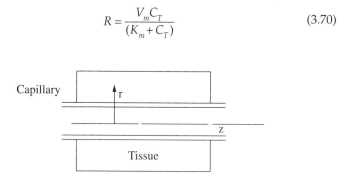

**Figure 3.14**   An idealized sketch of capillary bed and surrounding tissue layer.

where $C_T$ is the concentration of the solute in the tissue space. For consumption of the solute, $R$, it will have a positive value and for solute production, it will have a negative value. $V_m$ represents the maximum reaction rate. The maximum reaction rate occurs when $<C> >> K_m$. The reaction rate is then in zero order in solute concentration. The blood flows through the capillary with an average velocity of $V$. A steady-state shell balance on the solute in the blood from $z$ to $z + \Delta z$ can be written as:

$$-V \frac{dC}{dz} = \frac{2}{r_c} K_0 (C - C_{T(rc+tm)})$$

(3.71)

where $K_0$ is represented by an overall mass transfer coefficient. The overall mass transfer coefficient represents the combined resistance of fluid flowing through the capillary $k_m$ and the permeability of the solute in the capillary wall $P_m$. A steady-state shell balance at a given value of $z$ from $r$ to $r + \Delta r$ may also be written for the solute concentration in the tissue space:

$$\frac{D_T}{r} \frac{d}{dr} \left( r \frac{dC_T}{dr} \right) - R = 0$$

(3.72)

The boundary conditions for Eqs. (3.71) and (3.72) are:

$$z = 0, C = C_o$$

(3.73)

$$r = r_c + t_m, C_T = C_T|_{rc + tm}$$

(3.74)

$$r = r_T, dC_T/dr = 0$$

(3.75)

The axial diffusion is neglected in the tissue space in comparison with the radial diffusion. From the zero-order rate of reaction, $R = R_o$ is a constant. Solving for Eq. (3.72) with the boundary conditions given in Eqs. (3.73) and (3.75):

$$C_T - C_T|_{rc + tm} = (r^2 - (r_c + t_m)^2) \frac{R_o}{4D_T} - \frac{r_T^2 R_o}{2D_T} \ln \left( \frac{r}{r_c + t_m} \right)$$

(3.76)

The variation of concentration as a function of $z$ can be calculated by equating the change in solute concentration within the blood to the consumption of solute in the tissue space:

$$C = C_o - \frac{R_0}{Vr_c^2} (r_T^2 - (r_c + t_m)^2) z$$

(3.77)

Equation (3.76) is combined with Eq. (3.77):

$$C_T|_{rc + tm} - C_o = -\frac{R_0}{Vr_c^2} (r_T^2 - (r_c + t_m)^2) z - \frac{R_o}{2r_c} K_o (r_T^2 - (r_c + t_m)^2)$$

(3.78)

Combining Eqs. (3.76), (3.77), and (3.78):

$$C_T - C_o = (r^2 - (r_c + t_m)^2)\frac{R_o}{4D_T} - r_T^2 \frac{R_o}{2D_T} \ln\left(\frac{r}{r_c + t_m}\right) \tag{3.79}$$

$$- \frac{R_0}{Vr_c^2}(r_T^2 - (r_c + t_m)^2)z - \frac{R_o}{2r_c}(r_T^2 - (r_c + t_m)^2)$$

It can be deduced that under certain conditions some regions may not receive any solute. A critical radius of tissue can be identified, $r_{critical}$, and defined as the distance beyond which no solute is present in the tissue.

$$\text{At } r = r_{critical}, \ dC_T/dr = 0 \tag{3.80}$$

and

$$C_T = 0 \tag{3.81}$$

This can be solved for from Eq. (3.79) after replacing $r_T$ with $r_{critical}$. The equation is nonlinear.

## 3.12  Cartesian Coordinates

Idealize Fig. 3.14 in the Cartesian coordinates and obtain the solution for the concentration of the solute in the tissue space. The governing equations for the concentration of the solute in the capillary and in the tissue can be written after taking the $r$ in Fig. 3.14 as $x$ [22–23]:

$$-V\frac{dC}{dz} = \frac{2}{r_c}K_0\left(C - C_{T(r_c + t_{mem})}\right) \tag{3.82}$$

Considering the effects of diffusion in $x$ direction only in the tissue and assuming a zero-order reaction rate:

$$D_{AB}\frac{\partial^2 C_T}{\partial x^2} = r''' \tag{3.83}$$

Integrating, and substituting for the boundary conditions:

$$x = x_c + t_m, \ C_T = C_T|_{xc + tm} \tag{3.84}$$

$$x = x_T, \ dC_T/dx = 0 \tag{3.85}$$

$$-r'''x_T/D_{AB} = c_1 \tag{3.86}$$

$$C_T - C_T|_{xc + tm} = (r'''/2D_{AB})(x^2 - (x_c + t_m)^2)$$
$$- r'''x_T/D_{AB}(x - (x_c + t_m)) \tag{3.87}$$

The variation of concentration as a function of $z$ can be calculated by equating the change in solute concentration within the blood to the consumption of solute in the tissue space:

$$VAC_0 - VAC = r'''zA_T \tag{3.88}$$

$$C = C_0 - r'''zA_T/VA \tag{3.89}$$

Equation (3.89) is combined with Eq. (3.87):

$$r'''A_T/A = 2/x_c\,K_0\,(C - C_T|_{rc+tm}) \tag{3.90}$$

$$C_T|_{rc+tmem} = C - K_0 x_c\,r'''A_T/2A \tag{3.91}$$

Therefore:

$$CT - Co = r'''zAT/VA + K0xc\,r'''AT/2A + (r'''/2D_{AB}) \\ \times (x^2 - (x_c + t_m)^2) - r'''xT/D_{AB}\,(x - (x_c + t_m)) \tag{3.92}$$

At a critical distance from the capillary wall, the concentration in the solute will become zero. This can be solved for from Eq. (3.92). At and beyond the critical distance:

$$dC_T/dx = 0 = C_T \tag{3.93}$$

replacing $x_T$ with $x_{critical}$:

$$0 = C_0 + r'''zA_T/VA + K_0 x_c\,r'''A_T/2A + (r'''/2D_{AB})(x^2 - (x_c + t_{mem})^2) \\ - r'''x_{critical}/D_{AB}\,(x - (x_c + t_{mem})) \tag{3.94}$$

$$x_{critical}^2\,(-r'''/2D_{AB}) = C_0 + r'''zA_T/VA + K_0 x_c\,r'''A_T/2A - (r'''/2D_{AB}) \\ \times (x_c + t_m)^2 - r'''x_{critical}/D_{AB}\,(x - (x_c + t_m)) \tag{3.95}$$

The quadratic equation in $x_{critical}$ is then:

$$Ax_{critical}^2 + Bx_{critical} + C = 0 \tag{3.96}$$

where:

$$A = -\frac{R_0}{2D_{AB}} \tag{3.97}$$

$$B = \frac{(x_c + t_m)R_0}{D_{AB}} \tag{3.98}$$

$$C = C_0 + \frac{R_0 z A_T}{VA} + \frac{K_0 x_c R_0 A_T}{2A} - \left(\frac{R_0}{2D_{AB}} x_c + t_m\right)^2 + \frac{R_0(x_c + t_m)}{D_{AB}} \tag{3.99}$$

When the solution of the quadratic expression for the critical distance in the tissue is real and found to be less than the thickness of the tissue, the onset of zero concentration will occur prior to the periphery of the tissue. This zone can be seen as the anorexic or oxygen-depleted regions in the tissue.

## Summary

The three important developments that gave impetus to the emergence of the field of biofluid transport phenomena are the discovery of osmosis and osmotic pressure, permeability for a solvent across the membrane and Starling's law, and diffusion of solute across the membrane. Osmosis is the flow of solvent from a region of low solute concentration to a region of high solute concentration. The pressure difference that creates flow is caused by the presence of solutes, and is called the colloid osmotic pressure. If a cell is placed within a solution that has a lower concentration of solutes, water flows into the cell and the system is considered hypotonic. When the cell is placed in a solution that has a higher concentration of solutes, water flows out of the cell and the system is considered hypertonic. When the system isotonic, there is neither swelling nor shrinking of cells. Dutrochet discovered the phenomena of osmosis in the 1800s.

Van't Hoff's law can be used to determine the osmotic pressure in terms of the concentration of the solution. It can be derived by equating fugacities of the solvent and solution. The Poynting correction factor and Taylor series expansion are used in the derivation.

Porous membranes are solid materials with an internal pore structure comprised of macropores and micropores. Pores can be classified into passing penetrable pores, nonpassing penetrable pores, isolated pores, and tortuous channels. Pores are characterized by the surface area and porosity. When the length of the pore is greater than the thickness of the specimen, the pore is said to be tortuous. Darcy's law may be written for fluid passage through the pores, relating the discharge rate, $Q$, to the permeability of medium, $\kappa$; cross-sectional area, $A$; viscosity of fluid, $\mu$; and pressure gradient, $\partial P/\partial z$.

The combined effect of osmotic pressure and hydrostatic pressure is described by Starling's law, which provides the relation between the flow of fluid across the capillary wall or a porous membrane and the pressure difference across the capillary. Both the hydrostatic and osmotic pressure drops are accounted for. A hydraulic conductance, $L_p$, can be defined for flowing fluid as a function of pore radius, peripheral area, and wall thickness. Starling's law can be derived from pressure and chemical potential considerations. Starling's law applicability is not universal, but it has been improved upon.

Oftentimes, solvent moving across the membrane will carry with it some solute molecules. A sieving coefficient, $S_e$, has been developed to account for solute separation based on molecular size. Analytical expressions based on the motion of a spherical solute through a cylindrical pore have been derived to estimate $S_e$. The mathematical expression given by Deen can be written as a seventh-degree polynomial for the sieving coefficient in terms of the ratio of the solute radius to the capillary pore radius [Eq. (3.20)]. Formation of the concentration polarization layer was also discussed.

There are several kinds of diffusion of solutes through membranes: viscous flow, molecular flow, surface diffusion, capillary condensation, molecular sieving diffusion, and solubility diffusion. The solute size can be calculated as a function of molecular weight directly from Eq. (3.24). Renkin and Curry [7,8] presented an expression for diffusion coefficient as a function of the molecular weight of the migrating species in Eq. (3.25). There are several diffusion coefficients, such as solute diffusion coefficient in blood and tissue, $D_{bl}$; solute diffusion coefficient in plasma, $D_{pl}$; solute diffusion within the pores of a capillary wall, $D_{pore}$; solute diffusion in the interstitial fluid, $D_{int}$; solute diffusion within the cell, $D_{cell}$; and solute diffusion in water, $D$. Expressions for steric exclusion and hindered diffusion were provided. The Renkin equation gives the ratio of pore diffusivity to that of bulk diffusivity [Eq. (3.27)]. The Renkin euation is a seventh-degree polynomial in $\lambda$, the ratio of the solute radius with pore mouth radius

The mathematical expression developed by Maxwell for diffusion through blood and cells is given by Eq. (3.30). Expression for diffusion coefficient in blood developed using Monte Carlo simulations as a function of the volume fraction of cells in blood is given by Eq. (3.31). The effect of molecular weight as evidenced in polymers developed by Brinkman is given by Eq. (3.32). The effective diffusivities of different sizes of solutes through tumor and normal tissue developed by Jain [10] is given by Eq. (3.33). The solute diffusion can also be accounted for by use of a solute permeability [Eq. (3.35)]. Renkin and Curry [8] developed empirical relations [Eqs. (3.36) and (3.37)] to solute permeability as a function of solute size.

The combined effect of hydrostatic pressure and osmotic pressure and the solute flux is captured by the Kadem-Katchalsky equation [Eqs. (3.38) and (3.39)]. They were derived by the application of irreversible thermodynamics. A Staverman reflection coefficient gives the ratio of the hydraulic conductance of a solute to that of a solvent. Oftentimes, the three parameters need to be solved for: $L_p$, the hydraulic conductance of solvent; $P_m$, the permeability of the solute; and $\sigma$, the Staverman reflection coefficient. Anderson and Quinn [13] showed that the sieving coefficient, $S_e = 1 - \sigma$, using hydrodynamic equation accounting for hindered particle motion is small pores.

Eight worked examples illustrating the use of the theory described were presented. The human anatomy is comprised of three types of body fluids: extracellular fluids, intracellular fluids, and transcellular fluids. The Nernst equation also was derived. It can be used to calculate the equilibrium membrane potential.

Dialysis is a membrane separation process that lets solute diffuse across the membrane and from the permeate and retentate at desired concentrations. A mass exchanger is used in hospitals to treat human patients with kidney disease by removing toxic metabolites from urea. Industrial dialysis and reclamation of alcohol from beer was shown with examples from commercial operations used throughout the world. The electrodialysis apparatus was shown with a schematic as a method to reduce common salt concentrations in alternate compartments of anode and cathode.

Oxygen-depleted regions were identified by the theory of Krogh using mathematical modeling in both cylindrical and Cartesian coordinates. Simultaneous metabolic reactions and diffusion leads to the zone of null transfer after a critical length. A numerical solution is needed for cylindrical coordinates. Closed form analytical solution is derived for Cartesian coordinates.

## References

[1]   A. G. Ogston, "The spaces in uniform random suspension of fibers," *Trans. Farad. Soc.*, 54 (1958), 1754–1757.
[2]   H. Darcy, *Les Fontaines Publiques de la Ville de Dijon* (The Public Fountains of the Town of Dijon), Dalmont, Paris: 1856.
[3]   A. M. Katz, "Ernest Henry Starling, his predecessors and the 'law of the heart,'" *Circulation*, 106(23), 2002, 2986–2992.
[4]   W. M. Deen, "Hindered transport of large molecules in liquid-filled pores," *AIChE Journal*, 33 (1987), 1409–1425.
[5]   A. L. Zydney, "Therapeutic apheresis and blood fractionation," in *Biomedical Engineering Handbook*, J. D. Bronzino (ed.), Boca Raton, FL: CRC Press, 1951.
[6]   E. I. Cussler, *Diffusion: Mass Transfer in Fluid Systems*, Cambridge, UK: Cambridge University Press, 1997.
[7]   E. M. Renkin, "Filtration, diffusion and molecular sieving through porous cellulose membranes," *J. Gen. Physiology*, 38 (1954), 225–243.
[8]   E. M. Renkin and F. E. Curry, "Transport of water and solutes across capillary endothelium," in *Membrane Transport in Biology*, vol. 4, G. Giebisch and D. C. Tosteson (eds.), New York: Springer-Verlag, 1979.
[9]   L. J. Gaydos and H. Brenner, "Field-flow fractionation: Extensions to nonspherical particles and wall effects," Separation *Science and Techology*, 13(3), 1978, 215–240.
[10]  R. L. Fournier, *Basic Transport Phenomena in Biomedical Engineering*, Philadelphia, PA: Taylor & Francis, 1999.
[11]  G. A. Truskey, F. Yuan, and D. F. Katz, *Transport Phenomena in Biological Systems*, Upper Saddle River, NJ: Pearson Prentice Hall, 2009.
[12]  R. K. Jain, "Transport of molecules in the tumor interstitium: A review," *Cancer Rev.*, 47 (1987), 3039–3051.
[13]  F. E. Curry, "The Eugene M. Landis Award Lecture 1993: Regulation of water and solute exchange in microvessel endothelium: studies in single perfused capillaries," *Microcirculation*, 1(1), 1994, 11–26.
[14]  O. Kedem and A. Katchalsky, "Thermodynamic analysis of the permeability of biological membranes to non-electrolytes," *Biochem. Biophys. Acta.*, 27 (1958), 229.
[15]  A. J. Staverman, "The theory of measurement of osmotic pressure," *Rec. Trav. Chim.*, 70 (1951), 344–352.

[16]  C. C. Michel and M. E. Philips, "Steady-state fluid filtration at different capillary pressures in perfused frog mesenteric capillaries," *J. Physiol.*, 388 (1987), 421–435.

[17]  E. S. Avgoustiniatos, K. E. Dionne, D. F. Wilson, M. L. Yarmush, and C. K. Colton, "Measurements of the effective diffusion coefficient in pancreatic islets," *IEC Res.*, 46 (2007), 6157–6163.

[18]  C. E. Jennings, "Implantable human kidney replacement unit," U.S. Patent 7,083,653 (2006), Houston, TX.

[19]  Y. Nose, P. S. Malchesky, and T. Horiuchi, "Thermofiltration of plasma," U.S. Patent 5,080,796 (1992), The Cleveland Clinic Foundation, Cleveland, OH.

[20]  K. R. Sharma, *Principles of Mass Transfer*, N. Delhi, India: Prentice Hall of India, 2007.

[21]  A. Krogh, "The number and distribution of capillaries in muscles with calculations of the oxygen pressure head necessary for supplying the tissue," *J. Physiology*, 52 (1919), 409–415.

[22]  K. R. Sharma, "Oxygen-depleted regions by the theory of Krogh," in *32nd Northeast Bioengineering Conference*, Easton, PA, April 2006.

[23]  K. R. Sharma, "Anorexic region in tissue in Cartesian coordinates," in *CHEMCON 2005*, N. Delhi, India, Indian Institute of Technology, December 2005.

# Exercises

## Questions for Discussion

**1.0**   Can the osmotic pressure be used to cause a flow that can operate a turbine and generate electricity? Why?

**2.0**   Can Kedem-Katchalsky equations violate the second law of thermodynamics? Can transport occur from low concentration to high concentration of the said species?

**3.0**   Will recycling in sea water reverse osmosis (SWRO) plants reduce the pressure needed at the pump?

**4.0**   What is the reason for the higher water flux in the membranes developed later, such as a polyamide membrane, compared with the cellulose acetate membrane?

**5.0**   Should there be a temperature difference between the feed and permeate sides? What will happen to the predictions of van't Hoff's law?

**6.0**   Can osmotic pressure be extended to other systems such as gases?

**7.0**   Consider a layered solution. A higher concentration of solute is found in the bottom and a lower concentration of solute is found at the top. Will the solvent flow from the top of the jar to the bottom by osmosis? Why?

**8.0**   What are the energetic considerations during osmotic flow?

**9.0**   Can the osmotic pressure and hydrostatic pressure cancel out each other, resulting in zero net flow?

**10.0**  Why are not a lot of trees found near the beach or coastline?

**11.0**  Can 100 percent separation be effected by use of semipermeable membrane technology?

**12.0**  Is osmotic pressure accounted for in the development of Darcy's law? Why?

**13.0**  Sketch the different mechanisms of pore diffusion: viscous flow, Knudsen diffusion, capillary condensation, surface diffusion, molecular sieving, and diffusion solubility mechanism.

**14.0**  How would you write a generalized Starling's law, taking into account the observations that rendered it not universal?

**15.0**  What are some of the important considerations during transient solvent filtration and solute diffusion?

## Problems That Require Analysis—Reverse Osmosis

Reverse osmosis (RO), ultrafiltration, molecular sieves, and electrodialysis can be used to separate liquid solutions. Reverse osmosis and ultrafiltration differ in the solute size rejected. Conventional filtration rejects particles of the size of 10 μm and above, the ultrafiltration rejects solute sizes of 10 nm to 10 μm, and the reverse osmosis rejects solute sizes in the ionic range of 1 pm to 0.1 nm. The driving force for RO, ultrafiltration, and molecular sieve operations comes from a pressure difference. No filter cake is allowed to form. The driving force in electrodialysis is an induced electric field and polarization of the ions to the anode and cathode. The compartments are made from alternating anode and cathode, and the ions are segregated from the diluate water in alternating compartments. It is used to desalinate brackish water and is found to be the most economical method at low salt concentration in the feed, at little over 1 wt %. Any material that can exclude molecular species by size is referred to as a molecular sieve. These are made up of inorganic materials that possess uniform pores with diameters less than 2 nm (microrange) or 2 to 20 nm (mesorange).

In reverse osmosis, the solvent from the solution is pumped across a semipermeable membrane, opposing the osmotic pressure difference with the solute largely rejected by the membrane (Fig. 3.16). For example, in sea water desalination by reverse osmosis, the osmotic pressure will cause a flow from the region of low solute concentration to a region of higher solute concentration. Pressure is supplied in order to overcome this so that the solvent from the sea water flows across a semipermeable membrane that permits only the solvent and not the solute. Although the membrane rejects the solute, some will diffuse across the pores.

The drinking water needs of major cities around the world can be met using the reverse osmosis method for desalination, as the drought conditions in major cities have become an important concern. In the Middle East gulf region, where the energy is cheap and river water scarce, sea water reverse osmosis

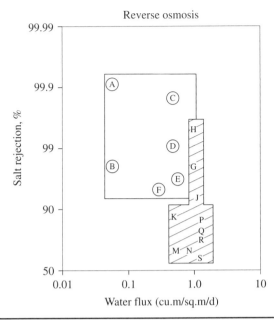

FIGURE **3.15**    Water flux and salt rejection for various SWRO semipermeable membranes.

**(A–F. Sea water membranes operate at 5.5 Mpa and 25°C. H–J. Brackish water membranes operate at 1,500 mg/L NaCl feed, 1.5 Mpa, and 25°C. K–S. Nanofiltration membranes operate at 500 mg/L NaCl feed, 0.74 Mpa, and 25°C.)**

**Cellulose acetate/triacetate**
**Linear aromatic polyamide**
**Cross-linked polyether**
**Cross-linked fully aromatic polyamide**
**I- Du Pont A-15**
**J, K, L- Nitto-Denko, NTR-739HF NTR-729HF, NTR-7250**
**O- Toray, UTC-40HF**

**E- Other thin-film composites**
**F- Asymmetric membranes**
**G- FilmTec, BW 30**
**H- Toray, SU –700**
**Q- Toray, UTC 60**
**R- Toray, UTC 20HF**
**S- FilmTec, NF50**
**P- FilmTec, NF70**

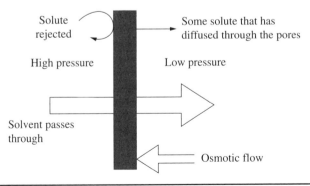

FIGURE **3.16**    Transport processes during reverse osmosis.

(SWRO) plants are increasingly relied upon to obtain potable drinking water (Fig. 3.17). Reverse osmosis is also used in the food industry to concentrate fruit juices. Reverse osmosis is dominated by desalination—that is, by removing salt from water. It is used in the treatment of sea water, brackish water, and the reclamation of municipal waste water. It is used in boiler feed water pretreatment in order to avoid scale formation later on. The market for ultra pure water is growing; thus, reverse osmosis is used in injectable pharmaceuticals and for semiconductors, water for domestic use, sweetener concentration, and fruit juice concentration and fermentation product recovery.

In the 1950s, it was shown that cellulose acetate RO membranes were capable of separating salt from water. But the water fluxes obtained were too small to be feasible in practice. Since then, RO membrane technology has improved a great deal. Recently DuPont patented a polyamide membrane made from interfacial polymerization that has a high solvent flux, and turbulence is used to reduce the concentration polarization layer thickness formed during the operation of the RO during desalination. Advances in thin-film composite membranes and polymer materials have widened the applications of RO from desalination to treatment of hazardous wastes, material recovery in electroplating industries, production of ultra pure water, water softening, food processing, dairy, and the semiconductor industry.

The advantages of these systems over traditional separation processes, such as distillation, extraction, ion exchange, and adsorption, are that RO is a pressure-driven process, there are no energy-intensive phase changes, and expensive solvents or adsorbents are not needed. RO processes have a simple design and are easy to operate compared with other traditional methods.

Membrane properties play a pivotal role in the performance of RO technology and depend on the chemical structure of the membrane. An ideal RO membrane is low in cost, resistant to chemical and microbial attack, possesses high mechanical and structural stability over long period of operation and a wide range of temperature, and has the desired separation characteristics for the given system.

RO membranes are classified into asymmetric membranes (containing one polymer) and thin-film composite membranes. Asymmetric RO membranes have a thin permselective skin layer of about 100 nm thickness supported on a more porous sublayer of the same polymer. The dense skin layer determines the fluxes and selectivities of these membranes. The porous layer serves as a mechanical support for the skin layer. Asymmetric membranes are formed by a phase-inversion polymer precipitation process. In this process, a polymer solution is precipitated into a polymer-rich solid phase that forms the membrane and a polymer-poor liquid phase that forms the membranes or void spaces. Composite RO and ultrafiltration membranes are thin films consisting of a thin polymer barrier layer formed on one or more porous support layers, which is a different polymer compared with the barrier layer. The surface layer determines the flux and separation characteristics of the membrane. The porous backing is intended largely to support the barrier layer. The barrier layer is extremely thin, thus allowing for high water fluxes. The most important thin-film composite membranes are made by interfacial polymerization, a process in which a highly porous membrane such as polysulfone is coated with an aqueous solution of a

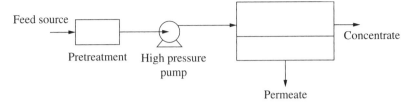

Feed source

Pretreatment    High pressure
pump

Concentrate

Permeate

**FIGURE 3.17**    Schematic of an SWRO semipermeable membrane process.

polymer or monomer and then reacted with a cross-linking agent in a water-immiscible solvent.

**16.0**    *Sea water reverse osmosis semipermeable membrane process.* Prepare a preliminary estimate of an SWRO plant where the inlet NaCl composition is 3.6 percent by weight and the outlet composition is 100 ppm. A typical RO process schematic is given in Fig. 3.16. What should the size of the pump be? The RO membrane used is a linear aromatic polyamide. The expected water supply for the township is 6.3 million liters per day. Spiral-wound modules are used with an interfacial area of 5 m²/gm. How much membrane is needed? The operating temperature is 25°C. What would be the reduction in pump pressure if recycling is used (Sharma [20])?

**17.0**    *Effect of solute concentration.* Van't Hoff's law, as given by Eq. (3.5), was derived assuming that the solute concentration was small. For not-so-dilute systems, what would be *the* expression for osmotic pressure? What effect does it have in the solution in Exercise 16.0? Use the $\ln(x_w)$ in the van't Hoff's law instead of the Taylor series approximation. What effect does it have?

In the Taylor series expansion of $\ln(1 - x_s)$, another two terms are taken:

$$\ln(x_w) = \ln(1 - x_s) \sim -x_s + \frac{x_s^2}{2} - \frac{x_s^3}{6} \qquad (3.100)$$

Equation (3.3) becomes, on substituting for Eq. (3.100):

$$\pi = \frac{RT}{V_w} x_s \left(1 - \frac{x_s}{2}\right) = RTC_s \left(1 - \frac{C_s V_w}{2} + \frac{\left(C_s V_w\right)^3}{6}\right) \qquad (3.101)$$

**18.0**    *Use of reflection coefficient.* A brackish water body is desalinated using the reverse osmosis method. The concentration of the NaCl feed is 15 gm/liter. A FilmTec NF40HF semipermeable membrane is used. How much water flux can be handled when a spiral module with interfacial area of 8 m2/gm is used and exit water at a concentration of 100 ppm is produced? The operating temperature is 37°C. How much membrane is needed? What is the pressure needed at the high pressure pump?

**19.0**    *Hydraulic conductance in the human body.* Calculate the normal rate of net filtration for the human body. Assume that the capillaries have a total surface area of 413 m² and that the slit-pore surface area is 3/1000 of the total capillary surface area. Assume that the porous structure of the capillary wall is a series of parallel cylindrical pores with a diameter of 7 nm. Plasma filtrate may be considered as Newtonian fluid with a viscosity of 1 cp. The mean net filtration pressure for the capillary was just calculated to be 0.3 mmHg. The capillary characteristics are as follows:

| | |
|---|---|
| Inside diameter: 10 μm | Length ($L$): 0.1 cm |
| Wall thickness, tm: 0.5 μm | Average blood velocity: 0.05 cm/s |
| Pore fraction: 0.001 | Wall pore diameter: 6–7 nm |
| Inlet pressure: 30 mmHg | Outlet pressure: 10 mmHg |
| Mean pressure: 17.3 mmHg | Colloid osmotic pressure: 28 mmHg |
| Interstitial fluid pressure: –3 mmHg | |
| Interstitial fluid colloid osmotic pressure: 8 mmHg | |

**20.0**  *Blood storage in the army.* For transfusion purposes, in the army, donated human blood is stored for a month. There is interest in improving the storage procedure such as concentrating red cells, white cells, platelets, vitamins, proteins, sugars, minerals, hormones, and enzymes by water removal. Ultrafiltration devices are sought by the army to remove water, with 50 percent by volume water, to levels low enough to effect significant volume reduction during blood storage. The temperatures have to be kept low during separation to prevent hemolysis. Use an ultrafiltration membrane and for a solute rejection of 80 percent, find the pore radius of the membrane from the expression for the sieving coefficient. The solute radius can be taken as an effective radius of the different ions present in the blood. For the given pore size of the membrane, and for a membrane thickness of 1 mm, what is the hydraulic conductance during flow of water across the membrane? The interfacial area of the membrane, the amount of water that can be treated, and the volume and weight of the membrane needed can be related to the information in Fig. 3.16. Choose a membrane for the given solute rejection, and by trial and error in a Microsoft Excel spreadsheet, obtain the throughput that can be handled during ultrafiltration of blood and the pressure needed at the high pressure pump.

**21.0**  *Reverse osmosis to separate acrylonitrile from water.* In the manufacture of Acrylonitrile Butadiene Styrene (ABS) engineering thermoplastics using a continuous polymerization process, for every pound of product manufactured, a little over a pound of water is generated that contains acrylonitrile ($CH_2 = CHCN$). A 5 percent Acrylonitrile (AN) solution of water needs to be separated by reverse osmosis, and the product needs to have AN less than 1 ppm. A membrane made of cross-linked polyether resin at an interfacial area of 10 m$^2$/gm and 1 mm thickness is used. How much membrane is needed for a solute rejection of 99.9 percent? What is the pore size of the membrane? What is the throughput of water it can handle? What is the pressure at the high pressure pump on the feed side?

**22.0**  *Effect of concentration polarization layer.* During reverse osmosis in sea water desalination, the salt is rejected by the semipermeable membrane and is accumulated near the feed side of the membrane (Fig. 3.16). Perform a mass balance of the solvent in the region of the concentration polarization layer and the membrane, and show that the flux will decrease with time—*that is:*

$$J/L_pS = (\Delta P - RTC_{sf0}/\delta S)/(1/L_pS + RTC_{sf0}t/\delta S) \qquad (3.102)$$

where $\delta$ is the polarization layer and $t$ is the time of operation.

**23.0** *Hydraulic conductance.* Consider water that contains polychlorinated biphenyls (PCBs) and tetrachlorinated ethylenes (TCEs) at 4.5 percent and 2.0 percent by weight, respectively. A reverse osmosis membrane of 0.5 mm thickness is used. The membrane is Toray, SU-700. What is the rejection rate for SU-700? What is the flux rate the Toray membrane can handle per day with an interfacial area of 13 $m^2/gm$? How much membrane is needed to produce a filtrate with a concentration of 1 ppb? For a throughput of 22,500 liters/day, what is the osmotic pressure? What is the hydraulic conductance ($L_pS$) of the system? What should the pressure at the pump be?

**24.0** *Starch removal.* A new membrane on the market was tested and found to have permeability, $L_pS$, of 1 E – 4 $m^4s/kg$ under a pressure difference of 10 atm. The membrane handles 4.8 percent of partially hydrolyzed starch (MW 17,000) as feed and puts out a product at 175 ppm. What is the Staverman reflection coefficient? What is the throughput of water the membrane can handle? What is the solute rejection rate of the membrane? Can you provide the pore diameter and length of the membrane for a thickness of the membrane of 500 $\mu$m when 100 cc of the membrane is used?

**25.0** *Tallest tree in the world.* What is the limit on the height of a tree? Include the Bernoulli law as well as the osmotic pressure drop. Assume that the leaves on the tree top have a starch concentration of 10 wt %.

**26.0** *Porous membrane.* A copolymer with a high acrylonitrile content of styrene acrylonitrile (SAN) is tested for use as a reverse osmosis membrane. The pore radius is 200 nm. For a solution of 1 percent by weight of polyethylene glycol of 18,000 molecular weight in water, what is the sieving coefficient? Using this as the Staverman reflection coefficient, what is the pressure at the pressure pump to reduce the water content to less than 1 ppb in the filtrate side? What is the hydraulic conductance? Show that the effect of the molecular weight has reduced the pressure needed at the pump in such as fashion that the solvent filtration pressure drop is the limiting factor, compared with most RO processes where the osmotic pressure is the limiting factor. Show that the water flux rate this membrane can handle is 26,000 liters/day for a 1 mm thick membrane and effective volume of 1 cc.

**27.0** *Effect of molecular weight of the solute.* In some applications of RO technology, such as the desalination of sea water, the cost-limiting step is the osmotic pressure that needs to be overcome in order to achieve the desired degree of separation. This leads to a large pump size. As the molecular weight of the solute increases, for the same solute concentration by weight, say, 3.6 wt %, what is the molecular weight of the solute when the pressure drop needed for the hydraulic motion of the solvent alone is greater than or equal to the osmotic pressure from the solute? Make suitable assumptions about the Staverman reflection coefficient.

**28.0**   *Effect of operating temperature.* In Exercise 16.0, what happens to the pressure needed at the high pressure pump during winter? Consider an operating temperature at 4°C, an inlet salt concentration of 4.4 wt %, and a product expected salt concentration of 42 ppm. For the same membrane used in the worked example, what is the reduction in pressure needed at the pump?

**29.0**   *Salt precipitation by freezing.* Based on the results in Exercise 16.0, estimate the energy needed to pump the sea water using a high pressure pump to achieve the desired separation objectives. Make a comparative study of the decrease in solubility of NaCl with temperature. As you compress the sea water, at what point does it become ice, or at what point does it become favorable for salt precipitation? In which route is less energy required?

**30.0**   *Virial expansion of osmotic potential.* The osmotic pressure for albumin was developed by fitting a curve to experimental data and presented as:

$$\pi = 0.345\rho_s + 2.657E - 3\rho_s^2 + 2.26E - 5\rho_s^3 \qquad (3.103)$$

where $\rho_s$ is in mgm/mL and $\pi$ is in cm of water. Show that this form of the equation can be derived from van't Hoff's law for nondilute solutions by a series expansion of the concentration of solute in absolute mass units instead of molar units.

**31.0**   *Effect of gravity in fluid flow through porous medium.* Modify Darcy's law in order to take into account gravity forces and show that:

$$v = -K(\nabla P - \rho g) \qquad (3.104)$$

**32.0**   Apply the principle of conservation of momentum, and derive the Laplace equation for the case of zero source and sink of fluid.

**33.0**   In Worked Example 3.2, the data point for BSA deviated from the fit of equation to data. Can the equation for solute radius, given the molecular weight, *be applied to* BSA? Discuss.

**34.0**   *Treatment of type 1 diabetes.* Nanoporous biocapsules can be used *to transport* insulin-secreting cells by providing an immunoisolating microenvironment. For immunoprotection of pancreatic cells, the immunoprotection membrane ought to allow permeation of glucose, insulin, oxygen, and other metabolic products to ensure islet functionality and therapeutic effectiveness. The sieving coefficient for *vitamin* $B_{12}$ may be taken as 0.7. What ought to be the size of the pores in the membrane should the molecular mass of vitamin $B_{12}$ be 1355? The molecular formula for the antioxidant is $C_{63}H_{88}CoN_{14}O_{14}P$. It is involved in the metabolism of every cell of the body. It affects DNA synthesis and regulation, energy production, and fatty acid synthesis.

**35.0**   *Islets of Langerhans.* Islets of Langerhans are spheroidal aggregates of cells that are located in the pancreas (Fig. 3.18). Islets may be viewed as a suspension of tissue spheres. Some islets were isolated from male rats under a dissecting microscope, as discussed in Worked Example 3.3, and cultured. Rotation of tubes prevented settling and aggregation of islets. Oxygen uptake measurements were conducted. The oxygen uptake chamber was equipped for measurement of the oxygen-dependent lifetime of Pd-coproporphyrin

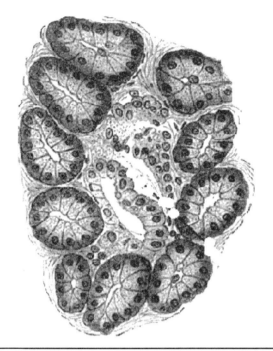

**Figure 3.18**   Islets of Langerhans.

phosphorescence to enable rapid and accurate measurements of oxygen concentration up to low concentrations. The measured effective diffusivity of oxygen through the islets was found to be 5.0 E – 5 cm²/sec. Given *that* the volume fraction of the cells was 50 percent, calculate the diffusion coefficient of oxygen in interstitial fluid. The diffusion coefficient in the cell *is* 2.5 E – 5 cm²/s.

**36.0**   Calculate the Brinkman parameter in Problem 21.0.

**37.0**   The Renkin equation takes into account the effects of hindered diffusion of the solute, especially the hydrodynamic drag experienced by the solute. The equation gives the ratio of pore diffusivity to that of bulk diffusivity:

$$\frac{D_{pore}}{D} = 1 - 4.1\lambda + 5.2\lambda^2 - 0.01\lambda^3 - 4.18\lambda^4 + 1.14\lambda^5 + 1.9\lambda^6 - 0.95\lambda^7 = K\omega \qquad (3.105)$$

where $\lambda$ is the ratio of the solute radius *to* the pore mouth radius. The partition coefficient K captures the steric exclusion. $\omega$ accounts for the hydrodynamic drag faced by the diffusing solute through the pore. What is the error involved in neglecting the cubic term in the seventh-degree polynomial expression—that is, $- 0.01\,\lambda^3$?

**38.0**   In the Maxwell's expression for diffusion coefficients in suspensions, the diffusion coefficient in the blood can be calculated given the volume fraction of cells, the diffusion coefficient in the interstitium, and the diffusion coefficient in the cell. Given the diffusion coefficient in the suspension and the diffusion coefficient in the volume fraction of cells, can the diffusion coefficient in the interstitium be estimated? Is a numerical solution necessary?

**39.0**   Blood cells can be isolated from plasma using a blood separation technique called *plasmapheresis,* as discussed in Worked Example 3.4. Given that the hydraulic conductance of the membrane is 2.0 E–6 m²s/kg and the thickness of the membrane is 90 μm, find the radius of the pore given that the porosity is 0.5.

**40.0**   *Intravenous therapy.* Liquid substances are introduced into the veins using IV therapy, or intravenous therapy. Rather than using a catheter, a novel device pumps the fluid across a semipermeable membrane into the bloodstream. Suppose the membrane thickness is 500 nm and has a pore size of 80 nm. What is the desired pump pressure to inject 0.9 perent NaCl solution at 3.6 mL/hr? Assume that the saline concentration in the blood is zero. In practice, the intravenous therapy is by transonic flow, as the saline concentration in the bloodstream is also 0.9 percent by weight. The pump pressure has to overcome the osmotic pressure and filter through the pores in accordance with Starling's law.

**41.0**   *Sieving coefficient of dialyzer filter.* As discussed in Worked Example 3.6, hemodialysis machines are stationary hydromechanical devices. They are used to prolong life temporarily for patients with end-stage renal disease. A dialyzer filter made of cellulose acetate has a pore size of 125 nm. Creatinine and urea have to pass through the filter and phosphorous has to be retained. Calculate the sieving coefficient of the filter.

**42.0**   *Glucose transport across a capillary.* Calculate the transport rate of glucose across the capillary wall. Glucose is a water-soluble and lipid-insoluble solute. The mean pressure of blood in *a* capillary is 17.3 mmHg, *and* interstitial blood pressure is –3 mmHg. The colloid osmotic pressure inside the capillary and the interstitial fluid are 28 and 8 mmHg, respectively. The capillary length is 900 μm and the inside diameter is 9 μm. It may be assumed that glucose is rapidly consumed by the cells upon transport. The average concentration of glucose in the plasma is 6 μmole/mL. The Staverman reflection coefficient can be taken as 0.424, and the pore radius can be taken as 15 nm.

**43.0**   *Concentration of protein.* A small bag containing albumin protein solution is dropped into a bath containing water. The molecular weight of albumin is 69,000 gm/mol. The bag wall is made up of a semipermeable membrane. What is the osmotic pressure developed? Will the bag increase or decrease in size?

**44.0**   *"Pot hole" in a membrane.* Suppose a "pot hole" forms in a dialysis membrane used to filter toxic solutes from the bloodstream. What will happen to the expression for flow rate of fluid across the membrane?

**45.0**   *Hydraulic conductance of ultrafiltration membrane.* Ultrafiltration membranes are used to clean a sanitation pond. For a clean water flow rate of 1 mL/min/cm² at 50 psi, calculate the hydraulic conductance. Should the thickness of the membrane be 500 nm? Assuming a porosity of 0.6, what is the pore radius?

**46.0**  *Combined hydrostatic and osmotic flow.* In a given system, *the* solvent filtration rate is in *an* opposite direction compared with solute diffusion across the membrane. Show that $N_s$, the transport rate of solute, would be zero when $J = -J_s$.

**47.0**  *Derive Darcy's law permeability coefficient.* Given that the Patlak, Goldstein et al. [14] account for the diffusion as follows:

$$N_s = C_s J(1-\sigma) - D_e S \frac{\partial C_s}{\partial z} \qquad (3.106)$$

Incorporate Darcy's law by realizing that $J = -\frac{\kappa}{\mu} \frac{S \partial P}{\partial z}$ in Eq. (3.97). At steady state, the principle of conservation of mass can be applied and:

$$-\frac{\partial N_s}{\partial z} = 0 \qquad (3.107)$$

Transform the governing equation [Eq. (3.103)]. Write pressure in terms of concentration by using $P = h\rho g$, and derive an expression for the permeability $\kappa$. Express the permeability coefficient as a function of the Staverman reflection coefficient, viscosity of the solvent, diffusion coefficient of the solute, density of the solute, and density of the solvent. What is the physical significance of this derivation?

**48.0**  *Permeability coefficient of membrane to solute.* Derive an expression for the permeability coefficient of the membrane to the solute comprise*d* of cylindrical pores. The population of pores is $N$, pore radius is $r_{pore}$, membrane thickness *is* $t_{mem}$, and diffusion coefficient of solute in the pore. Consider either one dimension diffusion of solute in pores or apply Kedem-Katchalsky equations at $J = 0$, called the zero convection velocity condition.

**49.0**  What would be the hydraulic conductance of the membrane to the solvent described in Exercise 34.0?

**50.0**  *Ultrafiltration of starch solution.* An ultrafiltration membrane recently developed by Omega Membranes, Inc., rejects 90 percent of a 3.6 wt % of partially hydrolyzed starch. The molecular weight of starch is 17,000 gm/mole. The permeability of the membrane to water is 0.63 m/day under a pressure drop of 4 atm. The volumetric flow is zero when 5 sq cm of membrane separates 66 cc of starch solution from the same volume of pure water. The osmotic pressure difference is 80 percent of the original value in one week. Find the Darcy's law permeability, solute permeability, and the reflection coefficient.

**51.0**  *Removal of aluminum from blood.* Aluminum (Al) has been found to be deposited in the bones of dialysis patients. In healthy individuals, Al is excreted through the kidneys. Patients accumulate Al in the body when the intake is greater than the excretion rate during dialysis treatment. The sources of Al are aluminum-hydroxide–based phosphate binders, drinking water and food storage containers made of Al, and a range of prescription drugs. In a modern, open-pore, high-flux dialyzer, only a reduction of 40 percent of the

Al level is achieved during passage through the dialyzer. The remaining Al settles in bones. This may cause bone cysts and osteomalacia. After five years of dialysis treatment, significant Al deposits in bones and tissues, causing the patient's health to deteriorate.

A complexing agent is used to react with Al and then filtered. The dosage of complexing agents should be in sufficient quantities to facilitate removal of Al ions. Ion-protein complexes are allowed to form. It is then filtered across a semipermeable membrane filter. The coagulant molecular weight is 20,000 gm/mole. What is the pore size of the semipermeable membrane for a rejection of 85 percent of Al coagulant by the membrane?

**52.0** Hemodialysis and ultrafiltration. Hemodialysis is an example of a dialysis process that is assisted using ultrafiltration. A hemodialyzer is used to remove waste products such as urea, creatinine, and uric acid from blood. The patient's blood is introduced into the hemodialyzer under the patient's perfusion pressure and flows past a semipermeable membrane. The blood solutes containing the wastes permeate through the membrane into the dialysate. The dialysate is a sterilized solution formulated to regulate solute permeation through the membrane. Osmosis can result in water from the dialysate flowing into the blood, causing edema. To avoid this, hemodialysis is used in conjunction with ultrafiltration to remove excess water. The dialysate is prepared using pure water obtained from the reverse osmosis process. Using Fig. 3.15, select a suitable membrane for RO to generate pure water for hemodialysis and *u*ltrafiltration. Use the information given in Worked Example 3.6.

**53.0** Apply the theory of Krogh and develop a mathematical model for diffusion of *o*xygen from blood capillaries into the tissue space along with simultaneous metabolic reactions in spherical coordinates. Is a numerical solution needed?

**54.0** In the food processing industry, during the concentration of orange juice, water needs to be removed. A plastic bag containing orange juice at 1 wt % sucrose concentration is dropped into a brine solution. Calculate the concentration of NaCl by % weight for hypotonic state, hypertonic state, and transonic state.

**55.0** Islets of Langerhans are spheroidal aggregates of cells that are located in the pancreas and secrete hormones that are involved in glucose metabolism [17]. Type 1 diabetes can be cured by transplanting isolated islets. Islets removed from the pancreas lose their internal vascularization and are dependent on the diffusion of oxygen from the external environment and through the oxygen-consuming islet tissue to satisfy the metabolic requirements of the cells. Islets can be viewed as a suspension of tissue spheres. The diffusivity of oxygen *is* measured. The islets were isolated from male rats using a modified digestion and purification technique under a dissecting microscope and cultured for a day in nonattacking polystyrene Petri dishes *containing* 5.6 mm of glucose, 50 U/mL penicillin, 50 µg/mL streptomycin, and 10 % newborn calf serum. The material was placed in an incubator at a temperature of oxygen uptake measurements. A known number of islets were placed in a tube that contained 45 mL of culture medium and 5 mL of air. The tubes were intermittently rotated to prevent settling and aggregation of the islets and to enhance oxygen transfer. The oxygen uptake chamber was equipped for measuring the oxygen-dependent lifetime of Pd-coproporphyrin phosphorescence to provide rapid and accurate measurements of oxygen concentration down to values as low as 0.05 μM. The chamber was a glass cuvette that contained a small Teflon-coated magnetic stirring bar *that* rotated at a speed of 1,200 rpm. A sample of 1,500 islets *was* loaded in the curette, which was filled with phosphate-buffered saline (pH 7.4) containing 0.35 gm/lit HEPES buffer, 0.5 gm/lit bovine serum albumin, and 300 mg/L glucose supplemented with 0.01 μM palladium coproporphyrin and 1-5 U/mL catalase. The cuvette was capped with a ground-glass stopper to eliminate the gas phase. The measured effective diffusivity of oxygen through the islets was found to be 2.31 E – 6 cm$^2$/s. Take the diffusivity of oxygen in interstitial fluid to be 4.1 E – 6 cm$^2$/s. Should the diffusivity in the cells be 3.5 E – 6 cm$^2$/s, calculate the volume fraction of the islets in the suspension.

**56.0**  *Plasmapheresis* is a blood separation procedure *used* to isolate blood cells from plasma. In hemofiltration, the "cut-off" for *the* passage of molecules through the membrane is 1,000 *to* 50,000 Dalton molecular weight and the cut-off in molecular weight of species in *plasmapheresis* is 3 million Dalton. A German manufacturer developed a polycarbonate membrane with a*n* average pore diameter of 200 nm and a porosity of 0.45. The membrane thickness was about 10 μm. A polycarbonate polymer solution was cast onto a smooth surface and contacted with a gel medium, followed by precipitation of membrane and gelled layer to form the membrane. Calculate the hydraulic conductance of capillary flow.

**57.0**  What ought to be the pump pressure to inject 1.6 ml/hr of saline water at a 15 wt % NaCl into the human blood stream across a membrane of thickness 50 nm. The membrane has a porous structure with a pore radius of 5 nm.

**58.0**  Healthy kidneys in the average adult person process about 125 ml/min or 180 liters of blood/day and filter out about 2 liters of waste product and extra water in the urine. The kidneys remove excess minerals and wastes and regulate the composition of such inorganic ions as sodium, phosphorous and chloride in the blood plasma at a nearly constant level. Blood urea nitrogen, BUN, a waste product produced in the liver as the end product of protein metabolism is removed from the blood by the kidneys in the Bowman's capsule along with creatinine, a waste product of creatinine phosphate as energy storing molecule produced largely from muscle breakdown. When a person kidneys fail, harmful wastes build up in their body, their blood pressure elevates and the blood retains fluid. The person will soon die unless their life is temporarily prolonged by either a kidney transplant their immune system attacks the foreign kidney requiring that the patient take immunosuppressant the rest of their life.

With hemodialysis the patient must be dialyzed three times a week: each treatment lasting from 3-4 hrs. Although the dialyzers are removing poisons there are side-effects caused primarily by the dialyzers themselves. Dialyzer filters are made of cellulose acetate, polysulfone or similar materials and sterilized with a solution of ethylene oxide, bleach or formaldehyde. Dialyzer filters have a membrane pore size of 90 nm. Find the radius of the solute that will be cut-off or rejected by the filter.

**59.0** The transport of glucose across the capillary wall is $5.0*10^{-5}$ μmole/hr. Glucose is a water soluble and lipid insoluble solute. The mean pressure of blood in capillary is 17.3 mm Hg, interstitial pressure of blood is −3 mm Hg. The colloid osmotic pressure inside the capillary and the interstitial fluid are 28 and 8 mm Hg respectively. The capillary length is 1 mm and inside diameter is 10 μm. It can be assumed that all the glucose transported to the extra capillary space is consumed rapidly by the cells. All plasma protein is retained by the capillary wall. Average concentration of glucose in plasma is 4 μmole/ml. Filtration rate can be taken as 10.75 E − 6 μL/hr. Calculate the Staverman reflection coefficient and the pore radius through which the solute transfers.

**60.0** Diseases usually have undesirable elevated levels of plasma solutes such as toxins, excessive antibodies and other metabolites. Plasma filtration has been used to separate undesirable solutes from blood plasma. Successful treatment of such diseases involves removal of undesirable plasma solutes from the blood plasma using membrane filtration.

Cholesterol has been determined to be an important component of arterial plague formation in atherosclerosis as well as in hypercholesterolemia. Cholesterol circulates in the blood linked to large protein molecules. One form of cholesterol carrying protein called low-density lipoprotein, LDL is known to promote atherosclerosis. About $2/3^{rd}$ or more of the total blood cholesterol is transported in LDL. Another form, called high density lipoprotein, HDL, is known to the protective against the disease process. Therefore the selective removal of LDL and maintenance of HDL is important in the treatment of atherosclerosis and the therapeutic control of hypercholesterolemia.

The observed sieving coefficients for different solutes are given in the following table. What would be the pore diameter used in filtration.

| Solute | Sieving Coefficient at 25°C |
|---|---|
| Albumin | 0.51 |
| Fibrinogen | 0.03 |
| LDL Cholesterol | 0.1 |
| HDL Cholesterol | 0.94 |

# CHAPTER 4

# Rheology of Blood and Transport

## Learning Objectives

- Better understand the Fahraeus-Lindqvist effect
- Develop marginal zone theory
- Explicit relation for plasma layer thickness
- Determine the manifestation of ballistic transport
- Discuss Casson's equation, Bingham equation, and damped wave and momentum transfer equation
- Womersley flow
- Apply the Bernoulli equation to cardiovascular flow work done by the heart

Blood is a colloidal dispersion system. This fluid system consists of cells and plasma. Major proteins found in the blood are albumin, globulin, and fibrinogen. The three main cells present in the blood are red blood cells (RBCs), white blood cells (WBCs), and platelets. The RBCs, or erythrocytes, tend to occupy 95 percent of the cellular component of the blood. They play a critical role in the transport of oxygen through hemoglobin contained within the RBCs. The density of RBCs is higher than that of plasma. The RBC volume fraction is called the *hemotocrit* and typically varies between 40 and 50 wt %. The true hemotocrit $H$ is about 96 percent of the measured hematocrit, Hct. RBCs can form stacked-coinlike structures called rouleaux. Rouleaux tend to clump together to form aggregates. They often break up in conditions of high shear or increased volumetric flow rate. About 5 percent of the blood consists of platelets. They are responsible for blood coagulation and homeostasis. The leukocytes, or WBCs, form the basis of the cellular component of the immune system. The effect of blood platelets and WBCs on the flow characteristics of the blood can be expected to be low on account of their low volume fraction.

## 4.1    Marginal Zone Theory

The marginal zone theory proposed by Haynes [1] can explain the *Fahraeus-Lindquist* effect. When the viscosity of blood was attempted to be measured using cylindrical tube viscometers, it was found that the viscosity measurements changed with changes in tube diameter! This happened at high shear rates. Viscosity measurements during tube flow at high shear rates greater than 100 sec$^{-1}$ were found to depend on the diameter of the tube. When the diameter of the tube is less then 500 μm, the viscosity of the blood will decrease accordingly, down to tube diameters of 4 to 6 μm.

This effect is attributed to the existence of a cell-free layer adjacent to the tube wall referred to as the *plasma skimming layer*. The occurrence of layered blood flow in capillaries and the existence of a cell-free layer in flowing blood have been confirmed using high-speed video photography. Simultaneously, an axial accumulation of the cells near the center of the tube results in a *core layer*—an expression for the apparent viscosity in terms of plasma layer thickness, tube diameter, and the hemotocrit in the marginal zone theory.

The blood flow within a tube or some other vessel is divided into two regions: a central core that contains cells with a viscosity of $\mu_c$ and a cell-free plasma peripheral layer with thickness δ and a viscosity of plasma denoted by $\mu_p$ (Fig. 4.1) In each region, the flow is considered Newtonian and at steady state. For the core region, the governing equation neglecting the ballistic effects can be written as [2]:

$$\tau_{rz} = \frac{\Delta P r}{2L} = -\mu_c \frac{\partial v_z^c}{\partial r} \tag{4.1}$$

The boundary conditions can be written as:

$$r = R - \delta, \ \tau_{rz}^{\text{core}} = \tau_{rz}^{\text{plasma}} \tag{4.2}$$

$$r = 0, \ \frac{\partial v_z^c}{\partial r} = 0 \tag{4.3}$$

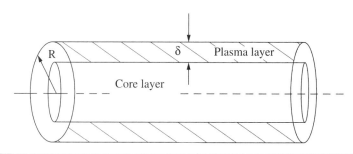

**FIGURE 4.1**    Layered blood flow in circular conduits.

The first boundary condition stems from the continuity of the transfer of momentum across the interface between the core and plasma layers. The second boundary condition derives from the realization that axial velocity reaches a maximum value at the center of the tube. This ought to be the case based on symmetry arguments. In a similar fashion, for the plasma layer, the governing equation and boundary conditions can be written as:

$$\tau_{rz}^{plasma} = \frac{\Delta P r}{2L} = -\mu_p \frac{\partial v_z^p}{\partial r} \tag{4.4}$$

The boundary conditions of the blood flow in the plasma layer can be written as:

$$r = R, \ v_z^p = 0 \tag{4.5}$$

$$r = R - \delta, \ v_z^p = v_z^c \tag{4.6}$$

At the wall, the fluid is considered to be at rest and at zero velocity. At the interface of the plasma and core layers, the velocity needs to be the same from continuity considerations, without any accumulation. Equations (4.1) and (4.4) can be integrated and the integration constants solved for using the previously mentioned four boundary conditions. The discharge rate of the plasma and core layers can be found to be [3]:

$$Q_p = \frac{\pi \Delta P}{8\mu_p L}(R^2 - (R - \delta)^2)^2 \tag{4.7}$$

$$Q_c = \frac{\pi \Delta P R^2}{8\mu_p L}\left[(R - \delta)^2 - \left(1 - \frac{\mu_p}{\mu_c}\right)\left(\frac{(R - \delta)^4}{R^2} - \frac{\mu_p}{\mu_c}\left(\frac{(R - \delta)^4}{2R^2}\right)\right)\right] \tag{4.8}$$

The total discharge rate of the blood is equal to the sum of the flow rates in the core and plasma regions, and is given by:

$$Q = \frac{\pi \Delta P R^4}{8\mu_p L}\left[1 - \left(1 - \frac{\delta}{R}\right)^4\left(1 - \frac{\mu_p}{\mu_c}\right)\right] \tag{4.9}$$

## 4.2  Slit Limit of Layered Flow

The blood flow within a rectangular conduit, as used in a dialysis machine, is divided into two regions: a central core that contains cells with a viscosity of $\mu_c$ and a cell-free plasma peripheral layer with thickness $\delta$ and a viscosity of plasma denoted by $\mu_p$ (Fig. 4.2). In each

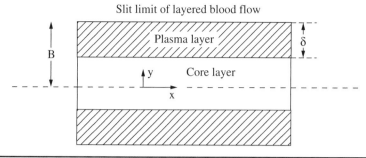

**Figure 4.2** Plasma layer and core layer of blood flowing across a rectangular slit of width 2B, length L, and plasma layer thickness δ.

region, the flow is considered to be Newtonian and at a steady state. For the core region, the governing equation neglecting the ballistic effects can be written as:

$$\tau_{xy} = \frac{\Delta P y}{2L} = -\mu_c \frac{\partial v_z^c}{\partial y} \tag{4.10}$$

The boundary conditions can be written as:

$$y = R - \delta, \ \tau_{xy}^{core} = \tau_{xy}^{plasma} \tag{4.11}$$

$$y = 0, \ \frac{\partial v_z^c}{\partial y} = 0 \tag{4.12}$$

The first boundary condition stems from the continuity of the transfer of momentum across the interface between the core and plasma layers. The second boundary condition is derived from the realization that axial velocity reaches a maximum speed at the center of the tube. This ought to be the case based on symmetry considerations. In a similar fashion, for the plasma layer, the governing equation and boundary conditions can be written as:

$$\tau_{xy}^{plasma} = \frac{\Delta P y}{2L} = -\mu_p \frac{\partial v_y^p}{\partial y} \tag{4.13}$$

The boundary conditions of the blood flow in the plasma layer can be written as:

$$y = B, \ v_y^p = 0 \tag{4.14}$$

$$y = B - \delta, \ v_z^p = v_z^c \tag{4.15}$$

At the wall, the fluid is considered to be at rest and at zero velocity. At the interface of the plasma and core layers, the velocity needs to be the same from continuity considerations, without any accumulation.

Equations (4.10) and (4.13) can be integrated and the integration constants solved for using the previously mentioned four boundary conditions. The fluid velocity of the plasma and core layers can be found to be:

$$v_x^P = \frac{\Delta P B^2}{2\mu_p L}\left(1 - \left(\frac{y}{B}\right)^2\right)$$  (4.16)

$$v_x^c = \frac{\Delta P B^2}{2\mu_p L}\left(1 - \left(1 - \frac{\delta}{B}\right)^2\left(1 - \frac{\mu_p}{\mu_c}\right) - \left(\frac{y}{B}\right)^2\mu_c\right)$$  (4.17)

The total discharge rate of the blood is equal to the sum of the flow rates in the core and plasma regions, and is given by:

$$Q = \frac{2B^3 W \Delta P}{3\mu_p L}\left(1 - \left(1 - \frac{\delta}{B}\right)^3\left(1 - \frac{\mu_p}{\mu_c}\right)\right)$$  (4.18)

## 4.3    Explicit Expression for Plasma Layer Thickness

Plasma layer thickness $\delta$ can be obtained by fitting apparent viscosity data and using Eq. (4.9) or Eq. (4.18), depending on whether the conduit is circular or rectangular. The core hemotocrit variation as a function of the tube diameter or slit width can also be obtained. A relation between the core hematocrit, $H_c$, and the feed hematocrit, $H_F$, and the thickness of the plasma layer is needed. An equation is needed to describe the dependence of the blood viscosity on the hematocrit since the value of $H_c$ will be larger than $H_F$ because of the axial accumulation of the RBCs. This relative increase in the core hematocrit will make the equation in the core have a higher viscosity than the blood in the feed. The following equation developed by Charm and Kurland [4] may be used to express the dependence of the viscosity of the blood at high shear rates on the hematocrit and temperature:

$$\mu = \mu_p\left(\frac{1}{(1 - \alpha H)}\right)$$  (4.19)

or  $$\alpha H = 1 - \frac{\mu_p}{\mu}$$  (4.20)

where  $$\alpha = 0.070\exp\left[2.49H + \frac{1107}{T}\exp(-1.69H)\right]$$  (4.21) where the temperature is in kelvins. These equations are valid to a hematocrit of 0.6 with an accuracy of 10 percent. The viscosity of the blood $\mu$ and the core layer hematocrit, $H_c$, can be related as:

$$\mu = \mu_c \frac{1}{(1 - \alpha_c H_c)} \tag{4.22}$$

defining $\sigma' = 1 - (\delta/R)$.

The solution for the plasma layer thickness $\delta$ is implicit and requires solving for two equations and two unknowns:

$$\frac{\mu_{app}}{\mu_F} = \frac{(1 - \alpha_F H_F)}{(1 - \sigma^4 \alpha_c H_c)} \tag{4.23}$$

$$\frac{H_c}{H_F} = 1 + \frac{(1 - \sigma^2)^2}{\sigma^2 \left[ 2(1 - \sigma^2) + \sigma^2 \dfrac{\mu_p}{\mu_c} \right]} \tag{4.24}$$

The plasma layer thickness $\sigma$ is implicit in Eqs. (4.23) and (4.24). An explicit expression for $\sigma$ is desirable. This was developed by Sharma [5]. Equation (4.21) was examined using a Microsoft Excel spreadsheet (Fig. 4.3).

Upon examination of Eq. (4.21), it was found that the temperature parameter used to describe the variation of viscosity of blood varied linearly with $H$ when checked against $\alpha H$ at a given temperature. This is shown in Fig. 4.3 at 300 K. This can be expressed mathematically for the core layer and plasma layer respectively as:

$$\alpha_c H_c = m H_c + c \tag{4.25}$$

$$\alpha_T H_T = m H_T + c \tag{4.26}$$

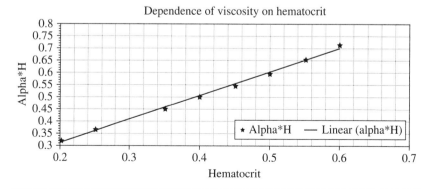

FIGURE 4.3 Variation of temperature parameter $\alpha H$ with hematocrit $H$.

The slope and intercept may be obtained by the least squares regression line between $\alpha H$ and $H$, as shown in Fig. 4.3 $m$ and $c$ would vary with the temperature. It is interesting that $m$ and $c$ are independent of the layer—that is, the core or plasma layer or the feed!

A material balance can be written over the two phases in the tube, and it can be realized that:

$$H_T = \sigma^2 H_c \tag{4.27}$$

Equation (4.25) is divided with Eq. (4.26):

$$\frac{\alpha_c H_c}{\alpha_T H_T} = \frac{mH_c + c}{mH_T + c} \tag{4.28}$$

It can be seen from Eq. (4.28) that:

$$\alpha_T H_T = 1 - \frac{\mu_p}{\mu_{app}} \tag{4.29}$$

It can also be seen upon minor rearrangement that:

$$\alpha_c H_c \sigma^4 = 1 - \frac{\mu_p}{\mu_{app}} = \alpha_T H_T = \alpha_T \sigma^2 H_c \tag{4.30}$$

or

$$\sigma = \sqrt{\frac{\alpha_T}{\alpha_c}} \tag{4.31}$$

Combining Eqs. (4.30) and (4.31):

$$\sigma^4 = \frac{mH_T + c}{\dfrac{mH_T}{\sigma^2} + c} \tag{4.32}$$

Let $\sigma^2 = s$, and it can be seen that Eq. (4.32) is quadratic in $s$:

$$s^2 + \left(\frac{mH_T - 1}{c}\right)s + 1 = 0 \tag{4.33}$$

A solution to Eq. (4.33) can readily be obtained. Thus, an explicit expression for the plasma layer thickness in terms of the tube hematocrit has been developed. The tube hematocrit, $H_T$, can be read from the linear regression line between $\alpha H$ and $H$ at a given temperature once the apparent viscosity of the tube is known.

## 4.4    Constitutive Relations—Yield Stress Myth

As was briefly discussed in Chap. 1, several equations have been proposed for use as constitutive relations to describe blood rheology.
One of them is the *Bingham* yield stress fluid [6].

Some examples of fluids whose rheology has been described
using the Bingham model are listed in Table 4.1 [7].

| S. No. | Fluid | S. No. | Fluid |
|--------|-------|--------|-------|
| 1 | Acrylic rubber/polyethyl acrylate | 24 | Metal oxides/water |
| 2 | Applesauce | 25 | Orange juice concentrate |
| 3 | Blood | 26 | Paint |
| 4 | Borite/water | 27 | Plastic rocket propellant |
| 5 | Butter | 28 | Polymer latex/water |
| 6 | Carbon black/oil | 29 | Printing ink |
| 7 | Cement raw mix | 30 | PVC/organic liquids |
| 8 | Cement/clay/water | 31 | Rubber/benzene |
| 9 | Chemical-mechanical polishing | 32 | Sewage sludge |
| 10 | Clay methanol | 33 | Silica/Newtonian liquid |
| 11 | Clay water | 34 | Styrene-co-DVB/PS/DEP |
| 12 | Coal/Newtonian liquid | 35 | Sulfur/water |
| 13 | Drilling mud | 36 | Sweet potato puree |
| 14 | Explosives—Water/gelling agent and oxidizer | 37 | Thorium oxide/methanol |
| 15 | Fermentation broth | 38 | Tomato puree |
| 16 | Glass/glycerol | 39 | Tomato sauce |
| 17 | Glass/polymer | 40 | Toothpaste |
| 18 | Graphite/water | 41 | Tritolyl phosphate/ castor oil |
| 19 | Grease | 42 | Water/animal wastes— fertilizer |
| 20 | Inorganic solid/polymer/ solvent | 43 | Water/bentonite |
| 21 | Iron oxide/ethylene glycol | 44 | Water/benzene |
| 22 | Mayonnaise | 45 | Wood pulp/water |
| 23 | Meat extract | 46 | Xanthan gum/water |

**TABLE 4.1**    List of Viscoplastic Fluids

Most of the examples listed in Table 4.1 are slurries, pastes, and suspensions. The Bingham equation can be written as shown in Fig. 1.3 as:

$$\tau_{xy} = \pm\tau_0 - \mu_0 \frac{\partial v_x}{\partial y}, \text{ for } |\tau_{xy}| \geq \tau_0 \tag{4.34}$$

where $\tau_0$ is the yield stress and $\mu_0$ is the plastic viscosity. For values of shear stress less than the yield stress, the velocity gradient or shear rate would be zero. The Bingham model is a special case of a model suggest by Schwedoff many years ago:

$$\tau_{xy} + \tau_r \frac{\partial \tau_{xy}}{\partial t} = \pm\tau_0 - \mu_0 \frac{\partial v_x}{\partial y} \tag{4.35}$$

As discussed in Chap. 1, experimental evidence presented by Barnes and Walters from constant stress rheometers indicate that the yield stress concept is an *idealization*. Given accurate measurements, no yield stress exists. The non-Newtonian power law models are adequate to describe the rheology of the "Bingham" fluids. Yield stress is a British standard that represents the stress below which the substance behaves as an elastic solid and above which it is a liquid with a plastic viscosity. With the advent of better instrumentation, Eq. (4.34) is only an idealization that is valid at high shear rates. The yield stress was extrapolated and never directly measured. A range of constant stress instruments was developed as the Deer rheometer, which allows for accurate stress measurements to be made at shear rates as low as 1 E-6 sec$^{-1}$. Conventional rheometers, such as the Weissenberg rheogoniometer, can provide viscosity measurements above shear rates of 0.01 sec$^{-1}$. They found that lower the specifications of the instrument of shear rate lower were the measured yield stress for the same fluid at the same temperature and pressure!

Hartnett and Hu, as discussed in Chap. 1, argue that yield stress is an engineering reality. Nylon and Teflon balls were placed in carbopol solutions to study the yield stress myth. Photographs obtained every week for several months after 14 weeks revealed that Teflon ball A dropped a distance equal to approximately half the ball diameter was interpreted as "no appreciable movement." Reviewers called attention to the Harntett and Hu's reference which predicts that a sphere will fall in a viscoplastic medium only for values of dimensionless parameter $\left[1.5\tau_0/(\rho_s - \rho)g\right]$ less than 0.143, where $\tau_0$ is the yield stress. Since the Teflon ball moved in the carbopol solution, the value of the yield stress must be less than 107 dynes/cm$^2$. With the emergence of nanotechnology, if the yield stress of fluids is low, they ought to be measured directly. Higher values inferred by extrapolation may not be sufficient. Should the non-Newtonian models be better suited for the Bingham fluids, those ought to be used.

This kind of rationale may be applicable to the Casson and Herschel-Buckley models discussed in the literature to describe blood flow.

## 4.5   Generalized Newton's Law of Viscosity

Newton's law of viscosity relates the shear stress to the shear rate with the constant of proportionality, the viscosity of the fluid, which offers resistance to flow:

$$\tau_{xy} = -\mu \frac{\partial v_x}{\partial y} \tag{4.36}$$

where $\tau_{yx}$ is the shear stress, $\partial v/\partial y$ is the velocity gradient, and $\mu$ is the absolute viscosity. The negative sign in Eq. (4.1) is written to normalize the momentum transfer direction. Consider a plate at $y = l$, pulled at a constant velocity $V$ atop a stationary liquid. The layer of the liquid adjacent to the plate is also subjected to motion. The layer adjoining the bottom surface is stationary. The velocity gradient can be calculated as $V/l$. This, multiplied by the absolute viscosity, gives the shear stress in magnitude. If the force acting on the plate is $F$ and the area of the plate is $A$, then:

$$\tau_{xy} = \frac{F}{A} = -\mu \frac{V\rho}{l} \tag{4.37}$$

The right-hand side of Eq. (4.37) represents the rate of momentum transfer. $\gamma$ is the kinematic viscosity with units of $m^2/s$. The direction of momentum transfer is in the downward direction, from atop the liquid towards the origin. Hence, in order to render $F/A = \tau$ positive, the negative sign is added to Eq. (4.1). Consequently, $\tau$ may be viewed as the momentum flux in the $y$ direction. In addition the momentum flux and velocity gradient must have opposite signs to stay within the bounds of the second law of thermodynamics. Momentum transfers from the high velocity region to a low velocity region by molecular transfer, and the other direction is not allowed. The shear stress expression, when combined with the equation of momentum, results in a PDE that can be solved and the solution expressed as an infinite Fourier series. The singularity in Fourier series representation can be addressed by the use of the damped wave momentum and relaxation equation:

$$\tau_{xy} = -\mu \frac{\partial v_x}{\partial y} - \tau_r \frac{\partial \tau_{xy}}{\partial t} \tag{4.38}$$

The *damped wave momentum transfer and relaxation equation* can arise from the accumulation term in the kinetic theory of gases and derivation of physical properties of monatomic gases from molecular properties. From a molecular view, the viscosity can be derived and the momentum transport mechanism can be illustrated [2]. This derivation is revisited here. Consider molecules to be rigid, nonattracting spheres of mass $m$ and diameter $d$. The gas is assumed to be at rest, and the molecular motion is considered. The following results of kinetic theory for a rigid sphere dilute gas in which temperature, pressure, and velocity gradients are small are used:

$$\text{Mean molecular speed } <u> = \sqrt{\frac{8\kappa T}{\pi m}} \qquad (4.39)$$

Wall collision frequency per unit area:

$$Z = \tfrac{1}{4}\, n'<u> \qquad (4.40)$$

$$\text{Mean free path } \lambda = \frac{1}{\sqrt{2\pi d^2 n'}} \qquad (4.41)$$

The molecules reaching any plane in the gas have, on average, had their last collision at a distance $a$ from the plane where:

$$a = \frac{2}{3}\lambda \qquad (4.42)$$

In order to determine the viscosity of a dilute monatomic gas, consider the gas when it flows parallel to the $x$ axis with a velocity gradient $\partial v_x/\partial z$. Assuming the relations for the mean free path of the molecule, wall collision frequency, distance to collision, and mean velocity of the molecule are good during the nonequilibrium conditions, the flux of momentum in the $x$ direction across any plane $z$ is found by summing the $x$ momentum of the molecules that cross in the positive $y$ direction and subtracting the $x$ momentum of those that cross in the opposite direction. Thus:

$$\tau_{zx} = Z\, mv_x\big|_{z-a} - Z\, mv_x\big|_{z+a} \qquad (4.43)$$

It may be assumed that the velocity profile is essentially linear for a distance of several mean free paths. Molecules have a velocity representative of their last collision. Accordingly:

$$v_x\big|_{z-a} = v_x\big|_z - \frac{2\lambda}{3}\frac{\partial v_x}{\partial z} \qquad (4.44)$$

$$v_x\big|_{z+a} = v_x\big|_z + \frac{2\lambda}{3}\frac{\partial v_x}{\partial z} \qquad (4.45)$$

Substituting Eqs. (4.9) and (4.10) into Eq. (4.8):

$$\tau_{zx} = -1/3n\, m\, <u>\, \lambda\, \frac{\partial v_x}{\partial z} \qquad (4.46)$$

Equation (4.46) corresponds to Newton's law of viscosity, with the viscosity given by:

$$\mu = 1/3 \, \rho <u> \lambda \tag{4.47}$$

Maxwell obtained this relation in 1860. It can be seen that prior to writing Eq. (4.8), the *accumulation of momentum* was neglected. This may be a good assumption at steady state, but not at short-time transient events. Thus, considering a time increment $t^*$, the momentum is the momentum of molecules in minus momentum of molecules out minus the accumulation of momentum in the incremental volume under consideration near the surface. The accumulation of momentum may be written in terms of:

$$\tau_{zx} = Z \, mv_x\big|_{z-a} - Z \, mv_x\big|_{z+a} - t^*\partial/\partial t\{Z \, mv_x\big|_{z-a} - Z \, mv_x\big|_{z+a}\} \tag{4.48}$$

where $t^*$ is some characteristic time constant. To simplify matters, Eq. (4.13) is used in Eq. (4.8) to give:

$$\tau_{zx} = -\mu \frac{\partial v_x}{\partial z} - t^* \frac{\partial \tau_{zx}}{\partial t} \tag{4.49}$$

### 4.5.1 Flow Near a Horizontal Wall Suddenly Set in Motion

Consider a fluid with constant density $\rho$ and constant viscosity $\mu$ atop a horizontal plate. The fluid medium is assumed to be in a continuum. In most cases, at the macroscopic scale, the molecular structure of the fluid is not taken into account. Mass is concentrated in the nuclei of atoms and is far from uniformly distributed over the volume occupied by the liquid. Nonuniform distribution can be seen in other variables, such as composition and velocity, when viewed on a microscopic scale. The continuum supposition is that the behavior of fluids is the same as if they were perfectly continuous in structure and physical quantities such as mass and momentum associated with matter contained within a given volume will be regarded as being spread uniformly over that volume, instead of being concentrated in a small fraction of it. Atop the horizontal plate is a semi-infinite medium of fluid. The fluid is stationary at time zero. For times greater than zero, the plate is set at a constant velocity $V$. The velocity in the $z$ direction as a function of space and time is of interest. An error function results when Newton's law of viscosity is used for the fluid. The spatiotemporal velocity of the fluid is obtained from the damped wave momentum transfer and relaxation equation. From the equation of motion, neglecting convection effects:

$$\tau_{mom} \frac{\partial^2 v_x}{\partial t^2} + \frac{\partial v_x}{\partial t} = \gamma \frac{\partial^2 v_x}{\partial z^2} - 2g \tag{4.50}$$

Let $\quad u = v_x/V; \tau = t/\tau_{mom}; Z = z/\text{sqrt}(\gamma \tau_{mom})$ (4.51)

The one time and two space conditions are:

$$\tau = 0, u = 0 \tag{4.52}$$

$$Z = 0, u = 1 \tag{4.53}$$

$$Z = \infty, u = 0 \tag{4.54}$$

$$\frac{\partial^2 u}{\partial \tau^2} + \frac{\partial u}{\partial \tau} = \frac{\partial^2 u}{\partial Z^2} - 2\text{Acc} \tag{4.55}$$

where $\text{Acc} = (g\tau_{mom}/V)$ is a dimensionless number that represents the ratio of gravity forces to the ballistic "force" that corrects for the accumulation of momentum and can be called the accumulation number.

Sharma [8] has developed a closed-form analytical solution to the governing equation presented in Eq. (4.55) by the method of relativistic transformation. The solution is:

For $\tau > Z$:

$$u = \frac{I_0(1/2\sqrt{\tau^2 - Z^2})}{I_0\left(\dfrac{\tau}{2}\right)} \tag{4.56}$$

$$\frac{\tau_{zx}}{\sqrt{\dfrac{\rho V^2 \mu}{\tau_{mom}}}} = e^{-\frac{\tau}{2}} I_0\left(\sqrt{\frac{\tau^2 - Z^2}{4}}\right) \tag{4.57}$$

For $Z > \tau$:

$$u = \frac{J_0(1/2\sqrt{Z^2 - \tau^2})}{I_0\left(\dfrac{\tau}{2}\right)} \tag{4.58}$$

$$\frac{\tau_{zx}}{\sqrt{\dfrac{\rho V^2 \mu}{\tau_{mom}}}} = e^{-\frac{\tau}{2}} J_0\left(\sqrt{\frac{Z^2 - \tau^2}{4}}\right) \tag{4.59}$$

The solution exhibits some *space-time symmetry* with respect to the negative values as well as with each other. It can be seen that for a plate at some point in the interior of the semi-infinite medium, the shear force exerted by the fluid on the plate is bifurcated. In fact, it has *four different regimes.* The first is the thermal inertia regime. In this regime there is no action of the fluid on the plate. In the second regime, the shear stress is given by Eq. (4.94), which is a product of the decaying exponential and a Bessel composite function of the first kind and zeroth order. The third regime after the wave front is represented by Eq. (4.93), and is a product of a modified composite Bessel function of the first and zeroth orders. The momentum inertia can be calculated as:

$$\tau_{inertia} = sqrt(Z_p^2 - 23.1323) \qquad (4.60)$$

The fourth regime is the wave front where

$$u = e^{-\frac{\tau}{2}} = e^{-\frac{Z}{2}} \qquad (4.61)$$

In the first regime, the shear force may be negative should the Bessel function's negative sign have meaning. This could be the first few ripples that the plate sees from the disturbance from the surface. The shear force can be in the opposite direction, and eventually, after the thermal time lag has elapsed, the force is in the right direction. The shear stress undergoes a maxima. The second regime is a steep rise (Fig. 4.4). The first regime is an inertial time of up to 3.597 in dimensionless quantities. The third regime is a tailed fall. The curvature changes from convex to concave. There is an inflection point in the third regime. There is a skew to the right, and the kurtosis may be compared to the Maxwell distribution.

On examining Eq. (4.92), it can be seen that when $Z > \tau$, the expression for the dimensionless velocity becomes a Bessel composite function. This is because when $Z > \tau$, the argument in the modified Bessel composite function within the square root sign becomes negative. The square root of $-1$ is $i$. Furthermore:

$$J_0(x) = I_0(ix) \qquad (4.62)$$

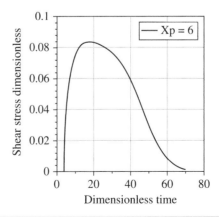

**Figure 4.4** Three regimes of dimensionless shear stress in the interior of a semi-infinite fluid suddenly pulled by a plate at a constant velocity from the bottom.

Hence, Eq. (4.92) becomes, for $Z > \tau$:

$$u = \frac{J_0(1/2\sqrt{Z^2 - \tau^2})}{I_0\left(\dfrac{\tau}{2}\right)} \qquad (4.63)$$

It is generally realized in analysis such as in boundary layer theory that after a finite region from the moving plate the fluid will be at the initial state or will have zero velocity. The first zero of the Bessel function occurs at 2.4048. Beyond that, the velocity predicted will be negative. Although the denominator in Eq. (4.63) will dampen the oscillations, why would the velocity of fluid be negative after a said distance from the moving flat plate at the boundary for a given time instant under consideration? Since it is damped oscillatory, the effect of the surface disturbance for distances further than sqrt($23.13 + \tau^2$) acts differently on account of the ballistic transport. It can be taken as zero from an analogy from heat wave conduction and relaxation or mass wave diffusion and relaxation. If it is taken as zero, the boundary layer thickness for a given instant in time greater than zero is given by:

$$\delta(\tau) = \text{sqrt}(23.13 + \tau^2) \qquad (4.64)$$

Beyond this distance, the fluid velocity can be taken to be zero from the analogy from heat or mass diffusion and relaxation.

The model prediction of Eq. (4.63) gives negative values for velocity beyond the boundary layer thickness. Velocity is a vector. In the momentum balance equation from which the solution is derived, the velocity is preserved through the analysis. Hence, a negative velocity could mean that the velocity of the fluid is in the opposite direction compared with the velocity of the flat plate. Up to the first root of the Bessel function, the second regime for the dimensionless velocity profile holds good. For a given instant of time, for values of $Z$ smaller than the instant of dimensionless time, the third regime, or the modified Bessel composite function solution is applicable. The negative values for the velocity can be due to the ballistic transport mechanism. The disturbance swims back from the region beyond the boundary layer. This is a type of ripple effect and backflow phenomena that needs to be borne out by experiment. It can be seen in graphical form as follows. For large values of the argument, the Bessel function can be approximated with a cosinuous function as follows:

$$u = \text{sqrt}(4/(\pi(Z^2 - \tau^2)^{1/2})) \cos[\tfrac{1}{2}(Z^2 - \tau^2)^{1/2} - \pi/4]/(I_0(\tau/2) \qquad (4.65)$$

In Figs. 4.5 and 4.6 are plotted the dimensionless velocity for a given instant in time ($\tau = 5$) as a function of dimensionless distance. In

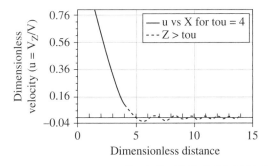

**FIGURE 4.5**    Dimensionless velocity of fluid in a semi-infinite fluid from a moving flat plate.

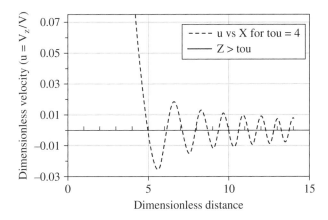

**FIGURE 4.6**    Damped oscillatory behavior of dimensionless velocity far from the flat plate in semi-infinite medium.

Fig. 4.4, it can be seen that close to the flat plate the dimensionless velocity obeys Eq. (4.98) and is valid for dimensionless distances less than the instant in time under consideration. This is given by the modified composite Bessel spatiotemporal function of the first kind and zeroth order divided by the modified Bessel function in time of the first kind and first order. For dimensionless distances greater than the time instant under consideration, the Bessel composite spatiotemporal function of the first kind and zeroth order divided by the modified Bessel function in time of the first kind and zeroth order gives the dimensionless velocity profile. Beyond the first zero of the Bessel function, the solution predicts damped oscillations for the dimensionless velocity. Up to the first zero of the Bessel function, the velocity of the fluid is positive. In this case, this value can be calculated to be sqrt(22.21 + 16) = 6.18. Beyond 6.18, the velocity is in the negative direction. This is the subcritical damped oscillatory regime, and it needs to be verified by experiment. The ballistic transport mechanism gives credence to some wave motion for certain conditions. This was seen in model predictions in the heat and mass transfer sections as well. However, for the fluid problem, it manifests as a vector, and a minus sign indicates a reversal of flow in the direction opposite to the movement of the flat plate.

### 4.5.2 Transient Vertical Darcy Flow

As discussed in earlier sections, during the study of emptying a pipe, a filled tube with porous packing was considered. Darcy's law can be used to relate the pressure gradient to the flow velocity. The solution for the resulting governing equation and for the vertical component of the velocity of the fluid is obtained.

The equation of motion considering the ballistic transport effects for the vertical component of the velocity of the fluid can be written as:

$$\tau_{mom}\rho\left(\frac{\partial^2 v_z}{\partial t^2} + v_z\frac{\partial^2 v_z}{\partial z\,\partial t} + \frac{\partial v_z}{\partial t}\frac{\partial v_z}{\partial z}\right) + \rho\frac{\partial v_z}{\partial t} + v_z\frac{\partial v_z}{\partial z} + \tau_{mom}\nabla\left(\frac{\partial p}{\partial t}\right)$$

$$= -\frac{\partial p}{\partial z} - \rho g \tag{4.66}$$

From Darcy's law:

$$v_z = -\left(\frac{\kappa}{\mu}\right)\left[\frac{\partial p}{\partial z} - \rho g\right] \tag{4.67}$$

$$-\frac{\partial^2 p}{\partial t\,\partial z} = \left(\frac{\mu}{\kappa}\right)\frac{\partial v_z}{\partial t} \tag{4.68}$$

where $\kappa$ is the permeability of the porous medium. Equation (4.66) becomes:

$$\tau_{mom}\rho\left(\frac{\partial^2 v_z}{\partial t^2} + v_z\frac{\partial^2 v_z}{\partial z\,\partial t} + \frac{\partial v_z}{\partial t}\frac{\partial v_z}{\partial z}\right) + \left(\rho - \frac{\mu\tau_{mom}}{\kappa}\right)\frac{\partial v_z}{\partial t}$$

$$+ v_z\frac{\partial v_z}{\partial z} - \frac{\mu}{\kappa}v_z = 0 \tag{4.69}$$

It can be assumed that during the drainage the velocity of liquid in the tube, $v_z$, is independent of $z$ and is only a function of time. Or this can be arrived at by writing the equation of continuity. This is true for wide reservoirs where the height in the container does not change appreciably. Thus, at constant density, the equation of momentum becomes:

$$\tau_{mom}\left(\frac{\partial^2 v_z}{\partial t^2}\right) + \left(1 - \frac{\gamma\tau_{mom}}{\kappa}\right)\frac{\partial v_z}{\partial t} - \frac{\gamma}{\kappa}v_z = 0 \tag{4.70}$$

Let
$$\tau = \frac{t}{\tau_{mom}}$$

$$\left(\frac{\partial^2 v_z}{\partial \tau^2}\right) + \left(1 - \frac{\gamma \tau_{mom}}{\kappa}\right)\frac{\partial v_z}{\partial \tau} - \frac{\gamma \tau_{mom}}{\kappa} v_z = 0 \qquad (4.71)$$

Let
$$Pb = \frac{\gamma \tau_{mom}}{\kappa}$$

$Pb$ is a sort of a permeability number that gives the ratio of the kinematic viscosity times the relaxation time of momentum divided by the Darcy permeability of the medium. It may represent the ratio of viscous forces and ballistic transport "forces" to the permeability forces.

Let
$$u = \frac{v_z}{g\tau_{mom}} \qquad (4.72)$$

$$\frac{\partial^2 u}{\partial \tau^2} + (1 - Pb)\frac{\partial u}{\partial \tau} - Pb(u) = 0 \qquad (4.73)$$

The solution to Eq. (4.73), which is a second-order ODE with constant coefficients, can be written as:

$$u = c_1 e^{-\tau} + c_2 e^{Pb\tau} \qquad (4.74)$$

From the constraint at infinite time—that is, $u = 0$, $c_2$ can be seen to be zero.

The initial condition can be written assuming a pseudo-steady state and using Torricelli's theorem:

$$\sqrt{2gH} = c_1 \qquad (4.75)$$

Thus:

$$u = \sqrt{2gH}e^{-\tau} \qquad (4.76)$$

## 4.5.3  Transient Vertical Darcy Flow under Reduced Gravity

In Sec. 4.5.2, during the study of emptying a pipe filled with liquid, a tube with porous packing was considered, and the apparatus was taken in a Space Shuttle and into the galaxy. Darcy's law is then used to relate the pressure gradient with the flow velocity as follows. The resulting governing equation is considered and a solution is obtained for the vertical component of the velocity of the fluid.

The equation of motion considering the ballistic transport effects for the vertical component of the velocity of the fluid can be given by

Eq. (4.66). From Darcy's law in a new gravitational field in the Space Shuttle in the galaxy:

$$v_z = \frac{\kappa}{\mu}\left(\frac{\partial p}{\partial z} + \rho g\right) \qquad (4.77)$$

where $\kappa$ is the permeability of the porous medium. After neglecting the pressure changes with time, Eq. (4.77) becomes:

$$\tau_{mom}\rho\left(\frac{\partial^2 v_z}{\partial t^2} + v_z\frac{\partial^2 v_z}{\partial z \partial t} + \frac{\partial v_z}{\partial t}\frac{\partial v_z}{\partial z}\right) + (\rho)\frac{\partial v_z}{\partial t} + v_z\frac{\partial v_z}{\partial z} + \frac{\mu}{\kappa}v_z = 0 \qquad (4.78)$$

It can be assumed that during the drainage, the velocity of liquid in the tube, $v_z$, is independent of $z$ and is only a function of time. Or this can be arrived at by writing the equation of continuity. This is true for wide reservoirs where the height in the container does not change appreciably. Thus, at constant density, the equation of momentum becomes:

$$\tau_{mom}\left(\frac{\partial^2 v_z}{\partial t^2}\right) + \frac{\partial v_z}{\partial t} + \frac{\gamma}{\kappa}v_z = 0 \qquad (4.79)$$

Let $\tau = t/\tau_{mom}$

$$\left(\frac{\partial^2 v_z}{\partial \tau^2}\right) + \frac{\partial v_z}{\partial \tau} + \frac{\gamma\tau_{mom}}{\kappa}v_z = 0 \qquad (4.80)$$

where $Pb = \gamma\tau_{mom}/\kappa$.

$Pb$ is a sort of a permeability number that gives the ratio of the kinematic viscosity times the relaxation time of momentum divided by the Darcy permeability of the medium. It may represent the ratio of viscous forces and ballistic transport "forces" to the permeability forces. Let $u = v_z/g\tau_{mom}$

$$\frac{\partial^2 u}{\partial \tau^2} + \frac{\partial u}{\partial \tau} + Pb(u) = 0 \qquad (4.81)$$

Equation (4.81) is a second-order ODE with constant coefficients and is homogeneous. The solution to Eq. (4.81) may be written as:

$$u = e^{-\frac{\tau}{2}}\left(c_1 e^{-\frac{\tau}{2}\sqrt{1-4Pb}} + c_2 e^{\frac{\tau}{2}\sqrt{1-4Pb}}\right) \qquad (4.82)$$

From the constraint at infinite time, that is, $u\exp(\tau/2) = 0$, $c_2$ can be seen to be zero. The initial condition can be written assuming a pseudo-steady state and using Torricelli's theorem:

$$\sqrt{2gH} = c_1$$

$$u = \sqrt{2gH}\left(e^{-\tau\sqrt{1-4Pb}}\right) \qquad (4.83)$$

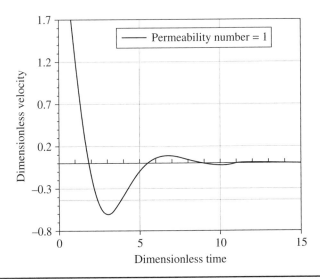

FIGURE **4.7**   Subcritical damped oscillations with positive permeability.

It can be seen that for small values of the permeability number, that is, when $Pb > 1/4$:

$$u = \sqrt{2gH}e^{-\frac{\tau}{2}}\cos\left(\frac{\tau\sqrt{4Pb-1}}{2}\right) \tag{4.84}$$

The positive gradient of pressure dependence of velocity of flow through a porous medium can happen in packings in which the channel size changes on account of pressure. When the channel size decreases with increased pressure, the flow velocity through the porous medium also will decrease. The dimensionless velocity as a function of dimensionless time is shown in Fig. 4.7, and subcritical damped oscillations can be seen. After a time, the velocity changes direction on account of the added consideration of the ballistic transport, which takes into consideration the accumulation of momentum in the momentum flux expression.

### 4.5.4   Shear Flow between Two Plates Moving in Opposite Directions at Constant Velocity with Separation Distance 2a

Consider two flat plates (Fig. 4.8) pulled in opposite directions at a constant velocity $V$ with confined fluid. Let the separation distance between the plates be $2a$. The initial velocity of the fluid is zero. Define the axes in a fashion so that the plate velocity is in the $\pm x$ direction and the shear stress acts in and imparts the momentum transfer in the

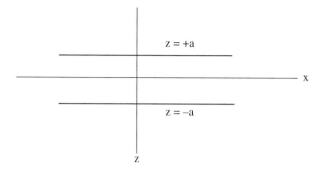

FIGURE **4.8**   Two plates pulled at constant velocity in opposite directions with confined fluids.

$z$ direction. The governing equation for the velocity for the fluid at constant density and viscosity, neglecting pressure and gravity effects in one dimension, including the ballistic transport term for correcting for the accumulation of momentum, can be written as:

$$\tau_r \frac{\partial^2 v_x}{\partial t^2} + \frac{\partial v_x}{\partial t} = \gamma \frac{\partial^2 v_x}{\partial z^2} \tag{4.85}$$

Let    $\tau = t/\tau_r;\ Z = z/\text{sqrt}(\gamma\tau_r);\ u = (v_x - V)/V$ \tag{4.86}

The governing equation in the dimensionless form is then:

$$\frac{\partial u}{\partial \tau} + \frac{\partial^2 u}{\partial \tau^2} = \frac{\partial^2 u}{\partial Z^2} \tag{4.87}$$

The solution can be assumed to consist of a steady-state part and a transient part, that is, $u = u^t + u^{ss}$. The steady-state part and boundary conditions can be selected in such a fashion that the transient portion becomes homogeneous:

$$0 = \frac{\partial^2 u^{ss}}{\partial Z^2} \tag{4.88}$$

The boundary conditions are:

$$Z = 0, u = -1 \quad \text{from symmetry} \tag{4.89}$$

$$Z = \frac{a}{\sqrt{\gamma\tau_r}}, u = 0 \tag{4.90}$$

$$Z = -\frac{a}{\sqrt{\gamma\tau_r}}, u = -2 \tag{4.91}$$

Solving for Eq. (4.88):

$$u^{ss} = c_1 Z + c_2 \qquad (4.92)$$

From the boundary condition given in Eq. (4.89), $c_2$ can be seen to be $-1$. From the boundary condition given in Eq. (4.90), $c_1$ can be seen to be $1/Z_a$. Equation (4.91) is obeyed by Eq. (4.92). Thus:

$$u^{ss} = \frac{Z}{Z_a} - 1 \qquad (4.93)$$

The equation and time and space conditions for the transient portion of the solution can be written as:

$$\frac{\partial u^t}{\partial \tau} + \frac{\partial^2 u^t}{\partial \tau^2} = \frac{\partial^2 u^t}{\partial Z^2} \qquad (4.94)$$

The initial condition:   $\tau = 0,\ u^t = -1$ \qquad (4.95)

The final condition:    $\tau = \infty,\ u^t = 0$ \qquad (4.96)

The boundary conditions are now homogeneous after the expression of the result as a sum of steady-state and transient parts, and are:

$$u^t = 0,\ Z = 0 \qquad (4.97)$$

$$Z = \pm Z_a,\ u^t = 0 \qquad (4.98)$$

The solution is obtained by the method of separation of variables. Initially, the damping term is eliminated using a substitution such as $u^t = W \exp(-n\tau)$. Equation (4.94) then becomes at $n = \frac{1}{2}$:

$$\frac{\partial^2 W}{\partial X^2} = \frac{\partial^2 W}{\partial \tau^2} - \frac{W}{4} \qquad (4.99)$$

Equation (4.99) also can be solved by the method of separation of variables:

Let $\qquad\qquad W = g(\tau)\phi(Z) \qquad (4.100)$

Equation (4.99) becomes:

$$g(\tau)\phi''(Z) = -g(\tau)\phi(Z)/4 + g''(\tau)\phi(Z) \qquad (4.101)$$

$$\frac{\phi''(Z)}{\phi(Z)} = -\frac{1}{4} + \frac{g''(\tau)}{g(\tau)} = -\lambda_n^2 \qquad (4.102)$$

The space domain solution is:

$$\phi(Z) = c_1 \sin(\lambda_n Z) + c_2 \cos(\lambda_n Z) \qquad (4.103)$$

From the boundary conditions:

$$X = 0, u = 0, \text{ it can be seen that } c_2 = 0 \qquad (4.104)$$

$$\phi(Z) = c_1 \sin(\lambda_n Z) \qquad (4.105)$$

From the boundary condition given by Eq. (4.98):

$$0 = c_1 \sin(\lambda_n Z_a) \qquad (4.106)$$

$$n\pi = \lambda_n Z_a \qquad (4.107)$$

$$\lambda_n = \frac{n\pi \sqrt{\gamma \tau_r}}{a} \qquad (4.108)$$

$$a = \frac{\sqrt{\gamma \tau_r}(n\pi)}{\lambda_n} \qquad (4.109)$$

Since $a$ is a nonzero quantity, $n$ can take on the values 1, 2, 3, …. The time domain solution would be:

$$g = c_3 e^{\tau \sqrt{\frac{1}{4} - \lambda_n^2}} + c_4 e^{-\tau \sqrt{\frac{1}{4} - \lambda_n^2}} \qquad (4.110)$$

From the final condition given by Eq. (4.96), not only does the transient velocity have to decay out to zero, but also the wave velocity. Because $W = u^t \exp(\tau/2)$, at time infinity, the transient velocity is zero and any number multiplied by zero is zero even if it is infinity, $W = 0$ at the final condition. Applying this condition in the solution in Eq. (4.110), it can be seen that $c_3 = 0$. Thus:

$$u^t = \sum_1^{\infty} -\frac{2}{n\pi}(1 - (-1)^n) e^{-\frac{\tau}{2}} e^{-\tau \sqrt{\frac{1}{4} - \lambda_n^2}} \sin(\lambda_n Z) \qquad (4.111)$$

$\lambda_n$ is described by Eq. (4.108). $C_n$ can be derived using the orthogonality property and can be shown to be $-(2/n\pi)(1 - (-1)^n)$. It can be seen that the model solutions given by Eq. (4.111) is bifurcated, that is, the characteristics of the function change considerably when a parameter such as the separation distance of the plates is varied. Here, a decaying exponential becomes an exponentially damped cosine function. This is referred to as subcritical damped oscillatory behavior.

For $a < 2\pi \sqrt{(\gamma \tau_r)}$, all the terms in the infinite series will pulsate. This is when the argument within the square root sign in the exponentiated time domain expression becomes negative and the result becomes imaginary. Using De Moivre's theorem and taking a real part for the small width of the slab:

$$u^t = \sum_1^\infty c_n e^{-\frac{\tau}{2}} \cos\left(\tau\sqrt{\lambda_n^2 - \frac{1}{4}}\right) \sin(\lambda_n Z) \qquad (4.112)$$

At $Z = Z_a/2$, the dimensionless velocity:

$$u^t = \sum_1^\infty c_n e^{-\frac{\tau}{2}} \cos\left(\tau\sqrt{\lambda_n^2 - \frac{1}{4}}\right) \qquad (4.113)$$

This is shown in Figs. 4.9 and 4.10. The maximum velocity can be expected to undergo subcritical damped oscillations. The oscillations are overdamped by the decaying exponential in terms of time.

### 4.5.5    Vertical Flow between Plates Moving in Opposite Directions

The governing equation for the fluid between two moving plates in the opposite direction (Fig. 4.11) is obtained after considering the additional forces of pressure and gravity. A permeability law between pressure gradient and flow velocity where the velocity increases as the pressure gradient becomes lower is assumed. This can be seen during elutriation in gas-solid flow and pneumatic conveying under certain conditions. The dimensionless velocity variation of the fluid spatiotemporally is discussed. The two plates are moved in opposite directions along the $z$ axis. Thus, the shear stress or momentum transfer is in the $x$ direction, the horizontal axis. Let the separation between the two vertical plates be given by $2a$.

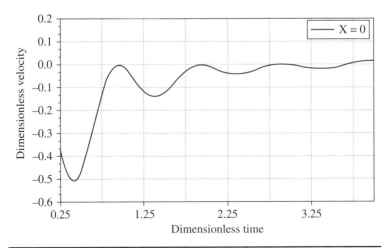

**Figure 4.9**   Subcritical damped oscillations in a fluid between two moving plates in opposite direction $Z = a/2(\gamma \tau_r)$.

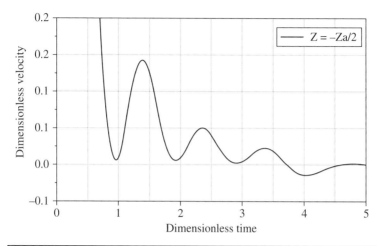

**FIGURE 4.10**    Subcritical damped oscillations in a fluid between two moving plates in opposite direction $Z = -a/2\text{sqrt}(\gamma\tau_{mom})$.

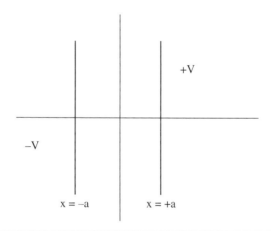

**FIGURE 4.11**    Fluid in between two vertical plates moving in opposite directions with constant velocity.

The governing equation, including the pressure gradient and gravity forces, and after writing the pressure gradient in terms of the velocity of flow and neglecting the changes in pressure gradient with time, can be written as:

$$\left(\frac{\partial^2 v_z}{\partial \tau^2}\right) + \frac{\partial v_z}{\partial \tau} + \frac{\gamma\tau_{mom}}{\kappa}v_z = \gamma\frac{\partial^2 v_z}{\partial z^2} \qquad (4.114)$$

Let    $\tau = \dfrac{t}{\tau_r};\ Z = \dfrac{z}{\sqrt{\gamma\tau_r}};\ u = \dfrac{(v_z - V)}{V};\ Pb = \dfrac{\gamma\tau_{mom}}{\kappa} \qquad (4.115)$

The governing equation in the dimensionless form is then:

$$\frac{\partial^2 u}{\partial \tau^2} + \frac{\partial u}{\partial \tau} + Pb(u) + Pb = \frac{\partial^2 u}{\partial Z^2} \qquad (4.116)$$

The solution can be assumed to consist of a steady-state part and a transient part, that is, $u = u^t + u^{ss}$. The steady-state part and boundary conditions can be selected in such a fashion that the transient portion becomes homogeneous:

$$Pb(u^{ss}) + Pb + \frac{\partial^2 u^{ss}}{\partial Z^2} = 0 \qquad (4.117)$$

The boundary conditions are:

$$Z = 0, u = -1 \quad \text{from symmetry} \qquad (4.118)$$

$$Z = \frac{R}{\sqrt{\gamma \tau_r}}, u^{ss} = 0 \qquad (4.119)$$

$$Z = -\frac{R}{\sqrt{\gamma \tau_r}}, u^{ss} = -2 \qquad (4.120)$$

Solving for Eq. (4.117):

$$u^{ss} = c_1 \sin(Pb^{1/2}Z) + c_2 \cos(Pb^{1/2}Z) + c_3 \qquad (4.121)$$

From the boundary condition given in Eq. (4.118):

$$c_2 + c_3 = -1 \qquad (4.122)$$

$$0 = c_1 \sin(Pb\, Z_R) + c_2 (\cos(Pb^{1/2}Z_R) - 1) + 1 \qquad (4.123)$$

$$-2 = -c_1 \sin(Pb\, Z_R) + c_2 (\cos(Pb^{1/2}Z_R) - 1) + 1 \qquad (4.124)$$

$$\text{or } c_1 = -c_2(\cot(Pb^{1/2}Z_R) - 1/\sin(Pb^{1/2}Z_R) + 1/\sin(Pb^{1/2}Z_R) \qquad (4.125)$$

$$c_2 = -2/(\sin(Pb^{1/2}Z_R) - 1/\sin^2(Pb^{1/2}Z_R)) \qquad (4.126)$$

The boundary condition given in Eq. (4.119) assumes that the viscous effects predominate over the gravitational and pressure effects. The equation and time and space conditions for the transient portion of the solution can be written as:

$$\frac{\partial^2 u^t}{\partial \tau^2} + \frac{\partial u^t}{\partial \tau} + Pb(u^t) + Pb = \frac{\partial^2 u^t}{\partial Z^2} \qquad (4.127)$$

The initial condition: $\tau = 0, u^t = -1$ \qquad (4.128)

The final condition: $\tau = \infty, u^t = 0$ \qquad (4.129)

The boundary conditions are now homogeneous after the expression of the result as a sum of steady-state and transient parts, and are:

$$u^t = 0, Z = 0 \tag{4.130}$$

$$Z = \pm Z_{a'}, u^t = 0 \tag{4.131}$$

The solution is obtained by the method of separation of variables. Initially, the damping term is eliminated using a substitution such as $u^t = W \exp(-n\tau)$. Equation (4.127) then becomes at $n = \frac{1}{2}$:

$$\frac{\partial^2 W}{\partial Z^2} = \left(Pb - \frac{1}{4}\right)\frac{W}{4} + \frac{\partial^2 W}{\partial \tau^2} \tag{4.132}$$

For large permeability numbers, Eq. (4.132) can be transformed into a Bessel equation with the following substitution:

$$\text{For } \tau > Z, \text{ and } Pb > \frac{1}{4}, \eta = \tau^2 - Z^2 \tag{4.133}$$

As shown in earlier sections, Eq. (4.132) is transformed into:

$$4\eta \frac{\partial^2 W}{\partial \eta^2} + 4\frac{\partial W}{\partial \eta} + \left(Pb - \frac{1}{4}\right)W = 0 \tag{4.134}$$

The solution is a Bessel composite function and can be written by:

$$d = \frac{1}{4}(Pb - \frac{1}{4})$$

$$W = c J_0 \sqrt{\left(Pb - \frac{1}{4}\right)(\tau^2 - Z^2)} \tag{4.135}$$

$$u = c e^{-\frac{\tau}{2}} J_0 \sqrt{\left(Pb - \frac{1}{4}\right)(\tau^2 - Z^2)} \tag{4.136}$$

From the boundary condition:

$$-1 = u = c e^{-\frac{\tau}{2}} J_0 \tau \sqrt{\left(Pb - \frac{1}{4}\right)} \tag{4.137}$$

Eliminating $c$ between Eqs. (4.136) and (4.137):

$$u = -\frac{J_0 \sqrt{\left(Pb - \frac{1}{4}\right)(\tau^2 - Z^2)}}{J_0 \left(\tau \sqrt{Pb - \frac{1}{4}}\right)} \tag{4.138}$$

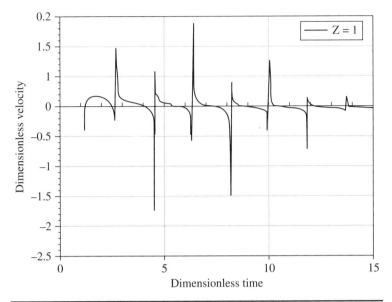

FIGURE **4.12**  Transient velocity of viscous fluid flow in porous medium with a positive permeability coefficient between vertical plates.

It can be seen that Eq. (4.138) describes the velocity profile in between the vertical plates considering the viscous, gravitational, and permeability effects. The spatiotemporal velocity is given by a Bessel composite function of the first kind and zeroth order (Fig. 4.12). This is expected to be valid for permeability numbers greater than ¼. The expression exhibits space symmetry and subcritical damped oscillations can be expected. The Bessel function can be approximated as:

$$u = -\frac{\tau}{(\tau^2 - Z^2)^{1/4}} \frac{\cos\left(\sqrt{Pb - \frac{1}{4}}(\tau^2 - Z^2) - \frac{\pi}{4}\right)}{\cos\left(\tau\sqrt{Pb - \frac{1}{4}} - \frac{\pi}{4}\right)} \tag{4.139}$$

### 4.5.6  Transient Laminar Flow in a Circular Conduit

Consider the laminar flow in a circular pipe of narrow dimensions in transience. The damped wave momentum transfer and relaxation equation is written as:

$$\tau_{xy} = -\mu \frac{\partial v}{\partial r} - \tau_{mom} \frac{\partial \tau_{xy}}{\partial t} \tag{4.140}$$

The governing equation for the axial velocity as a function of the radial direction can be written after combining the modified Newton's law of viscosity, including the relaxation term, with the momentum balance equation to yield:

$$\tau_{mom}\frac{\partial^2 v}{\partial t^2}+\frac{\partial v}{\partial t}=\frac{\Delta P}{\rho L}+\frac{v}{r}\frac{\partial}{\partial r}\left(r\frac{\partial v}{\partial r}\right)\qquad(4.141)$$

Let the dimensionless variables be:

$$\tau=\frac{t}{\tau_{mom}};\ X=\frac{r}{\sqrt{v\tau_{mom}}};\ u=\frac{v}{v_{max}};\ P^*=\frac{\Delta P\tau_{mom}}{L\rho v_{max}}\qquad(4.142)$$

The dimensionless governing equation then becomes:

$$\frac{\partial^2 u}{\partial\tau^2}+\frac{\partial u}{\partial\tau}=P^*+\frac{1}{X}\frac{\partial u}{\partial X}+\frac{\partial^2 u}{\partial X^2}\qquad(4.143)$$

The solution is assumed to be a sum of two parts, that is, the steady-state and transient parts. Let $u=u^{ss}+u^t$. Then the governing equation becomes:

$$0=P^*+\frac{1}{X}\frac{\partial u^{ss}}{\partial X}+\frac{\partial^2 u^{ss}}{\partial X^2}\qquad(4.144)$$

$$\frac{\partial^2 u^t}{\partial\tau^2}+\frac{\partial u^t}{\partial\tau}=P^*+\frac{1}{X}\frac{\partial u^t}{\partial X}+\frac{\partial^2 u^t}{\partial X^2}\qquad(4.145)$$

Integrating the equation for the steady-state component of the velocity with respect to $X$:

$$X\frac{\partial u^{ss}}{\partial X}=-\frac{P^*X^2}{2}+c'\qquad(4.146)$$

At $X=0$, the gradient of the velocity is zero, as a condition of extremama can be expected from symmetry considerations. So it can be seen that $C'=0$. Integrating the resulting equation, again with respect to $X$:

$$u^{ss}=-\frac{P^*X^2}{4}+d'\qquad(4.147)$$

From the boundary condition at $X=X_R$:

$$d'=P^*X_R^2/4\qquad(4.148)$$

and
$$u^{ss}=\frac{P^*R^2}{4v\tau_{mom}}\left(1-\left(\frac{r}{R}\right)^2\right)\qquad(4.149)$$

The above relation is the Poiseuille distribution. Specifically, is the Hagen-Poiseuille flow in laminar pipes at steady state where $P^*$ is given by Eq. (4.142). The rest of the problem obtains the transient part. First, the damping term is removed by a substitution: $u^t = W \times \exp(-n\tau)$. At $n = \frac{1}{2}$, the governing equation reduces to:

$$\frac{\partial^2 W}{\partial X^2} + \frac{1}{X}\frac{\partial W}{\partial X} = -\frac{W}{4} + \frac{\partial^2 W}{\partial \tau^2} \tag{4.150}$$

The basis for this substitution is to recognize that the damped wave conduction and relaxation equation, which is of the hyperbolic type, has a damping component and wave component to it. In order to better study the characteristics of the wave component, it would be desirable to remove the damping component from the governing equation. The transformation given in Eq. (4.145) was selected in order to delineate the damping component and the wave component of the transient temperature. Furthermore, it is realized that transient temperatures decay out in time exponentially. This leads to the negative exponent in the exponentiated term. At $n = \frac{1}{2}$, it can be seen that the governing equation in transient temperature reverts to an equation for wave temperature. This happens to be a Bessel special differential equation.

The method of separation of variables can be used to obtain the exact solution to the Eq. (4.150). The boundary and time conditions for the transient portion of the velocity are then:

$$\tau = 0, u^t = 1 \tag{4.151}$$

$$\tau = \text{infinity}, u^t = 0 \tag{4.152}$$

$$\tau > 0, X = X_R, u^t = 0 \tag{4.153}$$

$$X = 0, \text{symmetry considerations} \tag{4.154}$$

Let $\qquad W = V(\tau)\phi(X) \tag{4.155}$

The wave equation becomes:

$$\frac{V''}{V} - \frac{1}{4} = \frac{\phi'' + \dfrac{\phi'}{X}}{\phi} = -\lambda_n^2 \tag{4.156}$$

Thus:

$$X^2\phi'' + X\phi' + X^2\lambda_n^2\phi = 0 \tag{4.157}$$

This can be recognized as a Bessel equation of the first order (App. A) and the solution can be written as:

$$\varphi = c_1 J_0(\lambda_n X) \tag{4.158}$$

$c_2$ can be seen to be zero, as $\phi$ is finite at $X = 0$. From the boundary condition at $X = R/\text{sqrt}(\nu\tau_{\text{mom}})$:

$$c_1 J_0(\lambda_n X_R) = 0 \tag{4.159}$$

$$\lambda_n = \frac{\sqrt{\nu\tau_{\text{mom}}}}{R}(2.4048 + (n-1)\pi) \tag{4.160}$$

$n = 1, 2, 3, \ldots$.

Now the time domain part of the wave is obtained by solving the second-order ODE:

$$\frac{V''}{V} - \frac{1}{4} = -\lambda_n^2 \tag{4.161}$$

Thus:

$$ue^{\frac{\tau}{2}} = W = \left(c_3 e^{\tau\sqrt{\frac{1}{4}-\lambda_n^2}} + c_4 e^{-\tau\sqrt{\frac{1}{4}-\lambda_n^2}}\right)\phi \tag{4.162}$$

At infinite times, the right-hand side becomes infinitely large. The left-hand side is zero multiplied by infinity and is zero. At steady state, the velocity is bounded and hence, the constant $c_3$ is set to zero. It can be seen that for small channel dimensions, that is, when $R < 4.8096 \text{ sqrt}(\nu\tau_{\text{mom}})$, the solution is periodic with respect to time. The general solution for such cases can be written as:

$$u = \sum_1^\infty C_n e^{-\frac{\tau}{2}} \cos\left(\tau\sqrt{\lambda_n^2 - \frac{1}{4}}\right) J_0(\lambda_n X) \tag{4.163}$$

The initial condition can be taken to be the maximum velocity in plug flow. So the initial superficial velocity essentially at plug flow becomes a periodic profile, as in Hagen-Poiseuille flow when channels are formed. The transient portion is governed for small channels by the generalized Newton's law of viscosity. So the initial condition is:

$$1 = \sum_1^\infty C_n J_0(\lambda_n X) \tag{4.164}$$

The constant can be solved for by the orthogonality property:

$$C_n = -\frac{\int_0^{X_R} J_0(\lambda_m X)dX}{\int_0^{X_R} J_0^2(\lambda_m X)dX} \tag{4.165}$$

The maximum transient velocity is given by that at the center of the circular tube:

$$u_{\text{max}} = \sum_1^\infty C_n e^{-\frac{\tau}{2}} \cos\left(\tau\sqrt{\lambda_n^2 - \frac{1}{4}}\right) \tag{4.166}$$

### 4.5.7 Oscillations in a U-Tube Manometer

Consider the oscillations in a U-tube manometer. The additional ballistic term is used to discuss the velocity and the height in the manometer.

The governing equation in the $z$ direction in the U-tube manometer integrated with respect to $z$ between the two points in the manometer, 1 and 2, on either side may be written as follows:

$$\tau_{mom} \frac{\partial^2 v_x}{\partial t^2} + \frac{\partial v_x}{\partial t} + \frac{2gz}{L} = 0 \tag{4.167}$$

$\partial v_z / \partial t$ is independent of $z$, $L$, where $L$ is the length of the column:

$$P_1 = P_2 \tag{4.168}$$

$$V_{z1} = -V_{z2} \text{ (from continuity)} \tag{4.169}$$

$$V_{z1}^2 = V_{z2}^2 \tag{4.170}$$

Writing $v_z$ as $dz/dt$, Eq. (4.167) can be written as:

$$\tau_{mom} \frac{\partial^3 z}{\partial t^3} + \frac{\partial^2 z}{\partial t^2} + \frac{2gz}{L} = 0 \tag{4.171}$$

Let the oscillation number

$$\text{Osc} = (g\tau_{mom}^2 / L); \tau = t/\tau_{mom}; Z = z/L \tag{4.172}$$

Equation (4.171) becomes:

$$\frac{\partial^3 Z}{\partial \tau^3} + \frac{\partial^2 Z}{\partial \tau^2} + 2\text{Osc} = 0 \tag{4.173}$$

The third-order ODE with constant coefficients is homogeneous and can be solved for as follows:

$$r^3 + r^2 + 2\text{Osc} = 0 \tag{4.174}$$

Equation (4.174) can be compared with the general form of the cubic equation:

$$r^3 + a_2 r^2 + a_1 r + a_0 = 0 \tag{4.175}$$

Let $e$ and $f$ be defined as:

$$e = 1/3\, a_1 - 1/9 a_2^2 = -1/9 \tag{4.176}$$

$$f = 1/6(a_1 a_2 - 3a_0) - 1/27 a_2^3 = -(\text{Osc} + 1/27) \tag{4.177}$$

where $\qquad\qquad a_0 = 2\text{Osc}; \; a_1 = 0; \; a_2 = 1$ $\qquad\qquad$ (4.178)

Consider $\qquad\qquad e^3 + f^2 > 0$ $\qquad\qquad$ (4.179)

This is when:

$$\text{Osc} > -2/27 \qquad\qquad (4.180)$$

The oscillation number is the ratio of the gravitational force divided by the relaxation frequency normalized by the length of the column in the U-tube. The oscillation number will always be positive; hence, Eq. (4.173) will be valid for real systems. In such cases, the cubic equation solution results in one real root and two imaginary roots. Let:

$$
\begin{aligned}
s_1 &= (f + (e^3 + f^2)^{1/2})^{1/3}\\
&= (\text{Osc}[(1 + 2/27\text{Osc})^{1/2} - 1] - 1/27)^{1/3}
\end{aligned}
\qquad (4.181)
$$

$$
\begin{aligned}
s_2 &= (f - (e^3 + f^2)^{1/2})^{1/3}\\
&= -(\text{Osc}[(1 + 2/27\text{Osc}))^{1/2} - 1] + 1/27)^{1/3}
\end{aligned}
\qquad (4.182)
$$

The cubic roots are then:

$$r_1 = (s_1 + s_2) - a_2/3 \qquad\qquad (4.183)$$

$$r_2 = 1/2(s_1 + s_2) - a_2/3 + i\text{sqrt}(3)/2\,(s_1 - s_2) \qquad\qquad (4.184)$$

$$r_3 = 1/2(s_1 + s_2) - a_2/3 + i\text{sqrt}(3)/2\,(s_1 - s_2) \qquad\qquad (4.185)$$

Thus, for a finite oscillation number, the displacement will pulsate:

$$Z = c_1\exp(r_1 t) + c_2\exp(r_2 t) + c_3\exp(r_3 t) \qquad\qquad (4.186)$$

The imaginary roots can be seen to predict the oscillations (Fig. 4.13) that are subcritical and damped. Using De Moivre's theorem and

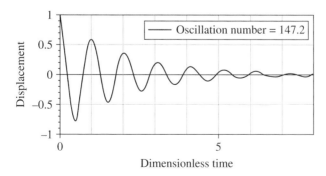

**Figure 4.13** Subcritical damped oscillations in a U-tube manometer, including the ballistic transport term.

obtaining the real parts, the term that contributes the subcritical damped oscillations can be written as:

$$Z = c'e^{-\tau\left(\frac{(s_1+s_2)}{2}-\frac{1}{3}\right)}\cos\left(\tau\left(\frac{s_1+s_2}{2}-\frac{1}{3}\right)\right) \tag{4.187}$$

$C'$ can be solved for using the initial condition for velocity. The other conditions are the velocity at zero time and infinite time to solve for the other two integration constants.

### 4.5.8  Tangential Flow Induced by a Rotating Cylinder

The velocity distributions and pressure distributions during the tangential laminar flow of an incompressible fluid induced by a sudden rotating cylinder at constant velocity is examined in this section. The rotation of the cylinder of radius $R$ is at a tangential velocity of $V_\theta$ (Fig. 4.14). The radial and azimuthal velocity components are zero. The equation of motion can be written as:

$$\rho\tau_{mom}\left(\frac{\partial^2 v_\theta}{\partial t^2}+\frac{v_\theta}{r}\frac{\partial^2 v_\theta}{\partial t\,\partial\theta}\right)+\rho\frac{\partial v_\theta}{\partial t}=\mu\frac{\partial}{\partial r}\left(\frac{1}{r}\frac{\partial(r v_\theta)}{\partial r}\right) \tag{4.188}$$

$\theta$ component:

$$\rho\tau_{mom}\left(-\frac{2v_\theta}{r}\frac{\partial v_\theta}{\partial t}\right)+\tau_{mom}\frac{\partial^2 p}{\partial t\,\partial r}-\frac{\rho v_\theta^2}{r}=-\frac{\partial p}{\partial r} \tag{4.189}$$

$r$ component:

$$0=-\frac{\partial p}{\partial z}+\rho g \tag{4.190}$$

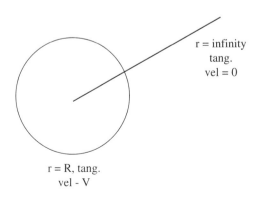

r = infinity
tang.
vel = 0

r = R, tang.
vel - V

**FIGURE 4.14**  Tangential flows past a cylinder.

It can be seen that the pressure gradient in the $z$ direction does not change with time; hence, $\tau_{mom}\partial^2 p/\partial t\,\partial z$ is zero. Once $v_\theta$ is solved for from Eq. (4.189), the radial pressure distribution can then be calculated using Eq. (4.190). The tangential velocity is assumed to be a function of time and the radial space coordinate. Equation (4.188) is made dimensionless by the following substitutions:

$$\tau = \frac{t}{\tau_{mom}}; \; X = \frac{r}{\sqrt{v\tau_{mom}}}; \; u = \frac{v_\theta}{V_\theta} \tag{4.191}$$

The dimensionless governing equation then becomes:

$$\frac{\partial^2 u}{\partial \tau^2} + \frac{\partial u}{\partial \tau} = +\frac{\partial}{\partial X}\left(\frac{1}{X}\frac{\partial(uX)}{\partial X}\right) \tag{4.192}$$

Let $$uX = V \tag{4.193}$$

The dimensionless governing equation then becomes:

$$\frac{\partial^2 V}{\partial \tau^2} + \frac{\partial V}{\partial \tau} = -\frac{1}{X}\frac{\partial V}{\partial X} + \frac{\partial^2 V}{\partial X^2} \tag{4.194}$$

The damping term is removed from the governing equation. This is done realizing that the transient velocity decays with time in an exponential fashion. The other reason for this maneuver is to study the wave equation without the damping term. Let $V = w\exp(-n\tau)$. By choosing $n = \frac{1}{2}$, the damping component of the equation is removed.

Thus, for $n = \frac{1}{2}$:

$$\frac{\partial^2 W}{\partial X^2} - \frac{1}{X}\frac{\partial W}{\partial X} = -\frac{W}{4} + \frac{\partial^2 W}{\partial \tau^2} \tag{4.195}$$

Equation (4.195) can be solved by using the method of relativistic transformation of coordinates. Consider the transformation variable $\eta$ for $\tau > X$:

$$\eta = \tau^2 - X^2 \tag{4.196}$$

Equation (4.195) becomes:

$$4\eta^2\frac{\partial^2 W}{\partial \eta^2} + 2\eta\frac{\partial W}{\partial \eta} - \eta\frac{W}{4} = 0 \tag{4.197}$$

or $$\eta^2\frac{\partial^2 W}{\partial \eta^2} + \frac{\eta}{2}\frac{\partial W}{\partial \eta} - \eta\frac{W}{16} = 0 \tag{4.198}$$

Comparing Eq. (4.198) with the generalized Bessel equation, the solution is $a = 1/2$; $b = 0$; $c = 0$; $d = -1/16$; and $s = \frac{1}{2}$. The order $p$ of the solution is then $p = 2 \ \text{sqrt}(1/16) = \frac{1}{2}$:

$$W = c_1 (\tau^2 - X^2)^{1/4} I_{1/2}\left(\frac{1}{2}\sqrt{(\tau^2 - X^2)}\right)$$

$$+ c_2 (\tau^2 - X^2)^{1/4} I_{-1/2}\left(\frac{1}{2}\sqrt{\tau^2 - X^2}\right) \tag{4.199}$$

$c_2$ can be seen to be zero, as $W$ is finite and not infinitely large at $\eta = 0$. An approximate solution can be obtained by eliminating $c_1$ between the equation derived from the boundary condition at $X = X_R$ and Eq. (4.199) after setting $c_2$ to be zero. It can be noted that this is a mild function of time, however, as the general solution of a PDE consists of $n$ arbitrary functions when the order of the PDE is $n$ compared with $n$ arbitrary constants for an ODE. From the boundary condition at $X = X_R$:

$$X = e^{-\frac{\tau}{2}} c_1 (\tau^2 - X^2)^{1/4} I_{1/2}\sqrt{(\tau^2 - X_R^2)} \tag{4.200}$$

or

$$u = \frac{(\tau^2 - X^2)^{1/4}}{(\tau^2 - X_R^2)^{1/4}} \frac{I_{1/2}\left(\frac{1}{2}\sqrt{\tau^2 - X^2}\right)}{I_{1/2}\left(\frac{1}{2}\sqrt{\tau^2 - X_R^2}\right)} \tag{4.201}$$

In terms of elementary functions, Eq. (4.201) can be written as:

$$u = \frac{\sinh\left(\frac{1}{2}\sqrt{\tau^2 - X^2}\right)}{\sinh\left(\frac{1}{2}\sqrt{\tau^2 - X_R^2}\right)} \tag{4.202}$$

In the limit of $X_R$ going to zero, the expression becomes for $\tau > X$:

$$u = \frac{\sinh\left(\frac{1}{2}\sqrt{\tau^2 - X^2}\right)}{\sinh\left(\frac{\tau}{2}\right)} \tag{4.203}$$

For $X > \tau$:

$$u = \frac{(X^2 - \tau^2)^{1/4}}{(\tau^2 - X_R^2)^{1/4}} \frac{J_{1/2}\left(\frac{1}{2}\sqrt{X^2 - \tau^2}\right)}{I_{1/2}\left(\frac{1}{2}\sqrt{\tau^2 - X_R^2}\right)} \tag{4.204}$$

Equation (4.204) can be written in terms of trigonometric functions as:

$$u = \frac{\sin\left(\frac{1}{2}\sqrt{X^2 - \tau^2}\right)}{\sinh\left(\frac{1}{2}\sqrt{\tau^2 - X_R^2}\right)} \tag{4.205}$$

Four different regimes can be seen. The first regime is that of the thermal lag and consists of no change from the initial velocity. The second regime is when:

$$\tau_{lag}^{\ 2} = X^2 - 4\pi^2$$

or 
$$\tau_{lag} = \text{sqrt}(X_p^{\ 2} - 4\pi^2) = 3.09 \qquad \text{when } X_p = 7 \tag{4.206}$$

For times greater than the time lag and less than $X_p$, the dimensionless velocity is given by Eq. (4.204). For dimensionless times greater than 7, for example, the dimensionless velocity is given by Eq. (4.201). For distances *closer to the surface compared with* $2\pi$, *the time lag will be zero*. The fourth regime is at the wavefront. Here, $u = \exp(-X/2) = \exp(-\tau/2)$. The radial pressure distribution can be estimated from the following equation:

$$\rho\left(-\frac{2v_\theta}{r}\frac{\partial v_\theta}{\partial \tau}\right) + \frac{\partial^2 p}{\partial \tau \partial r} - \frac{\rho v_\theta^2}{r} = -\frac{\partial p}{\partial r} \tag{4.207}$$

It can be seen that for materials with relaxation greater than a certain threshold value, the instantaneous pressure values will be pulsation values.

### Tangential Flow at Small Distances

In writing Eq. (4.194), the assumption made that large distances were involved was valid. Obtain the governing equation from the tangential shear term and obtain the solution for the tangential flow problem by relaxing the assumption at large distances:

$$\rho\tau_{mom}\left(\frac{\partial^2 v_\theta}{\partial t^2} + \frac{v_\theta}{r}\frac{\partial^2 v_\theta}{\partial t \partial \theta}\right) + \rho\frac{\partial v_\theta}{\partial t} = \mu\frac{\partial}{\partial r}\left(\frac{1}{r}\frac{\partial (r v_\theta)}{\partial r}\right) \tag{4.208}$$

Let $V = u/X$ where $u$, $X$, and $\tau$ are defined using the dimensionless variables in Eq. (4.191). Equation (4.208) becomes:

$$\frac{\partial^2 V}{\partial \tau^2} + \frac{\partial V}{\partial \tau} = \frac{3}{X}\frac{\partial V}{\partial X} + \frac{\partial^2 V}{\partial X^2} \tag{4.209}$$

The damping term is removed from the governing equation. This is done realizing that the transient velocity decays with time in an exponential fashion. The other reason for this maneuver is to study the wave equation without the damping term. Let $V = w\exp(-n\tau)$. By choosing $n = \frac{1}{2}$, the damping component of the equation is removed. Thus, for $n = \frac{1}{2}$:

$$\frac{\partial^2 W}{\partial X^2} + \frac{3}{X}\frac{\partial W}{\partial X} = -\frac{W}{4} + \frac{\partial^2 W}{\partial \tau^2} \tag{4.210}$$

Equation (4.210) can be solved using the method of relativistic transformation of coordinates. Consider the transformation variable $\eta$ for $\tau > X$ as:

$$\eta = \tau^2 - X^2 \tag{4.211}$$

Upon applying relativistic transformation, Eq. (4.210) becomes:

$$4\eta^2 \frac{\partial^2 W}{\partial \eta^2} + 10\eta \frac{\partial W}{\partial \eta} - \eta\frac{W}{4} = 0 \tag{4.212}$$

or

$$\eta^2 \frac{\partial^2 W}{\partial \eta^2} + \frac{5\eta}{2}\frac{\partial W}{\partial \eta} - \eta\frac{W}{16} = 0 \tag{4.213}$$

Comparing Eq. (4.213) with the generalized Bessel equation, the solution is $a = 5/2$; $b = 0$; $c = 0$; $d = -1/16$; and $s = \frac{1}{2}$. The order $p$ of the solution is then $p = 3/2 \, \text{sqrt}(1/16) = \frac{1}{2}$:

Or

$$W = c_1(\tau^2 - X^2)^{3/4} I_{3/2}\left(\frac{1}{2}\sqrt{(\tau^2 - X^2)}\right)$$

$$+ c_2(\tau^2 - X^2)^{3/4} I_{-3/2}\left(\frac{1}{2}\sqrt{\tau^2 - X^2}\right) \tag{4.214}$$

$c_2$ can be seen to be zero, as $W$ is finite and not infinitely large at $\eta = 0$. An approximate solution can be obtained by eliminating $c_1$ between the boundary condition and the Eq. (4.214) after setting $c_2$ to be zero. It can be noted that this is a mild function of time, however, as the general solution of a PDE consists of $n$ arbitrary functions when the order of the PDE is $n$, compared with $n$ arbitrary constants for the ODE. From the boundary condition at $X = X_R$:

$$\frac{1}{X} = e^{-\frac{\tau}{2}} c_1(\tau^2 - X_R^2)^{1/4} I_{1/2}\sqrt{(\tau^2 - X_R^2)} \tag{4.215}$$

or

$$u = \frac{(\tau^2 - X_R^2)^{3/4}}{(\tau^2 - X^2)^{3/4}} \frac{I_{3/2}\left(\frac{1}{2}\sqrt{\tau^2 - X^2}\right)}{I_{3/2}\left(\frac{1}{2}\sqrt{\tau^2 - X_R^2}\right)} \tag{4.216}$$

For $X > \tau$:

$$u = \frac{(\tau^2 - X_R^2)^{3/4}}{(X^2 - \tau^2)^{3/4}} \frac{J_{3/2}\left(\frac{1}{2}\sqrt{X^2 - \tau^2}\right)}{I_{3/2}\left(\frac{1}{2}\sqrt{\tau^2 - X_R^2}\right)} \tag{4.217}$$

### 4.5.9  Transient Flow Past a Sphere

Consider a solid sphere settling in an infinite fluid at terminal settling velocity. Of interest is the transient velocity in the tangential direction of the sphere as a function of $r$. Defining the stream function as follows and neglecting the $\phi$ and $r$ component velocities:

$$v_\theta = \frac{1}{r \sin\theta} \frac{\partial \psi}{\partial r} \tag{4.218}$$

The $\theta$ component of the velocity, considering only its dependence in $r$ direction, becomes:

$$\tau_{mom} \frac{\partial^3 \psi}{\partial t^2 \partial r} + \frac{\partial^2 \psi}{\partial t \partial r} = \frac{v}{r} \frac{\partial}{\partial r}\left(r^2 \frac{\partial}{\partial r}\left(\frac{1}{r}\frac{\partial \psi}{\partial r}\right)\right) \tag{4.219}$$

Let

$$\tau = \frac{t}{\tau_{mom}}; \ X = \frac{r}{\sqrt{v\tau_{mom}}} \tag{4.220}$$

Then

$$v_\theta = \frac{1}{X \sin\theta \ v\tau_{mom}} \frac{\partial \psi}{\partial r} \tag{4.221}$$

$$\frac{\partial^3 \psi}{\partial \tau^2 \partial X} + \frac{\partial^2 \psi}{\partial \tau \partial X} = \frac{1}{X}\frac{\partial}{\partial X}\left(X^2 \frac{\partial}{\partial X}\left(\frac{1}{X}\frac{\partial \psi}{\partial X}\right)\right) \tag{4.222}$$

Integrating with respect to $X$:

$$\frac{\partial^2 \psi}{\partial \tau^2} + \frac{\partial \psi}{\partial \tau} = \frac{\partial^2 \psi}{\partial X^2} \tag{4.223}$$

Let $\psi = \chi \exp(-\tau/2)$. This will remove the damping term in the governing equation to give:

$$\frac{\partial^2 \chi}{\partial X^2} = -\frac{\chi}{4} + \frac{\partial^2 \chi}{\partial \tau^2} \tag{4.224}$$

Using the relativistic transformation [8]:

$$\eta = \tau^2 - X^2 \tag{4.225}$$

For $\tau > X$ the governing equation without the damping term becomes:

$$4\eta\, \partial^2\chi/\partial\eta^2 + 4\,\partial\chi/\partial\eta - \chi/4 = 0 \qquad (4.226)$$

or

$$\eta^2\partial^2\chi/\partial\eta^2 + \eta\,\partial\chi/\partial\eta - \eta\chi/16 = 0 \qquad (4.227)$$

Comparing Eq. (4.227) with the generalized Bessel equation, the solution can be written as:

$$\chi = c_1 e^{-\frac{\tau}{2}} I_0\left(\frac{1}{2}\sqrt{\tau^2 - X^2}\right) \qquad (4.228)$$

$$v_\theta = \frac{1}{\nu\tau_{mom}\sin\theta} c_1 e^{-\frac{\tau}{2}} \frac{I_1\left(\frac{1}{2}\sqrt{\tau^2 - X^2}\right)}{\sqrt{\tau^2 - X^2}} \qquad (4.229)$$

From the boundary condition at $r = R$:

$$v_\theta = U_t = \frac{1}{\nu\tau_{mom}\sin\theta} c_1 e^{-\frac{\tau}{2}} \frac{I_1\left(\frac{1}{2}\sqrt{\tau^2 - X_R^2}\right)}{\sqrt{\tau^2 - X_R^2}} \qquad (4.230)$$

For $X > \tau$:

$$\frac{v_\theta}{U_t} = \sqrt{\frac{\tau^2 - X_R^2}{X^2 - \tau^2}} \frac{J_1\left(\frac{1}{2}\sqrt{X^2 - \tau^2}\right)}{I_1\left(\frac{1}{2}\sqrt{\tau^2 - X_R^2}\right)} \qquad (4.231)$$

Eliminating $c_1$ between Eqs. (4.230) and (4.229) for $\tau > X$:

$$\frac{v_\theta}{U_t} = \sqrt{\frac{\tau^2 - X_R^2}{\tau^2 - X^2}} \frac{I_1\left(\frac{1}{2}\sqrt{\tau^2 - X^2}\right)}{I_1\left(\frac{1}{2}\sqrt{\tau^2 - X_R^2}\right)} \qquad (4.232)$$

## 4.5.10   Radial Flow between Two Concentric Spheres

Consider the radial flow between two concentric spheres of an incompressible, isothermal liquid. The transient velocity distribution is examined using the damped wave momentum transfer and relaxation equation. Let the radii of the two spheres be $R$ and $mR$, respectively. The governing equation for the radial component of the velocity can be written as:

$$\tau_{mom}\left(\frac{1}{r^2}\frac{\partial^2(r^2v_r)}{\partial t^2}+\frac{v_r}{r^2}\frac{\partial^2(r^2v_r)}{\partial t\,\partial r}\right)+\left(\frac{\tau_{mom}}{r^2}\frac{\partial(r^2v_r)}{\partial t}+v_r\right)\left(\frac{1}{r^2}\frac{\partial(r^2v_r)}{\partial r}\right)$$

$$+\tau_{mom}\left(\frac{\partial^2 p}{\partial t\,\partial r}\right)+\frac{\rho}{r^2}\frac{\partial(r^2v_r)}{\partial t}=-\frac{1}{\rho}\frac{\partial p}{\partial r}+\frac{v}{r^2}\frac{\partial^2(r^2v_r)}{\partial r^2} \tag{4.233}$$

In order to obtain the dimensionless form of the governing equation, the substitution given in Eq. (4.233) is used, and after neglecting the nonlinear term using the creeping flow assumption:

$$\left(\frac{1}{X^2}\frac{\partial^2(X^2u)}{\partial\tau^2}\right)+\left(\frac{1}{X^2}\frac{\partial(X^2u)}{\partial\tau}+u\right)\left(\frac{1}{X^2}\frac{\partial(X^2u)}{\partial X}\right)+\left(\frac{\partial^2 P^*}{\partial\tau\,\partial X}\right)+\frac{1}{X^2}\frac{\partial(X^2u)}{\partial\tau}$$

$$=-\frac{\partial P^*}{\partial X}+\frac{1}{X^2}\frac{\partial^2(X^2u)}{\partial X^2} \tag{4.234}$$

where $$P^*=\frac{p}{\rho v_R v_{mom}}; u=\frac{v_r-v_R}{v_R} \tag{4.235}$$

The space boundary conditions can be written as

$$r=R, v_r=v_R \tag{4.236}$$

From the equation of continuity for a constant density system:

$$\frac{1}{r^2}\frac{\partial(\rho r^2v_r)}{\partial r}=0 \tag{4.237}$$

$$r^2v_r=c_1=R^2v_R \tag{4.238}$$

The velocity at $mR$ is then:

$$v_{mR}=v_R/m^2 \tag{4.239}$$

Thus,

$$u^{ss}=(R/r)^2-1 \tag{4.240}$$

$$=\frac{X_R^2}{X^2}-1 \tag{4.241}$$

The time conditions are:

$$\tau=0, u=-1 \tag{4.242}$$

Let the velocity consist of steady-state and transient components:

$$u=u^{ss}+u^t \tag{4.243}$$

$$-\frac{\partial P^*}{\partial X} = \frac{2}{X} + \frac{2}{X^2} + \frac{2X_R^2}{X^3} \tag{4.244}$$

$$P^* = -X_R^2/X^2 + 2\ln X - 2/X + c_3 \tag{4.245}$$

$$X = 0, \; P^* \text{ is finite and, therefore, } c_3 = 0 \tag{4.246}$$

$$u^{ss} = (X_R^2/X^2) \tag{4.247}$$

$$X^2 \, u^{ss} = X_R^2 \tag{4.248}$$

Thus, $P^* = c_2$

The transient dimensionless velocity can be written as:

$$\left(\frac{1}{X^2}\frac{\partial^2(X^2u)}{\partial\tau^2}\right) + \left(\frac{1}{X^2}\frac{\partial(X^2u)}{\partial\tau} + u\right)\left(\frac{1}{X^2}\frac{\partial(X^2u)}{\partial X}\right) + \frac{1}{X^2}\frac{\partial(X^2u)}{\partial\tau} \tag{4.249}$$

$$= -\frac{1}{X^2}\frac{\partial^2(X^2u)}{\partial X^2} \tag{4.250}$$

Let $V = X^2u$:

$$\frac{\partial^2 V}{\partial\tau^2} + X^2\frac{\partial V}{\partial X}\left(V + \frac{\partial V}{\partial\tau}\right) + \frac{\partial V}{\partial\tau} = \frac{\partial^2 V}{\partial X^2} \tag{4.251}$$

The method of separation of variables can be used to solve Eq. (4.251):

$$\text{Let } V = g(X)\,\theta(\tau) \tag{4.252}$$

$$\theta'' g + X^2 \left[\theta' g + g\theta\right] g'\theta + g\theta' = g''\theta \tag{4.253}$$

Dividing by $g\theta$ throughout:

$$\frac{\theta'' + \theta'}{\theta} = \frac{g''}{g} - X^2 g'(\theta + \theta') = -\lambda_n^2 \tag{4.254}$$

$[\theta' + \theta]$ can be set to zero to obtain a separation of the time and space variables. From this constraint:

$$\theta' + \theta = c \tag{4.255}$$

Then
$$\theta''/\theta = 1 - \lambda_n^2 \tag{4.256}$$

Equation (4.256) can be used to obtain the $\theta$. The $\theta$ obtained from this constraint may not meet the $\theta' + \theta = 0$ requirement. Hence, the solution is an approximation:

$$g''/g = -\lambda_n^2 \tag{4.257}$$

$$\theta = c_1 e^{\tau\sqrt{1-\lambda_n^2}} + c_2 e^{-\tau\sqrt{1-\lambda_n^2}} \tag{4.258}$$

At steady state, the velocity profile is given by Eq. (4.247). Hence, $c_1$ can be taken as zero, as $\theta$ is not infinite at infinite time:

$$\theta = c_2 e^{-\tau\sqrt{1-\lambda_n^2}} \tag{4.259}$$

In order to solve for the constants $c_3$ and $c_4$, redefine $X$ as follows:

$$Y = X - X_R \tag{4.260}$$

Equation (4.254) then becomes

$$g''(Y)/g = -\lambda_n^2 \tag{4.261}$$

$$g = c_3 \sin(\lambda_n Y) + c_4 \cos(\lambda_n Y) \tag{4.262}$$

at $\qquad\qquad Y = 0,\, u^t = g\theta = 0 \tag{4.263}$

Hence, $\quad c_4 = 0 \tag{4.264}$

$$g = c_3 \sin(\lambda_n Y) \tag{4.265}$$

at $\qquad Y = Y_{mR},\, u^t = u - u^{ss} = 1/m^2 - 1 - 1/m^2 + 1 = 0 \tag{4.266}$

Hence, $\qquad \lambda_n = \dfrac{n\pi}{Y_{mR}} \qquad \text{for } n = 1, 2, 3.\dots \tag{4.267}$

$$= \dfrac{\sqrt{\nu\tau_{mom}}\, n\pi}{mR} \tag{4.268}$$

Thus:

$$V = \sum_1^\infty c_n e^{-\tau\sqrt{1-\lambda_n^2}} \sin(\lambda_n(X - X_R)) \tag{4.269}$$

$c_n$ can be solved using the orthogonal property and the initial condition, and shown to be $(2/n\pi)(1 - (-1)^n)$. It can be seen that when:

$$\dfrac{\sqrt{\nu\tau_{mom}}\,\pi}{mR} > 1 \tag{4.270}$$

$$\text{dor } \tau_{mom} > \dfrac{m^2 R^2}{\pi^2 \nu} \tag{4.271}$$

The solution is given after using the De Moivre's theorem and obtaining the real parts:

$$X^2 u = V = \sum_{n=1}^\infty c_n \cos\tau\sqrt{1-\lambda_n^2} \sin(\lambda_n(X - X_R)) \tag{4.272}$$

The maximum velocity is obtained when $\lambda_n X = \pi/2$. The velocity is sustained periodically. When the velocity changes in sign, the radial velocity becomes inward in direction. This occurs when the distance between the spheres is small. The energy for the oscillations is provided by the kinetic energy from the inflow of the fluid from the surface at $R$. The pressure drop at steady state can be calculated from Eq. (4.234).

### 4.5.11  Squeeze Flow between Parallel Disks

Consider the outward radial squeeze flow between two parallel circular disks (Fig. 4.15). A potential application is in a lubricant system consisting of two circular disks between which a lubricant flows radially. The flow takes place because of a pressure drop, $\Delta p$, between the inner and outer radii, $r_1$ and $r_2$. Perform the analysis, including the transient effects.

The equation of motion for $v_r$ can be written as:

$$
\begin{aligned}
\tau_{mom} & \left( \frac{1}{r} \frac{\partial^2 (rv_r)}{\partial t^2} + \frac{v_r}{r^2} \frac{\partial^2 (rv_r)}{\partial t\, \partial r} \right) + \left( \frac{\tau_{mom}}{r} \frac{\partial (rv_r)}{\partial t} + v_r \right) \left( \frac{1}{r} \frac{\partial (rv_r)}{\partial r} \right) \\
& + \tau_{mom} \left( \frac{\partial^2 p}{\partial t\, \partial r} \right) + \frac{1}{r} \frac{\partial (rv_r)}{\partial t} + \left( \tau_{mom} \frac{\partial v_\theta}{\partial t} + v_\theta \right) \left( \frac{1}{r^2} \frac{\partial (rv_r)}{\partial \theta} \right) \\
& + \frac{\partial v_r}{\partial z} \left( v_z + \tau_{mom} \frac{\partial v_z}{\partial t} \right) \\
& = -\frac{\partial p}{\partial r} + \mu \frac{\partial}{\partial r} \left( \frac{1}{r} \frac{\partial (rv_r)}{\partial r} \right) + \frac{\partial^2 v_r}{\partial z^2}
\end{aligned}
\qquad (4.273)
$$

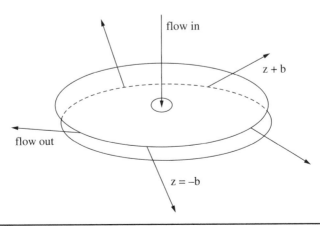

**Figure 4.15**  Radial outflow between two circular parallel disks.

Considering only the radial component of velocity from the equation of continuity:

$$\frac{1}{r}\frac{\partial(\rho r v_r)}{\partial r} = 0 \qquad (4.274)$$

Obtain the dimensionless form of Eq. (4.273) using the following substitutions:

Dimensionless velocity: $\qquad u = \dfrac{v_r}{v_{ref}}$ $\qquad\qquad$ (4.275)

Dimensionless time: $\qquad \tau = \dfrac{t}{\tau_{mom}}$ $\qquad\qquad$ (4.276)

Dimensionless distance: $\quad X = \dfrac{z}{\sqrt{\gamma \tau_{mom}}}$ $\qquad\quad$ (4.277)

Dimensionless radius: $\qquad Y = \dfrac{r}{\sqrt{\gamma \tau_{mom}}}$ $\qquad\quad$ (4.278)

Dimensionless pressure: $\quad P^* = \dfrac{p}{\rho v_{rb} v_{mom}}$ $\qquad$ (4.279)

Peclet (momentum): $\qquad Pe_{mom} = \dfrac{v_{rb}}{v_{mom}}$ $\qquad$ (4.280)

$$\left(\frac{1}{Y}\frac{\partial^2(Yu)}{\partial\tau^2}\right) + \left(\frac{\partial^2 P^*}{\partial\tau\partial Y}\right) + \frac{1}{Y}\frac{\partial(Yu)}{\partial\tau} + uPe_{mom}\left(\frac{1}{Y}\frac{\partial(Yu)}{\partial Y}\right)$$

$$= -\frac{\partial P^*}{\partial Y} + \frac{\partial^2 u}{\partial Y^2} \qquad (4.281)$$

Let $\qquad\qquad\qquad\qquad Yu = V \qquad\qquad\qquad\qquad (4.282)$

Then Eq. (4.281) becomes:

$$\frac{\partial^2 V}{\partial\tau^2} + \frac{\partial V}{\partial\tau} - \frac{\partial^2 V}{\partial X^2} = -Y\frac{\partial P^*}{\partial Y} - Y\frac{\partial^2 P^*}{\partial\tau\partial Y} \qquad (4.283)$$

Equation (4.283) is obeyed when the right-hand side and left-hand side go to zero or constant.

The solution of $(\partial^2 V/\partial\tau^2) + (\partial V/\partial\tau) - (\partial^2 V/\partial X^2) = 0$ is sought as follows:

Let $\qquad\qquad\qquad\qquad V = V^{ss} + V^t \qquad\qquad\qquad (4.284)$

The solution is assumed to consist of a steady-state and transient part. At steady state:

$$\frac{1}{Y}\frac{\partial^2 V^{ss}}{\partial X^2} = \frac{\partial P^*}{\partial Y} \tag{4.285}$$

Integrating with respect to $Y$:

$$\frac{\partial^2 V^{ss}}{\partial X^2} = \frac{\Delta P^*}{\ln\left(\dfrac{Y_2}{Y_1}\right)} \tag{4.286}$$

or

$$V^{ss} = \frac{\Delta P^*}{\ln\left(\dfrac{Y_2}{Y_1}\right)}\frac{X^2}{2} + c_1 X + c_2 \tag{4.287}$$

At

$$X = 0, \ \partial u/\partial X = 0 \tag{4.288}$$

so $c_1$ can be set to zero.

Solving for $c_2$ from the boundary condition at

$$X = \pm X_b, \ u = 0 \tag{4.289}$$

$$V^{ss} = \frac{\Delta P^*\left(X^2 - X_b^2\right)}{2\ln\left(\dfrac{Y_2}{Y_1}\right)} \tag{4.290}$$

The transient part of the solution can be obtained by solving the following equation:

$$\frac{\partial^2 V}{\partial \tau^2} + \frac{\partial V}{\partial \tau} - \frac{\partial^2 V}{\partial X^2} = 0 \tag{4.291}$$

The damping term can be removed from Eq. (4.291) by $V = W\exp(-\tau/2)$. Equation (4.291) becomes, as shown in Ref. 8:

$$\frac{\partial^2 W}{\partial X^2} = -\frac{W}{4} + \frac{\partial^2 W}{\partial \tau^2} \tag{4.292}$$

The solution to Eq. (4.292) can be obtained by the method of separation of variables:

Let

$$W = f(\tau)g(X) \tag{4.293}$$

Equation (4.292) then becomes:

$$f''/f - \tfrac{1}{4} = g''/g = -\lambda_n^2 \tag{4.294}$$

$$g = c_1 \sin(\lambda_n X) + c_2 \cos(\lambda_n X) \tag{4.295}$$

From the boundary condition at $X = 0$, that is, $g' = 0$, it can be seen that $c_1$ can be set to zero:

$$g = c_2 \cos(\lambda_n X) \qquad (4.296)$$

From the boundary condition at $X = \pm X_b$, $g = 0$:

$$\lambda_n X_b = \frac{(2n-1)\pi}{2}, \; n = 1, 2, 3, \ldots \qquad (4.297)$$

$$f = c_3 e^{\tau\sqrt{1-\lambda_n^2}} + c_4 e^{-\tau\sqrt{1-\lambda_n^2}} \qquad (4.298)$$

At infinite time, $u = 0$ and $V = uX = 0$. $W$ is $V\exp(\tau/2)$ and is zero at infinite time. Hence, $c_3$ needs to be set to zero. Thus:

$$uY = V = \sum_{1}^{\infty} e^{-\tau\sqrt{\frac{1}{4}-\lambda_n^2}} \cos\left(\frac{(2n-1)\pi z}{2b}\right) \qquad (4.299)$$

$c_n$ can be solved from the initial condition, which is when the fluid is at the reference velocity. $c_n$ can be shown to be $4(-1)^{n+1}/(2n-1)\pi$.

For large values of relaxation times, the characteristic nature of the solution changes from decaying exponential to subcritical damped oscillatory. This happens when:

$$\tau_{\text{mom}} > \frac{b^2}{\pi v} \qquad (4.300)$$

## 4.5.12   Periodic Boundary Condition

Blood flow in and out of the heart is periodic. Blood flows from the right ventricle into the pulmonary artery upon contraction of the heart muscle and from the left ventricle into the aorta. Backflow of blood is prevented by heart valves. Pressure and flow vary with time over the period of contraction and relaxation of the heart. During the phase of *systole,* the blood is pumped from the heart; during the phase of *diastole,* no blood is pumped from the heart and the ventricles fill with blood. An infinite Fourier series can be used to represent the pressure and velocity waveforms. Pressure waveform representation requires a minimum of 6 harmonics, and 10 harmonics are needed to describe the velocity waveform. The velocity profile in response to an oscillating pressure field was determined by Womersley [9].

For transient flow, based on dimensionless analysis, two dimensionless groups to describe the flow and velocity fields have been identified. The two dimensionless groups are:

1. Reynolds number ($\rho v 2R/\mu$). It is the ratio of the inertial forces to the gravity forces in the system. It is a quantitative criterion

to delineate the laminar flow and turbulent flow, as discussed in Chap. 1.

2. Strouhal number $St = L_z / t^* <v_z>$. The characteristic time $t^*$ for oscillatory flow is proportional to the reciprocal of the frequency $\omega$. The Womersley number is sometimes used in place of the Strouhal number:

$$W = R \sqrt{\frac{\rho \omega}{\mu}} \qquad (4.301)$$

The *dominant frequency* in blood flow used in Eq. (4.301) is the one that arises from the heart beat. Womersley numbers for different vessels are given in Table 4.2.

The velocity distribution can be obtained as follows. Consider the laminar flow of a Newtonian fluid in a rigid, long, circular blood vessel. This vessel is subjected to an oscillating pressure field $v_r = v_\theta = 0$. The azimuthal velocity, $v_z$, is a function of radial position only. The equation of momentum, or the Navier-Stokes equation, that governs the profile of $v_z$ as a function of space and time can be written as follows:

$$\rho \frac{\partial v_z}{\partial t} = -\frac{\partial p}{\partial z} + \mu \frac{\partial}{\partial r}(r \frac{\partial v_z}{\partial r}) + \rho g \qquad (4.302)$$

The pressure gradient imposed is periodic and:

$$-\frac{\partial p}{\partial z} = \text{Real}(Ae^{i\omega t}) = \frac{\Delta p}{L \cos(\omega t)} \qquad (4.303)$$

The method of complex velocity can be used to analyze the manifestation of the oscillating pressure gradient in the velocity flow field. Let:

$$v_z = \phi e^{i\omega t} \qquad (4.304)$$

| Vessel | Radius (mm) | W |
|---|---|---|
| Artery (Femoral) | 2.7 | 3.9 |
| Artery (Left coronary) | 4.25 | 6.15 |
| Artery (Right coronary) | 0.97 | 1.82 |
| Artery (Anterior disc) | 1.7 | 2.4 |
| Artery (Terminal) | 0.5 | 0.72 |
| Aorta (Proximal) | 15 | 21.7 |

**TABLE 4.2**  Womersley Number for Different Blood Vessels

where $\phi$ is a function of $z$ only. Substituting Eq. (4.304) into Eq. (4.302), Eq. (4.302) becomes, neglecting gravity effects:

$$\rho i \omega e^{i\omega t} \phi - \mu e^{i\omega t} \frac{\partial}{\partial r}\left(r \frac{\partial \phi}{\partial r}\right) = -A e^{i\omega t} \tag{4.305}$$

or
$$\rho i \omega \phi - \mu \left(\frac{d\phi}{dr} + r \frac{d^2\phi}{dr^2}\right) + A = 0 \tag{4.306}$$

Equation (4.306) is compared with the generalized Bessel equation. The boundary conditions are:

$$r = 0, \; \frac{\partial v_z}{\partial r} = 0 \tag{4.307}$$

$$r = R, \; v_z = 0 \tag{4.308}$$

The solution can be seen to be:

$$v_z = \frac{A}{i\omega\rho}\left(1 - \frac{J_0\left(i^{3/2}\dfrac{Wr}{R}\right)}{J_0(i^{3/2}W)}\right) e^{i\omega t} \tag{4.309}$$

In the time span of the heartbeat, the non-Newtonian finite momentum transfer effects may become significant and cannot be ignored. Consider the oscillations of a fluid in a tube of radius $R$ about a mean position where an oscillating pressure gradient is imposed on the system using the momentum transfer and relaxation equation. The pressure gradient imposed is periodic with respect to time with frequency $\omega$:

$$-\frac{\partial p}{\partial z} = a_0 \; \text{real part of } \exp(i\omega t) \tag{4.310}$$

The equation of motion for the vertical component of the velocity in the pipe, taking into account the finite speed momentum transfer effects, can be written as:

$$\rho \tau_{mom}\left(\frac{\partial^2 v_z}{\partial t^2} + v_z \frac{\partial^2 v_z}{\partial t \partial z}\right) + \left(\tau_{mom} \frac{\partial v_z}{\partial t} + v_z\right)\left(\frac{\partial v_z}{\partial z}\right)$$

$$+\tau_{mom}\left(\frac{\partial^2 p}{\partial t \partial z}\right) + \rho \frac{\partial v_z}{\partial t} = -\frac{\partial p}{\partial z} + \mu \frac{\partial}{\partial r}(r \frac{\partial v_z}{\partial r}) + \frac{\partial^2 v_z}{\partial z^2} + \rho g \tag{4.311}$$

From the equation of continuity, neglecting the radial and angular component of velocities and considering an incompressible fluid:

$$\frac{\partial(\rho v_z)}{\partial z} = 0 \tag{4.312}$$

and $v_z = \phi(r)$

Hence, after non-dimensionalizing and neglecting the gravity effects and keeping the tube horizontal, the equation of motion can be written as:

$$\tau = \frac{t}{\tau_{mom}}; \ X = \frac{r}{\sqrt{v\tau_{mom}}}; \ u = -\frac{v_z - v_{z\max}}{v_{ref}}; \ P^* = \frac{P}{\rho v_{mom} v_{ref}} \tag{4.313}$$

$$\frac{\partial^2 u}{\partial \tau^2} + \frac{\partial u}{\partial \tau} + \frac{\partial^2 P^*}{\partial X \partial \tau} = -\frac{\partial P^*}{\partial Z} + \frac{\partial}{\partial X}\left(X\frac{\partial u}{\partial X}\right) \tag{4.314}$$

The nonhomogeneity in the boundary condition can be removed by supposing that the solution can consist of two parts, one transient and the other a steady-state part:

Let
$$u = \phi(X)\exp(i\omega^*\tau) + u^{ss} \tag{4.315}$$

$$\frac{\Delta P^*}{L^*} = f(X)\exp(i\omega^*\tau)$$

where
$$\omega^* = \omega\tau_{mom} \tag{4.316}$$

The steady-state part of the solution can be written as:

$$\frac{\partial P^*}{\partial Z} = \frac{\partial}{\partial X}\left(X\frac{\partial u}{\partial X}\right) \tag{4.317}$$

Integrating both sides with respect to $Z$:

$$\frac{\Delta P^*}{L} = \frac{\partial}{\partial X}\left(X\frac{\partial u}{\partial X}\right) \tag{4.318}$$

Integrating with respect to $X$:

$$c' + X\frac{\Delta P^*}{L} = X\frac{\partial u}{\partial X} \tag{4.319}$$

$$c' = 0 \text{ as at } X = 0, \ \partial u/\partial X = 0 \tag{4.320}$$

$$c'' + X\Delta P^*/L = u^{ss} = v_{z\max}/v_{ref} \tag{4.321}$$

$$\text{At } X = X_{R'} \tag{4.322}$$

$$C'' = v_{zmax}/v_{ref} - X_R \Delta P^*/L \tag{4.323}$$

$$u^{ss} = v_{zmax}/v_{ref} - (X_R - X) f(X) \cos(\omega^* \tau) \tag{4.324}$$

The transient part of the velocity profile can be solved for as follows by assuming that the velocity also has a periodic component with the same frequency $\omega$:

$$-\omega^{*2} \phi + \phi i \omega^* + i \omega^* f' = -f' + \phi' i \omega^* - X\phi'' \omega^{*2} \tag{4.325}$$

Assuming $f$ and $\phi$ are the same:

$$X^2 \phi'' + \frac{X\phi'}{\omega^{*2}} + X\phi\left(\frac{i}{\omega^*} - 1\right) = 0 \tag{4.326}$$

Comparing Eq. (4.326) with the generalized Bessel equation:

$$a = 1/\omega^{*2}; \; b = 0; \; c = 0; \; s = \frac{1}{2}; \; d = i/\omega^* - 1; \; p = (\omega^{*2} - 1)/\omega^{*2} \tag{4.327}$$

For high dimensionless frequency, the order of the Bessel solution can be taken as 1. The solution can be presented as:

$$\phi = c' X^{1/2} I_1\left(2\sqrt{X\left(\frac{i}{\omega^*} - 1\right)}\right) + c'' X^{1/2} I_1\left(2\sqrt{X\left(\frac{i}{\omega^*} - 1\right)}\right) \tag{4.328}$$

$c''$ can be set to be zero because $\phi$ is finite at $X = 0$. Realizing that $I(ix) = J(x)$:

$$\phi = c' X^{1/2} J_1\left(2\sqrt{X\left(\frac{1}{i\omega^*} + 1\right)}\right) \tag{4.329}$$

$c'$ can be solved from the boundary condition of zero velocity at $r = R$ or $X = X_R$:

$$-\frac{v_{zmax}}{v_{rel}} = (c' X_R^{1/2} J_1\left(2\sqrt{X\left(\frac{1}{i\omega^*} + 1\right)}\right) \tag{4.330}$$

$$c' = -\frac{v_{zmax}}{v_{rel}} X_R^{1/2} J_1\left(2\sqrt{X\left(\frac{1}{i\omega^*} + 1\right)}\right) \tag{4.331}$$

$$\frac{v_{zmax} - v_z}{v_{zmax}} = \sqrt{\frac{X}{X_R}} J_1\left(2\sqrt{X\left(\frac{1}{i\omega^*} + 1\right)}\right) \cos(\omega^* \tau) \tag{4.332}$$

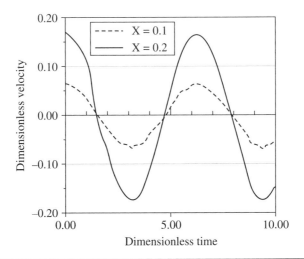

**Figure 4.16**  Dimensionless velocity responses to an oscillating pressure gradient imposed on tube flow by momentum transfer and relaxation equation.

At high frequency:

$$\frac{v_{z\max} - v_z}{v_{z\max}} = \sqrt{\frac{X}{X_R}} \frac{J_1(2\sqrt{X})}{J_1(2\sqrt{X_R})} \cos(\omega^*\tau) \qquad (4.333)$$

Equation (4.333) is shown in Fig. 4.16.

## 4.6  Friction Factors

Many engineering flow problems fall under the categories of flow in circular pipes and flow past spherical objects. The "Peace Pipeline Project" is designed to construct a pipeline to meter out oil from the Gulf region, such as from Iran to North Delhi, India, through a host of countries, such as Turkmenistan, Afghanistan, Pakistan, and India. The optimal number of pipes needed is derived in Sharma [10]. Examples of flow in pipes are piping oil in pipes, the flow of water in channels, the extrusion of polymer through a die, the flow of fluid through a filter, pulsatile flow from the lungs to the nostrils, blood flow through the capillaries, and flow during reverse osmosis in desalination. Examples of flow around submerged objects are the flow of air around the airplane wing, motion of fluid around particles, fluid flow in fluidized bed combustors, reactors, and heat exchangers circulating fluidized beds. In such problems, there is a relationship between the pressure drop and the volume rate of flow. In flow past submerged objects, the drag force is important. Sometimes experimental data are utilized to obtain correlations for the drag coefficients for the appropriate geometry and flow situation. In this section, the notion of friction factor is introduced and charts can be found in chemical engineering, bioengineering, mechanical engineering, and civil engineering handbooks. Often times, friction factors are defined for steady-state scenarios. In this section, friction factor for transient flow driven by damped wave momentum transfer and relaxation is introduced.

A force is exerted by the flowing fluid on the solid surface it is in contact with. This has been calculated at steady-state for incompressible flow under various conditions. This force consists of two parts: one that would act even when fluid is stationary and additional force associated with the kinetic behavior of the fluid:

$$F_k = AK_e f \tag{4.334}$$

$A$ is the characteristic area, $K$ is a characteristic energy per unit volume, and a dimensionless quantity, $f$, is called the friction factor. For circular tubes of radius $R$ and length $L$, $f$ is defined as:

$$F_k = (2\pi RL)(1/2\rho <v>^2)f \tag{4.335}$$

For a fully developed pipe flow, a force balance on the fluid between 0 and $L$ in the direction of flow yields:

$$F_k = \pi R^2 \, (\Delta p + \rho g \, (h_0 - h_L) = \pi R^2 \, (\Delta P) \tag{4.336}$$

Comparing Eqs. (4.326) and (4.327):

$$f = \frac{1}{4}\frac{D}{L}\frac{\Delta P}{\frac{1}{2}\rho <v>^2} \tag{4.337}$$

This is usually referred to as the Fanning friction factor. For flows around submerged objects, the characteristic area $A$ is usually taken to be the area obtained by projecting the solid onto a plane perpendicular to the velocity of approach of the fluid. Then $k$ is taken to be $1/2\rho \, v_\infty^2$, where $v_\infty$ is the approach velocity of the fluid at a large distance from the object. Thus, for flow past a sphere at steady state:

$$F_k = (\pi R^2)(1/2\rho \, v_\infty^2)f \tag{4.338}$$

The resultant force of gravity and buoyancy driving the motion of the sphere is given by:

$$F_s = \frac{\pi D^3}{6}\frac{(\rho_s - \rho)}{\rho} \tag{4.339}$$

Comparing Eqs. (4.329) and (4.330), for the net force to be zero, as it is at the terminal settling velocity of the fluid:

$$f = \frac{4gD(\rho_s - \rho)}{3\rho v_\infty^2} \tag{4.340}$$

The friction factor in Eq. (4.331) is referred to as drag coefficient, and is represented by $C_D$. For a long, smooth, horizontal pipe of length $L$ at steady state for fluid with constant $\rho$ and $\mu$, the force exerted by the fluid on the inner pipe wall for either laminar or turbulent flow is:

$$F_k = \int_0^L \int_0^{2\pi} -\mu \frac{\partial v_z}{\partial r}\bigg|_{r=R} R\, d\theta\, dz \tag{4.341}$$

Comparing Eqs. (4.340) and (4.341):

$$f = \frac{\int_0^L \int_0^{2\pi} -\mu \frac{\partial v_z}{\partial r}\bigg|_{r=R} d\theta\, dz}{\pi R L \rho <v>^2} \tag{4.342}$$

Thus, the friction factor is a function of the Reynold's number and the $L/D$ ratio.

For laminar, steady, incompressible flow in circular pipe, the Hagen-Poiseuille flow distribution is given by:

$$v^{ss} = \frac{\Lambda P R^2}{4\mu L}\left(1 - \left(\frac{r}{R}\right)^2\right) \tag{4.343}$$

Substituting Eq. (4.343) in Eq. (4.342), $f$ can be calculated as:

$$f = 16/\mathrm{Re} \tag{4.344}$$

where $\mathrm{Re} = \rho <v_z> 2R/\mu$.

Equation (4.335) has been found to be valid for Reynold's numbers less than 2100. For turbulent flow in a smooth, circular tube, the friction factor is given by the Blasius formula:

$$f = \frac{0.0791}{\mathrm{Re}^{1/4}} \tag{4.345}$$

The Blasius formula is valid for a Reynold's number less than $10^5$. The friction factor during flow between two plates moving in opposite directions when governed by the momentum transfer and relaxation equation is given by:

for 
$$\tau_{mom} > \frac{a^2}{4\nu\pi^2} \tag{4.346}$$

$$u^t = \sum_1^\infty c_n e^{-\frac{\tau}{2}} \cos\left(\tau\sqrt{\lambda_n^2 - \frac{1}{4}}\right) \sin(\lambda_n Z) \qquad (4.347)$$

$$\partial u^t / \partial Z\big|_b = \sum_1^\infty \lambda_n (-1)^n c_n e^{-\frac{\tau}{2}} \cos\left(\tau\sqrt{\lambda_n^2 - \frac{1}{4}}\right) \qquad (4.348)$$

The definition used for the friction factor at steady state is retained and:

$$f = -v_{max} / \mathrm{sqrt}(\gamma\tau_{mom})\, \mu\partial u / \partial Z / (\rho/2 <v>^2) \qquad (4.349)$$

$$<u^t> = \sum_1^\infty -(1-(-1)^n)^2 \exp(-\tau/2) \cos(\mathrm{sqrt}((\lambda_n^2 - 1/4)\tau) / n^2\pi^2 \qquad (4.350)$$

Combining Eqs. (4.347), (4.348), and (4.349):

$$f = -v_{max} / \mathrm{sqrt}(\gamma\tau_{mom})\, \mu\partial u / \partial Z / (\rho/2 <v>^2) \qquad (4.351)$$

Defining $\mathrm{Re} = (\rho <v> b/\mu)$

$$f = \frac{2}{\mathrm{Re}} \left| \frac{\sum_1^\infty 2(1-(-1)^n)(-1)^{n+1} e^{-\frac{\tau}{2}} \cos\left(\tau\sqrt{\lambda_n^2 - \frac{1}{4}}\right)}{\sum_1^\infty -(1-(-1)^n)^2 e^{-\frac{\tau}{2}} \dfrac{\cos\left(\tau\sqrt{\lambda_n^2 - \frac{1}{4}}\right)}{n^2\pi^2}} \right| \qquad (4.352)$$

For example, taking the first few terms in the infinite series in Eq. (4.352):

$$f = \frac{2}{\mathrm{Re}} \left| \frac{\cos(\tau\sqrt{\lambda_1^2 - \frac{1}{4}}) + \cos\left(\tau\sqrt{\lambda_3^2 - \frac{1}{4}}\right)}{\dfrac{\cos\left(\tau\sqrt{\lambda_1^2 - \frac{1}{4}}\right)}{\pi^2} + \dfrac{1}{9\pi^2}\cos\left(\tau\sqrt{\lambda_3^2 - \frac{1}{4}}\right)} \right| \qquad (4.353)$$

For the flow around the sphere, the friction factor can be shown to be:

$$f = \frac{24}{\mathrm{Re}} \qquad (4.354)$$

This can be derived from Stokes' law. This has been found valid for Re < 0.1 in the creeping flow regime.

## 4.7  Other Constitutive Relations

Notable in the literature among the equations used as a constitutive relation to describe blood rheology is Casson's equation. The shear stress and shear rate dependence for a Casson fluid is show in Fig. 4.17.

As the shear rate increases, the apparent viscosity decreases in Casson fluids. This could mean that the particulate aggregates become smaller, and at some point the fluid reverts to Newtonian behavior. In blood, the aggregates are formed by RBCs. At low shear rates, the apparent viscosity of Casson fluid is high, indicating aggregates of RBCs. Casson's equation may be written as:

$$\sqrt{\tau} = \sqrt{\tau_0} + \mu\sqrt{\frac{\partial v_x}{\partial y}} \tag{4.355}$$

The pressure drop and flow data from blood in tubes can be used in a log graph to obtain the transition point where the fluid reverts from Casson to Newtonian behavior. The momentum balance equation and the Casson equation can be combined to obtain the governing equation for fluid flow in a circular pipe. This can be solved and the velocity profile obtained as:

$$v_z = \frac{R\tau_w}{2\mu^2}\left\{\left(1-\left(\frac{r}{R}\right)^2\right) - \frac{8}{3}\sqrt{\frac{\tau_0}{\tau_w}}\left(1-\left(\frac{r}{R}\right)^{3/2}\right) + 2\left(\frac{\tau_0}{\tau_w}\right)\left(1-\frac{r}{R}\right)\right\} \tag{4.356}$$

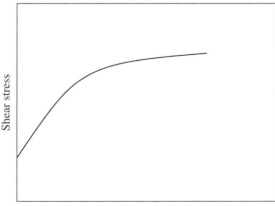

**Figure 4.17**  Casson fluid shear stress-shear rate relation.

When the shear stress equals the yield stress, the corresponding radius $r^*$ can be denoted as the critical radius.

## 4.8  Bernoulli Equation for Blood Pumped by the Heart

The generalized mechanical energy balance equation can be written to account for the pumping work performed by the heart in the human anatomy to cause blood to flow through the arteries. This equation is also referred to as the Bernoulli equation for work done by the heart. The changes in fluid pressure, potential energy, and kinetic energy during the flow of blood can be accounted for by this equation. The flow is assumed to be at a steady state. The density of the fluid is also assumed to be constant, or the flow is said to be incompressible. The Bernoulli equation, which was introduced in Chap. 1, can be written between two locations in the bloodstream in human anatomy, 1 and 2, as follows:

$$\frac{p_1}{\rho} + gz_1 + \frac{v_1^2}{2} + \eta W_p = \frac{p_2}{\rho} + gz_2 + \frac{v_2^2}{2} + h_{\text{friction}} \qquad (4.357)$$

Each term in the equation has units of energy per unit mass and represents the pressure head, potential energy, and kinetic energy at locations 1 and 2 in the blood flow stream. The work done by the heart is given by $W_p$, and $\eta$ is the efficiency of the pump work. The friction losses can also be accounted for by $h_{\text{friction}}$.

## Summary

Blood is a colloidal dispersion. It consists of RBCs (90 percent), WBCs (1 percent), platelets (5 percent). The RBC volume fraction is called the hematocrit and typically varies between 10 and 50 wt %. The Fahraeus-Lindqvist effect is the change observed in the viscosity of blood during flow with a change in the diameter of the circular conduits. The marginal zone theory was developed to explain the Fahraeus-Lindqvist effect. The discharge rate of blood flow as a function of plasma layer thickness, viscosity of the core layer, viscosity of the plasma layer, radius of the conduit, and pressure drop was derived from first principles.

The blood flow in a dialysis machine is through rectangular conduits. The discharge rate of blood through a narrow slit $2B$ distance apart as a function of the core layer, plasma layer viscosities, and pressure drop was derived for Newtonian fluids. Expression for plasma layer thickness as a function of core layer hematocrit, $H_c$, and feed hematocrit, $H_F$, can be developed. The mathematical expression was derived by Charm and Kurland [4] to capture the dependence of the viscosity of the blood at high shear rates on hematocrit and temperature. The solution for plasma layer thickness $\delta$ is implicit and requires solving two simultaneous nonlinear equations and two unknowns. An explicit method for obtaining plasma layer thickness was developed by Sharma [5]. The temperature parameter used to describe the variation of blood viscosity is expressed as a linear function of hematocrit for core and plasma layers. This leads to the simplification of the system of two simultaneous nonlinear equations into a single quadratic expression for the square of the plasma layer thickness.

A list of 46 viscoplastic fluids was given in Table 4.1. The yield stress concept is an idealization and has not been measured directly. It has been found by extrapolation from high shear data. Other constitutive relations that may be applicable to describe blood flow, such as Casson fluid, were discussed briefly. An expression to represent the shear stress versus shear rate behavior that may be applicable to describe blood flow is the generalized Newton's law of viscosity. This expression can be used to account for the finite speed of propagation of momentum. This is significant in transient applications. The damped wave momentum transfer and relaxation equation can be derived from the kinetic theory of gases or by analogy to molecular diffusion from the Stokes-Einstein equation, taking into account the acceleration of the molecules; or by analogy to molecular conduction from the free electron theory, taking into account the acceleration of the electrons due to a collision with an obstacle.

The transient velocity profile that arises when a flat plate is suddenly subject to velocity $V$ is derived using the damped wave momentum transfer and relaxation equation for a semi-infinite medium of fluid. A novel procedure called the method of relativistic transformation was used to obtain closed-form analytical solutions for the velocity profile for a initial, stagnant, semi-infinite fluid. Four regimes of solution for velocity profile can be recognized: (a) an inertial regime; (b) a regime at long times at a given location, $\tau > X$, characterized by a modified Bessel composite function in space and time of the zeroth order and first kind; (c) a regime at shorter times at a given location, $\tau_{inertia} < \tau < X$, characterized by a Bessel composite function in space and time of the zeroth order and first kind; and (d) a regime at the wavefront, $\tau = X$. Some interesting features can be seen from the model solutions. These include space-time symmetry, point of inflection in the velocity profile, zero curvature at $X = 0$, and subcritical damped oscillations under certain conditions. Mathematical expressions for inertial lag time and penetration distance were derived.

The transient velocity profile during vertical flow subject to a Darcy pressure gradient was developed using damped wave momentum transfer and relaxation equation. The permeability number is an important dimensionless group that governs the flow characteristics. For permeability numbers, $Pb > \frac{1}{4}$, the solution changed in character from monotonic decay to cosinuous oscillatory. These oscillations were found to be subcritical and damped.

The transient velocity profile between two plates moving in opposite directions at velocity $V$ and $-V$ separated by distance $2a$ was studied using the damped wave momentum transfer and relaxation equation. The method of separation of variables was used to obtain closed-form analytical solutions expressed as infinite Fourier series. The solution was found to be bifurcated. For certain fluids with large momentum relaxation times, the velocity is expected to transition from monotonic exponential decay to cosinuous damped oscillatory. Subcritical damped oscillations can be expected under certain conditions. The solution was found to be in accordance with Clausius inequality. The final condition in time at steady state was used, leading to dropping a growing exponential term in the time domain.

The transient velocity profile of fluid under vertical flow between two plates moving in opposite directions at velocity $V$ and $-V$ separated by distance $2a$ was studied using the damped wave momentum transfer and relaxation equation. For large permeability numbers, the governing equation for wave velocity was seen to be a Bessel differential equation. The velocity profile of fluid under vertical flow between two plates considering the viscous, gravitational, and permeability effects was derived. The profile was characterized by a Bessel composite function in space and time of zeroth order and first kind. The expression was valid for $Pb > \frac{1}{4}$. It is subcritical damped oscillatory.

The transient velocity profile of a viscous fluid in a circular conduit using damped wave momentum transfer and relaxation equation was derived using the method of separation of variables. An infinite Fourier series expression for velocity profile was developed. At large relaxation times, subcritical damped oscillations in velocity can be expected.

The oscillations in a U-tube manometer were modeled using the damped wave momentum transfer and relaxation equation. The third-order ODE was analyzed. The conditions where oscillations can be expected were also derived. This is when the oscillation number is $> -2/27$. A mathematical expression for displacement of fluid that is damped oscillatory was obtained.

Transient tangential flow in an infinite medium, transient flow past a sphere, flow between concentric spheres, and radial flow between parallel disks were studied using the damped wave momentum transfer and relaxation equation. The velocity profile under periodic boundary conditions also was derived. The use of friction factors under transient flow was reviewed. The Bernoulli equation was written, taking into account the work done by the heart causing blood to flow.

# References

[1]   R. F. Haynes, "Physical basis of the dependence of blood viscosity on tube radius," *Am. J. Physiol.*, vol. 198 (1960), 1193–1200.

[2]   R. B. Bird, W. E. Stewart, and E. N. Lightfoot, *Transport Phenomena,* 2nd ed., New York: John Wiley & Sons, 2007.

[3]   R. L. Fournier, *Basic Transport Phenomena in Biomedical Engineering,* Philadelphia, PA: Taylor & Francis, 1999.

[4]   S. E. Charm and G. S. Kurland, *Blood Flow and Microcirculation,* New York: John Wiley & Sons, 1974.

[5]   K. R. Sharma, "Layered annular flow of blood in capillaries and explicit relation for plasma-layered thickness," in *32nd Northeast Bioengineering Conference,* Easton, PA, April 2006.

[6]   E. C. Bingham, *Fluidity and Plasticity,* New York: McGraw-Hill, 1922, Chapter 8.

[7]   R. B. Bird, G. C. Dai, and B. J. Yarusso, "The rheology and flow of viscoplastic materials," *Reviews in Chemical Engineering* (1983), I(1).

[8]   K. R. Sharma, *Damped Wave Transport and Relaxation,* Amsterdam, Netherlands: Elsevier, 2005.

[9]   J. R. Womersley, "Method for the calculation of velocity, rate of flow and viscous drag in arteries when the pressure gradient is known," *J. Physiol.* (1955), 127, 553–563.

[10]  K. R. Sharma, *Fluid Mechanics and Machinery,* Chennai, India: Anuradha Publishers, 2003.

# Exercises

## Review Questions

**1.0**   What does hematocrit mean?

**2.0**   What are the differences between RBCs and WBCs?

**3.0**   What happens when the diameter of the blood-flowing capillary is decreased?

**4.0**   What are the key parameters of the marginal zone theory?

**5.0**   What would be the key parameters should the assumption of Newtonian fluid be replaced with non-Newtonian fluid in the marginal zone theory?

**6.0**   Is there allowance for boundary layer formation in the marginal zone theory?

**7.0**   What is the difference between plasma layer thickness and boundary layer thickness?

**8.0**   Will the plasma layer thickness change with a change in temperature?

**9.0**   What are the advantages of using the explicit method for calculation of plasma layer thickness?

**10.0**   Can a viscoplastic fluid that is homogeneous and made of one component be found?

**11.0**   Why is yield stress an idealization?

**12.0** What is the physical significance of the relaxation time of momentum?

**13.0** How many regimes were found in the transient velocity profile governed by the damped wave momentum transfer and relaxation equation?

**14.0** What is penetration distance?

**15.0** What is momentum inertial lag time?

**16.0** What is a permeability number? What happens at $Pb > 0.25$ during vertical flow subject to a Darcy pressure gradient governed by the damped wave momentum transfer and relaxation equation?

**17.0** What are the differences in a transient velocity profile for a fluid moving between two vertical plates at constant velocity when the plates are moving in the same direction and when the plates are moving in opposite directions? The governing equation is the damped wave momentum transfer and relaxation equation.

**18.0** What are the differences in the transient velocity profile during vertical flow when the fluid is governed by the parabolic momentum transfer equation and the hyperbolic momentum transfer equation?

**19.0** What are subcritical damped oscillations in velocity?

**20.0** What is the significance of the oscillation number during oscillations of a fluid in a U-tube manometer?

**21.0** Discuss the transient velocity profile during tangential flow past a sphere governed by the damped wave momentum transfer and relaxation equation.

**22.0** Discuss the transient velocity profile during flow between two concentric spheres governed by the damped wave momentum transfer and relaxation equation.

**23.0** Discuss the transient velocity profile during radial flow between parallel disks governed by the damped wave momentum transfer and relaxation equation.

**24.0** Discuss the transient velocity profile when the periodic boundary condition is used when the fluid is governed by the damped wave momentum transfer and relaxation equation.

**25.0** What is the term-by-term significance of the Bernoulli equation for the pump work done by the heart?

**26.0** Why are friction factors important in the flow of blood in human anatomy?

**27.0** What is meant by rheology of blood flow?

**28.0** How will you measure the relaxation time of momentum of blood?

**29.0**  Will there be two relaxation times of momentum of blood, one for the core layer and one for the plasma layer?

**30.0**  What are the issues in a system described by a third-order ODE?

## Problems

**31.0**  A flat plate is moved suddenly between two fluids of kinematic viscosity, $\gamma_1$ and $\gamma_2$, and relaxation times $\tau_{mom1}$ and $\tau_{mom2}$, respectively. The fluid at the two surfaces that binds them is stationary. Compute the force exerted on the plate, and obtain the distance of the plate from the wall for balance.

Show that the governing equations for the two fluids may be written as:

$$\partial^2 u/\partial\tau^2 + \partial u/\partial\tau = \partial^2 u/\partial X^2$$

$$\kappa\partial^2 u/\partial\tau^2 + \partial u/\partial\tau = \beta\partial^2 u/\partial X^2$$

where $u = v_x/V$
$\tau = t/\tau_{mom}$
$X = z/(\gamma\tau_{mom1})$
$\beta = \gamma_2/\gamma_1$
$\kappa = \tau_{mom2}/\tau_{mom1}$

The nonhomogeneity in the boundary condition can be removed by superposing the steady-state and transient solutions:

Let the distance of the plate from the top surface be $b$ and the distance from the bottom surface be $a$:

Let $u = u^{ss} + u^t$

Show that at steady state:

$$u^{ss} = 1 - X/X_b \text{ (fluid on top of the plate)}$$

$$u^{ss} = X/X_a + 1 \text{ (fluid below the plate)}$$

After removing the damping terms, show that the governing equations for the transient component of the fluid atop the plate and for the fluid below the plate can be written after a $u = w\exp(-\tau/2)$ substitution:

$$\partial^2 w/\partial\tau^2 - w/4 = \partial^2 w/\partial X^2$$

The $u = w\exp(-\tau/2\kappa)$ substitution for the governing equation for the fluid below the plate:

$$\partial^2 w/\partial\tau^2 - w/4\kappa^2 = (\beta/\kappa)\partial^2 w/\partial X^2$$

The space and time conditions are:

$$X = 0, u^t = 0$$

$X = X_a, u^t = 0$; $X = X_b, u^t = 0, \tau = 0, u = 0$. Use the method of separation of variables to solve for the transient component of the dimensionless velocity. Show that $u^t = \sum_1^\infty c_n \exp(-\tau/2)\exp(-\tau\mathrm{sqrt}(1/4 - \lambda_n^2)) \sin(\lambda_n X)$ where $\lambda_n = n\pi/X_b$, $n = 1, 2, 3, \ldots$ and where $c_n = 2(1 - (-1)^n)/n\pi$. Furthermore, for a small distance between the plate and the bounded surface, show that $u^t = \sum_1^\infty c_n \exp(-\tau/2)\cos(\tau\text{sqrt}(\lambda_n^2 - 1/4)\times \sin(\lambda_n X)$.

In a similar fashion, for the fluid below the plate, show that $u^t = \sum_1^\infty c_n \times \exp(-\tau/2)\exp(-\tau\text{sqrt}(1/4\kappa^2 - \zeta_n^2))\sin((\kappa/\beta)^{1/2}\zeta_n X)$ where $\zeta_n = n\pi/X_a$, $n = 1, 2, 3\ldots$ and where $c_n = 2(1 - (-1)^n)/n\pi$.

Derive the force on the plate during transient flow and equate the contributions of the fluid on top of the plate and from the bottom of the plate. Make use of the integrating factor if necessary. Is the force oscillatory for small separation distances between the late and the bounded surface?

**32.0**   Microlayer composites were coextruded in up to 3,713 alternating layers. The interdiffusion of two miscible layers of polycarbonate and copolyester was studied at temperatures from 200 to 230°C at the polymer laboratory at Case Western Reserve University. Extend the analysis in Problem 1.0 to $n$ layers.

**33.0**   From the analysis in Exercise 32.0, can the layer rearrangement be predicted? What secondary flows can be predicted?

**34.0**   Consider $n$ layers of $n$ different fluids on top of each other in a vertical container. Extend Torricelli's theorem to obtain the efflux velocity of the fluid from the bottom of the container. Derive the azimuthal velocity as a function of space and time.

**35.0**   The flow of blood in a circular pipe for the plasma layer and core layer was studied, and the discharge rate as a function of the intrinsic viscosity of the plasma and core layers and the thickness of the plasma layer radius of the capillary was derived. Derive the discharge rates as a function of pressure drop and other parameters of flow for the slit flow limit.

Let the width of the flat plates be $2W$, the area of the cross-section be $A$, and the thickness of the core layer be $2\delta_c$. The boundary conditions are:

$$x = 0, \partial v_x^c/\partial z = 0$$
$$x = W, v_x^p = 0$$
$$x = \delta_c, \tau_{zx}^c = \tau_{zx}^c \,;\, v_x^p = v_x^c$$

Show that at steady state:

$$\partial p/\partial x = \mu_c(\partial^2 v_x/\partial z^2) = c_1 = -(p_0 - p_L)/L$$
$$v_x^c = (\Delta p/2\mu_c L)(W^2 - z^2) - \Delta p\delta_c^2/2L(1/\mu_p - 1/\mu_c)$$
$$v_x^p = (\Delta p/2\mu_p L)(W^2 - z^2)$$

Show that the average flow rates in the core and plasma layers can be calculated as:

$$<v_x^c> = 2/\delta_c^2 \int_0^{\delta_c} z v_x^c \, dz = (\Delta p/\mu_c L)(W^2/2 - \delta_c^2/4) - \Delta p\delta_c^2/L(1/\mu_p - 1/\mu_c)$$

$$Q^c = (\Delta pW\delta_c^3/\mu_c L)\,(2(W/\delta_c)^2 - 1) - 4\Delta pW(\delta_c^3)/L(1/\mu_p - 1/\mu_c)$$
$$Q^p = 4W(\Delta p/\mu_p L)(W^2 - \delta_c^2)^2/(W + \delta_c)$$

**36.0**   For the geometry and space conditions shown in Problem 5.0, derive the transient pressure and velocity distributions as a function of $z$ and time. What is the critical thickness prior to the onset of subcritical oscillations in the flow rate?

**37.0**   Repeat the analysis in Problem 1.0 for a vertical plate. Study the response to an oscillating velocity introduced by the vertical plate.

**38.0**   Consider the coaxial flow between two cylinders. The free stream velocity approaching the coaxial cylinders is constant at $V$. Develop the transient velocity profile in the annulus using the damped wave momentum transfer and relaxation equation. What is the average velocity? Where is the location of maximum velocity? Obtain the pressure drop versus discharge rate relationships at steady state and transient state. Use the method of separation of variables, and let the inner and outer radii be $R$ and $\kappa R$, respectively, and develop the conditions where subcritical oscillations in the velocity can be observed. What is the force exerted by the fluid on the surface? Defining the friction factor $f$ as:

$$F_k = f(1/2\rho <v>^2)A_k$$

Obtain the friction factor at a steady state and transient state for large pipes and at a transient state for small pipes. $A_k$ may be taken as the wetted surface area, which is $2\pi LR\,(1 + \kappa)$.

**39.0**   *Bulk flow effect.* Consider a one-dimensional flow due to a constant pressure drop along with bulk flow. Show that after neglecting the viscous effects, the governing equation can be written as:

$$\tau_{mom}\,(\partial^2 v_x/\partial t^2 + v_x\partial^2 v_x/\partial t\,\partial x) + (\tau_{mom}\,\partial v_x/\partial t + v_x)(\partial v_x/\partial x) + \partial v_x/\partial t = \Delta p/\rho L$$

At steady state:

$$v_x^2/2 = c + x(\Delta p/L)$$

Let $v_x = v_x^s + v_x^t$

$$\tau = t/\tau_{mom};\ u = v_x/V_{ref}\ ;\ X = x/V\tau_{mom};\ P^* = \Delta p/(\rho V_{ref}\,L/\tau_{mom})$$

Show that the transient portion of the solution will obey:

$$(\partial^2 u/\partial\tau^2 + u\partial^2 u/\partial\tau\,\partial X) + (\partial u/\partial\tau + u)(\partial u/\partial X) + \partial u/\partial\tau = 0$$

Let $u = V(\tau)g(X)$

$$V''g + Vg\,V'g' + g'V(V'g + Vg) + V'g = 0$$
$$V''/V + V'/V\,/(2V' + V) = -g' = c^2$$
$$g(x) = -c^2x + d$$
$$V'' + V'(1 - 2c^2V) - V^2c^2 = 0$$

Seek a solution for $V$ using the Frobenius method.

**40.0** In the chapter, the radial flow between two concentric spheres of an incompressible, isothermal liquid was derived. The transient velocity distribution is examined using the damped wave momentum transfer and relaxation equation. Let the radii of the two spheres be $R$ and $mR$, respectively. The governing equation for the radial component of the velocity can be written. The velocity is assumed to consist of steady-state and transient parts. From the steady-state part of the solution to the velocity profile obtain the friction factor as a function of the Reynold's number for laminar flow. The transient velocity profile is derived by the method of separation of variables. Obtain the friction factor for transient flow for large spheres and small spheres.

**41.0** A conical thrust bearing idealized as a cone of vertex angle $2\theta$ and maximum cone radius $R$ rests and revolves over a uniform fluid layer of thickness $\delta$ at a constant angular velocity $\omega$. Derive the transient and steady-state velocity profileand obtain expressions for the torque required and the rates of heat dissipation in the bearing at steady state and transient state using the damped wave momentum transfer and relaxation equation.

**42.0** In the falling ball viscometer, the shear rate is given by the terminal settling velocity of the sphere over the radius of the falling ball and the shear stress by $2/9gR(\rho_s - \rho)$. Consider the acceleration regime of the settling sphere. Develop the friction factor and Reynold's number relationship during acceleration. Show how the falling ball viscometer can be used to obtain the viscosity and relaxation time information from experiments.

**43.0** Examine the rotating cylinder viscometer in transient and steady-state conditions. The radii of the cylinders are 3.2 cm and 3 cm, and the outer cylinder is suddenly rotated at 180 rpm. For a liquid filled in the annular space to a depth of 8 cm, the torque produced on the inner cylinder is $10^{-4}$ Nm at steady state. Use the damped wave momentum transfer and relaxation equation, and obtain the spatiotemporal velocity profile. Calculate the viscosity of the liquid. Develop a procedure to obtain the relaxation time of the liquid using the transient torque data.

**44.0** There is interest in a "Peace Pipeline" to bring gas from Iran to North Delhi via different countries such as Afghanistan, Pakistan, and possibly Turkmenistan. Prepare a preliminary estimate of the pipe size required for a transcontinental pipeline between the Gulf region and North Delhi. It has to handle 1000 std $m^3$/hr of natural gas at an average pressure drop of 3 atm abs at an average temperature of 25°C. What is the maximum force exerted by the pipe? Is this during transience or steady state? What is the ideal friction factor relation to be used? What is the optimal number of pipes to minimize the total cost to achieve the objectives of the task? Using the damped wave momentum transfer and relaxation time equation, obtain the time it takes to reach steady state and the conditions needed to avoid subcritical damped oscillations.

**45.0** Consider a hot circular pipe through which a fluid is flowing in laminar flow. Obtain the transient velocity profile using the damped wave momentum transfer and relaxation equation. Obtaining an average velocity, use the governing equation for heat transfer and plug the derived expression for the velocity in the azimuthal direction.

$$\tau_r(\partial^2 T/\partial t^2 + <v_{z>}\partial^2 T/\partial t\,\partial z + \partial T/\partial t = \alpha(1/r\,\partial/\partial r\,(r\partial T/\partial r)$$

Obtain the transient temperature profile using the damped wave heat conduction and relaxation equation. Obtain the hydrodynamic boundary layer thickness to the thermal boundary layer thickness. Discuss the implications of Prandtl number in transience.

**46.0**   In Ref. 8, a general substitution was used to reduce the hyperbolic PDE in one space dimension into a parabolic PDE. Consider all three space dimensions, and seek a suitable general substitution to reduce the hyperbolic PDE in three space and time dimensions into a parabolic PDE.

**47.0**   Obtain the pressure and velocity distribution as a function of $z$ and $t$ in a hemispherical cup using the extended Euler equation in one dimension and two dimensions, respectively. What is different in the predictions of the efflux time, velocity, and pressure profile?

$$\text{Volume of a partially filled sphere} = \pi/6h_1\,(3r_2^2 + h_1^2)$$

**48.0**   Develop the friction factor for a bubble moving through a liquid. Obtain the transient and steady-state relationships. What assumptions are necessary?

**49.0**   Consider an elutriating bed of particles. Write the governing equation for $v_z$. Obtain the pressure drop versus flow rate in a circular pipe for materials with a positive permeability coefficient. Obtain the conditions where the velocity will exhibit subcritical damped oscillations using the extended Euler equation, making allowance for Darcy's law with a positive permittivity. In an elutriating bed, as the superficial velocity increases, the pressure drop decreases. This is in contrast to Darcy's law for packed beds when the pressure drop is increased for increased flow rates. Use the damped wave momentum transfer and relaxation equation.

**50.0**   *Intravenous infusion.* Gravity flow is used for a patient infusion system. The fluid is allowed to flow out an IV bag by gravity flow. A 400-mL IV bag containing an aqueous solution is connected to a vein in the forearm of a patient. Pressure at the veins is 1 atm. The IV bag is placed on a pedestal such that the entrance to the tube leaving the IV bag is 1.8 meters above the vein into which the IV fluid enters. The length of the IV bag is 26 cm. The IV is fed through an 16"-gauge tube, and the total length of the tube is 3 m. Compute the flow rate of the IV fluid. Estimate the time needed to completely deplete the contents of the bag.

**51.0**   The cardiac output in human anatomy is about 5 L/min. Blood enters the right side of the heart at a pressure of 1 atm. It flows via the pulmonary arteries to the lungs at a mean pressure of 1.0144 atm. Blood returns to the left side of the heart through the pulmonary veins at a mean pressure of 1.0105 atm. The blood is then ejected from the heart through the aorta at a mean pressure of atm. Apply the Bernoulli equation and estimate the power or rate of work done by the heart.

**52.0**    One type of a compact mass exchanger used for detoxification of blood uses tubes whose cross-sectional area is an equilateral triangle. Each of the sides of the triangle is $H$. When blood is forced to flow through these mass-exchanger elements, it can be expected that the core layer and plasma layer form according to the Fahraeus-Lindqvist effect. Calculate the total discharge rate of the blood as a function of the viscosity of the core layer, viscosity of the plasma layer, plasma layer thickness, length of the tube, pressure drop, and any other parameters needed.

**53.0**    Repeat Exercise 52.0 for a trapezoidal cross-section of width $a$ and $b$ separated by $H$.

**54.0**    Calculate the total discharge rate of the blood in a circular tube with radius $R$ as a function of viscosity of core layer, viscosity of plasma layer, plasma layer thickness, length of the tube, pressure drop, and any other parameter that is needed when the fluid is said to be non-Newtonian. For such fluids:

$$\tau_{rz} = -\mu \left( \frac{\partial v_z}{\partial r} \right)^{n-1}$$

**55.0**    Repeat Exercise 54.0 for flow through a rectangular narrow slit $2B$ apart and width $W$.

**56.0**    The marginal zone theory and the discharge rate of blood as a function of the applied pressure drop, length of the circular conduit, radius of the circular conduit, viscosity of the core and plasma layer, and plasma layer thickness was derived in Sec. 4.1. This was at steady state. In Sec. 4.5.6, the transient flow in a circular conduit was studied using the damped wave momentum transfer and relaxation equation. Now study the flow of blood in the core and plasma layers in transit using the damped wave momentum transfer and relaxation equation. What are the interesting features of the solution to the transient velocity profile? Under what conditions of the relaxation time of momentum and plasma layer thickness can subcritical damped oscillations in velocity be expected? Under these circumstances, what will happen to the plasma layer formation?

**57.0**    Repeat Exercise 56.0 for rectangular slit $2B$ apart and width $W$.

**58.0**    Repeat Exercise 56.0 for a triangular cross-section, as described in Exercise 56.0.

**59.0**    The cardiac output in human anatomy is about 5 lit/min. What would be the fluid velocity in the human arm, human leg, human spinal region, human stomach, etc.? What idealization of the geometry would you recommend for each region of the human anatomy? Assume steady state.

**60.0**    How will you design an electronic blood pressure monitor for human anatomy in a noninvasive manner using the theory developed in Sec. 4.5.12?

# CHAPTER 5

# Gas Transport

## Learning Objectives

- Learn Hill plot and equilibrium dissociation
- Simultaneous diffusion and reaction in Islets of Langerhans
- Michaelis-Menten kinetics
- Asymptotic limits at high and low concentrations
- Transient oxygen diffusion in capillary and tissue layers
- Oxygen concentration profile in cell-free plasma layer
- Continue discussion on Krogh tissue cylinder
- Include wave diffusion effects in transient conditions
- Kinetics of nitric oxide formation, diffusion, and transport

## 5.1 Oxygenation Is a Reversible Reaction

Oxyhemoglobin, $HbO_{2n}$, dissociates to hemoglobin, $Hb$, and oxygen via an equilibrium reaction. Hemoglobin binds to oxygen to form oxyhemoglobin at high partial pressures of oxygen, usually in the lungs. *Heme* means group and *globin* represents the globular protein. Hemoglobin is a metalloprotein with a molecular weight of 68,000 gm/mole. It consists of four polypeptide chains, two of them $\alpha$ type and two of them $\beta$ type. It is contained in the RBCs of vertebrates. Ninety-seven percent of RBCs is Hb. Hb transports oxygen from the lungs or gills to the rest of the anatomy, where the oxygen is released for use in cells. It possesses oxygen-binding capacity. Hemoglobin was discovered by Hunefeld in 1840. Funke grew Hb crystals in water and alcohol. In 1959, Perutz elucidated the structure of hemoglobin by x-ray crystallography. He was awarded the Nobel Prize for it in 1962. At full saturation, all erythrocytes are in the form of oxyhemoglobin. As the erythrocytes diffuse to tissues deficient in oxygen, the partial pressure of oxygen will decrease, resulting in the decrease of oxygen and hemoglobin from oxyhemoglobin.

$$HbO_{2n} \underset{k_b}{\overset{k_f}{\rightleftharpoons}} Hb + nO_2 \tag{5.1}$$

The equilibrium reaction is shown in Eq. (5.1), where $k_f$ is the forward reaction rate, and $k_b$ is the reverse reaction rate, and $n$ is the number of molecules of oxygen that bind with hemoglobin. Chemical equilibrium is the state at which the chemical activities, usually denoted concentrations of the reactants and products, are invariant with time. At this juncture, the forward and reverse reaction rates are equal. They are not zero, and the process is in a state of dynamic equilibrium.

An equilibrium rate constant, $K_{eq}$, can be defined as follows:

$$K_{eq} = \frac{k_f}{k_b} \tag{5.2}$$

The rate of oxygenation can be written as:

$$\frac{dC_{HbO_{2n}}}{dt} = -k_f C_{HbO_2} + k_b C_{Hb} C_{O_{2n}}^n \tag{5.3}$$

Equation (5.3) may be written provided the rate of the forward reactions is simple and obeys the first-order kinetics and the rate of reverse reaction is first order with respect to hemoglobin and $n$th order with respect to oxygen. At equilibrium, Eq. (5.3) becomes zero and:

$$C_{HbO_{2n}} = \frac{C_{Hb} C_{O_{2n}}^n}{K_{eq}} \tag{5.4}$$

The extent of oxygenation can be quantitated by a term called *saturation*. Defining saturated hemoglobin as $\phi$:

$$\phi = \frac{C_{HbO_{2n}}}{C_{Hb} + C_{HbO_{2n}}} = \frac{1}{1 + \dfrac{C_{Hb}}{C_{HbO_{2n}}}} \tag{5.5}$$

Combining Eqs. (5.5) and (5.4):

$$\phi = \frac{1}{1 + \dfrac{K_{eq}}{C_{O_{2n}}^n}} = \frac{C_{O_{2n}}^n}{C_{O_{2n}}^n + K_{eq}} \tag{5.6}$$

Assuming ideal gas law:

$$C_{O_{2n}} = \frac{p_{O_{2n}}^n}{(RT)^n} \tag{5.7}$$

Plugging Eq. (5.7) in Eq. (5.6):

$$\phi = \frac{p_{O_{2n}}^n}{\left(p_{O_{2n}}^n + K_{eq}(RT)^n\right)} \tag{5.8}$$

Equation (5.8) can be made simpler to use by defining a certain $p_{50}$—that is, the partial pressure of oxygen at which 50 percent of the oxygen-binding sites are filled. Thus, when $\phi = 0.5$:

$$0.5 = \frac{p_{50}^n}{p_{50}^n + K_{eq}(RT)^n} \tag{5.9}$$

Obtaining the reciprocal of Eq. (5.9) and rearranging it can show that:

$$p_{50} = K_{eq}^{1/n}(RT) \tag{5.10}$$

Substituting Eq. (5.10) in Eq. (5.8):

$$\phi = \frac{p_{O_{2n}}^n}{\left(p_{O_{2n}}^n + p_{50}^n\right)} \tag{5.11}$$

Equation (5.11) is referred to as the *Hill equation*. It is named after Archibald V. Hill, who was awarded the Nobel Prize in physiology or medicine in 1922 for his discovery relating to the production of heat in the muscle. A plot of $\phi$ versus the partial pressure of oxygen is called the Hill plot. It can be seen that as temperature varies, the plot will vary. The values of $n$ and $p_{50}$ can be obtained by a log-log plot of the experimental values of the partial pressure of oxygen and the saturation level of hemoglobin upon suitable modification.

This can be seen in Fig. 5.1 The curve is sigmoidal in shape. When the temperature is increased, as indicated in Eq. (5.10), the $p_{50}$ will increase, provided the equilibrium rate constant does not change significantly. This can result in the dissociation curve shifting to the right. This is represented by the dashed curve in Fig. 5.1. Another cause attributable to the shift of the curve to the right in Fig. 5.1 with an increase in temperature is the denaturing of the bond between the oxygen and hemoglobin.

The sigmoidal shape of the Hill plot is attributed to the cooperative binding of oxygen to the four polypeptide chains. Cooperative binding is the increased affinity for more oxygen to bind with hemoglobin once the first oxygen atom has attached itself to hemoglobin.

Other factors that can change the equilibrium rate constant $K_{eq}$ and $p_{50}$ are pH and organic phosphates. An increase in acidity or a decrease in pH results in what is referred to as the *Bohr shift.*

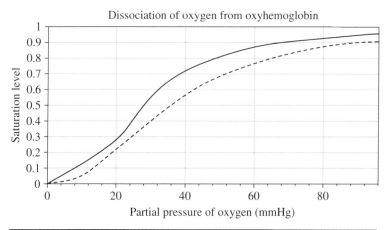

**FIGURE 5.1**  Oxygen dissociation from oxyhemoglobin—the Hill plot.

During Bohr shift, the curve in the Hill plot will shift to the right. Due to the increase in sensitivity to acid, more oxygen needs to be given up. 2.3-diphosphoglycerate (DPG) organic phosphate binds to hemoglobin. This decreases the affinity of oxygen, causing a shift of the curve to the right in the Hill plot [1].

Each heme group in hemoglobin contains one iron (Fe) atom that is capable of binding one oxygen molecule via ion-induced dipole forces. The oxidation state of Fe in oxyhemoglobin is 3, not 2. The polypeptide chains in Hb bind together by noncovalent interactions. The binding is said to be cooperative. When oxygen binds to the iron complex, the Fe atom moves back toward the center of the plane of the porphyrin ring. The imidazole side chain of the histidine residue interacting at the other pole of the iron is pulled toward the porphyrin ring. The binding of oxygen is a cooperative process. When one subunit in hemoglobin is attached to oxygen, the other subunits undergo a conformational change, resulting in an increase in affinity to oxygen. This is why the Hill plot is sigmoidal in shape (Fig. 5.1).

The oxygen-binding capacity of Hb is decreased in the presence of carbon monoxide (CO). This is because CO also competes for the available binding sites for oxygen. This effect can be seen in tobacco smokers. In heavy smokers, 20 percent of the active sites in Hb are blocked. The affinity of CO to Hb is 200 times greater than that with oxygen. Small amounts of CO can reduce the oxygen transport capability of hemoglobin. When Hb combines with CO, it forms a bright pink compound called carboxyhemoglobin. Due to pollution, air containing CO can cause headache, nausea, and even unconsciousness.

Henry's law can be applied to estimate the partial pressure of oxygen:

$$p_{O_2} = H C_{O_2} \tag{5.12}$$

$H$ is the Henry's law constant, and $C_{O_2}$ is the concentration of the dissolved oxygen concentration in the blood. The Henry's law constant for human blood at 37°C is 0.75 mmHg/μM.

The equilibrium rate constant, $K_{eq}$, changes in response to temperature. This is described by *van't Hoff's equation*:

$$\frac{dln(K_{eq})}{dT} = \frac{\Delta H}{RT^2} \tag{5.13}$$

where $\Delta H$ is the enthalpy change for the reaction. Equation (5.13) was integrated and an expression developed in terms of $K_{eq}$ at two different temperatures. Thus:

$$ln\left(\frac{K_{eq2}}{K_{eq1}}\right) = -\frac{\Delta H}{R}\left(\frac{1}{T_2} - \frac{1}{T_1}\right) \tag{5.14}$$

Furthermore, it is known that:

$$K_{eq} = e^{-\frac{\Delta G}{RT}} = e^{-\frac{\Delta H}{RT}}e^{\frac{\Delta S}{R}} \tag{5.15}$$

Equation (5.15) follows from the relation $\Delta G = \Delta H - T\Delta S$. $G$ is the Gibbs free energy, as discussed in Chap. 1.

## 5.2    Diffusion of Oxygen in Tissue and Blood

Oxygen availability becomes limited in some regions of the tissue. The metabolic rate in the cells and the demand for oxygen is greater than the oxygen that has diffused to that region. Oxygenation becomes a diffusion-limited process. Due to this phenomena, growth of multi-cellular systems over 100 μm does not happen. A condition called *hypoxia* has been observed in Brockmann bodies in fish. Oxygen-partial pressures were measured in the islet organs placed in culture. A microelectrode was used to detect oxygen-partial pressure in the surrounding region of an islet organ that is 800 μm in diameter and within the cells. Within a distance of 100 μm for the case of no convection [2], $p_{O_2}$ is close to zero. A condition called *necrosis* is reached where the cells begin to die without sufficient oxygen supply. The experiments with convection showed increased $p_{O_2}$ at the surface and core regions of the islet.

Oxygen supply, in addition to diffusion, comes about by the circulatory system and through the hemoglobin molecule. Oxygen is carried in the blood by convection to capillaries by the circulatory system.

Islets of Langerhans (Fig. 3.18) are spheroidal aggregates of cells that are located in the pancreas [2]. They secrete hormones that are involved in glucose metabolism, particularly insulin. Transplantation of isolated cells is a promising treatment for some forms of type 1 diabetes. Islets removed from the pancreas are devoid of their internal vascularization. The metabolic requirement of the cells requires oxygen to diffuse from the external environment and through the oxygen-consuming islet tissue. The oxygen supply is a critical limiting factor for the functionality and feasibility of islets that are encapsulated, placed in devices for *implantation,* cultured, and used in anaerobic conditions. Theoretical models are needed to describe the oxygen diffusion. The parameters of the model require knowledge of the consumption rate of oxygen, oxygen solubility, and the effective diffusion coefficient to oxygen in the tissue.

### 5.2.1  Fick Diffusion and Michaelis-Menten Kinetics in Spherical Coordinates

Colton et al. [3] developed an oxygen reaction and diffusion model. They assumed that the islet preparation is a suspension of tissue spheres that can be divided into $m$ groups. Each sphere in group $i$ $(1 < i < m)$ has the same equivalent radius, $R_i$, that varies from group to group. The tissue is assumed to be uniform, with constant physical properties that are invariant in space. The governing equation for oxygen diffusion and reaction in spherical coordinates with azimuthal symmetry, accounting for Fick's diffusion, can be written as:

$$\frac{\partial C_{O_2}}{\partial t} = D_T \left(\frac{1}{r^2}\right)\frac{\partial}{\partial r}\left(r^2 \frac{\partial C_{O_2}}{\partial r}\right) - \frac{C_{E0} C_{O_2}}{C_M + C_{O_2}} \tag{5.16}$$

where $D_T$ = the diffusion coefficient in the tissue
$\quad\ C_{E0}$ = the total enzyme or complexation species concentration
$\quad\ C_M$ = the Michaelis constant

The oxygen consumption rate is assumed to obey the Michaelis-Menten kinetics. Equation (5.16) describes the interplay of transient diffusion and metabolic consumption of oxygen in the tissue in spherical coordinates. The concentration of oxygen, $C_{O_2}$, can be expressed in terms of its partial pressure, $p_{O_2}$. This is obtained by using the Bunsen solubility coefficient, $\alpha_t$, such that:

$$C_{O_2} = \alpha_t p_{O_2} \tag{5.17}$$

Substituting Eq. (5.17) in Eq. (5.16), Eq. (5.16) becomes:

$$\alpha_t \frac{\partial p_{O_2}}{\partial t} = \alpha_t D_T \left(\frac{1}{r^2}\right)\frac{\partial}{\partial r}\left(r^2 \frac{\partial p_{O_2}}{\partial r}\right) - \frac{C_{E0} p_{O_2}}{C'_M + p_{O_2}} \tag{5.18}$$

The product $\alpha_t D_T$ can be seen to the product of solubility and diffu-sivity, and hence is the permeability of oxygen in the tissue. The Michaelis constant, $C_M$, is also modified: $C'_M$ expressed in units of mmHg. The initial condition can be written as:

$$p_{O_2} = p_{O_{20}}, \, t = 0 \qquad (5.19)$$

From symmetry at the center of the sphere:

$$\frac{\partial p_{O_2}}{\partial r} = 0 \qquad (5.20)$$

At the surface, the oxygen diffusive transport from within the sphere must be equal to the oxygen transport by convection across the boundary layer surrounding each sphere:

$$J_i = k_i \alpha_m (p_{O_{2m}} - p_{O_2}(R_i)) = \alpha_t D_T \frac{\partial p_{O_2}}{\partial r} \qquad (5.21)$$

where $p_{O_2}(R_i)$ = the partial pressure of oxygen at the surface,
$\quad k_i$ = the mass transfer coefficient between the surround-ing space and the surface of the sphere,
$\quad \alpha_m$ = the oxygen solubility in the surrounding space.

The total rate of oxygen transfer $N$ from the surrounding space to all of the spheres can be summed up as:

$$N = \sum_{i=1}^{m} J_i \left(4\pi R_i^2\right) n_s f_i = -V_m \alpha_m \frac{\partial p_m}{\partial t} \qquad (5.22)$$

where $V_m$ = the volume of the surrounding space is given by
$\quad n_s$ = the total number of spheres,
$\quad f_i$ = the fraction of spheres in group $i$.

The initial condition for the surrounding space is:

$$p_{O_{2m}} = p_{O_{2m}}(0), \, t = 0 \qquad (5.23)$$

The mass transfer coefficient can be obtained from suitable Sherwood number correlations. For instance, the mass transfer coefficient for spherical particles in an agitated tank in the islet size range of 100 to 300 µm can be written as:

$$Sh_i = \frac{k_i d_i}{D_m} = 2 + \left(\frac{\mu}{\rho D_m}\right)^{1/3} \left(\frac{d_{imp}}{d_{tank}}\right)^{0.15} f\left(\frac{\varepsilon \rho^3 d_i^4}{\mu^3}\right) \qquad (5.24)$$

where $\varepsilon$ is the power input per unit fluid mass and $f$ is the function that has to be obtained from experimental data.

Numerical methods are needed to obtain the solution to Eq. (5.18). This is because of the nonlinearity of Michaelis-Menten kinetics. Closed formed analytical solutions to Eq. (5.18) can be obtained in the asymptotic limits of the following:

1. High concentration of oxygen: The rate is independent of $p_{O_2}$ (zeroth order).

2. Low concentration of oxygen: The rate is first order with respect to $p_{O_2}$.

The reasons for choosing the asymptotic limits are elucidated in Fig. 5.2. It can be seen that at low reactant concentrations, the rate is linear [4]. At high enzyme or complexing agent concentrations, the rate is invariant with respect to concentration. Hence, a zeroth order can be assumed at high concentrations and a first order at low reactant concentrations.

Thus, at high reactant concentrations, Eq. (5.18) becomes:

$$\alpha_t \frac{\partial p_{O_2}}{\partial t} = \alpha_t D_T \left(\frac{1}{r^2}\right) \frac{\partial}{\partial r}\left(r^2 \frac{\partial p_{O_2}}{\partial r}\right) + r_{max} \tag{5.25}$$

Equation (5.18) can be nondimensionalized as follows:

$$\text{Let} \quad \tau = \frac{D_T t}{R_i^2}; u = \frac{p_{O_2} - p_{O_{2m}}}{p_{O_{20}} - p_{O_{2m}}}; X = \frac{r}{R_i}; r_{max}^* = \frac{r_{max} R_i^2}{(p_{O_{20}} - p_{O_{2m}})D_T \alpha_t} \tag{5.26}$$

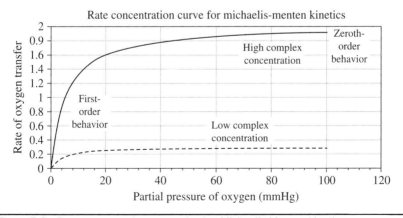

FIGURE 5.2    Rate-concentration curve obeying Michaelis-Menten kinetics.

Equation (5.25) becomes:

$$\frac{\partial u}{\partial \tau} = \frac{1}{X^2}\frac{\partial}{\partial X}\left(X^2\frac{\partial u}{\partial X}\right) + r^*_{max} \tag{5.27}$$

The zeroth-order reaction at high concentrations of oxygen is a heterogeneity in the partial differential equation. Systems such as this can be solved for by assuming that the solution consists of a steady-state part and a transient part:

$$\text{Let } u = u^{ss} + u^t \tag{5.28}$$

Substituting Eq. (5.28) in Eq. (5.27), Eq. (5.27) becomes:

$$\frac{\partial u^t}{\partial \tau} = \frac{1}{X^2}\frac{\partial}{\partial X}\left(X^2\frac{\partial(u^t + u^{ss})}{\partial X}\right) + r^*_{max} \tag{5.29}$$

Equation (5.29) holds good when:

$$-r^*_{max} = \frac{1}{X^2}\frac{\partial}{\partial X}\left(X^2\frac{\partial u^{ss}}{\partial X}\right) \tag{5.30}$$

and

$$\frac{\partial u^t}{\partial \tau} = \frac{1}{X^2}\frac{\partial}{\partial X}\left(X^2\frac{\partial u^t}{\partial X}\right) \tag{5.31}$$

Equation (5.30) can be integrated twice and the boundary condition given by Eq. (5.20) applied to yield:

$$u^{ss} = \frac{X^2 r^*_{max}}{6} + d \tag{5.32}$$

In order to obtain the solution of the integration constant $d$ in Eq. (5.32), the boundary condition given by Eq. (5.21) needs to be modified. Assuming that after attaining steady state, the surface concentration of the sphere would have reached the surrounding space concentration, $d$ can be solved for the solution for the $p_{O_2}$ at steady state, written as:

$$u^{ss} = \frac{\left(R_i^2 - r^2\right)r^*_{max}}{6R_i^2} \tag{5.33}$$

The solution to Eq. (5.31) may be obtained by separating the variables as follows:

$$\text{Let } u^t = V(\tau)g(X)$$

Then Eq. (5.31) becomes:

$$\frac{V'(\tau)}{V(\tau)} = \frac{1}{g(X)X^2}\frac{\partial}{\partial X}(X^2 g'(X)) = -\lambda_n^2 \tag{5.34}$$

Hence,
$$V = ce^{-\lambda_n^2 \tau} \tag{5.35}$$

$$X^2 g''(X) + 2Xg'(X) + X^2 \lambda_n^2 g(X) = 0 \tag{5.36}$$

Comparing Eq. (5.36) with the generalized Bessel function [5]:

$$a = 2; c = 0; s = 1; d = \lambda_n^2; p = 1/2$$

The solution to Eq. (5.36) can be seen to be:

$$g(X) = c_1 \frac{J_{1/2}(\lambda_n X)}{\sqrt{X}} + d_1 \frac{J_{-1/2}(\lambda_n X)}{\sqrt{X}} \tag{5.37}$$

From the boundary condition given by Eq. (5.20), it can be seen that $d_1$ can be set to zero and:

$$g(X) = c_1 \frac{J_{1/2}(\lambda_n X)}{\sqrt{X}} \tag{5.38}$$

The eigenvalues $\lambda_n$ can be solved for from the boundary condition given by Eq. (5.21). In the dimensionless form, Eq. (5.21) may be written as:

$$-\left(\frac{\alpha_m}{\alpha_t}\right)\left(\frac{k_i R_i}{D_T}\right) u = \frac{\partial u}{\partial X}, \; r = R_i \tag{5.39}$$

where $(k_i R_i / D_T) = Bi_m$, the Biot number (mass). This represents the ratio of mass transfer from the surrounding space and the diffusion within the sphere. To simplify matters from a mathematical standpoint, Eq. (5.38) can be written in terms of elementary trigonometric functions as:

$$g(X) = c_1 \sqrt{\frac{2}{\pi \lambda_n}} \frac{\sin(\lambda_n X)}{X} \tag{5.40}$$

The eigenvalues can be obtained from the solution of the following transcendental equation:

$$\tan(\lambda_n X) = \frac{\lambda_n X}{1 - \dfrac{\alpha_m}{\alpha_t} Bi_m X} \tag{5.41}$$

The general solution for the dimensionless $p_{O_2}$ can be written as:

$$u = \sum_0^\infty c_n e^{-\lambda_n^2 \tau} \frac{J_{1/2}(\lambda_n X)}{\sqrt{X}} \tag{5.42}$$

The eigenvalues are given by Eq. (5.41). The $c_n$ can be solved for from the initial condition given by Eq. (5.19) using the principle of orthogonality and:

$$c_n = \frac{\int_0^{R_i} \frac{J_{1/2}(\lambda_n X)}{\sqrt{X}} dX}{\int_0^{R_i} \frac{J_{1/2}^2(\lambda_n X)}{X} dX} \qquad (5.43)$$

Thus, the oxygen concentration profile at high oxygen concentration is obtained. At a low concentration of oxygen, the rate of consumption is first order. The governing equation, Eq. (5.18), can be written as:

$$\alpha_t \frac{\partial p_{O_2}}{\partial t} = \alpha_t D_T \left(\frac{1}{r^2}\right) \frac{\partial}{\partial r}\left(r^2 \frac{\partial p_{O_2}}{\partial r}\right) - k_p p_{O_2} \qquad (5.44)$$

Obtaining the dimensionless form of Eq. (5.44):

$$\frac{\partial u}{\partial \tau} = \frac{1}{X^2}\frac{\partial}{\partial X}\left(X^2 \frac{\partial u}{\partial X}\right) - \phi^2 u \qquad (5.45)$$

where

$$u = \frac{p_{O_2}}{p_{O_{2m}}}; X = \frac{r}{R_i}; \tau = \frac{D_T t}{R_i^2}; \phi^2 = \frac{k_p R_i^2}{\alpha_t D_T} \qquad (5.46)$$

It can be recognized that $\phi$ is the Thiele modulus. Equation (5.45) can be solved for by the method of separation of variables. Let $u = V(\tau)g(X)$:

$$\frac{V'}{V} = \frac{1}{gX^2}\frac{\partial}{\partial X}\left(X^2 \frac{\partial g}{\partial X}\right) = -\left(\phi^2 + \lambda_n^2\right) \qquad (5.47)$$

The solution in the time domain can be seen to be:

$$V = ce^{-\left(\lambda_n^2 + \phi^2\right)\tau} \qquad (5.48)$$

The solution in the space domain can be seen to be:

$$X^2 g''(X) + 2Xg'(X) + X^2\left(\phi^2 + \lambda_n^2\right)g(X) = 0 \qquad (5.49)$$

Comparing Eq. (5.49) with the generalized Bessel function [5]:

$$a = 2; c = 0; s = 1; d = \left(\phi^2 + \lambda_n^2\right); p = 1/2$$

The solution to Eq. (5.49) can be seen to be:

$$g(X) = c_1 \frac{J_{1/2}\left(\sqrt{\lambda_n^2 + \phi^2}\, X\right)}{\sqrt{X}} + d_1 \frac{J_{-1/2}\left(\sqrt{\lambda_n^2 + \phi^2}\right)}{\sqrt{X}} \tag{5.50}$$

From the boundary condition given by Eq. (5.20), it can be seen that $d_1$ can be set to zero and:

$$g(X) = c_1 \frac{J_{1/2}\left(\sqrt{\lambda_n^2 + \phi^2}\, X\right)}{\sqrt{X}} \tag{5.51}$$

The eigenvalues, $\lambda_n$, can be solved for from the boundary condition given by Eq. (5.21). In the dimensionless form, Eq. (5.21) may be written as:

$$-\left(\frac{\alpha_m}{\alpha_t}\right)\left(\frac{k_i R_i}{D_T}\right) u = \frac{\partial u}{\partial X}, \quad r = R_i \tag{5.52}$$

where $(k_i R_i / D_T) = Bi_m$, the Biot number (mass). This represents the ratio of mass transfer from the surrounding space of the diffusion within the sphere. To simplify matters from a mathematical standpoint, Eq. (5.51) can be written in terms of elementary trigonometric functions as:

$$g(X) = c_1 \sqrt{\frac{2}{\pi \sqrt{\lambda_n^2 + \phi^2}}} \frac{\sin\left(\sqrt{\lambda_n^2 + \phi^2}\, X\right)}{X} \tag{5.53}$$

The eigenvalues can be obtained by solving the following transcendental equation:

$$\tan\left(\sqrt{\lambda_n^2 + \phi^2}\, X\right) = \frac{\sqrt{\lambda_n^2 + \phi^2}\, X}{1 - \dfrac{\alpha_m}{\alpha_t} Bi_m X} \tag{5.54}$$

The general solution for the dimensionless $p_{O_2}$ can be written as:

$$u = \sum_0^\infty c_n e^{-\left(\phi^2 + \lambda_n^2\right)\tau} \frac{J_{1/2}\left(\sqrt{\lambda_n^2 + \phi^2}\, X\right)}{\sqrt{X}} \tag{5.55}$$

The eigenvalues are given by Eq. (5.54). The $c_n$ can be solved for from the initial condition given by Eq. (5.19) using the principle of orthogonality and:

$$c_n = \frac{\displaystyle\int_0^{R_i} \frac{J_{1/2}\left(\sqrt{\lambda_n^2 + \phi^2}\, X\right)}{\sqrt{X}}\, dX}{\displaystyle\int_0^{R_i} \frac{J_{1/2}^2\left(\sqrt{\lambda_n^2 + \phi^2}\, X\right)}{X}\, dX} \tag{5.56}$$

Thus, the oxygen concentration profile at low oxygen concentration is obtained.

## 5.2.2  Wave Diffusion Effects

In the times associated with the oxygen consumption, the finite speed of diffusion effects cannot be ignored. The damped wave diffusion and relaxation effects may be included (Sharma [6]) in the following manner.

At low oxygen concentration, a first-order rate of reaction can be assumed. A semi-infinite medium of tissue is considered. A step change in concentration is given at the surface. At times zero, the concentration of oxygen is at an initial value. At *infinite distances*, the concentration of oxygen would be unchanged at the initial value. The mass balance equation for oxygen can be written as:

$$-\frac{\partial J_{O_2}}{\partial x} - kC_{O_2} = \frac{\partial C_{O_2}}{\partial t} \tag{5.57}$$

where $k$ is the lumped first-order reaction rate constant. Combining Eq. (5.57) with the damped wave diffusion and relaxation equation:

$$J_{O_2} = -D_T \frac{\partial C_{O_2}}{\partial x} - \tau_{mr} \frac{\partial J_{O_2}}{\partial t} \tag{5.58}$$

the governing equation is obtained. $\tau_{mr}$ is the mass relaxation time. When it is zero, Eq. (5.58) reverts to Fick's law of diffusion. When the rate of mass flux is greater than an exponential rise, the *wave regime* is the more dominant mechanism of transport, compared with the Fick regime. Equation (5.58) is differentiated by $x$, Eq. (5.57) is differentiated by $t$, and the cross term $\partial^2 J / \partial t \partial x$ between the two equations is:

$$D_T \frac{\partial^2 C_{O_2}}{\partial x^2} = \tau_{mr} \frac{\partial^2 C_{O_2}}{\partial t^2} + (1 + k\tau_{mr}) \frac{\partial C_{O_2}}{\partial t} + kC_{O_2} \tag{5.59}$$

The governing equation for oxygen concentration in the tissue is obtained in the dimensionless form by the following substitutions:

$$k^* = k\tau_{mr} ; u = \frac{C_{O_2}}{C_{O_{2m}}} ; \tau = \frac{t}{\tau_{mr}} ; X = \frac{x}{\sqrt{D_T \tau_{mr}}} \tag{5.60}$$

The governing equation is a partial differential equation of the *hyperbolic* type. It is second order with respect to time and space:

$$\frac{\partial^2 u}{\partial X^2} = \frac{\partial^2 u}{\partial \tau^2} + (1 + k^*) \frac{\partial u}{\partial \tau} + k^* u \tag{5.61}$$

The space and time conditions are:

$$X = 0, u = 1; X = \infty, u = 0$$
$$\tau = 0, u = 0$$

(5.62)

Equation (5.61) can be solved by a recently developed method called *relativistic transformation of coordinates*. The damping term is first removed by Eq. (5.61) by $e^{n\tau}$. Choosing $n = (1 + k^*)/2$ and letting $W = ue^{n\tau}$, Eq. (5.61) becomes:

$$\frac{\partial^2 W}{\partial X^2} = \frac{\partial^2 W}{\partial \tau^2} - \frac{W(1 - k^*)^2}{4}$$

(5.63)

The significance of $W$ is that it can be recognized as the wave concentration. During the transformation of Eq. (5.61) to Eq. (5.62), the damping term has vanished.

Now let us define a spatiotemporal symmetric substitution:

$$\eta = \tau^2 - X^2 \text{ for } \tau > X$$

Equation (5.62) becomes:

$$\eta^2 \frac{\partial^2 W}{\partial \eta^2} + \eta \frac{\partial W}{\partial \eta} - \eta \frac{(1 - k^*)^2}{16} = 0$$

(5.64)

Comparing Eq. (5.64) with the generalized Bessel equation :

$$a = 1; b = 0; c = 0; s = \tfrac{1}{2}; d = -\frac{(1 - k^*)^2}{16}$$

The order $p = 0$.

$\sqrt{d}/s = \tfrac{1}{2} i(1 - k^*)$ and is imaginary. Hence, the solution is:

$$W = c_1 I_0 \left( \frac{|1 - k^*| \sqrt{\tau^2 - X^2}}{2} \right) + c_2 K_0 \left( \frac{|1 - k^*| \sqrt{\tau^2 - X^2}}{2} \right)$$

(5.65)

$c_2$ can be seen to be zero from the condition that at $\eta = 0$, $W$ is finite:

$$W = c_1 I_0 \left( \frac{|1 - k^*| \sqrt{\tau^2 - X^2}}{2} \right)$$

(5.66)

From the boundary condition at $X = 0$:

$$1 e^{\frac{\tau(1+k^*)}{2}} = c_1 I_0 \left( \frac{|1 - k^*| \tau}{2} \right)$$

(5.67)

$c_1$ can be eliminated between Eqs. (5.66) and (5.67) in order to yield:

$$u = \frac{I_0\left(\frac{1}{2}\sqrt{\tau^2 - X^2}\,|1-k^*|\right)}{I_0\left(\frac{\tau}{2}|1-k^*|\right)} \tag{5.68}$$

This is valid for $\tau > X$, $k^* \neq 1$. For $X > \tau$:

$$u = \frac{J_0\left(\frac{1}{2}\sqrt{X^2 - \tau^2}\,|1-k^*|\right)}{I_0\left(\frac{\tau}{2}|1-k^*|\right)} \tag{5.69}$$

At the wavefront, $\tau = X$:

$$u = e^{-\tau\frac{(1+k^*)}{2}} = e^{-X\frac{(1+k^*)}{2}} \tag{5.70}$$

The mass inertia can be calculated from the first zero of the Bessel function at 2.4048. Thus:

$$\tau_{\text{inertia}} = \frac{\sqrt{X_p^2 - 23.1323}}{|1-k^*|^2} \tag{5.71}$$

The concentration at an interior point in the semi-infinite medium is shown in Fig. 5.3. Four regimes can be identified. These are:

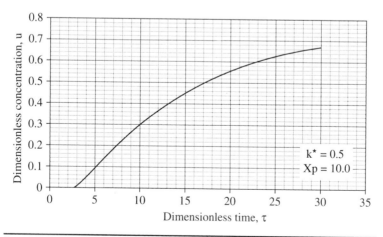

**FIGURE 5.3** Dimensionless concentration at an interior point $X_p = 10$ in a semi-infinite medium during simultaneous reaction and diffusion. $k^* = 0.5$.

1. Zero-transfer inertial regime: $0$ $0 \leq \tau \leq \tau_{inertia}$
2. Times greater than the inertial regime and less than at the wavefront: $X_p > \tau$
3. Wavefront: $\tau = X_p$
4. Open interval of times greater than at the wavefront: $\tau > X_p$

During the first regime of mass inertia, there is no transfer of mass up to a certain threshold time at the interior point $X_p = 10$. The second regime is given by Eq. (5.69), represented by a Bessel *composite function* of the first kind and zeroth order. The rise in dimensionless concentration proceeds from the dimensionless time 2.733 up to the wavefront at $X_p = 10.0$. The third regime is at the wavefront. The dimensionless concentration is described by Eq. (5.70).

The fourth regime is described by Eq. (5.68) and represents the decay in time of the dimensionless concentration. It is given by the modified Bessel composite function of the first kind and zeroth order. Figure 5.4 shows the three regimes of the concentration when $k^* = 2.0$. It can be seen from Fig. 5.4 that the mass inertia time has increased to 8.767. The rise is nearly a jump in concentration at the interior point $X_p = 10.0$. When $k^* = 0.25$, as shown in Fig. 5.5, the inertia time is 7.673. In Fig. 5.6, the three regimes for the case when $k^* = 0.0$ are plotted. In Table 5.1, the mass inertia time for various values of $k^*$ for the interior point $X_p = 10.0$ is shown. $k^*$ needs to be sufficiently far from 1 to keep the inertia time positive.

The steady-state solution for Eq. (5.61) can be written as:

$$u^{ss} = \exp(-(k^*)^{1/2} X) \tag{5.72}$$

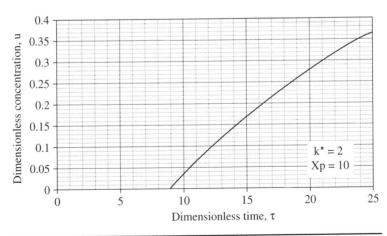

**FIGURE 5.4**   Dimensionless concentration at an interior point $X_p = 10$ in a semi-infinite medium during simultaneous reaction and diffusion. $k^* = 2.0$.

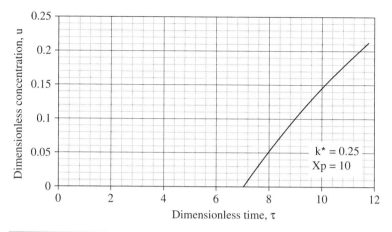

**FIGURE 5.5**   Dimensionless concentration at an interior point $X_p = 10$ in a semi-infinite medium during simultaneous reaction and diffusion. $k^* = 0.25$.

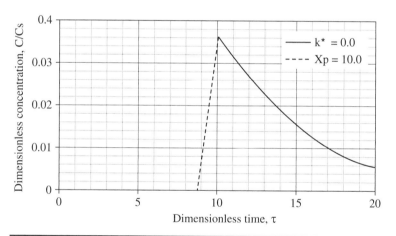

**FIGURE 5.6**   Dimensionless concentration at an interior point $X_p = 10$ in a semi-infinite medium during simultaneous reaction and diffusion. $k^* = 0.0$.

## 5.3   Krogh Tissue Cylinder

The Krogh tissue cylinder was briefly discussed in Sec. 3.11. In the early part of the 20th century, A. Krogh, a Danish physiologist, described oxygen delivery to tissues when the concentration of oxygen in the blood is uniform. More advanced models that include the variation of concentration along the length of the capillary have been developed in the literature. Krogh idealized the capillary and tissue region such that capillaries supply oxygen to a cylindrical region surrounding each capillary (Fig. 3.14). This pattern may be applicable in

| S. No. | $k^*$ ($k'''\tau_{mr}$) | Mass Inertia Time ($t/\tau_{mr}$) |
|--------|------------------------|-----------------------------------|
| 1. | 0.01 | 8.741 |
| 2. | 0.1 | 8.452 |
| 3. | 0.25 | 7.673 |
| 4. | 0.3 | 7.266 |
| 5. | 0.4 | 5.979 |
| 6. | 0.5 | 2.733 |
| 7. | 1.75 | 7.673 |
| 8. | 2.0 | 8.767 |
| 9. | 4.0 | 9.871 |
| 10. | 8.0 | 9.976 |
| 11. | 25.0 | 9.998 |
| 12. | 10.0 | 10.0 |

**TABLE 5.1**   Mass Inertia Time vs. $k^*$ for Interior Point $Xp = 10.0$.

some cases, but not in the brain tissue, where the capillary arrangement is more complex.

### 5.3.1   Transient Oxygen Fick Diffusion and Michaelis-Menten Kinetics

Consider the oxygen diffusion in the tissues at *transient state*. The governing equation for one-dimensional diffusion of oxygen in the tissue can be written considering only the Fick diffusion and Michaelis-Menten kinetics for the consumption of oxygen as:

$$\frac{\partial C_{O_2}}{\partial t} = D_T \left( \frac{1}{r} \right) \frac{\partial}{\partial r} \left( r \frac{\partial C_{O_2}}{\partial r} \right) - \frac{C_{E0} C_{O_2}}{C_M + C_{O_2}} \tag{5.73}$$

where $D_T$ is the diffusion coefficient of oxygen in the tissues. The boundary conditions are:

$$r = R_o, -D_T \frac{\partial C_{O_2}}{\partial r} = 0 \tag{5.74}$$

$$r = R_c, C_{O_2} = C_{R_c} \tag{5.75}$$

$$t = 0, C_{O_2} = 0 \tag{5.76}$$

Closed formed analytical solutions to Eq. (5.73) can be obtained in the asymptotic limits of the following:

1. High concentration of oxygen: The rate is independent of oxygen concentration (zeroth order).

2. Low concentration of oxygen: The rate is first order with respect to concentration of oxygen.

The reasons for choosing the asymptotic limits are elucidated in Fig. 5.2. It can be seen that at low reactant concentrations the rate is linear. At high enzyme or complexing agent concentrations the rate is invariant with respect to concentration. Hence, a zeroth order can be assumed at high concentrations and a first order at low reactant concentrations. Thus, at high reactant concentrations, Eq. (5.73) becomes:

$$\frac{\partial C_{O_2}}{\partial t} = D_T \left(\frac{1}{r}\right)\frac{\partial}{\partial r}\left(r\frac{\partial C_{O_2}}{\partial r}\right) + r_{max} \tag{5.77}$$

Equation (5.73) can be nondimensionalized as follows:

Let $\quad \tau = \frac{D_T t}{R_o^2}; u = \frac{C_{O_2} - C_{O_{2c}}}{-C_{O_{2c}}}; X = \frac{r}{R_o}; r_{max}^* = \frac{r_{max}R_o^2}{\left(-C_{O_{2c}}\right)D_T} \tag{5.78}$

Equation (5.77) becomes:

$$\frac{\partial u}{\partial \tau} = \frac{1}{X}\frac{\partial}{\partial X}\left(X\frac{\partial u}{\partial X}\right) + r_{max}^* \tag{5.79}$$

The zeroth-order reaction at high concentrations of oxygen is a heterogeneity in the partial differential equation. Systems such as this can be solved for by assuming that the solution consists of a steady-state part and a transient part:

Let $\qquad\qquad\qquad u = u^{ss} + u^t \tag{5.80}$

Substituting Eq. (5.80) in Eq. (5.79), Eq. (5.79) becomes:

$$\frac{\partial u^t}{\partial \tau} = \frac{1}{X}\frac{\partial}{\partial X}\left(X\frac{\partial(u^t + u^{ss})}{\partial X}\right) + r_{max}^* \tag{5.81}$$

Equation (5.81) holds good when:

$$-r_{max}^* = \frac{1}{X}\frac{\partial}{\partial X}\left(X\frac{\partial u^{ss}}{\partial X}\right) \tag{5.82}$$

and $\qquad\qquad \frac{\partial u^t}{\partial \tau} = \frac{1}{X}\frac{\partial}{\partial X}\left(X\frac{\partial u^t}{\partial X}\right) \tag{5.83}$

Equation (5.82) can be integrated twice and the boundary condition given by Eq. (5.74) applied to yield:

$$u^{ss} = -\frac{X^2 r_{max}^*}{4} + d + \frac{r_{max}^*}{2}\ln(X) \qquad (5.84)$$

In order to obtain the solution of the integration constant $d$ in Eq. (5.84), the boundary condition given by Eq. (5.75) is used. The solution for the $C_{O_2}$ concentration profile of oxygen in the tissue space surrounding the blood capillary at steady state can be written as:

$$u^{ss} = \frac{\left(X_c^2 - X^2\right) r_{max}^*}{4} + \frac{r_{max}^*}{2}\ln\left(\frac{X}{X_c}\right) \qquad (5.85)$$

Equation (5.85) is plotted for various values of dimensionless reaction rate $r_{max}^*$ in Fig. 5.7. A certain $X$ less than 1 can be calculated where the dimensionless concentration becomes 1. This is when the concentration of oxygen drops to zero and is the zone of zero transfer. This happens before the arrival of the tissue space boundary, where there is no flux, and can be expected at large reaction rates.

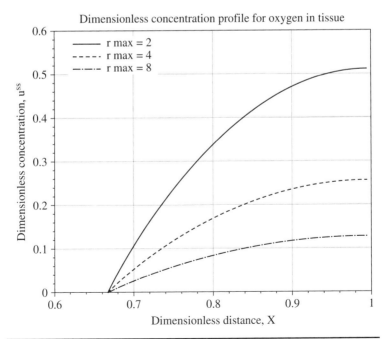

Dimensionless concentration profile for oxygen in tissue

**FIGURE 5.7** Dimensionless concentration profile of oxygen in tissue space for various values of maximum reaction rate.

The solution to Eq. (5.83) may be obtained by separating the variables as follows:

Let
$$u^i = V(\tau)g(X)$$

Then Eq. (5.83) becomes:

$$\frac{V'(\tau)}{V(\tau)} = \frac{1}{g(X)X}\frac{\partial}{\partial X}(Xg'(X)) = -\lambda_n^2 \tag{5.86}$$

Hence,
$$V = ce^{-\lambda_n^2 \tau} \tag{5.87}$$

$$X^2 g''(X) + Xg'(X) + X^2\lambda_n^2 g(X) = 0 \tag{5.88}$$

Comparing Eq. (5.88) with the generalized Bessel function:

$$a = 1; c = 0; s = 1; d = \lambda_n^2 \ p = 0 \tag{5.89}$$

The solution to Eq. (5.88) can be seen to be:

$$g(X) = c_1 J_0(\lambda_n X) + d_1 Y_0(\lambda_n X) \tag{5.90}$$

Based on the fact that at $X = 0$, the concentration of acid cannot be unbounded, it can be seen that $d_1$ can be set to zero and $g(X) = c_1 J_0(\lambda_n X)$ This is in the limit of the capillary radius tending to zero. The eigenvalues, $\lambda_n$, can be solved for from the boundary condition given by Eq. (5.91):

$$J_0(\lambda_n X_c) = 0, \ r = R_c \tag{5.91}$$

The eigenvalues can be obtained as:

$$X_c \lambda_n = 2.4048 + (n-1)\pi, \ n = 1, \ 2, \ 3.... \tag{5.92}$$

The general solution for the dimensionless concentration can be written as:

$$u = \sum_0^\infty c_n e^{-\lambda_n^2 \tau} J_0(\lambda_n X) \tag{5.93}$$

The eigenvalues are given by Eq. (5.92). The $c_n$ can be solved for from the initial condition given by Eq. (5.76) using the principle of orthogonality and:

$$c_n = \frac{\int_{R_o}^\infty J_0(\lambda_n X)dX}{\int_{R_o}^\infty J_0^2(\lambda_n X)dX} \tag{5.94}$$

Thus, the oxygen concentration profile at high oxygen concentration is obtained. At a low concentration of oxygen, the rate of oxygen generation may be approximated as obeying first-order kinetics. The governing equation for simultaneous diffusion and reaction of acid can be written as:

$$\frac{\partial C_{O_2}}{\partial t} = D_T \left(\frac{1}{r}\right) \frac{\partial}{\partial r}\left(r \frac{\partial C_{O_2}}{\partial r}\right) - kC_{O_2} \tag{5.95}$$

Obtaining the dimensionless form of Eq. (5.95):

$$\frac{\partial u}{\partial \tau} = \frac{1}{X} \frac{\partial}{\partial X}\left(X \frac{\partial u}{\partial X}\right) - \phi^2 u \tag{5.96}$$

where

$$u = \frac{C_{O_2} - C_{O_{2c}}}{-C_{O_{2c}}} ; X = \frac{r}{R_o} ; \tau = \frac{D_T t}{R_o^2} ; \phi^2 = \frac{kR_o^2}{D_T} \tag{5.97}$$

It can be recognized that $\phi$ is the Thiele modulus. The transient part of the concentration profile can be solved for by separating the variables. Let $u = V(\tau)g(X)$:

$$\frac{V'}{V} = \frac{1}{gX} \frac{\partial}{\partial X}\left(X \frac{\partial g}{\partial X}\right) = -\left(\phi^2 + \lambda_n^2\right) \tag{5.98}$$

The solution in the time domain can be seen to be:

$$V = ce^{-\left(\lambda_n^2 + \phi^2\right)\tau} \tag{5.99}$$

The solution in the space domain can be seen to be:

$$X^2 g''(X) + X g'(X) + X^2\left(\phi^2 + \lambda_n^2\right)g(X) = 0 \tag{5.100}$$

Comparing Eq. (5.100) with the generalized Bessel function:

$$a = 1; c = 0; s = 1; d = \left(\phi^2 + \lambda_n^2\right); p = 0$$

The solution to Eq. (5.100) can be seen to be:

$$g(X) = c_1 J_0\left(\sqrt{\lambda_n^2 + \phi^2}\, X\right) + c_2 Y_0\left(\sqrt{\lambda_n^2 + \phi^2}\, X\right) \tag{5.101}$$

Based on the realization that the concentration cannot become unbounded at $X = 0$, it can be seen that $c_2$ is 0. This is especially true for capillaries with a small radius:

$$g(X) = c_1 J_0\left(\sqrt{\lambda_n^2 + \phi^2}\, X\right) \tag{5.102}$$

The eigenvalues, $\lambda_n$, can be solved for from the boundary condition at $X = 1$:

$$-\sqrt{\lambda_n^2 + \phi^2}\, J_1\left(\sqrt{\lambda_n^2 + \phi^2}\right) = 0 \qquad (5.103)$$

and

$$3.8317 + (n - 1)\pi = \sqrt{\lambda_n^2 + \phi^2} \qquad (5.104)$$

The general solution for the dimensionless $C_{O_2}$ can be written as:

$$u = \sum_0^\infty c_n e^{-(\phi^2 + \lambda_n^2)\tau} J_0\left(\sqrt{\lambda_n^2 + \phi^2}\, X\right) \qquad (5.105)$$

The eigenvalues are given by Eq. (5.104). The $c_n$ can be solved for from the initial condition and using the principle of orthogonality:

$$c_n = \frac{\int_{R_c}^{R_o} J_0\left(\sqrt{\lambda_n^2 + \phi^2}\, X\right) dX}{\int_{R_c}^{R_o} J_0^2\left(\sqrt{\lambda_n^2 + \phi^2}\right) dX} \qquad (5.106)$$

Thus, the oxygen concentration profile at low oxygen concentration is obtained.

### 5.3.2  Anoxic Regions

The original Krogh model did not include the transient effects. It also did not include the axial variation of oxygen concentration. Axial variation in oxygen concentration can be accounted for by a steady-state mass balance on the bloodstream (Sec 3.11). This can be used to obtain the *anoxic region*. This is where the oxygen concentration becomes zero even before reaching the boundary. It is also called the lethal corner.

The Krogh model makes the following assumptions:

1. Axial diffusion of oxygen in the blood is insignificant.
2. Axial diffusion of oxygen in tissue is insignificant.
3. Other reactions with oxygen are insignificant.
4. Zeroth-order kinetics is assumed for oxygen uptake for large concentrations of oxygen (Fig. 5.2).
5. Capillaries are arranged in a regular array. The central capillary is surrounded by cylinders of tissue.
6. Oxyhemoglobin and oxygen reaction is reversible and occurs uniformly throughout the bloodstream.
7. Oxygen release in plasma is at a uniform rate. RBCs are not recognized as discrete bodies.

8. Mass transfer resistance in the endothelium and cell-free fluid layer are insignificant.

Since the Krogh model in the early part of the 20th century, other geometries have been considered for idealization of capillaries and tissue space. Examples are hexagonal, rectangular, triangular, etc. Complex geometries require numerical procedures for obtaining model solution(s). The organization of arteriolar and venular ends of adjacent capillaries has an effect on oxygen concentration in the tissue space.

### 5.3.3  Diffusion in the Cell-Free Plasma Layer

The oxygen concentration in the plasma layer in the absence of any reaction can be obtained as follows:

$$\frac{\partial C_{O_2}}{\partial t} = D_{Pl}\left(\frac{1}{r}\right)\frac{\partial}{\partial r}\left(r\frac{\partial C_{O_2}}{\partial r}\right) \tag{5.107}$$

The boundary conditions can be written as:

$$r = R_{RB}, \ C_{O_2} = C_{O_2^*} \tag{5.108}$$

The second boundary condition is derived from the oxygen balance between the flowing blood in convection and diffusion:

$$-2\pi R_c L_c D_{Pl}\frac{\partial C_{O_2}}{\partial r}\bigg|_{r=R_c} = rL_c\left(\pi R_o^2 - \pi R_c^2\right) \tag{5.109}$$

The initial condition is that at:

$$t = 0, \ C_{O_2} = 0$$

Obtaining the dimensionless form of Eq. (5.107):

$$\frac{\partial u}{\partial \tau} = \frac{1}{X}\frac{\partial}{\partial X}\left(X\frac{\partial u}{\partial X}\right) \tag{5.110}$$

where

$$u = \frac{C_{O_2} - C_{O_2^*}}{-C_{O_2^*}}; X = \frac{r}{R_c}; \tau = \frac{D_T t}{R_c^2} \tag{5.111}$$

The boundary condition given by Eq. (5.109) imposes a mathematical heterogeneity on the equations. Systems such as this can be solved for by assuming that the solution consists of a steady-state part and a transient part:

Let
$$u = u^{ss} + u^t \tag{5.112}$$

Substituting Eq. (5.112) in Eq. (5.110), Eq. (5.110) becomes:

$$\frac{\partial u^t}{\partial \tau} = \frac{1}{X}\frac{\partial}{\partial X}\left(X\frac{\partial(u^t + u^{ss})}{\partial X}\right) \qquad (5.113)$$

Equation (5.113) holds good when:

$$0 = \frac{1}{X}\frac{\partial}{\partial X}\left(X\frac{\partial u^{ss}}{\partial X}\right) \qquad (5.114)$$

with the boundary condition:

$$r = R_c, \frac{\partial u^{ss}}{\partial X} = r^* \qquad (5.115)$$

where

$$r^* = \frac{r\left(R_o^2 - R_c^2\right)}{2D_{pl}C_{O_2^*}} \qquad (5.116)$$

and

$$\frac{\partial u^t}{\partial \tau} = \frac{1}{X}\frac{\partial}{\partial X}\left(X\frac{\partial u^t}{\partial X}\right) \qquad (5.117)$$

with the boundary condition:

$$r = R_c, \frac{\partial u^t}{\partial X} = 0 \qquad (5.118)$$

Integrating Eq. (5.114) twice and solving for the integration constants from the boundary conditions:

$$C_{O_2} = C_{O_2^*} - \frac{r\left(R_0^2 - R_c^2\right)}{2D_{pl}}\ln\left(\frac{r}{R_{RB}}\right) \qquad (5.119)$$

The oxygen concentration profile in the cell-free layer at steady state is given by Eq. (5.119). This was developed by Groebe [7]. They also include a fraction of the length of the capillary occupied by the RBCs. This can go in the denominator with $2D_{pl}$. The fraction is usually 0.5, so $2 \cdot 0.5 = 1$. Equation (5.119) is valid for $R_{RB} < r < R_c$.

The transient part of the concentration profile of oxygen in the cell-free layer can be solved for by separating the variables. Let $u = V(\tau)g(X)$:

$$\frac{V'}{V} = \frac{1}{gX}\frac{\partial}{\partial X}\left(X\frac{\partial g}{\partial X}\right) = -\left(\lambda_n^2\right) \qquad (5.120)$$

The solution in the time domain can be seen to be:

$$V = ce^{-\left(\lambda_n^2\right)\tau} \tag{5.121}$$

The solution in the space domain can be seen to be:

$$X^2 g''(X) + X g'(X) + X^2\left(\lambda_n^2\right)g(X) = 0 \tag{5.122}$$

Comparing Eq. (5.122) with the generalized Bessel function:

$$a = 1; \; c = 0; \; s = 1; \; d = \left(\lambda_n^2\right); \; p = 0 \tag{5.123}$$

The solution to Eq. (5.122) can be seen to be:

$$g(X) = c_1 J_0(\lambda_n X) + c_2 Y_0(\lambda_n X) \tag{5.124}$$

Based on the realization that the concentration cannot become unbounded at $X = 0$, it can be seen that $c_2$ is 0. This is from the symmetry condition:

$$g(X) = c_1 J_0(\lambda_n X) \tag{5.125}$$

The eigenvalues, $\lambda_n$, can be solved for from the boundary condition given by Eq. (5.118). Hence:

$$3.8317 + (n - 1)\pi = \lambda n \tag{5.126}$$

The general solution for the transient dimensionless concentration can be written as:

$$u = \sum_0^\infty c_n e^{-\lambda_n^2 \tau} J_0(\lambda_n X) \tag{5.127}$$

The eigenvalues are given by Eq. (5.126). The $c_n$ can be solved for from the initial condition given by Eq. (5.109) using the principle of orthogonality and:

$$c_n = \frac{\displaystyle\int_{R_o}^\infty J_0(\lambda_n X)dX}{\displaystyle\int_{R_o}^\infty J_0^2(\lambda_n X)dX} \tag{5.128}$$

Thus, the oxygen concentration profile in the cell-free layer is obtained.

### 5.3.4   Wave Diffusion Effects during Diffusion in the Plasma Layer

Given that typical radius of the capillary is 5 μm, the residence time of blood in the capillary is on the order of a few seconds, and the nonhomogenous inner structure of blood, the relaxation time (mass) can be on the order of several seconds. The ballistic term in the generalized Fick's law of diffusion cannot be neglected. In this section, the wave diffusion effects in the cell-free plasma layer are attempted to be taken into account.

The oxygen concentration in the plasma layer in the absence of any reaction can be obtained from the governing equation for the concentration of oxygen, including the damped wave diffusion and relaxation effects, as follows:

$$\tau_{mr} \frac{\partial^2 C_{O_2}}{\partial t^2} + \frac{\partial C_{O_2}}{\partial t} = D_{Pl} \left( \frac{1}{r} \right) \frac{\partial}{\partial r} \left( r \frac{\partial C_{O_2}}{\partial r} \right) \qquad (5.129)$$

The boundary conditions can be written as:

$$r = R_{RB}, \; C_{O_2} = C_{O_2^*} \qquad (5.130)$$

where $R_{RB}$ is the radius of the core layer with aggregates of RBCs and it demarcates the plasma layer of interest.

The second boundary condition is derived from the oxygen balance between the flowing blood in convection and diffusion:

$$-2\pi R_c L_c D_{Pl} \frac{\partial C_{O_2}}{\partial r}\bigg|_{r=R_c} = rL_c \left( \pi R_o^2 - \pi R_c^2 \right) \qquad (5.131)$$

The initial condition is that at:

$$t = 0, \; C_{O_2} = 0 \qquad (5.132)$$

Obtaining the dimensionless form of Eq. (5.129):

$$\frac{\partial u}{\partial \tau} + \frac{\partial^2 u}{\partial \tau^2} = \frac{1}{X} \frac{\partial}{\partial X} \left( X \frac{\partial u}{\partial X} \right) \qquad (5.133)$$

where

$$u = \frac{C_{O_2} - C_{O_2^*}}{-C_{O_2^*}}; X = \frac{r}{\sqrt{D_T \tau_r}}; \tau = \frac{t}{\tau_r} \qquad (5.134)$$

The boundary condition given by Eq. (5.131) imposes a mathematical heterogeneity on the equations. Systems such as this can be solved for by assuming that the solution consists of a steady-state part and a transient part:

Let
$$u = u^{ss} + u^t \qquad (5.135)$$

Substituting Eq. (5.135) in Eq. (5.133), Eq. (5.133) becomes:

$$\frac{\partial u^t}{\partial \tau} + \frac{\partial^2 u^t}{\partial \tau^2} = \frac{1}{X} \frac{\partial}{\partial X}\left(X \frac{\partial(u^t + u^{ss})}{\partial X}\right) \tag{5.136}$$

Equation (5.136) holds good when:

$$0 = \frac{1}{X} \frac{\partial}{\partial X}\left(X \frac{\partial u^{ss}}{\partial X}\right) \tag{5.137}$$

with the boundary condition:

$$r = R_c, \frac{\partial u^{ss}}{\partial X} = r^* \tag{5.138}$$

where

$$r^* = \frac{r\left(R_o^2 - R_c^2\right)}{2D_{pl}C_{O_2^*}} \tag{5.139}$$

and

$$\frac{\partial u^t}{\partial \tau} + \frac{\partial u^t}{\partial \tau^2} = \frac{1}{X} \frac{\partial}{\partial X}\left(X \frac{\partial u^t}{\partial X}\right) \tag{5.140}$$

with the boundary condition:

$$r = R_c, \frac{\partial u^t}{\partial X} = 0 \tag{5.141}$$

Integrating Eq. (5.137) twice and solving for the integration constants from the boundary conditions:

$$C_{O_2} = C_{O_2^*} - \frac{r\left(R_0^2 - R_c^2\right)}{2D_{pl}} \ln\left(\frac{r}{R_{RB}}\right) \tag{5.142}$$

The oxygen concentration profile in the cell-free layer at steady state is given by Eq. (5.142). This was discussed in the previous section. The transient part of the concentration profile of oxygen in the cell-free layer can be solved for by separating the variables. Let $u = V(\tau)g(X)$:

$$\frac{V'}{V} + \frac{V''}{V} = \frac{1}{gX} \frac{\partial}{\partial X}\left(X \frac{\partial g}{\partial X}\right) = -\left(\lambda_n^2\right) \tag{5.143}$$

The solution in the time domain can be seen to be:

$$V = e^{-\frac{\tau}{2}}\left(ce^{-\frac{\tau\sqrt{1-4\lambda_n^2}}{2}} + de^{+\frac{\tau\sqrt{1-4\lambda_n^2}}{2}}\right) \tag{5.144}$$

It can be seen that the system reaches steady state after some time. At steady state, $Ve^{\tau/2}$ would become 0 multiplied by a countable large number tending to infinity. Zero multiplied by any finite number is zero. The time taken to reach steady state is usually not more than a few hours. So $e^{\tau/2}$ is large at steady state, yet finite or tending to countable infinity. In this case, zero multiplied by a large number is zero. At steady state, $V$ would be zero. This means that $d$ has to be set to zero in Eq. (5.144). This comes from the "final condition" in time.

The solution in the space domain can be seen to be:

$$X^2 g''(X) + X g'(X) + X^2 \left(\lambda_n^2\right) g(X) = 0 \tag{5.145}$$

Comparing Eq. (5.145) with the generalized Bessel function:

$$a = 1; \; c = 0; \; s = 1; \; d = \left(\lambda_n^2\right); \; p = 0 \tag{5.146}$$

The solution to Eq. (5.145) can be seen to be:

$$g(X) = c_1 J_0(\lambda_n X) + c_2 Y_0(\lambda_n X) \tag{5.147}$$

Based on the realization that the concentration cannot become unbounded at $X = 0$, it can be seen that $c_2$ is 0. This is from the symmetry condition:

$$g(X) = c_1 J_0(\lambda_n X) \tag{5.148}$$

The eigenvalues, $\lambda_n$, can be solved for from the boundary condition given by Eq. (5.130). Hence:

$$3.8317 + (n - 1)\pi = \lambda_n X_{RB} \tag{5.149}$$

The general solution for the transient dimensionless concentration can be written as:

$$u = \sum_0^\infty c_n e^{-\frac{\tau}{2}} e^{-\frac{\tau\sqrt{1-4\lambda_n^2}}{2}} J_0(\lambda_n X) \tag{5.150}$$

It can be seen that for large values of relaxation times of the fluid, Eq. (5.150) becomes *bifurcated*. For:

$$\lambda_n > 1/2 \tag{5.151}$$

Equation (5.150) would become:

$$u = \sum_0^\infty c_n e^{-\frac{\tau}{2}} \cos\left(\tau\sqrt{4\lambda_n^2 - 1}\right) J_0(\lambda_n X) \tag{5.152}$$

The eigenvalues are given by Eq. (5.126). Equation (5.151) is valid when:

$$\tau_r > \frac{R_{RB}^2}{58.7 D_T} \tag{5.153}$$

For a plasma layer thickness of 1 μ and the diffusion coefficient of $10^{-5}$ m$^2$/s, the threshold relaxation time of the fluid would be 27 ns, so there is a good chance that the concentration of oxygen in the plasma layer would exhibit *subcritical damped oscillations*.

The $c_n$ can be solved for from the initial condition given by Eq. (5.132) using the principle of orthogonality and:

$$c_n = \frac{\int_{R_{RB}}^{R_c} J_0(\lambda_n X) dX}{\int_{R_{RB}}^{R_c} J_0^2(\lambda_n X) dX} \tag{5.154}$$

Thus, the oxygen concentration profile in the cell-free layer is obtained, including the wave diffusion effects.

## 5.4   Nitric Oxide Formation and Transport in Blood and Tissue

Nitric oxide (NO) is a vasodilator. The widening of blood vessels when the surrounding smooth muscle cells relax is called vasodilation. *Au contraire*, the narrowing of blood vessels is called vasoconstriction. NO gas is an important signaling molecule and is involved in many physiological and pathological processes. In order to protect the liver from ischemic damage, a certain threshold level of *NO* is required. Excess *NO* can result in tissue damage and vascular collapse. Chronic expression of *NO* is associated with many carcinomas and inflammatory conditions, such as juvenile diabetes, multiple sclerosis, arthritis, and ulcerative colitis.

The *NO* has to diffuse through the blood vessel wall and then binds to the enzyme in smooth muscle cells. Cyclic guanine monophosphate (cGMP) is produced, and the smooth relaxation of muscle is completed. During *hypoxia*, which is a state of no or low oxygen concentration in the blood, *NO* is produced. *NO* production is sensitive to the presence of acetylcholine, histamine, adenosine triphosphate (ATP), and adenosine diphosphate (ADP), and is also stimulated by elevated stress levels. The triple-bonded structure of nitrogen and oxygen reacts with it, hemoglobin, and proteins, and binds with guanylate cyclase enzyme. The lifetime of *NO* is only a few seconds. Progression of atherosclerosis and septic shock are affected by changes in the release of *NO*. It participates in neurotransmission, smooth muscle cell formation and development, and changes the adhesion potential of leukocytes to endothelium. *NO* plays a role in the modulation of the hair cycle, synthesis of reactive nitrogen intermediates, penile erections, etc. Sildenafil, sold under the brand name Viagra, stimulates erections by increased signaling via *NO* pathways in the penis.

The concentration of *NO* can be measured using a reaction with ozone. Such reactions are called *chemiluminescent reactions*. The reaction produces light that can be detected using a photodetector:

$$NO + O_3 \rightarrow NO_2 + O_2 + light \tag{5.155}$$

The light can be detected using electron paramagentic resonance (EPR). The heat produced by the formation of *NO* is *endothermic*. *NO* production is elevated in people living at high altitudes, which aids them in avoiding hypoxia by increasing pulmonary vasculature vasodilation.

*NO* is produced by the reaction of nicotinamide adenine dinucleotide phosphate (NADPH ) with oxygen and L-arginine. Another by-product of this nitric oxide synthase (NOS), enzyme-catalyzed reaction is L-citrulline.

*NO*:

1. Reacts in parallel with proteins, such those found in hemoglobin, myoglobin, and soluble guanyl (sGC)

2. Reacts in parallel with cyclase and cytochrome. The reaction between *NO* and hemoglobin is complex. At first, *NO* reacts with the heme group of hemoglobin that is deoxygenated and forms a stable complex with iron. The reaction is first order with respect to *NO* and hemoglobin:

$$NO + Hb(Fe^{++})O_2 \xrightarrow{k_1} NO^{3+} + Hb(Fe^{2+}) \tag{5.156}$$

3. Reacts irreversibly with oxygen and reversibly with thiol groups (–SH):

$$4NO + O_2 + 2H_2O \xrightarrow{k_2} 4H^+ + 4NO_2^- \tag{5.157}$$

The reaction rates of the parallel reactions can be analyzed as follows:

$$r_{HbFe} = \frac{dC_{HbFe}}{dt} = k_1 C_{NO} C_{HbFeO_2} \tag{5.158}$$

$$r_{NO_2} = \frac{dC_{NO_2}}{dt} = -4k_2 C_{NO}^2 C_{O_2} \tag{5.159}$$

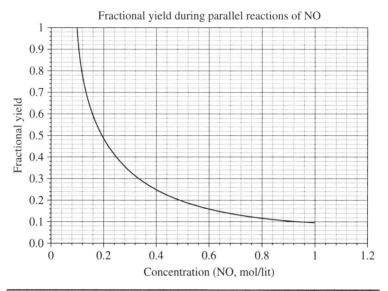

**Figure 5.8** Instantaneous fractional yield of heme complex during parallel reactions of *NO*.

The product distribution can be obtained by defining an *instantaneous fractional yield* (Figure 5.8). Thus, at any $C_{NO}$:

$$\varphi = \frac{dC_{HbFe}}{-dC_{NO}} = \frac{k_1 C_{NO} C_{HbFeO_2}}{k_1 C_{NO} C_{HbFeO_2} + 4k_2 C_{NO}^2 C_{O_2}} \tag{5.160}$$

Assuming plug flow reactor behavior, the final product yield can be calculated as:

$$\phi_p = -\frac{1}{C_{NOi} - C_{NOf}} \int_{C_{NOi}}^{C_{NOf}} \frac{k_1 C_{HbFeO_2}}{k_1 C_{HbFeO_2} + 4k_2 C_{O_2} C_{NO}} dC_{NO} \tag{5.161}$$

or $\quad \phi_p = -\dfrac{k_1 C_{HbFeO_2}}{(4k_2 C_{NO}) C_{NOi} - C_{NOf}} \ln\left(\dfrac{k_1 C_{HbFeO_2} + 4k_2 C_{O_2} C_{NOf}}{k_1 C_{HbFeO_2} + 4k_2 C_{O_2} C_{NOi}}\right) \tag{5.162}$

The transport of *NO* is similar to the transport of oxygen, but not exactly the same. Unlike oxygen, *NO* is generated via a surface reaction in the surface of the endothelium. *NO* diffuses in the blood and tissue. Consider the schematic in Fig. 5.9.

Capillary radii through which blood flows are approximately 25 to 75 μm. A mathematical model is developed to describe the diffusion of *NO* in the blood and tissue. The distance over which *NO* acts

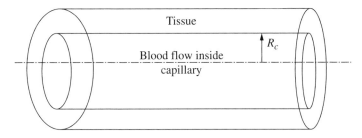

FIGURE 5.9   *NO diffusion in blood and tissue.*

in the tissue can be calculated. The governing equation for *NO* diffusion in the blood and tissue, including Fick and non-Fick transient diffusion effects, can be written for the blood and tissue space as:
   Blood space:

$$\frac{\partial^2 C_{NO}}{\partial \tau^2} + (1+k^*)\frac{\partial C_{NO}}{\partial \tau} + k^* C_{NO} = \frac{\partial^2 C_{NO}}{\partial X^2} + \frac{1}{X}\frac{\partial C_{NO}}{\partial X} \qquad (5.163)$$

where:

$$X = \frac{r}{\sqrt{D_{bl}\tau_r}}\; ; k^* = k\tau_r \; ; \tau = \frac{t}{\tau_r} \qquad (5.164)$$

Tissue space:

$$\frac{\partial^2 C_{NO}}{\partial \tau^2} + (1+k^*)\frac{\partial C_{NO}}{\partial \tau} + k^* C_{NO} = \left(\frac{D_T}{D_{bl}}\right)\left(\frac{\partial^2 C_{NO}}{\partial X^2} + \frac{1}{X}\frac{\partial C_{NO}}{\partial X}\right) \qquad (5.165)$$

where $D_T$ and $D_{bl}$ are the diffusion coefficients of *NO* in tissue space and blood space, respectively. $\tau_{rbl}$ and $\tau_{rT}$ are the relaxation times of *NO* in the blood and tissue space, respectively, and are assumed to be equal as a first approximation and are $\tau_r$.
   The time and blood and tissue space conditions for $C_{NO}$ can be written as follows:

$$r = 0, \frac{\partial C_{NO}}{\partial r} = 0 \qquad (5.166)$$

$$r = R_c, C_{NO}(tissue) = C_{NO}(blood) \qquad (5.167)$$

$$r = \infty, \frac{\partial C_{NO}}{\partial r} = 0 \qquad (5.168)$$

$$r = R_c, \dot{\Gamma} = D_{bl}\frac{\partial C_{NO}}{\partial r} - D_T\frac{\partial C_{NO}}{\partial r} \qquad (5.169)$$

where $\dot{\Gamma}$ is the metabolic production rate of NO at the surface of the capillary in the endothelium. Assuming no accumulation at the surface, the metabolic production rate of NO has to be equal to the sum of the rates of diffusion into the blood and tissue space. Due to opposite directions of NO movement in the blood and tissue spaces, there is a sign change in front of the Fick term in the blood space in Eq. (5.169):

$$t = 0, C_{NO} = C_{NOi} \tag{5.170}$$

Heterogeneity is introduced by Eqs. (5.169), (5.163), and (5.165), which can be solved for in the following manner:

Let the solution be assumed to consist of steady-state and transient parts;

$$C_{NO} = C_{NO}^t + C_{NO}^{ss} \tag{5.171}$$

The solutions to Eqs. (5.163) and (5.165) are same as the solutions to:

$$\frac{\partial^2 C_{NO}^t}{\partial \tau^2} + (1 + k^*)\frac{\partial C_{NO}^t}{\partial \tau} + k^* C_{NO}^t = \frac{\partial^2 C_{NO}^t}{\partial X^2} + \frac{1}{X}\frac{\partial C_{NO}^t}{\partial X} \tag{5.172}$$

and

$$k^* C_{NO}^{ss} = \frac{\partial^2 C_{NO}^{ss}}{\partial X^2} + \frac{1}{X}\frac{\partial C_{NO}^{ss}}{\partial X} \tag{5.173}$$

and

$$\frac{\partial^2 C_{NO}^t}{\partial \tau^2} + (1 + k^*)\frac{\partial C_{NO}^t}{\partial \tau} + k^* C_{NO}^t = \left(\frac{D_T}{D_{bl}}\right)\left(\frac{\partial^2 C_{NO}^t}{\partial X^2} + \frac{1}{X}\frac{\partial C_{NO}^t}{\partial X}\right) \tag{5.174}$$

and

$$k^* C_{NO}^{ss} = \left(\frac{D_T}{D_{bl}}\right)\left(\frac{\partial^2 C_{NO}^{ss}}{\partial X^2} + \frac{1}{X}\frac{\partial C_{NO}^{ss}}{\partial X}\right) \tag{5.175}$$

The boundary condition in Eq. (5.169) is applied in its entirety to the solution of the steady-state component of the solution. The transient part of the solution has a co-continuous derivative, making the equations homogeneous. The solution to Eq. (5.173) and the concentration profile for NO at steady state in the blood space can be written as:

$$C_{NO}^{ss} = cI_0(\sqrt{k^*}X) + dK_0(\sqrt{k^*}X) \tag{5.176}$$

Applying the boundary condition given by Eq. (5.173), it can be seen that $d$ has to be set to zero and:

$$C_{NO}^{ss} = cI_0(\sqrt{k^*}X) \tag{5.177}$$

The solution to Eq. (5.175) and the concentration profile of $NO$ in the tissue space can be written as:

$$C_{NO}^{ss} = eI_0\left(\sqrt{\left(\frac{k\tau_r D_{bl}}{D_T}\right)}X\right) + fK_0\left(\sqrt{\left(\frac{k\tau_r D_{bl}}{D_T}\right)}X\right) \tag{5.178}$$

The application of the boundary condition given by Eq. (5.168) leads to setting $e$ as zero in Eq. (5.168):

$$C_{NO}^{ss} = fK_0\left(\sqrt{\left(\frac{k\tau_r D_{bl}}{D_T}\right)}X\right) \tag{5.179}$$

The constants $c$ and $f$ in Eqs. (5.177), (5.178), and (5.179) can be obtained by applying the boundary conditions given by Eqs. (5.167) and (5.169). Thus, the steady-state concentration profile in the blood space can be given by:

$$C_{NO}^{ss} = \frac{\dot{\Gamma} K_0\left(\sqrt{\frac{k}{D_T}}R_c\right)I_0\left(\sqrt{\frac{k}{D_{bl}}}r\right)}{\sqrt{kD_{bl}}I_1\left(\sqrt{\frac{k}{D_{bl}}}R_c\right)K_0\left(\sqrt{\frac{k}{D_T}}R_c\right) - \sqrt{kD_T}K_1\left(\sqrt{\frac{k}{D_T}}R_c\right)I_0\left(\sqrt{\frac{k}{D_{bl}}}R_c\right)} \tag{5.180}$$

and the concentration profile in the tissue space can be given by:

$$C_{NO}^{ss} = \frac{\dot{\Gamma} I_0\left(\sqrt{\frac{k}{D_{bl}}}R_c\right)K_0\left(\sqrt{\frac{k}{D_T}}r\right)}{\sqrt{kD_{bl}}I_1\left(\sqrt{\frac{k}{D_{bl}}}R_c\right)K_0\left(\sqrt{\frac{k}{D_T}}R_c\right) - \sqrt{kD_T}K_1\left(\sqrt{\frac{k}{D_T}}R_c\right)I_0\left(\sqrt{\frac{k}{D_{bl}}}R_c\right)} \tag{5.181}$$

The solution to the transient portion of the concentration profile in the blood space can be obtained by separating the variables. Upon substituting $C_{NO}^{ss} = g(X)V(\tau)$, Eq. (5.172) becomes:

$$\frac{V''}{V} + (1+k^*)\frac{V'}{V} + k^* = \frac{g''}{g} + \frac{g'}{Xg} = -\lambda_n^2 \tag{5.182}$$

The space domain solution can be written as:

$$g(X) = cJ_0(\lambda_n X) + dY_0(\lambda_n X) \tag{5.183}$$

The boundary condition given by Eq. (5.168) can be applied to Eq. (5.183), and the integration constant $d$ has to be taken to be zero. Hence:

$$g(X) = cJ_0(\lambda_n X) \tag{5.184}$$

The solution to the transient portion of the concentration profile in the tissue space can be obtained by separating the variables. Upon substituting $C_{NO}^{ss} = g(X)V(\tau)$, Eq. (5.174) becomes:

$$\frac{V''}{V} + (1 + k^*)\frac{V'}{V} + k^* = \frac{g''}{g} + \frac{g'}{Xg} = -\lambda_n^2\left(\frac{D_{bl}}{D_T}\right) \tag{5.185}$$

The space domain solution can be written as:

$$g(X) = eJ_0\left(\frac{\lambda_n r}{\sqrt{D_T \tau_r}}\right) + fY_0\left(\frac{\lambda_n r}{\sqrt{D_T \tau_r}}\right) \tag{5.186}$$

The boundary condition given by Eq. (5.166) can be applied to Eq. (5.183), and the integration constant $e$ has to be taken to be zero. Hence:

$$g(X) = fY_0\left(\frac{\lambda_n r}{\sqrt{D_T \tau_r}}\right) \tag{5.187}$$

The boundary condition given by Eq. (5.167) can be applied to get $f$ in terms of integration constant $c$:

$$f = c\frac{J_0\left(\dfrac{\lambda_n R_c}{\sqrt{D_{bl}\tau_r}}\right)}{Y_0\left(\dfrac{\lambda_n R_c}{\sqrt{D_T\tau_r}}\right)} \tag{5.188}$$

The eigenvalues, $\lambda_n$, can be solved for by applying the boundary condition given by Eq. (5.169):

$$\sqrt{\frac{D_{bl}}{D_T}}J_1\left(\frac{\lambda_n R_c}{\sqrt{D_{bl}\tau_r}}\right) = Y_1\left(\frac{\lambda_n R_c}{\sqrt{D_T\tau_r}}\right)\frac{J_0\left(\dfrac{\lambda_n R_c}{\sqrt{D_{bl}\tau_r}}\right)}{Y_0\left(\dfrac{\lambda_n R_c}{\sqrt{D_T\tau_r}}\right)} \tag{5.189}$$

The solution to the time domain portion of the concentration profile in the blood space can be obtained by solving the following second-order ordinary differential equation (ODE):

$$V'' + (1+k^*)V' + V(k^* + \lambda_n^2) = 0 \tag{5.190}$$

The solution to Eq. (5.190) can be seen to be:

$$V = e^{-\frac{\tau}{2}} \left( c_1 e^{+\frac{\tau\sqrt{(1-k^*)^2 - 4\lambda_n^2}}{2}} + c_2 e^{-\frac{\tau\sqrt{(1-k^*)^2 - 4\lambda_n^2}}{2}} \right) \tag{5.191}$$

As the system reaches steady state, at large times, $Vge^{\tau/2}$ cannot be infinity or unbounded. Therefore, $c_1$ in Eq. (5.191) has to be set to zero. The general solution to the transient concentration profile in the blood space can be written as:

$$C_{NO}^t = \sum_0^\infty c_n e^{-\frac{\tau(1+k^*)}{2}} e^{-\frac{\tau\sqrt{(1-k^*)^2 - 4\lambda_n^2}}{2}} J_0(\lambda_n X) \tag{5.192}$$

The integration constant, $c_n$, can be solved for from the initial condition given by Eq. (5.170), and is obtained as:

$$c_n = \frac{C_{NOi} \int_0^{R_c} J_0(\lambda_n X) dX}{\int_0^{R_c} J_0^2(\lambda_n X) dX} \tag{5.193}$$

It can be realized that Eq. (5.192) is bifurcated. For large values of $\lambda_n$, the monotonic exponential decay behavior of the time domain portion of the solution will change in characteristic to a *subcritical damped oscillatory state*. This happens when:

$$\lambda_n > \frac{|1 - k^*|}{2} \tag{5.194}$$

When this happens, the general solution for the concentration profile of NO in the blood space can be seen to be:

$$C_{NO}^t = \sum_0^\infty c_n e^{-\frac{\tau(1+k^*)}{2}} \cos\left(\tau\sqrt{4\lambda_n^2 - (1-k^*)^2}\right) J_0(\lambda_n X) \tag{5.195}$$

## Summary

Oxyhemoglobin dissociation to hemoglobin and oxygen occurs through an equilibrium chemical reaction. The rate expression for oxygen was written. The Hill equation was derived. The Hill plot (Fig. 5.1) contains the saturated hemoglobin $\phi$ versus the partial pressure of oxygen. The curve is sigmoidal in shape. pH and organic phosphates can cause a change in the equilibrium rate constant, $K_{eq}$, values. During Bohr shift, the curve in the Hill plot shifts to the right. The change with temperature of the $K_{eq}$ rate constant is captured by van't Hoff's equation.

Oxygen availability becomes limited in some regions of the tissue. An oxygen reaction and diffusion model was developed by Colton [2]. The transient diffusion of oxygen in spherical coordinates that undergoes a simultaneous reaction by Michaelis-Menten kinetics was modeled. At the asymptotic limit of high reactant concentration, the reaction rate becomes zero order. The model solution consists of steady-state and transient parts, which removes the heterogeneity in the governing equation. The transient concentration profile is obtained by separating the variables. The solution is presented as an infinite Bessel series of 1/2th order and first kind. At a low concentration of oxygen, the kinetics of oxygenation reverts to a first-order expression. The transient and steady-state solutions are obtained. The eigenvalues have to be solved for from a transcendental equation. A Biot number (mass) was introduced.

During the course of oxygen consumption, the finite speed effects cannot be ignored. Simultaneous diffusion and reaction of the first order using the damped wave diffusion and reaction equation in a semi-infinite medium in Cartesian coordinates was studied. The closed-form analytical solution was obtained by the method of relativistic transformation. Four regimes in solution were identified: (a) an inertial regime of null transfer, (b) a regime at short times characterized by the Bessel composite function in space and time and rate constant, (c) a regime at long times characterized by a modified Bessel composite function in space and time and rate constant, and (d) a wavefront regime. The mass inertial lag time as a function of rate constant was obtained and tabulated in Table 5.1. Different characteristic behaviord at different dimensionless rate constants $k^*$ were found and shown in Figs. 5.3–5.6.

The Krogh tissue cylinder was modeled under transient conditions, and the kinetics obeyed the asymptotic limit of high concentration of oxygen (zeroth-order rate) and low concentration of oxygen (first-order rate). For intermediate values, a numerical solution is needed. An infinite Fourier series solution was obtained. Heterogeneity in the governing equation was removed by supposing that the concentration is a sum of steady and transient states.

The transient concentration profile of oxygen in the plasma layer in the absence of any reaction was modeled. The damped wave diffusion and relaxation effects were also accounted for in a separate section. The infinite Fourier series solution was seen to be bifurcated. At large relaxation times, $\tau_r > (R_{RB}^2/58.7D_T)$, the concentration of oxygen will undergo subcritical damped oscillations.

$NO$ is a vasodilator. The widening of blood vessels when the surrounding smooth muscle cells relax is called vasodilation. $NO$ has to diffuse through the blood vessel wall and then binds to the enzyme in smooth muscle cells. It participates in a set of reactions in parallel. The instantaneous fractional yield of heme complex during parallel reactions of *NO* was solved for and shown in Fig. 5.8.

*NO* diffusion in blood and tissue is similar to that of oxygen, but not exactly the same. The steady-state concentration profile for *NO* in blood and tissue was obtained from the model solutions and shown in Fig. 5.8. It is characterized by modified Bessel functions of the first and second kind and zeroth and first orders. The solution to the transient portion of the concentration profile in the blood space is obtained by separating the variables. The damped wave diffusion effects were included in the model. For materials with large relaxation times, the concentration of *NO* can be expected to be subcritical damped oscillatory in nature.

# References

[1]  R. L. Fournier, *Basic Transport Phenomena in Biomedical Engineering*, Boca Raton, FL: Taylor & Francis, CRC Press, 1999.

[2]  C. K. Colton and E. S. Avgoustiniatos, "Bioengineering in development of the hybrid artificial pancreas," *J. Biomech. Eng.* (1991), 113, 152–170.

[3]  E. S. Avgoustiniatos, K. E. Dionne, D. F. Wilson, M. L. Yarmush, and C. K. Colton, "Measurements of the effective diffusion coefficient of oxygen in pancreatic islets," *IEC, Res.*, (2007), 46, 6157–6163.

[4]  O. Levenspiel, *Chemical Reaction Engineering, Third Edition*, New York: John Wiley & Sons, 1999.

[5]  A. Varma and M. Morbidelli, *Mathematical Methods in Chemical Engineering*, Oxford, UK: Oxford University Press, 1997.

[6]  K. R. Sharma, *Damped Wave Transport and Relaxation*, Amsterdam, Netherlands: Elsevier, 2005.

[7]  K. Groebe, "An easy-to-use model for oxygen supply to red muscle: Validity of assumptions, sensitivity to errors in data," *Biophys. Journal*, (1995), 68, 1246–1269.

[8]  G. A. Truskey, F. Yuan, and D. F. Katz, *Transport Phenomena in Biological Systems*, Upper Saddle River, NJ: Pearson Prentice Hall, 2009.

# Exercises

## Review Questions

**1.0**  What is the difference between myoglobin and hemoglobin?

**2.0**  Does myoglobin undergo cooperative binding with oxygen?

**3.0**  Does hemoglobin undergo cooperative binding with oxygen?

**4.0**  When does the sigmoidal curve in the Hill plot shift to the right?

**5.0**  When does the sigmoidal curve in the Hill plot shift to the left?

**6.0**  At the time when the Krogh tissue cylinder model arrived, were transient effects studied?

**7.0**  What is the importance of hypoxia in the islets of Langerhans and the treatment of diabetes?

**8.0**   Is the diffusion of *NO* and oxygen the same in human anatomy?

**9.0**   What does vasoconstriction mean?

**10.0**   How can the oxygen-depleted regions be detected from the mathematical model solutions?

**11.0**   By mathematically modeling the diffusion of oxygen in the tissue space, it was found that after a critical distance from the blood capillary, there is a zone of null transfer. These are the oxygen-depleted regions in the tissue. Consider an idealized geometry of a sphere. What would be the shape of the zone of zero transfer?

**12.0**   In Exercise 11.0, consider an idealized geometry of a cylinder. What would be the shape of the zone of zero transfer?

**13.0**   In Exercise 11.0, consider an idealized geometry of a solid with a equilateral triangular cross-sectional area. What would be the shape of the zone of zero transfer?

**14.0**   Would carbon dioxide ($CO_2$) be transported in a similar manner compared with oxygen in the blood and tissue spaces?

**15.0**   Would there be $CO_2$-accumulated regions in the tissue in human anatomy?

**16.0**   What would happen to the model developed for the islets of Langerhans for diffusion and reaction of oxygen if the reversible reaction is accounted for instead of the Michaelis-Menten kinetics?

**17.0**   What would happen to the model developed in semi-infinite Cartesian coordinates in order to account for simultaneous diffusion and reaction if the rate expression is reversible rate instead of the first-order rate used?

**18.0**   Can the damped wave diffusion and relaxation equation violate the second law of thermodynamics?

**19.0**   How many regimes of solution are present in the solution of the damped wave diffusion and relaxation equation in a semi-infinite medium in Cartesian coordinates?

**20.0**   What happens during the inertial regime?

## Problems

**21.0**   *Parabolic diffusion of oxygen in islets of Langerhans in a finite slab.* The governing equation for oxygen diffusion and reaction in Cartesian coordinates in one dimension, accounting for Fick's diffusion, can be written as:

$$\frac{\partial C_{O_2}}{\partial t} = D_T \left( \frac{\partial^2 C_{O_2}}{\partial z^2} \right) - \frac{C_{E0} C_{O_2}}{C_M + C_{O_2}} \qquad (5.196)$$

where $D_T$ = the diffusion coefficient in the tissue
$\qquad C_{E0}$ = the total enzyme or complexation species concentration
$\qquad C_M$ = the Michaelis constant

The oxygen consumption rate is assumed to obey the Michaelis-Menten kinetics. The governing equation describes the interplay of transient diffusion and metabolic consumption of oxygen in the tissue in spherical coordinates. The concentration of oxygen, $C_{O_2}$, can be expressed in terms of its partial $p_{O_2}$. This is obtained by using the Bunsen solubility coefficient $\alpha_t$ such that:

$$C_{O_2} = \alpha_t p_{O_2} \tag{5.197}$$

Substituting Eq. (5.197) in Eq. (5.196), Eq. (5.196) becomes:

$$\alpha_t \frac{\partial p_{O_2}}{\partial t} = \alpha_t D_T \left( \frac{\partial^2 p_{O_2}}{\partial z^2} \right) - \frac{C_{E0} p_{O_2}}{C_M' + p_{O_2}} \tag{5.198}$$

The product $\alpha_t D_T$ can be seen to be the product of solubility and diffusivity, and hence is the permeability of oxygen in the tissue. The Michaelis constant, $C_M$, is also modified, $C_M'$, expressed in units of mmHg. The initial condition can be written as:

$$p_{O_2} = p_{O_{20}}, \, t = 0 \tag{5.199}$$

From the symmetry at the center of the slab:

$$\frac{\partial p_{O_2}}{\partial z} = 0 \tag{5.200}$$

At the surface ($z = \pm a$), the oxygen diffusive transport from within the tissue must be equal to the oxygen transport by convection across the boundary layer surrounding each tissue:

$$J_i = k_i \alpha_m (p_{O_{2m}} - p_{O_2}(R_i)) = \alpha_t D_T \left. \frac{\partial p_{O_2}}{\partial z} \right|_{z=a} \tag{5.201}$$

Obtain the concentration profile for oxygen in the finite slab at the asymptotic limit of high oxygen concentration.

**22.0** Obtain the concentration profile for oxygen in a finite slab at the asymptotic limit of low oxygen concentration in Problem 11.0.

**23.0** *Hyperbolic diffusion of oxygen in islets of Langerhans in a finite slab.* When the damped wave diffusion and relaxation effects are included, the governing equation for Problem 11.0 can be written in Cartesian coordinates as:

$$\alpha_t \tau_{mr} \frac{\partial^2 p_{O_2}}{\partial t^2} + \alpha_t \frac{\partial p_{O_2}}{\partial t} = \alpha_t D_T \left( \frac{\partial^2 p_{O_2}}{\partial z^2} \right) - \frac{C_{E0} p_{O_2}}{C_M' + p_{O_2}} \tag{5.202}$$

For the same set of time and space conditions as in Eqs. (5.199), (5.200), and (5.201), obtain the closed-form analytical solution for Eq. (5.202) in the asymptotic limit of high oxygen concentration.

**24.0**   Obtain the concentration profile for oxygen in a finite slab at the asymptotic limit of low oxygen concentration, taking into account the damped wave diffusion and relaxation effects. The governing equation is given by Eq. (5.202), and the time and space conditions are given by Eqs. (5.199), (5.200), and (5.201).

**25.0**   *Pulse boundary condition–finite slab–hyperbolic diffusion.* Obtain the closed-form analytical solution for Eq. (5.202) in the asymptotic limit of high oxygen concentration for the pulse boundary condition. In the time and space conditions given in Eqs. (5.199), (5.200). and (5.201), instead of Eq. (5.201), use the following boundary condition at $z = \pm a$ at the surface of a finite slab of width 2a:

$$t = 0, p_{O_2} = p^*\delta(a) \tag{5.203}$$

$$t > 0, \frac{\partial p_{O_2}}{\partial z} = 0 \tag{5.204}$$

**26.0**   *Pulse boundary condition–finite slab–hyperbolic diffusion.* Obtain the closed-form analytical solution for Eq. (5.202) in the asymptotic limit of low oxygen concentration for the pulse boundary condition. In the time and space conditions given in Eqs. (5.199), (5.200), and (5.201), instead of Eq. (5.201), use the boundary condition at $z = \pm a$ at the surface of a finite slab of width 2a given by Eqs. (5.203)and (5.204).

**27.0**   *Pulse boundary condition–finite slab–parabolic diffusion.* Obtain the closed-form analytical solution for Eq. (5.198) in the asymptotic limit of high oxygen concentration for the pulse boundary condition. In the time and space conditions given in Eqs. (5.199), (5.200), and (5.201), instead of Eq. (5.201), use the following boundary condition at $z = \pm a$ at the surface of a finite slab of width 2a given by Eqs. (5.203) and (5.204).

**28.0**   *Pulse boundary condition–finite slab–parabolic diffusion.* Obtain the closed-form analytical solution for Eq. (5.198) in the asymptotic limit of low oxygen concentration for the pulse boundary condition. In the time and space conditions given in Eqs. (5.199), (5.200) and (5.201), instead of Eq. (5.201), use the following boundary condition at $z = \pm a$ at the surface of a finite slab of width 2a given by Eqs. (5.203) and (5.204).

**29.0**   *Pulse boundary condition–finite cylinder–hyperbolic diffusion.* Obtain the closed-form analytical solution for the following governing equation that can be used to describe oxygen diffusion in a finite cylinder undergoing simultaneous reaction kinetics of the Michaelis-Menten type:

$$\tau_{mr}\alpha_t \frac{\partial^2 p_{O_2}}{\partial t^2} + \alpha_t \frac{\partial p_{O_2}}{\partial t} = \alpha_t D_T \left(\frac{1}{r}\right)\frac{\partial}{\partial r}\left(r\frac{\partial p_{O_2}}{\partial r}\right) - \frac{C_{E0}p_{O_2}}{C_M' + p_{O_2}} \tag{5.205}$$

in the asymptotic limit of high oxygen concentration for the pulse boundary condition. In the time and space conditions given in Eqs. (5.199), (5.200) and

(5.201), instead of Eq. (5.201), use the boundary condition at $r = R_i$ at the surface of a finite cylinder of radius $R_i$ given by Eqs. (5.203) and (5.204).

**30.0**  *Pulse boundary condition–finite cylinder–hyperbolic diffusion.* Obtain the closed-form analytical solution for Eq. (5.205) in the asymptotic limit of low oxygen concentration for the pulse boundary condition. In the time and space conditions given in Eqs. (5.199), (5.200) and (5.201), instead of Eq. (5.201), use the boundary condition at $r = R_i$ at the surface of a finite cylinder of radius $R_i$ given by Eqs. (5.203) and (5.204).

**31.0**  *Pulse boundary condition–finite cylinder–parabolic diffusion.* Obtain the closed-form analytical solution for Eq. (5.206) in the asymptotic limit of high oxygen concentration for the pulse boundary condition. In the time and space conditions given in Eqs. (5.199), (5.200) and (5.201), instead of Eq. (5.201), use the boundary condition at $r = R_i$ at the surface of a finite cylinder of radius $R_i$ given by Eqs. (5.203) and (5.204):

$$\alpha_t \frac{\partial p_{O_2}}{\partial t} = \alpha_t D_T \left( \frac{1}{r} \right) \frac{\partial}{\partial r} \left( r \frac{\partial p_{O_2}}{\partial r} \right) - \frac{C_{E0} p_{O_2}}{C'_M + p_{O_2}} \tag{5.206}$$

**32.0**  *Pulse boundary condition–finite cylinder–parabolic diffusion.* Obtain the closed-form analytical solution for Eq. (5.206) in the asymptotic limit of low oxygen concentration for the pulse boundary condition. In the time and space conditions given in Eqs. (5.199), (5.200) and (5.201), instead of Eq. (5.201), use the boundary condition at $r = R_i$ at the surface of a finite cylinder of radius $R_i$ given by Eqs. (5.203) and (5.204).

**33.0**  *Convective boundary condition–finite cylinder–hyperbolic diffusion.* Obtain the closed-form analytical solution for the governing equation given by Eq. (5.205) that can be used to describe oxygen diffusion in a finite cylinder undergoing simultaneous reaction kinetics of the Michaelis-Menten type:

$$\tau_{mr} \alpha_t \frac{\partial^2 p_{O_2}}{\partial t^2} + \alpha_t \frac{\partial p_{O_2}}{\partial t} = \alpha_t D_T \left( \frac{1}{r} \right) \frac{\partial}{\partial r} \left( r \frac{\partial p_{O_2}}{\partial r} \right) - \frac{C_{E0} p_{O_2}}{C'_M + p_{O_2}}$$

in the asymptotic limit of high oxygen concentration for the convective boundary condition. Use the time and space conditions given in Eqs. (5.199), (5.200), and (5.201).

**34.0**  *Convective boundary condition–finite cylinder–hyperbolic diffusion.* Obtain the closed-form analytical solution for Eq. (5.205) in the asymptotic limit of low oxygen concentration for the pulse boundary condition. Use the time and space conditions given in Eqs. (5.199), (5.200), and (5.201).

**35.0**  *Convective boundary condition–finite cylinder–parabolic diffusion.* Obtain the closed-form analytical solution for Eq. (5.206) in the asymptotic limit of high oxygen concentration for the pulse boundary condition. Use the time and space conditions given in Eqs. (5.199), (5.200) and (5.201).

**36.0**  *Convective boundary condition–finite cylinder–parabolic diffusion.* Obtain the closed-form analytical solution for Eq. (5.206) in the asymptotic limit of low oxygen concentration for the pulse boundary condition. Use the time and space conditions given in Eqs. (5.199), (5.200) and (5.201).

**37.0** *Pulse boundary condition–finite sphere–hyperbolic diffusion.* Obtain the closed-form analytical solution for Eq. (5.207) that can be used to describe oxygen diffusion in a finite cylinder undergoing simultaneous reaction kinetics of the Michaelis-Menten type:

$$\tau_{mr}\alpha_t \frac{\partial^2 p_{O_2}}{\partial t^2} + \alpha_t \frac{\partial p_{O_2}}{\partial t} = \alpha_t D_T \left(\frac{1}{r^2}\right)\frac{\partial}{\partial r}\left(r^2 \frac{\partial p_{O_2}}{\partial r}\right) - \frac{C_{E0}p_{O_2}}{C'_M + p_{O_2}} \qquad (5.207)$$

in the asymptotic limit of high oxygen concentration for the pulse boundary condition. In the time and space conditions given in Eqs. (5.199), (5.200), and (5.201), instead of Eq. (5.201), use the boundary condition at $r = R_i$ at the surface of a finite cylinder of radius $R_i$ given by Eqs. (5.201) and (5.202).

**38.0** *Pulse boundary condition–finite sphere–hyperbolic diffusion.* Obtain the closed-form analytical solution for Eq. (5.207) in the asymptotic limit of low oxygen concentration for the pulse boundary condition. In the time and space conditions given in Eqs. (5.199), (5.200), and (5.201), instead of Eq. (5.201), use the boundary condition at $r = R_i$ at the surface of a finite cylinder of radius $R_i$ given by Eqs. (5.203) and (5.204).

**39.0** *Pulse boundary condition–finite sphere–parabolic diffusion.* Obtain the closed-form analytical solution for Eq. (5.206) in the asymptotic limit of high oxygen concentration for the pulse boundary condition. In the time and space conditions given in Eqs. (5.199), (5.200) and (5.201), instead of Eq. (5.201), use the boundary condition at $r = R_i$ at the surface of a finite cylinder of radius $R_i$ given by Eqs. (5.203) and (5.204):

$$\alpha_t \frac{\partial p_{O_2}}{\partial t} = \alpha_t D_T \left(\frac{1}{r^2}\right)\frac{\partial}{\partial r}\left(r^2 \frac{\partial p_{O_2}}{\partial r}\right) - \frac{C_{E0}p_{O_2}}{C'_M + p_{O_2}} \qquad (5.208)$$

**40.0** *Pulse boundary condition–finite sphere–parabolic diffusion.* Obtain the closed-form analytical solution for Eq. (5.208) in the asymptotic limit of low concentration for the pulse boundary condition. In the time and space conditions given in Eqs. (5.199), (5.200), and (5.201), instead of Eq. (5.201), use the boundary condition at $r = R_i$ at the surface of a finite cylinder of radius $R_i$ given by Eqs. (5.201) and (5.204).

**41.0** *Periodic boundary condition–finite slab–hyperbolic diffusion.* Obtain the closed-form analytical solution for Eq. (5.202) in the asymptotic limit of high oxygen concentration for the pulse boundary condition. In the time and space conditions given in Eqs. (5.199), (5.200), and (5.201), instead of Eq. (5.20!), use the following boundary condition at $z = \pm a$ at the surface of a finite slab of width $2a$:

$$t > 0, p_{O_2} = p^* + A\sin\omega t \qquad (5.209)$$

**42.0** *Periodic boundary condition–finite slab–hyperbolic diffusion.* Obtain the closed-form analytical solution for Eq. (5.202) in the asymptotic limit of low oxygen concentration for the pulse boundary condition. In the time and space conditions given in Eqs. (5.199), (5.200), and (5.201), instead of Eq. (5.201), use the boundary condition at $z = \pm a$ at the surface of a finite slab of width $2a$ given by Eqs. (5.209).

**43.0** *Periodic boundary condition–finite slab–parabolic diffusion.* Obtain the closed-form analytical solution for Eq. (5.198) in the asymptotic limit of high oxygen concentration for the pulse boundary condition. In the time and space conditions given in Eqs. (5.199), (5.200), and (5.201), instead of Eq. (5.201), use the boundary condition at $z = \pm a$ at the surface of a finite slab of width $2a$ given by Eq. (5.209).

**44.0** *Periodic boundary condition–finite slab–parabolic diffusion.* Obtain the closed-form analytical solution for Eq. (5.198) in the asymptotic limit of low oxygen concentration for the pulse boundary condition. In the time and space conditions given in Eqs. (5.199), (5.200), and (5.201), instead of Eq. (5.201), use the boundary condition at $z = \pm a$ at the surface of a finite slab of width $2a$ given by Eq. (5.209).

**45.0** *Periodic boundary condition–finite cylinder–hyperbolic diffusion.* Obtain the closed-form analytical solution for the following governing equation that can be used to describe oxygen diffusion in a finite cylinder undergoing simultaneous reaction kinetics of the Michaelis-Menten type:

$$\tau_{mr}\alpha_t\frac{\partial^2 p_{O_2}}{\partial t^2} + \alpha_t\frac{\partial p_{O_2}}{\partial t} = \alpha_t D_T\left(\frac{1}{r}\right)\frac{\partial}{\partial r}\left(r\frac{\partial p_{O_2}}{\partial r}\right) - \frac{C_{E0}p_{O_2}}{C'_M + p_{O_2}} \qquad (5.210)$$

in the asymptotic limit of high oxygen concentration for the pulse boundary condition. In the time and space conditions given in Eqs. (5.199), (5.200), and (5.201), instead of Eq. (5.201), use the boundary condition at $r = R_i$ at the surface of a finite cylinder of radius $R_i$ given by Eq. (5.209).

**46.0** *Periodic boundary condition–finite cylinder–hyperbolic diffusion.* Obtain the closed-form analytical solution for Eq. (5.210) in the asymptotic limit of low oxygen concentration for the pulse boundary condition. In the time and space conditions given in Eqs. (5.199), (5.200), and (5.201), instead of Eq. (5.201), use the boundary condition at $r = R_i$ at the surface of a finite cylinder of radius $R_i$ given by Eq. (5.209).

**47.0** *Periodic boundary condition–finite cylinder–parabolic diffusion.* Obtain the closed-form analytical solution for Eq. (5.211) in the asymptotic limit of high oxygen concentration for the pulse boundary condition. In the time and space conditions given in Eqs. (5.199), (5.200), and (5.201), instead of Eq. (5.201), use the boundary condition at $r = R_i$ at the surface of a finite cylinder of radius $R_i$ given by Eq. (5.209):

$$\alpha_t\frac{\partial p_{O_2}}{\partial t} = \alpha_t D_T\left(\frac{1}{r}\right)\frac{\partial}{\partial r}\left(r\frac{\partial p_{O_2}}{\partial r}\right) - \frac{C_{E0}p_{O_2}}{C'_M + p_{O_2}} \qquad (5.211)$$

**48.0** *Periodic boundary condition–finite cylinder–parabolic diffusion.* Obtain the closed-form analytical solution for Eq. (5.211) in the asymptotic limit of low oxygen concentration for the pulse boundary condition. In the time and space conditions given in Eqs. (5.199), (5.200), and (5.201), instead of Eq. (5.201), use the boundary condition at $r = R_i$ at the surface of a finite cylinder of radius $R_i$ given by Eq. (5.209).

**49.0**  *Periodic boundary condition–finite sphere–hyperbolic diffusion.* Obtain the closed-form analytical solution for Eq. (5.212) that can be used to describe oxygen diffusion in a finite cylinder undergoing simultaneous reaction kinetics of the Michaelis-Menten type:

$$\tau_{mr}\alpha_t \frac{\partial^2 p_{O_2}}{\partial t^2} + \alpha_t \frac{\partial p_{O_2}}{\partial t} = \alpha_t D_T \left(\frac{1}{r^2}\right) \frac{\partial}{\partial r}\left(r^2 \frac{\partial p_{O_2}}{\partial r}\right) - \frac{C_{E0}p_{O_2}}{C_M' + p_{O_2}} \tag{5.212}$$

in the asymptotic limit of high oxygen concentration for the pulse boundary condition. In the time and space conditions given in Eqs. (5.199), (5.200), and (5.201), instead of Eq. (5.201), use the boundary condition at $r = R_i$ at the surface of a finite cylinder of radius $R_i$ given by Eq. (5.209).

**50.0**  *Periodic boundary condition–finite sphere–hyperbolic diffusion.* Obtain the closed-form analytical solution for Eq. (5.212) in the asymptotic limit of low oxygen concentration for the pulse boundary condition. In the time and space conditions given in Eqs. (5.199), (5.200), and (5.201), instead of Eq. (5.201), use the boundary condition at $r = R_i$ at the surface of a finite cylinder of radius $R_i$ given by Eq. (5.209).

**51.0**  *Periodic boundary condition–finite sphere–parabolic diffusion.* Obtain the closed-form analytical solution for Eq. (5.213) in the asymptotic limit of high oxygen concentration for the pulse boundary condition. In the time and space conditions given in Eqs. (5.199), (5.200), and (5.201), instead of Eq. (5.201), use the boundary condition at $r = R_i$ at the surface of a finite cylinder of radius $R_i$ given by Eq. (5.209):

$$\alpha_t \frac{\partial p_{O_2}}{\partial t} = \alpha_t D_T \left(\frac{1}{r^2}\right) \frac{\partial}{\partial r}\left(r^2 \frac{\partial p_{O_2}}{\partial r}\right) - \frac{C_{E0}p_{O_2}}{C_M' + p_{O_2}} \tag{5.213}$$

**52.0**  *Periodic boundary condition–finite sphere–parabolic diffusion.* Obtain the closed-form analytical solution for Eq. (5.213) in the asymptotic limit of low oxygen concentration for the pulse boundary condition. In the time and space conditions given in Eqs. (5.199), (5.200), and (5.201), instead of Eq. (5.201), use the boundary condition at $r = R_i$ at the surface of a finite cylinder of radius $R_i$ given by Eq. (5.209).

**53.0**  *Constant surface concentration condition–finite slab–hyperbolic diffusion.* Obtain the closed-form analytical solution for Eq. (5.202) in the asymptotic limit of high oxygen concentration for the pulse boundary condition. In the time and space conditions given in Eqs. (5.199), (5.200), and (5.201), instead of Eq. (5.201), use the following boundary condition at $z = \pm a$ at the surface of a finite slab of width $2a$:

$$t > 0, p_{O_2} = p^* \tag{5.214}$$

**54.0** *Constant surface concentration condition–finite slab–hyperbolic diffusion.* Obtain the closed-form analytical solution for Eq. (5.202) in the asymptotic limit of low oxygen concentration for the pulse boundary condition. In the time and space conditions given in Eqs. (5.199), (5.200), and (5.201), instead of Eq. (5.201), use the boundary condition at $z = \pm a$ at the surface of a finite slab of width 2a given by Eq. (5.214).

**55.0** *Constant surface concentration condition–finite slab–parabolic diffusion.* Obtain the closed-form analytical solution for Eq. (5.198) in the asymptotic limit of high oxygen concentration for the pulse boundary condition. In the time and space conditions given in Eqs. (5.199), (5.200), and (5.201), instead of Eq. (5.201), use the following boundary condition at $z = \pm a$ at the surface of a finite slab of width $2a$ given by Eq. (5.214).

**56.0** *Constant surface concentration condition–finite slab–parabolic diffusion.* Obtain the closed-form analytical solution for Eq. (5.198) in the asymptotic limit of low oxygen concentration for the pulse boundary condition. In the time and space conditions given in Eqs. (5.199), (5.200), and (5.201), instead of Eq. (5.201), use the following boundary condition at $z = \pm a$ at the surface of a finite slab of width $2a$ given by Eq. (5.214).

**57.0** *Constant surface concentration–finite cylinder–hyperbolic diffusion.* Obtain the closed-form analytical solution for the following governing equation that can be used to describe oxygen diffusion in a finite cylinder undergoing simultaneous reaction kinetics of the Michaelis-Menten type:

$$\tau_{mr}\alpha_t\frac{\partial^2 p_{O_2}}{\partial t^2} + \alpha_t\frac{\partial p_{O_2}}{\partial t} = \alpha_t D_T\left(\frac{1}{r}\right)\frac{\partial}{\partial r}\left(r\frac{\partial p_{O_2}}{\partial r}\right) - \frac{C_{E0}p_{O_2}}{C'_M + p_{O_2}} \qquad (5.215)$$

in the asymptotic limit of high oxygen concentration for the pulse boundary condition. In the time and space conditions given in Eqs. (5.199), (5.200), and (5.201), instead of Eq. (5.201), use the boundary condition at $r = R_i$ at the surface of a finite cylinder of radius $R_i$ given by Eq. (5.214).

**58.0** *Constant surface concentration–finite cylinder–hyperbolic diffusion.* Obtain the closed-form analytical solution for Eq. (5.205) in the asymptotic limit of low oxygen concentration for the pulse boundary condition. In the time and space conditions given in Eqs. (5.199), (5.200), and (5.201), instead of Eq. (5.201), use the boundary condition at $r = R_i$ at the surface of a finite cylinder of radius $R_i$ given by Eq. (5.214).

**59.0** *Constant surface concentration–finite cylinder–parabolic diffusion.* Obtain the closed-form analytical solution for Eq. (5.216) in the asymptotic limit of high oxygen concentration for the pulse boundary condition. In the time and space conditions given in Eqs. (5.199), (5.200), and (5.201), instead of Eq. (5.201), use the boundary condition at $r = R_i$ at the surface of a finite cylinder of radius $R_i$ given by Eq. (5.214):

$$\alpha_t\frac{\partial p_{O_2}}{\partial t} = \alpha_t D_T\left(\frac{1}{r}\right)\frac{\partial}{\partial r}\left(r\frac{\partial p_{O_2}}{\partial r}\right) - \frac{C_{E0}p_{O_2}}{C'_M + p_{O_2}} \qquad (5.216)$$

**60.0** *Constant surface concentration–finite cylinder–parabolic diffusion.* Obtain the closed-form analytical solution for Eq. (5.206) in the asymptotic limit of low oxygen concentration for the pulse boundary condition. In the time and space conditions given in Eqs. (5.199), (5.200), and (5.201), instead of Eq. (5.201), use the boundary condition at $r = R_i$ at the surface of a finite cylinder of radius $R_i$ given by Eq. (5.214).

**61.0** *Constant surface concentration–finite sphere–hyperbolic diffusion.* Obtain the closed-form analytical solution for Eq. (5.217) that can be used to describe oxygen diffusion in a finite cylinder undergoing simultaneous reaction kinetics of the Michaelis-Menten type:

$$\tau_{mr}\alpha_t \frac{\partial^2 p_{O_2}}{\partial t^2} + \alpha_t \frac{\partial p_{O_2}}{\partial t} = \alpha_t D_T \left(\frac{1}{r^2}\right) \frac{\partial}{\partial r}\left(r^2 \frac{\partial p_{O_2}}{\partial r}\right) - \frac{C_{E0}p_{O_2}}{C'_M + p_{O_2}} \qquad (5.217)$$

in the asymptotic limit of high oxygen concentration for the pulse boundary condition. In the time and space conditions given in Eqs. (5.199), (5.200), and (5.201), instead of Eq. (5.201), use the boundary condition at $r = R_i$ at the surface of a finite cylinder of radius $R_i$ given by Eq. (5.214).

**62.0** *Constant surface concentration–finite sphere–hyperbolic diffusion.* Obtain the closed-form analytical solution for Eq. (5.207) in the asymptotic limit of low oxygen concentration for the pulse boundary condition. In the time and space conditions given in Eqs. (5.199), (5.200), and (5.201), instead of Eq. (5.201), use the boundary condition at $r = R_i$ at the surface of a finite cylinder of radius $R_i$ given by Eq. (5.214).

**63.0** *Constant surface concentration–finite sphere–parabolic diffusion.* Obtain the closed-form analytical solution for Eq. (5.206) in the asymptotic limit of high oxygen concentration for the pulse boundary condition. In the time and space conditions given in Eqs. (5.199), (5.200), and (5.201), instead of Eq. (5.201), use the boundary condition at $r = R_i$ at the surface of a finite cylinder of radius $R_i$ given by Eq. (5.214):

$$\alpha_t \frac{\partial p_{O_2}}{\partial t} = \alpha_t D_T \left(\frac{1}{r^2}\right) \frac{\partial}{\partial r}\left(r^2 \frac{\partial p_{O_2}}{\partial r}\right) - \frac{C_{E0}p_{O_2}}{C'_M + p_{O_2}} \qquad (5.218)$$

**64.0** *Constant surface concentration–finite sphere–parabolic diffusion.* Obtain the closed-form analytical solution for Eq. (5.208) in the asymptotic limit of low oxygen concentration for the pulse boundary condition. In the time and space conditions given in Eqs. (5.199), (5.200), and (5.201), instead of Eq. (5.201), use the boundary condition at $r = R_i$ at the surface of a finite cylinder of radius $R_i$ given by Eq. (5.214).

**65.0** Equations (5.68) and (5.69) are not valid for $k^* \neq 1$. Derive a suitable expression for $k^* = 1$. The governing equation for simultaneous diffusion and reaction of oxygen with first-order kinetics, accounting for damped wave diffusion effects at $k^* = 1$, can be written as follows:

$$\frac{\partial^2 u}{\partial X^2} = \frac{\partial^2 u}{\partial \tau^2} + 2\frac{\partial u}{\partial \tau} + u \qquad (5.219)$$

Obtain the closed-form analytical solution by the method of Laplace transforms. The boundary conditions in space and time conditions for a semi-infinite medium in Cartesian coordinates can be the same as those given by Eq (5.62).

**66.0** Obtain the solution to the governing equation given by Eq. (5.61) for damped wave diffusion and relaxation of oxygen and simultaneous reaction of the first order by the method of Laplace transforms. The boundary conditions in space and time for a semi-infinite medium in Cartesian coordinates can be the same as those given by Eq. (5.62).

**67.0** *Semi-infinite medium-Cartesian-convective boundary–hyperbolic-first order.* Obtain the solution to the damped wave diffusion and relaxation and simultaneous reaction of oxygen of the first order in a semi-infinite medium. Use the governing equation in Eq. (5.61). The boundary conditions in space and time in Cartesian coordinates are as follows:

$$X = 0, D_T \frac{\partial C_{O_2}}{\partial x} = k''(C - C_b); X = \infty, C_{O_2} = 0$$

$$t = 0, C_{O_2} = 0$$

$$(5.220)$$

where $k''$ is the mass transfer coefficient between the bulk cross-flow with the surface of the semi-infinite slab.

**68.0** *Semi-infinite medium-Cartesian-constant concentration–parabolic-first order.* Obtain the solution to the diffusion and relaxation and simultaneous reaction of oxygen of the first order in a semi-infinite medium using the parabolic Fick's second law of diffusion. The boundary conditions in space and time for a semi-infinite medium in Cartesian coordinates are the same as in Eq. (5.62). The governing equation for the concentration of oxygen can be written as:

$$D_T \frac{\partial^2 C_{O_2}}{\partial x^2} = \frac{\partial C_{O_2}}{\partial t} + k C_{O_2}$$

$$(5.221)$$

**69.0** *Semi-infinite medium-Cartesian-constant concentration–hyperbolic-zeroth order.* Consider a semi-infinite medium at an initial concentration of zero. For times greater than zero, a step change in concentration is effected at one of the surfaces. The species reacts with a zeroth-order reaction as it comes in contact with the solid medium. Discuss the concentration profile as a function of space and time. The governing equation for damped wave diffusion and relaxation and simultaneous reaction of oxygen may be written as follows:

$$D_T \frac{\partial^2 C_{O_2}}{\partial x^2} = \tau_{mr} \frac{\partial^2 C_{O_2}}{\partial t^2} + \frac{\partial C_{O_2}}{\partial t} + k'''$$

$$(5.222)$$

The boundary conditions in space and time for a semi-infinite medium in Cartesian coordinates are the same as in Eq. (5.62).

**70.0** *Semi-infinite medium-Cartesian-convective boundary–hyperbolic-zeroth order.* Obtain the concentration profile of oxygen in space and time during damped wave diffusion and simultaneous reaction in a semi-infinite medium in Cartesian coordinates. The reaction order is of the zeroth order. The governing equation is given by Eq. (5.222). The boundary condition is of the convective type, as given by Eq. (5.220).

**71.0**  *Semi-infinite medium-Cartesian-constant concentration–parabolic-zeroth order.* Consider a semi-infinite medium at an initial concentration of zero. For times greater than zero, a step change in concentration is effected at one of the surfaces. The species reacts with a zeroth-order reaction as it comes in contact with the solid medium. Discuss the concentration profile as a function of space and time. The governing equation for Fick's parabolic diffusion and simultaneous reaction of oxygen may be written as follows:

$$D_T \frac{\partial^2 C_{O_2}}{\partial x^2} = \frac{\partial C_{O_2}}{\partial t} + k''' \tag{5.223}$$

The boundary conditions in space and time for a semi-infinite medium in Cartesian coordinates are the same as those given by Eq. (5.62).

**72.0**  *Semi-infinite medium-Cartesian-convective boundary–parabolic-zeroth order.* Obtain the concentration profile of oxygen in space and time during Fick's parabolic diffusion and simultaneous reaction in a semi-infinite medium in Cartesian coordinates. The reaction order can be taken to be of the zeroth order. The governing equation is given by Eq. (5.223). The boundary condition is of the convective type, as given by Eq. (5.220).

**73.0**  *Semi-infinite medium-Cartesian-periodic boundary–hyperbolic-first order.* Obtain the solution to the damped wave diffusion and relaxation and simultaneous reaction of oxygen of the first order in a semi-infinite medium. The governing equation is Eq. (5.61). The boundary conditions in space and time in Cartesian coordinates are of the periodic type as follows:

$$X = 0, D_T \frac{\partial C_{O_2}}{\partial x} = (k'' + A\sin \omega t)(C - C_b); X = \infty, C_{O_2} = 0$$

$$t = 0, C_{O_2} = 0 \tag{5.224}$$

where $k''$ is the mass transfer coefficient between the bulk cross-flow with the surface of the semi-infinite slab.

**74.0**  *Semi-infinite medium-Cartesian-periodic boundary–parabolic-first order.* Obtain the solution to the diffusion and simultaneous reaction of oxygen of the first order in a semi-infinite medium using the parabolic Fick's second law of diffusion. The boundary conditions in space and time for a semi-infinite medium in Cartesian coordinates are the same as in Eq. (5.224). The governing equation for the concentration of oxygen is Eq. (5.221).

**75.0**  *Semi-infinite medium-Cartesian-periodic boundary–parabolic-zeroth order.* Obtain the solution to the diffusion and simultaneous reaction of oxygen of the zeroth order in a semi-infinite medium using the parabolic Fick's second law of diffusion. The boundary conditions in space and time for a semi-infinite medium in Cartesian coordinates are given by Eq. (5.224). The governing equation for the concentration of oxygen is given by Eq. (5.221).

**76.0**  *Semi-infinite medium-Cartesian-periodic boundary–hyperbolic-zeroth order.* Obtain the solution to the damped wave diffusion and relaxation and simultaneous reaction of oxygen of the zeroth order in a semi-infinite medium. The boundary conditions in space and time for a semi-infinite medium in Cartesian coordinates are given by Eq. (5.224) and can be seen to be of the periodic type. The governing equation for the concentration of oxygen is given by Eq. (5.222).

**77.0**  *Semi-infinite medium-Cartesian-pulse boundary–hyperbolic-first order.* Obtain the solution to the damped wave diffusion and relaxation and simultaneous reaction of oxygen of the first order in a semi-infinite medium. The governing equation is Eq. (5.61). The boundary conditions in space and time in Cartesian coordinates are of the pulse type and are as follows:

$$X = 0, t = 0, C_{O_2} = C^*\delta(z); t > 0, \frac{\partial C_{O_2}}{\partial x} = 0; X = \infty, C_{O_2} = 0$$

$$t = 0, C_{O_2} = 0 \tag{5.225}$$

**78.0**  *Semi-infinite medium-Cartesian-pulse boundary–parabolic-first order.* Obtain the solution to the diffusion and simultaneous reaction of oxygen of the first order in a semi-infinite medium using the parabolic Fick's second law of diffusion. The boundary conditions in space and time for a semi-infinite medium in Cartesian coordinates are given by Eq. (5.225) and are seen to be of the pulse type. The governing equation for the concentration of oxygen is given by Eq. (5.221).

**79.0**  *Semi-infinite medium-Cartesian-pulse boundary–parabolic-zeroth order.* Obtain the solution to the diffusion and simultaneous reaction of oxygen of the zeroth order in a semi-infinite medium using the parabolic Fick's second law of diffusion. The boundary conditions in space and time for a semi-infinite medium in Cartesian coordinates are given by Eq. (5.225) and are seen to be of the pulse type. The governing equation for the concentration of oxygen is given by Eq. (5.221).

**80.0**  *Semi-infinite medium-Cartesian-pulse boundary–hyperbolic-zeroth order.* Obtain the solution to the damped wave diffusion and relaxation and simultaneous reaction of oxygen of the zeroth order in a semi-infinite medium. The boundary conditions in space and time for a semi-infinite medium in Cartesian coordinates are given by Eq. (5.225) and can be seen to be of the periodic type. The governing equation for the concentration of oxygen is given by Eq. (5.222).

**81.0**  *Infinite medium-cylindrical-convective boundary–hyperbolic-first order.* Obtain the solution to the damped wave diffusion and relaxation and simultaneous reaction of oxygen of the first order in an infinite medium in cylindrical coordinates. The governing equation can be derived as:

$$D_T \frac{1}{r} \frac{\partial}{\partial r}\left( r \frac{\partial C_{O_2}}{\partial r} \right) = \tau_{mr} \frac{\partial^2 C_{O_2}}{\partial t^2} + (1 + k\tau_{mr}) \frac{\partial C_{O_2}}{\partial t} + k C_{O_2} \tag{5.226}$$

The boundary conditions in space and time in cylindrical coordinates are as follows:

$$r = R_i, D_T \frac{\partial C_{O_2}}{\partial r} = k''(C - C_b); r = \infty, C_{O_2} = 0$$

$$t = 0, C_{O_2} = 0 \tag{5.227}$$

where $k''$ is the mass transfer coefficient between the flow within the cylinder of radius $R_i$.

**82.0**   *Infinite medium-cylindrical-constant concentration–parabolic-first order.* Obtain the solution to the diffusion and simultaneous reaction of oxygen of the first order in an infinite medium using the parabolic Fick's second law of diffusion. The boundary conditions in space and time for an infinite medium in cylindrical coordinates can be taken as:

$$r = R_i, C_{O_2} = C_{O_{2m}}; r = \infty, C_{O_2} = 0$$

$$t = 0, C_{O_2} = 0 \tag{5.228}$$

The governing equation for the concentration of oxygen can be written as:

$$D_T \frac{1}{r} \frac{\partial}{\partial r} \left( r \frac{\partial C_{O_2}}{\partial r} \right) = \frac{\partial C_{O_2}}{\partial t} + k C_{O_2} \tag{5.229}$$

**83.0**   *Infinite medium-cylindrical-constant concentration–hyperbolic-first order.* Obtain the solution to the damped wave diffusion and relaxation and simultaneous reaction of oxygen of the first order in an infinite medium using the governing equation given by Eq. (5.226). The boundary conditions in space and time for an infinite medium in cylindrical coordinates is given by Eq. (5.228).

**84.0**   *Infinite medium-cylindrical-constant concentration–hyperbolic-zeroth order.* Consider an infinite medium at an initial concentration of zero. For times greater than zero, a step change in concentration is effected at the surface of a cylinder of radius $R_i$. The species reacts with a zeroth-order reaction as it comes in contact with the solid medium. Discuss the concentration profile as a function of space and time. The governing equation for damped wave diffusion and relaxation and simultaneous reaction of oxygen may be written as follows:

$$D_T \frac{1}{r} \frac{\partial}{\partial r} \left( r \frac{\partial C_{O_2}}{\partial r} \right) = \tau_{mr} \frac{\partial^2 C_{O_2}}{\partial t^2} + \frac{\partial C_{O_2}}{\partial t} + k''' \tag{5.230}$$

The boundary conditions in space and time for an infinite medium in cylindrical coordinates are given by Eq. (5.228).

**85.0**   *Infinite medium-cylindrical-convective boundary–hyperbolic-zeroth order.* Obtain the concentration profile of oxygen in space and time during damped wave diffusion and simultaneous reaction in an infinite medium in cylindrical coordinates. The reaction order is of the zeroth order. The governing equation is given by Eq. (5.230). The boundary condition is of the convective type, as given by Eq. (5.227).

**86.0** *Infinite medium-cylindrical-constant concentration–parabolic-zeroth order.* Consider an infinite medium at a initial concentration of zero. For times greater than zero, a step change in concentration is effected at one of the surfaces. The species reacts with a zeroth-order reaction as it comes in contact with the solid medium. Discuss the concentration profile as a function of space and time in cylindrical coordinates. The governing equation for Fick's parabolic diffusion and simultaneous reaction of oxygen may be written as follows:

$$D_T \frac{1}{r} \frac{\partial}{\partial r}\left( r \frac{\partial C_{O_2}}{\partial r} \right) = \frac{\partial C_{O_2}}{\partial t} + k''' \tag{5.231}$$

The boundary conditions in space and time for an infinite medium in cylindrical coordinates are given by Eq. (5.228).

**87.0** *Infinite medium-cylindrical-convective boundary–parabolic-zeroth order.* Obtain the concentration profile of oxygen in space and time during Fick's parabolic diffusion and simultaneous reaction in an infinite medium in cylindrical coordinates. The reaction order is of the zeroth order. The governing equation is given by Eq. (5.231). The boundary condition is of the convective type, as given by Eq. (5.227).

**88.0** *Infinite medium-cylindrical-periodic boundary–hyperbolic-first order.* Obtain the solution to the damped wave diffusion and relaxation and simultaneous reaction of oxygen of the first order in an infinite medium. The governing equation is Eq. (5.226). The boundary conditions in space and time in cylindrical coordinates are of the periodic type and are as follows:

$$r = R_i, D_T \frac{\partial C_{O_2}}{\partial r} = (k'' + A\sin\omega t)(C - C_b); r = \infty, C_{O_2} = 0$$
$$t = 0, C_{O_2} = 0 \tag{5.232}$$

where $k''$ is the mass transfer coefficient between the flowing fluid within the cylinder of radius, $R_i$.

**89.0** *Infinite medium-cylindrical-periodic boundary–parabolic-first order.* Obtain the solution to the diffusion and simultaneous reaction of oxygen of the first order in an infinite medium using the parabolic Fick's second law of diffusion. The boundary conditions in space and time for an infinite medium in cylindrical coordinates are given by Eq. (5.232). The governing equation for the concentration of oxygen is given by Eq. (5.229).

**90.0** *Infinite medium-cylindrical-periodic boundary–parabolic-zeroth order.* Obtain the solution to the diffusion and simultaneous reaction of oxygen of the zeroth order in an infinite medium using the parabolic Fick's second law of diffusion. The boundary conditions in space and time for an infinite medium in cylindrical coordinates are given by Eq. (5.232). The governing equation for the concentration of oxygen is given by Eq. (5.229).

**91.0**  *Infinite medium-cylindrical-periodic boundary–hyperbolic-zeroth order.* Obtain the solution to the damped wave diffusion and relaxation and simultaneous reaction of oxygen of the zeroth order in an infinite medium. The boundary conditions in space and time for an infinite medium in cylindrical coordinates are given by Eq. (5.232) and are of the periodic type. The governing equation for the concentration of oxygen is given by Eq. (5.230).

**92.0**  *Infinite medium-cylindrical-pulse boundary–hyperbolic-first order.* Obtain the solution to the damped wave diffusion and relaxation and simultaneous reaction of oxygen of the first order in an infinite medium. The governing equation is Eq. (5.226). The boundary conditions in space and time in cylindrical coordinates are of the pulse type and are as follows:

$$r = R_i, t = 0, C_{O_2} = C^*\delta(z); t > 0, \frac{\partial C_{O_2}}{\partial r} = 0; r = \infty, C_{O_2} = 0$$

$$t = 0, C_{O_2} = 0$$

(5.233)

**93.0**  *Infinite medium-Cartesian-pulse boundary–parabolic-first order.* Obtain the solution to the diffusion and simultaneous reaction of oxygen of the first order in an infinite medium using the parabolic Fick's second law of diffusion. The boundary conditions in space and time for an infinite medium in cylindrical coordinates are given by Eq. (5.233) and are of the pulse type. The governing equation for the concentration of oxygen is given by Eq. (5.229).

**94.0**  *Infinite medium-cylindrical-pulse boundary–parabolic-zeroth order.* Obtain the solution to the diffusion and simultaneous reaction of oxygen of the zeroth order in an infinite medium using the parabolic Fick's second law of diffusion. The boundary conditions in space and time for an infinite medium in cylindrical coordinates are given by Eq. (5.233) and are of the pulse type. The governing equation for the concentration of oxygen is given by Eq. (5.231).

**95.0**  *Infinite medium-cylindrical-pulse boundary–hyperbolic-zeroth order.* Obtain the solution to the damped wave diffusion and relaxation and simultaneous reaction of oxygen of the zeroth order in an infinite medium. The boundary conditions in space and time for an infinite medium in cylindrical coordinates are given by Eq. (5.233) and are of the periodic type. The governing equation for the concentration of oxygenis given by Eq. (5.230).

**96.0**  *Infinite medium-spherical-convective boundary–hyperbolic-first order.* Obtain the solution to the damped wave diffusion and relaxation and simultaneous reaction of oxygen of the first order in an infinite medium in spherical coordinates. The governing equation is:

$$D_T \frac{1}{r^2} \frac{\partial}{\partial r}\left(r^2 \frac{\partial C_{O_2}}{\partial r}\right) = \tau_{mr} \frac{\partial^2 C_{O_2}}{\partial t^2} + (1 + k\tau_{mr})\frac{\partial C_{O_2}}{\partial t} + kC_{O_2}$$

(5.234)

The boundary conditions in space and time in spherical coordinates are as follows:

$$r = R_i, D_T \frac{\partial C_{O_2}}{\partial r} = k''(C - C_b); r = \infty, C_{O_2} = 0$$

$$t = 0, C_{O_2} = 0 \tag{5.235}$$

where $k''$ is the mass transfer coefficient between the flow within the cylinder of radius $R_i$.

**97.0**  *Infinite medium-spherical-constant concentration–parabolic-first order.* Obtain the solution to the diffusion and simultaneous reaction of oxygen of the first order in an infinite medium using the parabolic Fick's second law of diffusion. The boundary conditions in space and time for an infinite medium in spherical coordinates are:

$$r = R_i, C_{O_2} = C_{O_{2m}}; r = \infty, C_{O_2} = 0$$

$$t = 0, C_{O_2} = 0 \tag{5.236}$$

The governing equation for the concentration of oxygen can be written as:

$$D_T \frac{1}{r^2} \frac{\partial}{\partial r}\left(r^2 \frac{\partial C_{O_2}}{\partial r}\right) = \frac{\partial C_{O_2}}{\partial t} + kC_{O_2} \tag{5.237}$$

**98.0**  *Infinite medium-spherical-constant concentration–hyperbolic-first order.* Obtain the solution to the damped wave diffusion and relaxation and simultaneous reaction of oxygen of the first order in an infinite medium using the governing equation as given by Eq. (5.234). The boundary conditions in space and time for an infinite medium in spherical coordinates are given by Eq. (5.236).

**99.0**  *Infinite medium-spherical-constant concentration–hyperbolic-zeroth order.* Consider an infinite medium at an initial concentration of zero. For times greater than zero, a step change in concentration is effected at the surface of a cylinder of radius $R_i$. The species reacts with a zeroth-order reaction as it comes in contact with the solid medium. Discuss the concentration profile as a function of space and time. The governing equation for damped wave diffusion and relaxation and simultaneous reaction of oxygen may be written as follows:

$$D_T \frac{1}{r} \frac{\partial}{\partial r}\left(r \frac{\partial C_{O_2}}{\partial r}\right) = \tau_{mr} \frac{\partial^2 C_{O_2}}{\partial t^2} + \frac{\partial C_{O_2}}{\partial t} + k''' \tag{5.238}$$

The boundary conditions in space and time for an infinite medium in spherical coordinates are given by Eq. (5.236).

**100.0**  *Infinite medium-spherical-convective boundary–hyperbolic-zeroth order.* Obtain the concentration profile of oxygen in space and time during damped wave diffusion and simultaneous reaction in an infinite medium in spherical coordinates. The reaction order is of the zeroth order. The governing equation is given by Eq. (5.238). The boundary condition is of the convective type, as given by Eq. (5.235).

**101.0**    *Infinite medium-spherical-constant concentration–parabolic-zeroth order.* Consider an infinite medium at an initial concentration of zero. For times greater than zero, a step change in concentration is effected at one of the surfaces. The species reacts with a zeroth-order reaction as it comes in contact with the solid medium. Discuss the concentration profile as a function of space and time in spherical coordinates. The governing equation for Fick's parabolic diffusion and simultaneous reaction of oxygen may be written as follows:

$$D_T \frac{1}{r} \frac{\partial}{\partial r}\left(r \frac{\partial C_{O_2}}{\partial r}\right) = \frac{\partial C_{O_2}}{\partial t} + k''' \tag{5.239}$$

The boundary conditions in space and time for an infinite medium in spherical coordinates are given by Eq. (5.236).

**102.0**    *Infinite medium-spherical-convective boundary–parabolic-zeroth order.* Obtain the concentration profile of oxygen in space and time during Fick's parabolic diffusion and simultaneous reaction in an infinite medium in spherical coordinates. The reaction order is of the zeroth order. The governing equation is given by Eq. (5.239). The boundary condition is of the convective type, as given by Eq. (5.235).

**103.0**    *Infinite medium-spherical-periodic boundary–hyperbolic-first order.* Obtain the solution to the damped wave diffusion and relaxation and simultaneous reaction of oxygen of the first order in an infinite medium. The governing equation is Eq. (5.234). The boundary conditions in space and time in spherical coordinates are of the periodic type and are as follows:

$$r = R_i, D_T \frac{\partial C_{O_2}}{\partial r} = (k'' + A\sin\omega t)(C - C_b); r = \infty, C_{O_2} = 0$$
$$t = 0, C_{O_2} = 0 \tag{5.240}$$

where $k''$ is the mass transfer coefficient between the flowing fluid within the cylinder of radius, $R_i$.

**104.0**    *Infinite medium-spherical-periodic boundary–parabolic-first order.* Obtain the solution to the diffusion and simultaneous reaction of oxygen of the first order in an infinite medium using the parabolic Fick's second law of diffusion. The boundary conditions in space and time for an infinite medium in spherical coordinates are given by Eq. (5.240). The governing equation for the concentration of oxygen is Eq. (5.237).

**105.0**    *Infinite medium-spherical-periodic boundary–parabolic-zeroth order.* Obtain the solution to the diffusion and simultaneous reaction of oxygen of the zeroth order in an infinite medium using the parabolic Fick's second law of diffusion. The boundary conditions in space and time for an infinite medium in spherical coordinates are given by Eq. (5.240). The governing equation for the concentration of oxygen is given by Eq. (5.239).

**106.0** *Infinite medium-spherical-periodic boundary–hyperbolic-zeroth order.* Obtain the solution to the damped wave diffusion and relaxation and simultaneous reaction of oxygen of the zeroth order in an infinite medium. The boundary conditions in space and time for an infinite medium in spherical coordinates are given by Eq. (5.232) and are of the periodic type. The governing equation for the concentration of oxygen is given by Eq. (5.240).

**107.0** *Infinite medium-spherical-pulse boundary–hyperbolic-first order.* Obtain the solution to the damped wave diffusion and relaxation and simultaneous reaction of oxygen of the first order in an infinite medium. The governing equation is Eq. (5.234). The boundary conditions in space and time in spherical coordinates are of the pulse type and are as follows:

$$r = R_i, t = 0, C_{O_2} = C^*\delta(z); t > 0, \frac{\partial C_{O_2}}{\partial r} = 0; r = \infty, C_{O_2} = 0$$
$$t = 0, C_{O_2} = 0 \tag{5.241}$$

**108.0** *Infinite medium-Cartesian-pulse boundary–parabolic-first order.* Obtain the solution to the diffusion and simultaneous reaction of oxygen of the first order in an infinite medium using the parabolic Fick's second law of diffusion. The boundary conditions in space and time for an infinite medium in spherical coordinates are given by Eq. (5.241) and are of the pulse type. The governing equation for the concentration of oxygen is given by Eq. (5.237).

**109.0** *Infinite medium-spherical-pulse boundary–parabolic-zeroth order.* Obtain the solution to the diffusion and simultaneous reaction of oxygen of the zeroth order in an infinite medium using the parabolic Fick's second law of diffusion. The boundary conditions in space and time for an infinite medium in spherical coordinates are given by Eq. (5.239) and are of the pulse type. The governing equation for the concentration of oxygen is given by Eq. (5.231).

**110.0** *Infinite medium-spherical-pulse boundary–hyperbolic-zeroth order.* Obtain the solution to the damped wave diffusion and relaxation and simultaneous reaction of oxygen of the zeroth order in an infinite medium. The boundary conditions in space and time for an infinite medium in spherical coordinates are given by Eq. (5.241) and are of the periodic type. The governing equation for the concentration of oxygen is given by Eq. (5.238).

CHAPTER **6**

# Pharmacokinetic Study

## Learning Objectives

- Compartment models
- Drug distribution volumes
- Factors that affect drug distribution
- Nephron and glomerular filtration rate
- Single-compartment model with first-order absorption and elimination
- Zeroth-order, first-order, and second-order absorption
- Michaelis-Menten absorption
- Second-order absorption and elimination
- Subcritical damped oscillations
- Multiple compartment models
- PK tool

## 6.1   Introduction

The experimental, theoretical, and computational analysis of rate of change with time of concentration and volume distribution of compounds administered externally, such as drugs, metabolites, nutrients, hormones, and toxins, in various regions of the human physiology is called pharmacokinetics. Pharmacokinetics comes from the Greek words, *pharmacon*, which means drugs, and *kinetikos*, which means setting in motion. The science and techniques of chemical kinetics applied to biological systems, thus, is pharmacokinetics. The application of pharmacokinetics allows for the processes of *liberation, absorption, distribution, metabolism,* and *excretion* to be characterized mathematically. The development of the theory of kinetic processes in biological systems permits a quantitative prediction of the amounts and concentrations of a chemical in the anatomy as a function of time and dosing regimen.

The *absorption* of a drug can be affected by different methods. When drugs are administered through the gastrointestinal (GI) tract, it is referred to as an enteral route of entry. Parenteral routes refer to all other types of drug entry. Drug administration can be achieved in the following ways:

- Beneath the tongue (sublingual entry)
- Via the mouth (buccal entry)
- Through the stomach (gastric entry)
- Through the veins (IV therapy)
- Within the muscles (intramuscular therapy)
- Beneath the epidermal and dermal skin layers (subcutaneous therapy)
- Within the dermis (intradermal therapy)
- Applied to the skin (percutaneous therapy)
- Through the mouth, nose, pharynx, trachea, bronchi, bronchioles, alveolar sacs, or alveoli by inhalation
- Into an artery (intra-arterial route)
- Into cerebrospinal fluid (intrathecal route)
- Within the vagina (vaginal route)
- Through the eye (intraocular route)

*Systemic circulation* occurs when the drugs are absorbed from the buccal cavity and the lower rectum. The *splanchnic circulation* occurs when the drugs are absorbed from the stomach, intestines, colon, and upper rectum.

The drug is then circulated to the liver through the portal vein, and upon exit from the liver, enters the systemic circulation. During the first pass of the drug through the liver, a significant portion may be degraded by the various enzymes contained within the liver. The drug is then available for general circulation. The presence of a drug can be detected in the plasma, and the concentration of the drug changes with time. There can be three types of drug concentration as a function of time upon infusion in the body (Fig. 6.1).

The slow absorption of the drug is shown as curve A in Fig. 6.1. A maxima is reached in the concentration of the drug in the plasma. The fall in the drug concentration in the plasma is because of the drug's elimination through physiological processes. A rapid bolus of the drug intravenously is shown as curve B in Fig. 6.1. Peak concentration of the drug in the plasma is reached the next instant upon infusion of

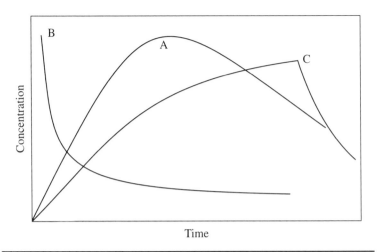

FIGURE **6.1**  Drug concentration in plasma: (A) absorption process;
(B) intravenous therapy; (C) intravenous continuous infusion.

the drug. This is followed by a decrease in the concentration of the
drug due to the elimination reactions. In some cases, the allowed dosage level is either exceeded or not met. This problem can be circum-
vented by the method of *controlled-rate drug delivery*. This is shown
as curve C in Fig. 6.1. The infusion is continuous, and after a
short period, the drug reaches a steady-state concentration in the
plasma. The fall in drug concentration in the plasma occurs
through elimination.

Pharmacokinetic studies can be performed by:

1. Noncompartmental method
2. Compartmental methods
3. Bioanalytical method
4. Mass spectrometry
5. Population pharmacokinetic methods

Noncompartmental methods are model-independent. The area
under the concentration-time graph of the drug is used to estimate
the exposure of the drug. Kinetic models may be used to obtain the
area under the concentration-time graph. These are used in bioequiv-
alence studies.

Compartmental methods [1] involve the development of mathematical models to describe the change in drug concentration with
time. These models are similar to those developed in chemical reaction engineering, thermodynamics, and biochemical kinetics. Compartmental models offer the advantage of being able to predict
drug concentration at any instant of time. There is a spectrum of

pharmacokinetic models and computer software, ranging from a simple one-compartmental pharmacokinetic mode with IV bolus administration with elimination, to complex models that rely on the use of physiological information to ease development and validation.

*Bioanalytical* methods can be used to construct the drug concentration versus time curve(s). The concentration of drugs in a biological matrix is measured using chemical techniques. These methods are designed to be selective and are sensitive. *Mass spectrometry* can be used in pharmacokinetic studies and offer high sensitivity with a low dosage of blood or urine. An LC-MS with a triple quadrapole mass spectrometer can be used for this purpose. High-sensitivity mass spectrometers for microdosing studies are becoming more popular and offer a better alternative to animal experimentation. Samples at different time points, including a $t = 0$ sample, can be obtained as a pharmaceutical is administered and then metabolized or cleared from the anatomy. Complex curve fitting, which is more advanced than linear, is used in mass spectrometry studies. *Population pharmacokinetics* involves the correlation of variability in drug concentration among individuals in the target population. Patient demographic pathophysiological, therapeutic features, such as body weight, excretory, and metabolic functions, can affect dose-concentration relationships. Measurable pathophysiological factors that affect the dose-concentration relationships are identified.

Mathematical models are proposed to describe the concentration profile of the drug in the plasma. Models will be allowed to contain parameters, some of which can be obtained by fitting experimental data to theory. In the compartmental method, the drug is assumed to be distributed into one or more "compartments," such as different organs, a group of tissues, or body fluids. The compartments are assumed to be "well mixed." Drug concentration, both within the compartment and with the exit of the compartment, is identical. Multiple compartments can be used to describe the distribution of the drug throughout the anatomy. The well-mixed assumption within the compartments stems from the fact that cardiac output is about 5 lit/min and the blood volume is bit more than a gallon, resulting in a residence time within the circulatory system of 1 minute. The filtration rate of the blood plasma and interstitial fluid is of the order of mm/min. The distribution of the drug is over hours, and the body fluids move over minutes. As a result, the well-mixed assumption is appropriate. Movement of a drug between compartments is described by simple irreversible or reversible first-order rate processes.

For example, the drug has to be modeled much like oxygen or nitric oxide diffusing through the blood and capillaries in the earlier chapters. The transport of a drug depends on the flow profile of blood through a desired organ or tissue. Rates of mass transport need be considered. The model parameters can be defined using the concentration of drug in blood and tissue. *Au contraire* to a noncompartmental study in the compartmental study, the physiological connection is made. According to Notari [2], the drug concentration can be denoted by a simple weighted summation of exponential decays.

## 6.2 Drug Distribution Issues

The factors that affect how a particular drug is distributed throughout the anatomy are as follows:

1. Rate of blood perfusion
2. Permeability of the capillary
3. Biological affinity of the drug
4. Rate of metabolism of the drug
5. Rate of renal excretion

The distribution volume of a drug when infused into the circulatory system alone would be about 3 lit; when allowed to penetrate the vascular walls, it would be about 15 lit; and when allowed to permeate the cell wall, it would be about 40 lit. Drug delivery rate would depend on the rate of the blood, that is, the rate of blood perfusion. The tissue blood perfusion rate and the rate of drug transport from the vascular system to extravascular space are important. The transport is said to be *perfusion-rate–limited* when the equilibrium between the concentration of drug in the blood and in the tissue is rapidly reached, such as when a drug is lipid-soluble. For cases when the drug is lipid-insoluble, the rate at which the drug is distributed between the blood and tissue is determined by the permeability of the capillary membrane. In such cases, the transport of the drug is said to be *diffusion-rate–limited*.

Drugs may sometimes bind to proteins found in the blood and tissue spaces. The distribution volume of the drug is then restricted. An *apparent distribution volume* of the drug may be defined in such cases. Let the volume of the blood space be denoted by $V_{plasma}$ and the volume of the tissue space be denoted by $V_T$. Let the concentration of the drug that is unbound be denoted by $C_{drug}$ and the concentration of the drug that binds to proteins be represented by $C_{Bplasma}$ and $C_{BT}$ in the plasma and tissue spaces, respectively. Thus:

$$C_{total}^T = C_{drug}^T + C_{BT} \tag{6.1}$$

Equation (6.1) represents the total concentration of drug in the tissue space. In a similar manner, in the blood space, the total concentration of the drug can be written as:

$$C_{total}^{plasma} = C_{drug}^{plasma} + C_{Bplasma} \tag{6.2}$$

The total drug concentration can be obtained by adding Eqs. (6.1) and (6.2):

$$C_{total} = C_{total}^T + C_{total}^{plasma} \tag{6.3}$$

The apparent distribution volume of the drug, $V_a$, can then be estimated as:

$$V_a = \frac{n_{total}}{C_{total}^u} \tag{6.4}$$

where $n_{total}$ is the total number of moles of drug within the anatomy. Let $\alpha$ and $\beta$ be the fraction of the drug that is bound to plasma protein and tissue protein, respectively. Then:

$$\alpha = \frac{C_{Bplasma}}{C_{total}^u + C_{Bplasma}} \; ; \; \beta = \frac{C_{BT}}{C_T^u + C_{BT}} \tag{6.5}$$

The unbound drug concentration in the plasma and tissue can be calculated as:

$$n_{total} = C_{Bplasma}V_{pl} + C_{BT}V_T + V_{true}C_{total}^u \tag{6.6}$$

The apparent volume can be calculated as:

$$V_a = \frac{\alpha}{1-\alpha}V_{pl} + \frac{\beta}{1-\beta}(V_{true} - V_{pl}) + V_{true} \tag{6.7}$$

where $C_{drug}^{plu}$ = the concentration of drug that is unbound and free in the plasma

$C_{drug}^{Tu}$ = the concentration of drug that is unbound and free in the tissue space

$C_{total}^u$ = the concentration of the drug that is unbound and free in the entire anatomy

Upon being absorbed and distributed in the body by circulation, a number of reactions will set in to degrade the drug. These metabolic reactions are catalyzed by enzymes. The depletion of the drug via enzymatic reactions will result in a decrease of pharmacological activity. The products tend to have higher water solubility and decreased permeability through the capillaries. They are removed from the anatomy through the kidneys. The kinetics of enzymatic reactions that deplete the drug in the tissues can be expected to obey the Michaelis-Menten kinetics:

$$r = -\frac{dC_{drug}}{dt} = k\frac{C_{EO}C_{drug}}{C_M + C_{drug}} \tag{6.8}$$

where $C_{EO}$ is the enzyme concentration and $C_M$ is the Michaelis constant. For cases when the drug concentration is much less than the Michaelis constant, Eq. (6.8) can be reduced to a first-order expression:

$$r = \frac{dC_{\text{drug}}}{dt} = -k'C_{\text{drug}} \qquad (6.9)$$

During the final stages of drug action, the rate can reduce to a zeroth-order expression. This is when the enzyme concentration is high and the rate becomes:

$$r = \frac{dC_{\text{drug}}}{dt} = -k''C_{E0} \qquad (6.10)$$

The elimination of the drug is affected by the kidneys in a big way by enzymatic degradation and the formation of water-soluble drug products. About 22 percent of the cardiac output is received by the kidneys. The organ most perfused is the kidneys. The human urinary system is made up of kidneys, the bladder, two ureters, and single urethra. The kidneys are a pair of organs resembling kidney beans measuring around 4 to 5 inches in length and 2 to 3 inches in width. They are situated against the rear wall of the abdomen in the middle of the back, on either side of the spine, beneath the liver on the right and the spleen on the left. Healthy kidneys in the average adult person process about 125 mL/min, or 180 liters of blood per day, and filter out about 2 liters of waste product and extra water in the urine. The kidneys remove excess minerals and wastes and regulate the composition of such inorganic ions as sodium, phosphorous, and chloride in the blood plasma at a nearly constant level. Potassium is controlled by the kidneys for proper functioning of the nerves and muscles, particularly those of the heart.

Blood urea nitrogen (BUN), a waste product produced in the liver as the end product of protein metabolism, is removed from the blood by the kidneys in the Bowman's capsule along with creatinine, a waste product of creatinine phosphate, an energy-storing molecule produced largely from muscle breakdown. Most kidney diseases, such as diabetes and high blood pressure, are caused by an attack on the nephrons, which causes them to lose their filtering capability. The damaged nephrons cannot filter out the poisons as they should. If the problems worsen and renal function drops below 10 to 15 percent, that person is diagnosed with end-stage renal disease. When a person's kidneys fail, harmful wastes build up in the body, the blood pressure elevates, and the blood retains fluid. The person will soon die unless his life is temporarily prolonged by a kidney transplant. To keep the immune system from attacking the foreign kidney, the person must take immunosuppressants for the rest of his life.

When the kidneys are functioning properly and the concentration of an ion in the blood exceeds its kidney threshold value, the excess ions and proteins in the filtrate are not reabsorbed but are released in the urine, thus maintaining near-constant levels. Maintaining constant levels is achieved by the mechanism of reverse osmosis, osmosis, and ion-exchange filtration.

The microscopic representation of the normal anatomy of the nephron is shown in Fig. 6.2. The kidney is comprised of more than a million nephrons. The nephron is comprised of a glomerulus, entering and exiting arterioles, and a renal tubule. The glomerulus consists of a tuft of 20 to 40 capillary loops protruding into *Bowman's capsule.* Bowman's capsule is a cup-shaped extension of the renal tubule and is the beginning of the renal tubule. The epithelial layer of Bowman's capsule is about 40 nm thick and facilitates passage of water into inorganic and organic compounds. The renal tubule has several distinct regions, which have different functions, such as the *proximal convoluted tubule, the loop of Henle, the distal convoluted tubule,* and the *collecting duct* that carries urine to the renal pelvis and the ureter.

There are two types of nephrons: cortical nephrons and juxta-medullary nephrons. About 85 percent of all nephrons in the kidney are cortical. They have glomeruli located in the renal cortex and short

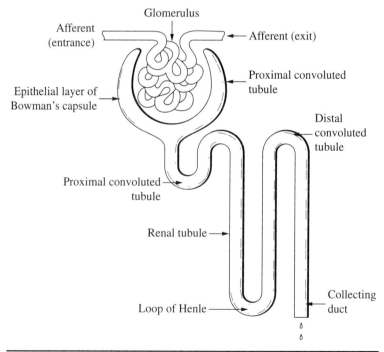

FIGURE **6.2**   Microscopic representation of the anatomy of a nephron [3].

loops of Henle that descend as far as the outer layer of the renal medulla. Long loops are found in juxtamedullary nephrons. These are located at the junction of the cortex and the medulla of the kidney. The long loops of Henle penetrate deep into the medulla and reach the tip of the renal papillae. Urine is concentrated in the kidneys by a countercurrent system of nephrons. About 2 million nephrons participate in making sure that the anatomy's internal environment is maintained at a constant level. As blood passes through the kidneys, the nephrons clear the plasma of unwanted substances, such as urea, while retaining others, such as water. Undesirable substances are removed by glomerular filtration and renal tubular secretion, and are passed into the urine. Substances that the anatomy needs are retained by renal tubular secretion and are returned to the blood by reabsorption.

Glomerular filtration is the amount of fluid movement from the capillaries into the Bowman's capsule. The glomerular filtration rate (GFR) is about 125 mL/min, or about 180 liters per day. GFR refers to the volume of glomerular filtrate formed each minute by all the nephrons in both the kidneys. The glomerular filtrate then passes along the renal tubule and is subject to the forces in the proximal convoluted tubule, the loop of Henle, the distal convoluted tubule, and finally the collecting duct. The renal tubule functions either to secrete or reabsorb organic and inorganic compounds into or from the glomerular filtrate. Both of these renal tubular functions involve active transport mechanisms, as opposed to passive transport mechanisms.

Glomerular filtration is proportional to the membrane permeability and to the balance between hydrostatic and oncotic forces. The hydrostatic pressure driving glomerular filtration is the gradient between the intrarenal blood pressure and the pressure within the Bowman's capsule, which is roughly atmospheric. The intrarenal pressure is, for all intents and purposes, equivalent to the systolic and diastolic blood pressures measured peripherally. Hydrostatic pressure can be conceptualized as the pressure driving fluid out of the glomerular capillary into Bowman's capsule. The colloid oncotic pressure gradient is the pressure driving fluid into the glomerular capillary. When the hydrostatic pressure exceeds the oncotic pressure, filtration occurs. Conversely, when the oncotic pressure exceeds the hydrostatic pressure, reabsorption occurs.

Renal clearance is the term reserved for elimination of a drug by the kidney. The volume of plasma that is totally cleared of the drug per unit time as a result of elimination reactions is called the renal clearance. This is useful when the elimination pathway of drugs is through the kidneys.

The concept of renal clearance is shown in Fig. 6.3. A drug is assumed to be uniformly distributed in the human anatomy with an apparent volume of $V_a$ and a total drug concentration in the plasma of $C_{\text{drug}}^{\text{plasma}}$. The renal plasma flow rate that is totally cleared of the

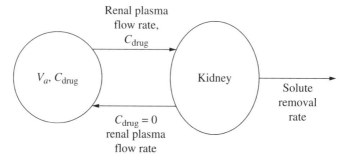

**Figure 6.3**    Renal clearance.

drug is denoted by $F_{renal}$. A mass balance on the drug in the anatomy's apparent distribution volume at transient state can be written as:

$$(rate)_{in} - (rate)_{out} = accumulation \tag{6.11}$$

$$0 - F_{renal}C_{drug}^{plasma} = V_a \frac{dC_{drug}^{plasma}}{dt} \tag{6.12}$$

Given an initial concentration of drug in the plasma as $C_{drug0}^{plasma}$, Eq. (6.12) can be integrated to yield the concentration of a drug as a function of time:

$$C_{drug}^{plasma} = C_{drug0}^{plasma} e^{-\left(\frac{F_{renal}t}{V_a}\right)} \tag{6.13}$$

Thus, the elimination of a drug in the urine is seen to be a *first-order* process. The first-order rate constant can be seen to be $k_{re} = (F_{renal}/V_a)$. Let the urine volumetric rate be given by $Q_u$ and the concentration of a drug in the urine given by $C_{drug}^{urine}$. A mass balance on the drug passing through the urine at steady state would yield:

$$Q_u C_{drug}^{urine} = \frac{F_{renal}}{V_a} C_{drug}^{plasma} \tag{6.14}$$

Equation (6.14) gives the formation of urine in terms of renal clearance. In a similar vein, the term *plasma clearance* represents all the drug-elimination processes of the body. The primary elimination processes are that of metabolism and glomerular filtration in the kidneys. The secondary processes can be from sweat, bile, respiration, and feces. The rate constant for each secondary process can be denoted as:

$$k_j = \frac{F_j}{V_a} \tag{6.15}$$

where $j$ represents the secondary processes. An overall rate constant can be defined and used to take into account all of the primary and secondary processes of the elimination of a drug in the human anatomy:

$$k_{\text{lumped}} = \sum_j k_j = \frac{1}{V_a} \sum_j F_j \qquad (6.16)$$

The change in concentration of drug with time can be written as:

$$C_{\text{drug}}^{\text{plasma}} = C_{\text{drug0}}^{\text{plasma}} e^{-\left(\frac{F_{\text{plasma}} t}{V_a}\right)} \qquad (6.17)$$

where $F_{\text{plasma}} = \sum_j F_j$. Equation (6.17) is an example of a pharmacokinetic model derived from first principles. The curve B in Fig. 6.1 can be explained using this model. The *half-life* of the drug is the time taken by the drug to reach one-half its initial concentration value. For a first-order process, half-life can be related to the rate constant by solving Eq. (6.13) and letting the drug reach half the initial concentration:

$$t_{1/2} = \frac{0.693}{k_{\text{lumped}}} = \frac{0.693 V_a}{F_{\text{plasma}}} \qquad (6.18)$$

The area under the concentration of drug versus time graph can be denoted by *Area* and can be seen to be:

$$\text{Area}^{0..\infty} = \int_0^\infty C_{\text{drug}} dt \qquad (6.19)$$

Combining Eqs. (6.17) and (6.19):

$$\text{Area}^{0..\infty} = \frac{C_{\text{drug0}}^{\text{plasma}}}{k_{\text{lumped}}} = \frac{\text{Dose}}{F_{\text{plasma}}} \qquad (6.20)$$

where Dose is the dose injected over the distribution volume $V_a$.

## 6.3 Single-Compartment Models

### 6.3.1 First-Order Absorption with Elimination

Drugs can take several routes into the human anatomy, as discussed in Sec. 6.1. Upon entry into the human anatomy, the drug finds its way to the plasma by diffusion. Upon infusion, the concentration of the drug gradually increases, reaches a maxima, and then decreases. The decrease in concentration of drug in the plasma is attributed to the elimination reactions, both primary and secondary, that tend to deplete the drug.

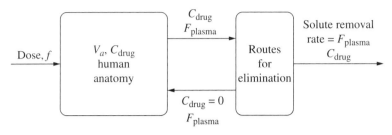

**Figure 6.4** Single compartmental model with first-order absorption and elimination.

A single compartmental model that accounts for the first-order absorption of the drug and its elimination is shown in Fig. 6.4.

The dose infused is given by Dose. A factor $f$ is introduced that represents the fraction of dose that is absorbable. A mass balance on the concentration of the drug within the human anatomy can be written for Fig. 6.4 as:

$$(\text{rate})_{in} - (\text{rate})_{out} = \text{accumulation} \tag{6.21}$$

$$k_{infusion}n_{anatomy} - F_{renal}C_{drug}^{plasma} = V_a \frac{dC_{drug}^{plasma}}{dt} \tag{6.22}$$

where $k_{infusion}$ is the first-order rate constant of the absorption process and $n_{antomy}$ is the amount of drug that is available for absorption. The first-order absorption of the drug process can be described by:

$$\frac{dn_{anatomy}}{dt} = -k_{infusion}n_{anatomy} \tag{6.23}$$

The solution to Eq. (6.23) can be written as:

$$n_{anatomy} = (f\,\text{Dose})e^{-k_{infusion}t} \tag{6.24}$$

Equations (6.24) and (6.22) can be combined and the concentration of drug as a function of time can be solved for from the first-order ordinary differential equation (ODE) by the method of Laplace transforms as:

$$C_{drug}^{plasma} = \left(\frac{f\,\text{Dose}}{V_a}\right)\left(\frac{k_{infusion}}{k_{infusion} - k_{lumped}}\right)\left(e^{-k_{lumped}t} - e^{-k_{infusion}t}\right) \tag{6.25}$$

This is valid for $k_{infusion} \neq k_{lumped}$. From Eq. (6.25), it can be seen that the concentration of the drug as a function of time varies inversely with the apparent volume of distribution of the drug within the human anatomy and is directly proportional to the amount of drug that is absorbable. In addition, it depends on the first-order rate constants of absorption and elimination. It can be seen that Eq. (6.25) exhibits a maxima. This occurs at:

$$\tau_m = \frac{1}{k_{\text{infusion}} - k_{\text{lumped}}} \ln\left(\frac{k_{\text{infusion}}}{k_{\text{lumped}}}\right) \tag{6.26}$$

The corresponding maximum concentration can be given by:

$$C_{\text{drug}}^{\text{plasma}(m)} = \left(\frac{f\,\text{Dose}}{V_a}\right)\left(\frac{k_{\text{infusion}}}{k_{\text{lumped}}}\right)^{\frac{k_{\text{lumped}}}{k_{\text{lumped}} - k_{\text{infusion}}}} \tag{6.27}$$

Eqs. (6.25) to (6.27) are valid only when $k_{\text{lumped}} \neq k_{\text{infusion}}$. For the special case when the overall rate constant of the primary and secondary elimination processes is equal to the rate constant of absorption, the following analysis would be applicable. Equation (6.22) can be written when $k = k_{\text{infusion}} = k_{\text{lumped}}$ as:

$$(kf\,\text{Dose})e^{-kt} - (V_a k)C_{\text{drug}}^{\text{plasma}} = V_a \frac{dC_{\text{drug}}^{\text{plasma}}}{dt} \tag{6.28}$$

Let

$$\tau = kt; u = \frac{C_{\text{drug}}^{\text{plasma}}}{\dfrac{f\,\text{Dose}}{V_a}} \tag{6.29}$$

Equation (6.28) becomes:

$$e^{-\tau} - u = \frac{du}{d\tau} \tag{6.30}$$

Obtaining the Laplace transforms of Eq. (6.30) and recognizing the initial concentration is $u_0$, the transformed variable can be solved for as:

$$\bar{u} = \frac{u_0}{(s+1)} + \frac{1}{(s+1)^2} \tag{6.31}$$

Obtaining the inverse of the Laplace transformed dimensionless concentration:

$$u = u_0 e^{-\tau} + \tau e^{-\tau} \tag{6.32}$$

The solution for the dimensionless concentration given by Eq. (6.32) is shown in Fig. 6.5 for the case of zero initial concentration. A maxima can be seen in the concentration versus time graphs. The Type A behavior, as shown in Fig. 6.1, can be accounted for from this model.

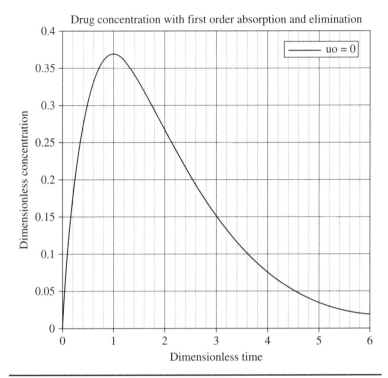

**Figure 6.5**   Drug concentration as a function of time for the special case when $k = k_{\text{lumped}} = k_{\text{infusion}}$.

In the dimensional form, Eq. 6.32 can be written as:

$$C_{\text{drug}}^{\text{plasma}} = \left(\frac{f\,\text{Dose}}{V_a}\right) kt e^{-kt} \tag{6.33}$$

Equation (6.25) can be integrated and given in terms of area under the concentration of the drug versus time graph as:

$$\text{Area}^{0..\infty} = \frac{f\,\text{Dose}}{V_a k_{\text{lumped}}} = \frac{f\,\text{Dose}}{F_{\text{plasma}}} \tag{6.34}$$

### 6.3.2   Second-Order Absorption with Elimination

A mass balance on the concentration of a drug within the human anatomy for the case of second-order absorption with elimination can be written for Fig. 6.4 as:

$$(\text{rate})_{\text{in}} - (\text{rate})_{\text{out}} = \text{accumulation} \tag{6.35}$$

$$k'' n_{\text{anatomy}}^2 - F_{\text{renal}} C_{\text{drug}}^{\text{plasma}} = V_a \frac{dC_{\text{drug}}^{\text{plasma}}}{dt} \tag{6.36}$$

where $k''$ is the second-order rate constant of the absorption process and $n_{anatomy}$ is the amount of drug that is available for absorption. The second-order absorption of the drug process can be described by:

$$\frac{dn_{anatomy}}{dt} = -k'' n_{anatomy}^2 \tag{6.37}$$

The solution to Eq. (6.37) can be written as:

$$\frac{1}{f\,Dose} + k''t = \frac{1}{n_{anatomy}} \tag{6.38}$$

Equations (6.38) and (6.36) can be combined and the concentration of drug as a function of time can be solved for from the first-order ODE and substituting the particular integral as;

$$C_{drug}^{plasma} = e^{-k_{lumped}t} - \frac{1}{\left(1 + \dfrac{k''t}{f\,Dose}\right)^2} \tag{6.39}$$

The initial condition is:

$$C_{drug0}^{plasma} = 0,\ t = 0 \tag{6.40}$$

Equation (6.39) can be seen to exhibit a maxima (Figure 6.6). The solution for the maxima needs numerical methods, as the resulting

**FIGURE 6.6**  Second-order absorption and elimination: $F_{renal}/V_a = 0.25$; $k''/f\,Dose = 1$.

equation is transcendental. Equation (6.39) can be seen to be an interplay of the rate of absorption and the rate of excretion. When the second-order absorption processes are rapid and excretion is slow, the drug tends to accumulate in the blood plasma. When the rate of excretion is rapid, the drug concentration tends to drop off rapidly.

**Worked Example 6.1**   *Pharmacokinetics of styrene in rats.* Styrene is a liquid at room temperature, with a vapor pressure of 4.5 mmHg at 20°C. It is widely used as a monomer in the production of polystyrene and various copolymers, such as Styrene Acrylonitrile (SAN), Acrylonitrile Butadiene Styrne (ABS), Methacrylate Butadiene Styrene (MBS), Styrene Maleic Anhydride (SMA), etc. Styrene is low in acute and chronic toxicity. It is metabolized in rats and humans by oxidation of the side chain to yield mandelic and phenylglyoxylic acids, among other things, that are rapidly excreted in the urine. Experimental data were obtained from inhalation exposure of male Sprague-Dawley rats to different styrene concentrations by Young et al. [4] for up to 24 hours. Rats were removed in groups of three from the exposure chambers at selected time intervals, up to 24 hours, and immediately decapitated to obtain whole heparinized blood for styrene analysis. Samples of epididymal fat, liver, and kidney were also obtained from each rat. A group of 27 rats was removed from the exposure chamber after 6 hours of exposure. These rats were killed in groups of three at selected time intervals up to 18 hours postexposure to examine the disappearance of styrene from blood and tissues. All samples were analyzed for styrene concentration by hexane extraction, followed by (Gas chromotograph) GC/mass spectrometry with selected ion monitoring. At each exposure level, the styrene concentration in the blood increased rapidly and approached a near-maximum value at 6 hours. The postexposure blood concentration curves revealed a biphasic log-linear shape at low concentrations. The data are given in the following table for one styrene concentration. Determine the pharmacokinetic parameters that describe the absorption and elimination of styrene.

| Time | Styrene Concentration (mg/lit) |
|------|-------------------------------|
| 0 | 1 |
| 2 | 10 |
| 4 | 1 |
| 6 | 0.2 |
| 8 | 0.025 |
| 12 | 5 E-4 |
| 16 | 1 E-5 |
| 20 | 1.5 E-7 |
| 24 | 3 E-9 |

TABLE **6.1**   Styrene Concentration in Blood

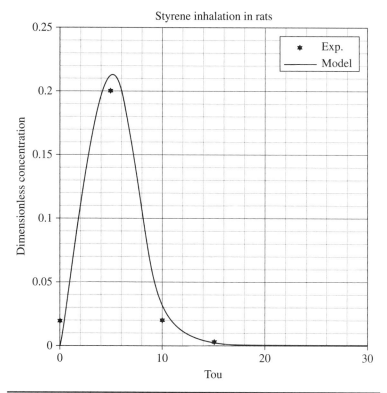

**Figure 6.7**  Styrene inhalation in rats—first-order absorption and elimination.

Equation (6.25) was made dimensionless and plotted against the experimental data points for various values of the absorption and elimination rate constants. As shown in Fig. 6.7, the following values of the rate constants gave the least error in a Microsoft Excel spreadsheet:

$$k_{\text{infusion}} = 2.5 \text{ hr}^{-1} \tag{6.41}$$

$$k_{\text{lumped}} = 1 \text{ hr}^{-1} \tag{6.42}$$

$$\frac{f\text{Dose}}{V_a} = 50 \tag{6.43}$$

### 6.3.3   Zeroth-Order Absorption with Elimination

A mass balance on the concentration of a drug within the human anatomy for the case of zeroth-order absorption with elimination can be written for Fig. 6.4 as:

$$r_{\text{zero}} - V_a k_{\text{lumped}} C_{\text{drug}}^{\text{plasma}} = V_a \frac{dC_{\text{drug}}^{\text{plasma}}}{dt} \tag{6.44}$$

where $r_{zero}$ is the zeroth-order reaction rate of the absorption process. Let $n_{anatomy}$ be the amount of drug that is available for absorption. The zeroth-order absorption of the drug process can be described by:

$$\frac{dn_{anatomy}}{dt} = -k_{zero} \tag{6.45}$$

where $k_{zero}$ is the zeroth-order reaction rate with units of moles/sec. The solution to Eq. (6.45) can be written as:

$$n_{anatomy} = (f\,Dose) - k_{zero}t \tag{6.46}$$

Equations (6.46) and (6.44) can be combined and the concentration of drug as a function of time can be solved for from the resulting equation by the method of Laplace transforms as follows:

$$k_{zero} - V_a k_{lumped} C_{drug}^{plasma} = V_a \frac{dC_{drug}^{plasma}}{dt} \tag{6.47}$$

Obtaining the Laplace transforms of Eq. (6.47):

$$\frac{k_{zero}}{s} - V_a k_{lumped} \bar{C} = V_a(s\bar{C} - 0) \tag{6.48}$$

or

$$\bar{C} = \frac{k_{zero}}{V_a(s)(s + k_{lumped})} \tag{6.49}$$

Obtaining the inverse of the Laplace transformed expression in Eq. (6.49), the concentration of the drug in the plasma can be seen to be:

$$C_{drug}^{plasma} = \frac{k_{zero}}{V_a k_{lumped}}(1 - e^{-k_{lumped}t}) \tag{6.50}$$

Equation (6.50) is valid for times from 0 up to $t = f\,Dose/k_{zero}$. For times $t > (f\,Dose/k_{zero})$, the zeroth-order absorption process concludes and the concentration of the drug has to be solved for from the following equation:

$$-k_{lumped} C_{drug}^{plasma} = \frac{dC_{drug}^{plasma}}{dt} \tag{6.51}$$

and

$$C_{drug}^{plasma} = \frac{k_{zero}}{V_a k_{lumped}}(1 - e^{-\frac{k_{lumped}(f\,Dose)}{k_{zero}}})e^{-k_{lumped}\left(t - \frac{f\,Dose}{k_{zero}}\right)} \tag{6.52}$$

Equations (6.50) and (6.52) for a set of rate constants, dose, and apparent volumes are shown in Fig. 6.8. This model can explain the Type C behavior shown in Fig. 6.1. For short times, the concentration profile is convex and the drug concentration reaches a maxima. After the

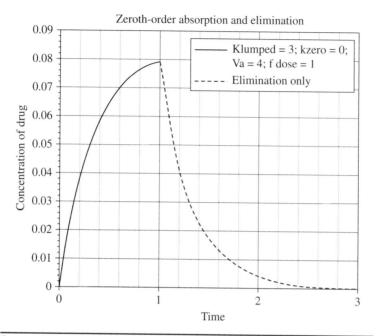

Zeroth-order absorption and elimination

Legend:
— Klumped = 3; kzero = 0; Va = 4; f dose = 1
---- Elimination only

(y-axis: Concentration of drug, 0 to 0.09; x-axis: Time, 0 to 3)

FIGURE 6.8    Zeroth-order absorption with elimination.

zeroth-order absorption processes ceases, upon complete consumption of the dose, the drug concentration falls in accordance with a monotonic exponential decay. This concentration profile is concave in shape, and the curve is asymmetrical. This model has the capability of predicting convex to concave changes in behavior of the functionality of the concentration with respect to time. In the convex portion of the curve, the rate of change in concentration decreases with time. In the concave portion of the curve, the rate of change in concentration is first-order decay.

## 6.3.4    Michaelis-Menten Absorption with Elimination

A mass balance on the concentration of a drug within the human anatomy for the case where the kinetics of absorption is in accordance with Michaelis-Menten kinetics with elimination can be written for Fig. 6.4 as:

$$\frac{kC_{E0}n_{anatomy}}{V_aC_M + n_{anatomy}} - V_a k_{lumped} C_{drug}^{plasma} = V_a \frac{dC_{drug}^{plasma}}{dt} \tag{6.53}$$

Let $n_{anatomy}$ be the amount of drug that is available for absorption. The absorption of the drug process can be described by Michaelis-Menten kinetics:

$$\frac{dn_{anatomy}}{dt} = -\frac{kV_aC_{E0}n_{anatomy}}{V_aC_M + n_{anatomy}} \tag{6.54}$$

where    $k$ = the infusion Michaelis-Menten rate constant
$C_{E0}$ = the total enzyme concentration
$C_M$ = the rate constant.

It can be seen [6] that the Michaelis-Menten kinetics becomes independent of concentration at high drug concentration and becomes zeroth order, and at the low concentration limit reverts to a simple first-order rate expression. An integration of Eq. (6.54) can be seen to be [7]:

$$C_M \ln\left(\frac{n_{\text{anatomy}}}{f\text{Dose}}\right) + \left(\frac{f\text{Dose}}{V_a} - \frac{n_{\text{anatomy}}}{V_a}\right) = kC_{E0}t \qquad (6.55)$$

It can be seen that Eq. (6.55) is in a form that is not readily usable in terms of a one-to-one mapping between the independent variable $t$ and dependent variable $n_{\text{anatomy}}$. In order to combine Eq. (6.55) with Eq. (6.53) and then solve for the concentration of drug in the plasma, a more usable form of Eq. (6.55) is sought. This can be done by realizing that any arbitrary function can be represented using the Taylor series. The Taylor series representation of any arbitrary function is an infinite series containing derivatives of the arbitrary function about a particular point. Prior to obtaining the Taylor series, Eq. (6.54) is made dimensionless as follows:

$$u = \frac{n_{\text{anatomy}}}{f\text{Dose}}; \tau = k_3 u_{E0}t; u_{E0} = \frac{C_{E0}V_a}{f\text{Dose}}; u_M = \frac{C_M V_a}{f\text{Dose}} \qquad (6.56)$$

Equation (6.54) becomes:

$$\frac{du}{d\tau} = -\frac{u}{u+u_M} \qquad (6.57)$$

The Taylor series in terms of derivatives of $u$ evaluated at the point $\tau = 0$ can be written as follows:

$$u = u(0) + \tau u'(0) + \frac{\tau^2}{2!}u''(0) + \frac{\tau^3}{3!}u'''(0) + \cdots\cdots \qquad (6.58)$$

from the initial condition,    $u(0) = 1$ \qquad (6.59)

From Eq. (6.57),    $u'(0) = \dfrac{1}{1+u_M}$ \qquad (6.60)

The initial value of the second derivative of the dimensionless concentration $u''$ can be seen to be:

$$\frac{d^2u}{d\tau^2} = \frac{-(u+u_M)u' + uu'}{(u_M+u)^2}\bigg|_{\tau=0} = \frac{u_M}{(u_M+1)^3} \qquad (6.61)$$

The initial value of the third derivative of the dimensionless concentration $u'''$ can be seen to be:

$$\frac{d^3u}{d\tau^3} = \frac{-(u+u_M)^2 u_M u'' + 2u_M u'(u+u_M)}{(u_M+u)^4}\bigg|_{\tau=0} = -\frac{u_M^2 - 2u_M}{(u_M+1)^5} \qquad (6.62)$$

Plugging Eqs. (6.59) to (6.62) in Eq. (6.58):

$$u = 1 - \frac{\tau}{u_M+1} + \frac{u_M \tau^2}{2!(u_M+1)^3} - \frac{\tau^3\left(u_M^2 - 2u_M\right)}{3!(u_M+1)^5} + \cdots\cdots \qquad (6.63)$$

Equations (6.63) and (6.55) are sketched for a particular value of $u_M = 16$ in Fig. 6.9. It can be seen that for times $t < (25 f \text{Dose}/kC_{E0}V_a)$, the Taylor series expression evaluated near the origin, up to the third derivative, is a reasonable representation of the integrated solution given in Eq. (6.55). More terms in the Taylor series expression can be added to suit the application and the apparent volume, dosage, enzyme concentration, Michaelis constant, and the desired accuracy level needed, as shown in Eq. (6.63).

Equations (6.63) and (6.53) can be combined and the concentration of drug as a function of time can be solved for from the resulting

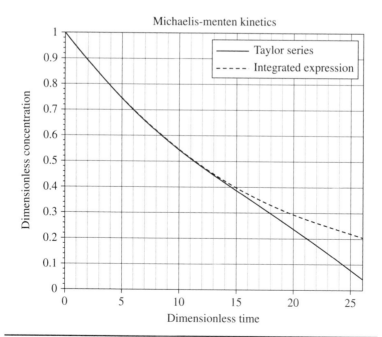

**FIGURE 6.9** Michaelis-Menten kinetics from the integrated and Taylor series expressions.

equation by the method of Laplace transforms as follows. The combined equation is also made dimensionless:

$$\frac{u}{u_M + u} - \left(\frac{k_{\text{lumped}}}{ku_{E0}}\right)u_{\text{drug}}^{\text{plasma}} = \frac{du_{\text{drug}}^{\text{plasma}}}{d\tau} \tag{6.64}$$

Combining Eqs. (6.63), (6.57), and (6.64):

$$\frac{1}{u_M + 1} - \frac{u_M \tau}{(u_M + 1)^3} + \frac{\tau^2 \left(u_M^2 - 2u_M\right)}{2(u_M + 1)^5} + \cdots - \left(\frac{k_{\text{lumped}}}{ku_{E0}}\right)u_{\text{drug}}^{\text{plasma}} = \frac{du_{\text{drug}}^{\text{plasma}}}{d\tau} \tag{6.65}$$

Obtaining the Laplace transform of the governing equation for dimensionless drug concentration in the compartment (plasma), Eq. (6.65):

$$\frac{1}{s(u_M + 1)} - \frac{u_M}{s^2(u_M + 1)^3} + \frac{\left(u_M^2 - 2u_M\right)}{2s^3(u_M + 1)^5} + \cdots - \left(\frac{k_{\text{lumped}}}{ku_{E0}}\right)\bar{u} = s\bar{u} - 0 \tag{6.66}$$

The transformed expression for dimensionless drug concentration in the compartment can be seen to be:

$$\frac{1}{s(u_M + 1)\left(s + \dfrac{k_{\text{lumped}}}{ku_{E0}}\right)} - \frac{u_M}{s^2(u_M + 1)^3\left(s + \dfrac{k_{\text{lumped}}}{ku_{E0}}\right)}$$

$$+ \frac{\left(u_M^2 - 2u_M\right)}{2s^3(u_M + 1)^5\left(s + \dfrac{k_{\text{lumped}}}{ku_{E0}}\right)} + \cdots - = \bar{u} \tag{6.67}$$

It can be seen that the inversion for each term in the infinite series is readily available from the tables (Appendix B). Thus, a nonlinear differential equation was transformed using Taylor series and some manipulations into an equation with a closed-form analytical solution. The term-by-term inversion of Eq. (6.67) can be seen from the tables in Appendix B as:

$$\left(\frac{k_{\text{lumped}}}{ku_{E0}}\right)\frac{1 - e^{-\frac{k_{\text{lumped}}\tau}{ku_{E0}}}}{(u_M + 1)} - \frac{u_M\left(\dfrac{k\tau}{k_{\text{lumped}}} + \left(\dfrac{k}{k_{\text{lumped}}}\right)^2\left(1 - e^{-\frac{k_{\text{lumped}}\tau}{k}}\right)\right)}{(u_M + 1)^3} + \cdots = u \tag{6.68}$$

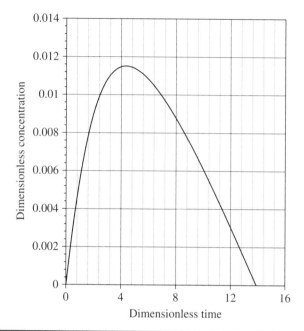

FIGURE **6.10**   Michaelis-Menten absorption and elimination.

The dimensionless drug concentration in the compartment or plasma is shown in Fig. 6.10.

It can be seen from Fig. 6.10 that the dimensionless drug concentration in the compartment goes through a maxima. The curve is convex throughout the absorption and elimination processes. The drug is completely depleted after a said time. The curve is asymmetrical with a right skew. The constants used to construct Fig. 6.10 using a Microsoft Excel spreadsheet were:

$$u_M = 16; k_{\text{lumped}} = 2\,\text{sec}^{-1}$$

$$k = 1\,\text{sec}^{-1}; u_{E0} = 5$$

## 6.4   Analysis of Simple Reactions in Circle

The mathematical model predictions for drug concentration, as discussed previously, depends on the nature of kinetics of absorption. It can be simple zeroth order, first order, second order, fractional order, and any order $n$. It can also be reversible in nature. It can obey Michaelis-Menten kinetics.

Sometimes in the absorption process, the Krebs cycle [8] may be encountered. Reactions such as these can be represented by a scheme of *reactions in circle* [9]. The essential steps in the Krebs cycle are the formation of:

- Oxalic acid
- Citric acid
- Isocitric acid
- α-ketoglutaric acid
- Succyl coenzyme A
- Succinic acid
- Fumaric acid
- Maleic acid

Other sets of reactions in metabolic pathways can be represented by a scheme of reactions in circle. Systems of reactions in series and reactions in parallel have been introduced [6].

Consider the following systems of reactions in circle:

- System of three reactants in circle
- System of four reactants in circle
- System of eight reactants in circle (such as in the Krebs cycle)
- General case

A scheme of reactants in circle is shown in Fig. 6.11.

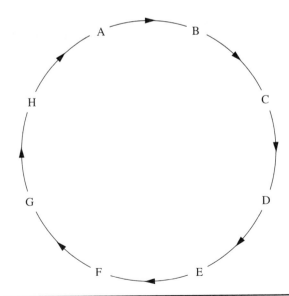

**FIGURE 6.11**   Simple reactions in circle representation of Krebs cycle.

## 6.4.1 Three Reactions in Circle

The simple first-order irreversible rate expressions for three reactants in circle can be written as:

$$\frac{dC_A}{dt} = -k_1 C_A + k_3 C_C \tag{6.69}$$

$$\frac{dC_B}{dt} = -k_2 C_B + k_1 C_A \tag{6.70}$$

$$\frac{dC_C}{dt} = -k_3 C_C + k_3 C_B \tag{6.71}$$

where $C_A$, $C_B$, and $C_C$ are the concentrations of the reactants A, B, and C at any instant in time $t$. Let the initial concentrations of reactants A, B, and C be given by $C_{A0}$, $C_{B0} = 0$, and $C_{C0} = 0$. The Laplace transforms of Eqs. (6.68) to (6.71) are obtained as:

$$(s + k_1)\bar{C}_A = C_{A0} + k_3 \bar{C}_C \tag{6.72}$$

$$(s + k_2)\bar{C}_B = k_2 \bar{C}_A \tag{6.73}$$

$$(s + k_3)\bar{C}_C = k_2 \bar{C}_B \tag{6.74}$$

Eliminating $C_B$ and $C_C$ between Eqs. (6.72) and (6.74), the transformed expression for the instantaneous concentration of reactant A can be written as:

$$(s + k_1)\bar{C}_A = C_{A0} + \frac{k_3 k_2 k_1 \bar{C}_A}{(s + k_3)(s + k_2)} \tag{6.75}$$

or

$$\bar{C}_A = \frac{C_{A0}(s + k_3)(s + k_2)}{s(s^2 + s(k_1 + k_2 + k_3) + k_1 k_2 + k_1 k_3 + k_2 k_3)} \tag{6.76}$$

The inversion of Eq. (6.76) can be obtained by using the residue theorem. The three simple poles can be recognized in Eq. (6.76). Furthermore, it can be realized that when the poles are complex, *subcritical damped oscillations* can be expected in the concentration of the reactant. This is when the quadratic $b^2 - 4ac < 0$. This can happen when:

$$(k_3 - k_2 - k_1)^2 - 4k_2 k_1 < 0 \tag{6.77}$$

or

$$(k_3 - k_2 - k_1) < 2\sqrt{k_1 k_2} \tag{6.78}$$

or

$$k_3 < (\sqrt{k_2} + \sqrt{k_1})^2 \tag{6.79}$$

This expression is *symmetrical* with respect to reactants $A$, $B$, and $C$. When the relation holds, that is, when one reaction rate constant is less than the square of the sum of the square root of the rate constants of the other two reactions, the subcritical damped oscillations can be expected in the reactant concentration.

### 6.4.2 Four Reactions in Circle

The equivalent Laplace transformed expression for concentration of reactant $A$ for a system of four reactions in circle, assuming that all the reactions in the cycle obey simple, first-order kinetics, can be derived as:

$$\bar{C}_A = \frac{C_{A0}(s+k_3)(s+k_2)(s+k_1)}{s(s^3 + s^2(k_1+k_2+k_3+k_4) + s(k_1k_2+k_1k_3+k_2k_3+k_1k_4+k_2k_4+k_3k_4) + k_1k_2k_3+k_1k_2k_4+k_1k_3k_4+k_2k_3k_4)}$$

(6.80)

The conditions where the concentration can be expected to exhibit subcritical damped oscillations when the roots of the following equation becomes complex are:

$$s^3 + \alpha s^2 + \beta s + \chi = 0 \tag{6.81}$$

where $\alpha = k_1 + k_2 + k_3 + k_4$ (6.82)

$$\beta = k_2k_3 + k_1k_3 + k_1k_2 + k_1k_4 + k_4k_3 + k_1k_4 \tag{6.83}$$

$$\chi = k_1k_2k_3 + k_1k_2k_4 + k_1k_3k_4 + k_2k_3k_4 \tag{6.84}$$

It can be seen that $\alpha$ is the sum of all four reaction rate constants, $\beta$ is the sum of the product of all possible pairs of the reaction rate constants, and $\chi$ is the sum of the product of all possible triple products of rate constants in the system of reactions in circle. Equation (6.81) can be converted to the depressed cubic equation by using the following substitution:

$$x = s - \frac{\alpha}{3} \tag{6.85}$$

This method was developed in the Renaissance period [10]. The depressed cubic without the quadratic term will then be:

$$x^3 + \left(\beta - \frac{\alpha^2}{3}\right)x + \chi + \frac{2\alpha^3}{27} - \frac{\alpha\beta}{3} = 0 \tag{6.86}$$

Let $\left(\beta - \frac{\alpha^2}{3}\right) = B; \ \left(\chi + \frac{2\alpha^3}{27} - \frac{\alpha\beta}{3}\right) = C$ (6.87)

Then Eq. (6.86) becomes:

$$x^3 + Bx + C = 0 \tag{6.88}$$

The complex roots to Eq. (6.88) shall occur when $D > 0$ where:

$$D = \frac{B^3}{27} + \frac{C^2}{4} \tag{6.89}$$

Thus, the conditions when subcritical damped oscillations can be expected for a system of four reactions in circle are derived.

### 6.4.3    General Case of *n* Reactions in Circle

For the general case, of which the Krebs cycle with eight reactions in circle is a particular case, it can be obtained by extending the expressions derived for three reactions in circle and four reactions in circle. Another way of doing this would be the method of eigenvalues and eigenvectors. The cases when $\lambda$ is imaginary is when the concentration of the species will exhibit subcritical damped oscillations given by the characteristic equation (6.90):

$$\text{Det} \left| K - \lambda I \right| = 0 \tag{6.90}$$

The size of the $K$ matrix depends on the number of reactions in circle. For $n$ reactions in circle, $K$ would be an $n \times n$ matrix. For the case of a Krebs cycle, it would be a $8 \times 8$ matrix.

Upon expansion, an eighth-order polynomial equation in $\lambda$ arises. Eight roots of the polynomial exist. Even if all the values in the characteristic matrix are real, some roots may be complex. When complex roots occur, they appear in pairs. The roots of the polynomial are called eigenvalues of the characteristic matrix. The polynomial equation is called the eigenvalue equation.

## 6.5    Subcritical Damped Oscillations

As was discussed in the previous section, the concentration of the drug during absorption on account of kinetics, such as the reactions in circle, can undergo subcritical damped oscillations. In such cases, how can the absorption with elimination process be modeled?

Let the solution for the dosage drug when absorbed by kinetics that result in subcritical damped oscillations be given by:

$$n_{\text{anatomy}} = \frac{f \text{Dose}}{V_a} e^{-k_{\text{infusion}} t} (2 - \cos(\omega_k t)) \tag{6.91}$$

A mass balance on the concentration of drug within the human anatomy for the case of kinetics of absorption resulting in subcritical damped oscillation can be written for Fig. 6.4 as:

$$\frac{k_{\text{infusion}}(f\text{Dose})e^{-k_{\text{infusion}}t}(2 - \cos(\omega_k t))}{V_a} - V_a k_{\text{lumped}} C_{\text{drug}}^{\text{plasma}} = V_a \frac{dC_{\text{drug}}^{\text{plasma}}}{dt}$$

(6.92)

Equation (6.92) is the governing equation for concentration of a drug in the single compartment. Equation (6.92) is made dimensionless by the following substitutions:

$$u = \frac{C_{\text{drug}}^{\text{plasma}}}{\left(\dfrac{f\text{Dose}}{V_a}\right)}; \tau = k_{\text{lumped}}t; \omega^* = \frac{\omega_k}{k_{\text{lumped}}}$$

(6.93)

Plugging Eq. (6.93) in, Eq. (6.92) becomes:

$$\frac{du}{d\tau} = \left(\frac{k_{\text{infusion}}}{k_{\text{lumped}}}\right)e^{-\tau}(2 + \cos(\omega^*\tau)) - u$$

(6.94)

The concentration of drug as a function of time can be solved for by the method of Laplace transforms as follows. Obtaining the Laplace transforms of Eq. (6.94):

$$\bar{u} = \left(\frac{k_{\text{infusion}}}{k_{\text{lumped}}}\right)\left(\frac{2}{(s+1)^2} + \frac{(s+1)}{(s+1)^2 + \omega^{*2}}\right)$$

(6.95)

Obtaining the inverse of the transformed expression by using the convolution property of Eq. (6.95):

$$u = e^{-\tau}\left(\frac{k_{\text{infusion}}}{k_{\text{lumped}}}\right)\left(\tau + \frac{\sin(\omega^*\tau)}{\omega^*}\right)$$

(6.96)

The solution for dimensionless concentration of the drug in the single compartment for different values of rate constants and dimensionless frequency are shown in Figs. 6.12 through 6.15. The drug profile reaches a maximum and drops to zero concentration after a given period. The fluctuations in concentration depend on the dimensionless frequency resulting from the subcritical damped oscillations during absorption. At low frequencies, the fluctuations are absent. As the frequency is increased, the fluctuations in concentration are pronounced. The frequency of fluctuations was found to increase along with an increase in the frequency of oscillations during absorption.

| | | | | | | | |
|---|---|---|---|---|---|---|---|
| $(-k_1 - \lambda)$ | 0 | 0 | 0 | 0 | 0 | 0 | $k_8$ |
| $k_1$ | $(-k_1 - \lambda)$ | 0 | 0 | 0 | 0 | 0 | 0 |
| 0 | $k_2$ | $(-k_3 - \lambda)$ | 0 | 0 | 0 | 0 | 0 |
| 0 | 0 | $k_3$ | $(-k_4 - \lambda)$ | 0 | 0 | 0 | 0 |
| 0 | 0 | 0 | $k_4$ | $(-k_5 - \lambda)$ | 0 | 0 | 0 |
| 0 | 0 | 0 | 0 | $k_5$ | $(-k_6 - \lambda)$ | 0 | 0 |
| 0 | 0 | 0 | 0 | 0 | $k_6$ | $(-k_7 - \lambda)$ | 0 |
| 0 | 0 | 0 | 0 | 0 | 0 | $k_7$ | $(-k_8 - \lambda)$ |

**FIGURE 6.12** Characteristic matrix for system of eight reactions in circle.

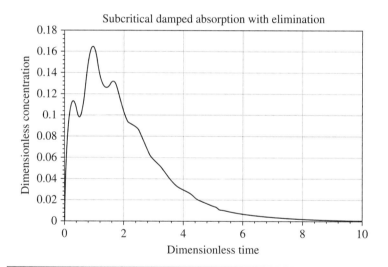

**FIGURE 6.13** Dimensionless concentration of drug in compartment $\omega^* = \omega / k_{\text{lumped}} = 8$; $k_{\text{infusion}} / k_{\text{lumped}} = 0.4$.

## 6.6 Multicompartment Models

Two- and three-compartment models are used when complex drug profiles need to be described. Such a need arises particularly when equilibrium between a central compartment and a peripheral tissue compartment to describe the concentration of drug in blood is not rapid.

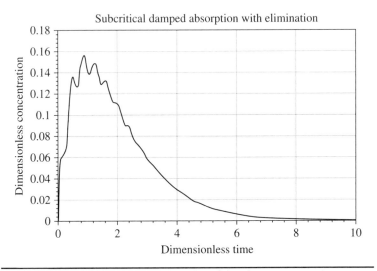

**Figure 6.14**   Dimensionless concentration of drug in compartment $\omega^* = \omega/k_{lumped} = 16$; $k_{infusion}/k_{lumped} = 0.4$.

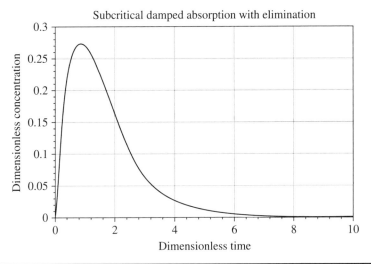

**Figure 6.15**   Dimensionless concentration of drug in compartment, $\omega^* = 1$, $k_{infusion}/k_{lumped} = 0.4$.

A two-compartment model to model the absorption process with elimination is shown in Fig. 6.17. The concentration that has diffused to the tissue region in the human anatomy is accounted for in addition to the concentration of drug in the blood plasma.

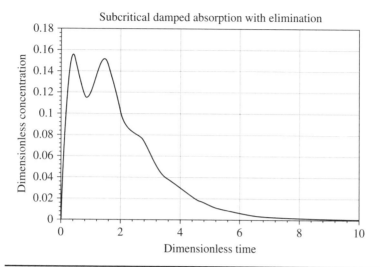

**FIGURE 6.16** Dimensionless concentration of drug in compartment $\omega^* = \omega / k_{\text{lumped}} = 5$; $k_{\text{infusion}} / k_{\text{lumped}} = 0.4$.

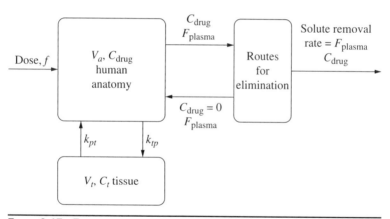

**FIGURE 6.17** Two-compartment model.

A bolus is administered intravenously. A mass balance on the concentration of drug within the human anatomy in the blood plasma and tissue compartments can be written for Fig. 6.17 as follows:

$$(\text{rate})_{\text{in}} - (\text{rate})_{\text{out}} = \text{accumulation} \tag{6.97}$$

$$-V_p C_{\text{drug}}^{\text{plasma}}(k_{pt} + F_{\text{renal}}) + k_{tp} V_T C_{\text{drug}}^{\text{tissue}} = V_p \frac{dC_{\text{drug}}^{\text{plasma}}}{dt} \tag{6.98}$$

$$V_p C_{\text{drug}}^{\text{plasma}}(k_{pt}) - k_{tp} V_T C_{\text{drug}}^{\text{tissue}} = V_T \frac{dC_{\text{drug}}^{\text{tissue}}}{dt} \tag{6.99}$$

where $V_p$ and $V_T$ are the apparent distribution volumes of the blood plasma and tissue compartments. The initial concentrations of the drug in the compartments are:

$$C_{drug}^{plasma} = \frac{Dose}{V_p}$$

$$C_{drug}^{tissue} = 0$$

(6.100)

Differentiating Eq. (6.98) with respect to time and eliminating the concentration of the drug in the tissue from Eq. (6.100), the governing equation for the concentration of drug in the blood plasma compartment can be written as:

$$\frac{d^2 C_{drug}^{plasma}}{dt^2} + (k_{pt} + F_{renal}) \frac{dC_{drug}^{plasma}}{dt} - (k_{tp} + k_{pt}) C_{drug}^{plasma} + \frac{k_{tp} Dose}{V_p} = 0$$

(6.101)

Equation (6.101) is an ODE of the second order with constant coefficients. This can be solved for by obtaining the roots of the complementary function and adding a particular solution. The solution to Eq. (6.101) can be written, after using the initial condition, as:

$$C_{drug}^{plasma} = \frac{Dose}{V_a} e^{\alpha t} + c_2 (e^{\alpha t} - e^{\beta t})$$

(6.102)

The concentration of the drug in the tissue can be written as:

$$C_{drug}^{tissue} = \frac{Dose}{V_a} (1 - e^{\alpha t}) - c_2 (e^{\alpha t} - e^{\beta t})$$

(6.103)

where $\alpha = \dfrac{-(k_{pt} + F_{renal}) + \sqrt{(k_{pt} + F_{renal})^2 + 4(k_{pt} + k_{tp})}}{2}$

(6.104)

$\beta = \dfrac{-(k_{pt} + F_{renal}) - \sqrt{(k_{pt} + F_{renal})^2 + 4(k_{pt} + k_{tp})}}{2}$

(6.105)

The solution of the integration constant needs another time condition in addition to the initial conditions given by Eq. (6.100). By mass balance at any given instant in time, the initial concentration of the drug in the plasma is the total of the concentration of the drug in the blood compartment plus the concentration of the drug in the tissue compartment. The fourth constraint can be that the initial rate of reaction in the tissue compartment is zero, that is:

$$0 = -\frac{\alpha Dose}{V_a} - c_2 (\alpha - \beta)$$

(6.106)

then
$$c_2 = \frac{\alpha \text{Dose}}{V_P (\alpha - \beta)}$$
(6.107)

Equation (6.107) is valid only when $\alpha \neq \beta$.

## 6.7    Computer Implementation of Models

With the advent of personal computers, the pharmacokinetic models are implemented on computers. Both linear and nonlinear pharmacokinetic models can be simulated in the computer. This is especially the case when drug concentration throughout the body or a particular location is high. The possible reason for this situation is when the capacity of a biochemical process to reduce the concentration of the drug becomes saturated. Michaelis-Menten kinetics are used to capture the nonlinear nature of the system. This involves mixtures of zeroth-order and first-order kinetics.

Experimental methods are deployed to collect data on the change in the concentration of drug with time from a patient who has been injected with a particular dose of a drug. This is followed by interpreting and analyzing the data. Data analysis involves plotting the points of concentration of a drug in a logarithmic graph. The slope and intercept of the best-fit, linear, and regression lines to the data can be used to obtain the rate constant and the initial concentration of the drug. These constants are used in the compartment models to describe the drug's time course for additional patients and dosing regimes.

Experimental methods to study drug profiles affected by Michaelis-Menten kinetics are similar to those used in standard compartment models. The drug profiles are usually nonlinear. However, at high concentrations, the drug concentration is linear. This is because the drug is eliminated at a maximal constant rate by a zeroth-order process. The data line then begins to curve in an asymptotic fashion with time until the drug concentration drops to a point where the rate process becomes proportional to the drug concentration via a first-order process. Nonlinear pharmacokinetics can be used to describe solvation of the therapeutic ingredient from a drug formulation, as well as metabolism and elimination processes. Toxicological events related to threshold dosing can be described using nonlinear pharmacokinetics.

Single-, two-, and three-compartment pharmacokinetic models require *in vivo* blood data to obtain rate constants and other relevant parameters that are used to describe drug profiles. Furthermore, what may work for one drug may not be suitable for another drug. Blood profile data need to be generated for each drug under scrutiny. In vivo state of a spectrum of drugs without experimental blood samples from animal testing cannot be predicted accurately using such models. Physiological pharmacokinetic models have been developed.

These integrate the basic physiology and anatomy with drug distribution and disposition. The compartments used correspond to anatomic entities, such as GI tract, liver, lungs, ocular, buttocks, etc., that are connected by the passage of blood. However, a large body of drug-specific physiological and physiochemical data are employed. The rate processes are lumped together in the physiological models.

Computer systems have been used in pharmacokinetics to provide easy solutions to complex pharmacokinetic equations and modeling of pharmacokinetic processes. Other uses of computers in pharmacokinetics include statistical design of experiments, data manipulation, graphical representation of data, projection of drug action, and preparation of written reports or documents.

Pharmacokinetic models are described by systems of differential equations. Computer systems and programming languages have been developed that are more amenable for the solution of differential equations. Graphics-oriented model development computer programs are designed for the development of multicompartment linear and nonlinear pharmacokinetic models. The user is allowed to interactively draw compartments and then link them with other iconic elements to develop integrated flow pathways using predefined symbols. The user assigns certain parameters and equations, relating the parameters to the compartments and flow pathways, and then the model development program generates the differential equations and interpretable code to reflect the integrated system in a computer-readable format. The resulting model can be used to simulate the system under scrutiny when input values for parameters corresponding to the underlying equations of the model, such as drug dose, etc., are used.

Tools are developed to implement pharmacokinetic models. However, the current state of the art does not permit predictability of the pharmacokinetic state of extravascularly administered drugs in a mammal from in vitro cell, tissue, or compound *structure-activity* relationship (SAR/QSAR) data. The predictability is poor when attempting to predict absorption of drug in one mammal from data derived from a second mammal. Different approaches to predict oral administration and fraction dose absorbed are presented in the literature [13–15]. There are lacunae in these models, as they make assumptions that limit the scope of prediction to a few specific compounds. These collections of compounds possess variable ranges of dosing requirements and of permeability, solubility, dissolution rates, and transport mechanism properties. Other deficiencies include the use of drug-specific parameters and values in pharmacokinetic models that limit the predictive capability of the models. Generation of rules that may be universally applicable to drug disposition in a complex physiological system, such as the GI tract is difficult.

The bioavailability of the drug includes the product price, patient compliance, and ease of administration. Failure to identify promising, problematic drug candidates during the discovery and preclinical stages of development is a significant consequence of problems with drug bioavailability. There is a need to develop a comprehensive, physiologically based, pharmacokinetic model and computer system capable of predicting drug bioavailability and variability in humans that utilizes relatively straightforward input parameters. Computer-based biopharmaceutical tools are needed for the medical community, which encounters new therapeutic alternates and the use of high-throughput drug screening for selecting drug candidates.

Lion Bioscience [16] has patented a pharmacokinetic-based design and selection PK tool. The tool can be used to predict absorption of a compound in a mammalian system of interest. The PK tool consists of an input/output system; a physiologic-based simulation model of one or more segments of a mammalian system of interest, with physiological barriers to absorption based on route of administration; and a simulation engine with a differential equation solver and a control statement module. The structure of the PK tool is shown in Fig. 6.18. The PK tool is a multicompartment mathematical model. Linked components include differential equations for fluid transport, fluid absorption, mass transit, mass solvation, mass solubility, and mass

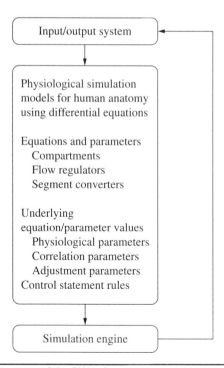

**FIGURE 6.18**   Structure of the PK tool.

absorption for one or more segments of the human anatomy; input parameter values for the differential equations corresponding to physiological parameters; and selectively optimized adjustable parameters for one or more segments of human anatomy and control statement rules for one or more transit, absorption, permeability, solubility, solvation, concentration, and mathematical error correction.

The dose, permeability, and solubility data of a drug or compound is received by the input/output system. The absorption profile for the compound is generated by applying physiological-based simulation models. The PK tool also has a database that includes physiological-based models; simulation model parameters; differential equations for fluid transport, fluid absorption, mass transport, mass solvation, and mass absorption for different parts of the human anatomy; initial parameter values for the differential equations; optimized adjustable parameters; regional correlation parameters; and control statement rules for transport, absorption, permeation, solvation, and mathematical error corrections for different parts of the human anatomy. The database also has a compartment-flow data structure that is portable into and readable by a simulation engine for calculating the rate of absorption, extent of absorption, and concentration of a compound at a sampling site across physiological barriers in different parts of the human anatomy as a function of time. The PK tool can be used to predict accurately one or more in vivo pharmacokinetic parameters of a compound in human anatomy. The method uses a curve-fitting algorithm to obtain the fit of the model with one or more input variables. Then adjustable parameters are generated. These steps are repeated until the adjustable parameters are optimized.

An example of simulation engine is the STELLA® program from High Performance Systems, Inc. It is an interpretive program that can use two different numerical schemes to evaluate differential equations: Euler's method or Runge-Kutta. The program KINETICA™ solves differential equations by evaluating the equations of the model. By translating the model from a STELLA-readable format to a KINETICA-readable format, physiological simulations can be constructed using KINETICA, which has various fitting algorithms.

## Summary

Pharmacokinetics comes from the Greek words *pharmacon*, which that means drugs, and *kinetikos*, which means setting in motion. The application of pharmacokinetics allows for the processes of *liberation, absorption, distribution, metabolism,* and *excretion* to be characterized mathematically. The absorption of drug can be affected by 13 different methods. The change with time of the concentration of the drug can be by three different types, as shown in Fig. 6.1: Slow absorption maxima and rapid bolus, and constant-rate delivery. Pharmacokinetic studies can be performed by five different methods, including compartment methods. The five different methods are noncompartment, compartment, bioanalytical, mass spectrometry, and population.

The factors that affect how a particular drug is distributed throughout the anatomy are rate of blood perfusion, permeability of the capillary, biological affinity of the drug, rate of metabolism of the drug, and rate of renal extraction. Drugs may bind to proteins, and the distribution volume is restricted. An apparent distribution volume of the drug is defined in such cases (Eq. 6.7). The kinetics of enzymatic reactions that deplete the drug in the tissues can be expected to obey the Michaelis-Menten kinetics. The elimination of a drug is affected by the kidneys to a large extent by enzymatic degradation and formation of water-soluble drug products. BUN, a waste product produced in the liver as the end product of protein metabolism, is removed from the blood by the kidneys in the Bowman's capsule, along with creatinine, a waste product of creatinine phosphate, an energy-storage molecule, produced largely from muscle breakdown. The microscopic representation of the anatomy of a nephron is shown in Fig. 6.2. The kidney is comprised of more than a million nephrons. The nephron is comprised of a glomerulus, entering and exiting arterioles, and a renal tubule. The glomerulus consists of a tuft of 20 to 40 capillary loops protruding into Bowman's capsule. Bowman's capsule is a cup-shaped extension of the renal tubule and is where the tubule begins. The epithelial layer of Bowman's capsule is about 40 nm in thickness and facilitates the passage of water into inorganic and organic compounds. The renal tubule has several distinct regions, which have different functions, such as the proximal convoluted tubule, the loop of Henle, the distal convoluted tubule, and the collecting duct that carries the final urine to the renal pelvis and the ureter. Glomerular filtration is the amount of fluid movement from the capillaries into the Bowman's capsule. GFR is about 125 mL/min, or about 180 liters per day.

The concept of renal clearance is introduced by performing a mass balance on the drug in the human anatomy's apparent distribution volume at transient state. The elimination of a drug in urine is seen to be a first-order process [Eq. (6.13)]. The term plasma clearance represents all the drug elimination processes of the body. The primary elimination processes are from metabolism and GFR. The secondary processes are from sweat, bile, respiration, and feces. The rate constant for each secondary process is denoted by $k_j = F_j / V_a$.

An overall rate constant is defined that can be used to account for all the primary and secondary processes of the elimination of a drug in human anatomy [Eq. (6.16)]. Equation (6.17) is an example of a pharmacokinetic model derived from first principles. Curve B in Fig. 6.1 can be explained using this model. The area under the concentration of drug versus time can be denoted by area [Eq. (6.19)].

Single-compartment models were developed for first-order absorption with elimination. The dose infused is given by *Dose*. A fraction *f* represents the fraction of dose that is absorbable. The model equation is solved for by the method of Laplace transforms [Eq. (6.25)]. The concentration of the drug as a function of time varies inversely and is proportional to the apparent distribution volume of the drug within the human anatomy. It is directly proportional to the amount of drug that is absorbable and depends on the first-order rate constants of absorption and elimination. The time where the concentration reaches a maxima is identified [Eq. (6.26)], and the corresponding maximum concentration is given by Eq. (6.27). A separate model solution for the special case when the overall rate constant of the primary and secondary elimination processes is equal to the rate constant of absorption [Eq. (6.32)] was developed. Curve A in Fig. 6.1 can be explained by this model solution: second-order absorption with elimination. The model solution was obtained by the method of particular integrals for first-order ODE, and the model solution was given by Eq. (6.39). The concentration profile exhibited a maxima. The solution for the time at which the maxima occurs needs a numerical solution, as in the resulting transcendental equation in Eq. (6.39). This equation can be seen to be an interplay of the rate of absorption and rate of excretion. When the second-order absorption processes are rapid and excretion is slow, the drug tends to accumulate in the blood plasma. When the rate of excretion is rapid, the drug concentration tends to drop off rapidly. The zeroth-order absorption with elimination model solution was obtained by the method of Laplace transforms. The model solution given by Eq. (6.50) is valid for $0 < t < (f\,\text{Dose}/k_{\text{zero}})$. For time $t > (f\,\text{Dose}\,/\,k_{\text{zero}})$, the model solution is given by Eq. (6.52). This model solution (Fig. 6.8) can explain the behavior of Curve C, as shown in Figure 6.1. The concentration profile changes from convex at small times to concave at long times. The Michaelis-Menten kinetics, when integrated (Eq. 6.55), results in a transcendental equation. However, it is not in a form that is readily usable. A more usable form of Eq. (6.55) is developed using a Taylor series expansion of dimensionless concentration *u* in terms of its derivatives. The infinite series expression for dimensionless concentration is given by Eq. (6.63). It can be seen that for times $t < (25 f\,\text{Dose}/kC_{E0}V_a)$, the Taylor series expression evaluated near the origin, up to the third derivative, is a reasonable representation of the integrated solution given in Eq. (6.55). More terms in the Taylor series expression can be added to suit the application and the apparent volume, dosage, enzyme concentration, Michaelis constant, and the desired accuracy level needed, as shown previously. The single-compartment model solution was obtained by the method of Laplace transforms (Eq. 6.68). It can be seen from Fig. 6.10 that the dimensionless drug concentration in the compartment goes through a maxima. The curve is convex throughout the absorption and elimination processes. The drug is completely depleted after a given period. The curve is asymmetrical with a right skew. With reactions in circle and absorption with elimination, a system of $n$ simple reactions in circle was considered. The concentration profile of the reactants was obtained by the method of Laplace transforms. The conditions when subcritical damped oscillations can be expected were derived. A model was developed for cases when absorption kinetics exhibiting subcritical damped oscillations can be expected. The solution was developed by the method of Laplace transforms (Eq. 6.96). The solutions for dimensionless concentration of the drug in a single compartment for different values of rate constants and dimensionless frequency were shown in Figs. 6.12 through 6.15. The drug profile reaches a maximum and drops to zero concentration after a given period. The fluctuations in concentration depend on the dimensionless frequency resulting from the subcritical damped oscillations during absorption. At low frequencies, the fluctuations are absent. As the frequency increases, the fluctuations in concentration are pronounced. The frequency of fluctuations was found to increase along with an increase in frequency of oscillations during absorption.

A two-compartment model for absorption with elimination was shown in Fig. 6.17. The concentration that has diffused to the tissue region in the human anatomy is accounted for in addition to the concentration of drug in the blood plasma. The model equation for the concentration of drug in the tissue is found to be an ODE of the second order with constant coefficients (Eq. 6.101). The model solution is given in Eq. (6.103) and is obtained by the method of complementary function and particular integrals.

The implementation of the pharmacokinetic models on personal computers was discussed. Software has been developed, with wide-ranging capabilities, from regression fit of experimental data to projection of drug action to preparation of written reports/documents. Computer systems and programming languages have been developed that are more amenable for solving differential equations. Tools have been developed to implement pharmacokinetic models. However, the current state of the art does not permit the predictability of the pharmacokinetic state of extravascularly administered drugs in a mammal from in vitro cell, tissue, or compound SAR/QSAR data. Lion Bioscience [16] has patented a pharmacokinetic-based design and selection PK tool. The tool can be used to predict absorption of a compound in a mammalian system of interest. The PK tool consists of an input/output system; a physiologic-based simulation model of one or more segments of a mammalian system of interest, with physiological barriers to absorption based on route of administration; and a simulation engine with a differential equation solver and a control statement module. The structure of the PK tool was shown in Fig. 6.17.

# References

[1]  D. O. Cooney, *Biomedical Engineering Principles*, New York: Marcel Dekker, 1976.

[2]  R. E. Notari, *Biopharmaceutics and Clinical Pharmacokinetics, Fourth Edition*, New York: Marcel Dekker, 1987.

[3]  E. F. Ikeguchi, "Method of inducing negative pressure in the urinary collecting system and apparatus therefore," U.S. Patent 6,500,158 (2002), New York: Columbia University.

[4]  J. D. Young, J. C. Ramsey, G. E. Blau, G. E. Karbowski, R. J. Nitschke, K. D. Slauter, and W. H. Braun, "Pharmacokinetics of inhaled or intraperitoneally administered styrene in rats," in *Toxicology and Occupational Medicine. Proc. of 10th Inter-America Conference on Toxicology and Occupational Medicine*, Amsterdam, Netherlands: Elsevier, 1979.

[5]  H. S. Mickley, T. K. Sherwood, and C. E. Reed, *Applied Mathematics in Chemical Engineering*, New York: McGraw Hill Professional, 1957.

[6]  O. Levenspiel, *Chemical Reaction Engineering, Third Edition*, New York: John Wiley & Sons, 1999.

[7]  L. Michaelis and M. L. Menten, "Die kinetik der intertinwerkung," *Biochem. Z.* (1913), 49, 333–369.

[8]  Sir Hans A. Krebs, "Citric acid cycle," Nobel prize lecture (1953), Retrieved from http://nobel.se.

[9]  K. R. Sharma, "Subcritical damped oscillatory kinetics of simple reactions in circle" in *CHEMCON 2003*, Bubanewar, Orissa, India, December 2003.

[10] Ans. Magna, "Solution to the cubic equation, Renaissance Mathematics. *Ars magna or The Rules of Algebra*, Dover (published 1993), ISBN 0-486-67811-3

[11] A. Varma and M. Morbidelli, *Mathematical Methods in Chemical Engineering*, Oxford, UK: Oxford University Press, 1997.

[12] S. Tang and Y. Xiao, "One-compartment model with Michaelis-Menten elimination kinetics and therapeutic window: An analytical approach," *J. Pharmacokinet. Pharmacodyn.* (2007), 34, 807–827.

[13] P. J. Sinko and G. L. Amidon, "Characterization of the oral absorption of β-lactam antibiotics. I. Cephalosporins: Determination of intrinsic membrane absorption parameters in the rate intestine *in situ*," *Pharmaceutical Res.* (1988), 5(10), 645–650.

[14] P. J. Sinko, G. D. Leesman, and G. L. Amidon, "Predicting fraction dose absorbed in humans using a macroscopic mass balance approach," *Pharmaceutical Res.* (1991), 8(8), 979–988.

[15] B. Levet-Trafit, M. S. Gruyer, M. Marjanovic, and R. C. Chou, "Estimation of oral drug absorption in man based on intestine permeability in rats," *Life Sciences* (1996), 58, 359–363.

[16] G. M. Grass, G. D. Leesman, D. A. Norris, P. J. Sinko, and J. E. Wehrli, "Pharmacokinetic-based drug design tool and method," U.S. Patent 6,542,858 (2003), Lion Bioscience AG, Heidelberg, DE.

# Exercises

## Review Questions

**1.0**  Does the concentration of a drug reach zero after a given period?

**2.0**  What is the difference between compartment methods and noncompartment methods?

**3.0**  What is the difference between the bioanalytical method and the mass spectrometry method?

**4.0**   What are the characteristics of the typical drug concentration profiles, Curve *A*, Curve *B*, and Curve *C*, as shown in Fig. 6.1?

**5.0**   What is the difference between absorption and elimination?

**6.0**   What is the difference between liberation and metabolism?

**7.0**   What is the effect of the permeability of a capillary on the concentration profile of a drug?

**8.0**   What is the effect of biological affinity on the concentration profile of a drug?

**9.0**   What are the differences in drug administration by sublingual entry and buccal entry?

**10.0**   What are the differences in drug administration by gastric entry and IV therapy?

**11.0**   What are the differences in drug administration by intramuscular therapy and subcutaneous therapy?

**12.0**   What are the differences in drug administration by intradermal therapy and percutaneous therapy?

**13.0**   What are the differences in drug administration by inhalation and by intra-arterial route?

**14.0**   What are the differences in drug administration by intraocular route and gastric entry?

**15.0**   What is Bowman's capsule?

**16.0**   What are the different regions of the renal tubule?

**17.0**   What are the differences between renal clearance and plasma clearance?

**18.0**   The overall rate constant given by Eq. (6.16) assumed that the processes considered act in series. What if they act in parallel?

**19.0**   What are the issues in developing a single-compartment model with absorption kinetics that is reversible?

**20.0**   What are the features of a PK tool in the PC?

## Problems

**21.0**   *Accumulation of drug in urine.* During the elimination of the drug by primary and secondary processes, some drug is passed along with the urine. In order to supplement the pharmacokinetic analysis, some additional data on the amount of the drug can be used. The rate at which drug accumulates in the urine at any given instant in time can be written as:

$$\frac{dm_{\text{urine}}}{dt} = k_{re} V_a C_{\text{drug}}^{\text{plasma}} \qquad (6.108)$$

where $m_{\text{urine}}$ is the mass of the urine at any given instant in time. Substitute Eq. (6.17) in Eq. (6.40). Show that at infinite time:

$$m_{\text{urine}} = \text{Dose}\, \frac{k_{re}}{k_{\text{lumped}}} = \text{Area}^{0..\infty}F_{\text{renal}} \tag{6.109}$$

Eq. (6.41) would be a relationship between the total amount of drug collected in the urine, the renal clearance, and the area under the concentration of the drug versus time curve.

**22.0**   *Drug concentration during continuous intravenous injection.* Curve C in Fig. 6.1 can be explained by the following pharmacokinetic model (Fig. 6.19).

The drug is infused at a constant rate, $q_0$, as shown in Fig. 6.4. Write the mass balance for the drug and show that the concentration of the drug as a function of time can be given by:

$$C_{\text{drug}}^{\text{plasma}} = \frac{q_0}{k_{\text{lumped}}V_a}(1 - e^{-k_{\text{lumped}}t}) \tag{6.110}$$

Show that Eq. (6.93) can describe the behavior of Curve C as depicted in Fig. 6.1.

**23.0**   *Accumulation of drug in urine with first-order absorption and elimination.* During the elimination of the drug by primary and secondary processes, some drug is passed along with the urine. In order to supplement the pharmacokinetic analysis, some additional data on the amount of drug can be used. The rate at which drug accumulates in the urine at any given instant in time can be written as:

$$\frac{dm_{\text{urine}}}{dt} = k_{re}V_a C_{\text{drug}}^{\text{plasma}} \tag{6.111}$$

where $m_{\text{urine}}$ is the mass of the urine at any given instant in time. Substitute Eq. (6.25) in Eq. (6.94). Show that:

$$m_{\text{urine}} = \left(\frac{f\,\text{Dose}\,k_{re}}{k_{\text{lumped}}}\right)\left(1 - \frac{1}{k_{\text{infusion}} - k_{\text{lumped}}}(k_{\text{infusion}}e^{-k_{\text{lumped}}t} - k_{\text{lumped}}e^{-k_{\text{infusion}}t})\right)$$

$$\tag{6.112}$$

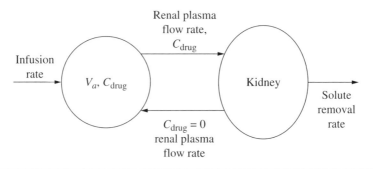

**FIGURE 6.19**   Continuous infusions with elimination.

This is true for $k_{\text{infusion}} \neq k_{\text{lumped}}$. Show that at infinite time;

$$m_{\text{urine}}^{\infty} = \frac{f\,\text{Dose}\,k_{re}}{k_{\text{lumped}}} \qquad (6.113)$$

Why are the limits at $t = 0$ and $t = \infty$ equal to each other?

**24.0** The equations shown in Exercise 3.0 are only valid for $k_{\text{infusion}} \neq k_{\text{lumped}}$. For the special case when $k_{\text{infusion}} = k_{\text{lumped}} = k$ show that:

$$m_{\text{urine}} = \left(\frac{k_{re}}{k}(f\,\text{Dose})\right)(1 - e^{-kt}(1 + kt)) \qquad (6.114)$$

Show that at infinite time:

$$m_{\text{urine}}^{\infty} = \left(\frac{k_{re}}{k}(f\,\text{Dose})\right) \qquad (6.115)$$

What happens to the expression in the limit of $t \to 0$?

**25.0** *Accumulation of drug in urine with zeroth-order absorption and elimination.* During the elimination of the drug by primary and secondary processes, some drug is passed along with the urine. In order to supplement the pharmacokinetic analysis, some additional data on the amount of drug can be used. The rate at which drug accumulates in the urine at any given instant in time can be written as:

$$\frac{dm_{\text{urine}}}{dt} = k_{re}V_a C_{\text{drug}}^{\text{plasma}} \qquad (6.116)$$

where $m_{\text{urine}}$ is the mass of the urine at any given instant in time. Substitute Eq. (6.25) in Eq. (6.99). Show that:

$$m_{\text{urine}} = \left(\frac{k_{\text{zero}}k_{re}}{k_{\text{lumped}}}\right)\left(t + \frac{e^{-k_{\text{lumped}}t}}{k_{\text{lumped}}}\right) \qquad (6.117)$$

This is true for $t \leq (f\,\text{Dose}/k_{\text{zero}})$. What happens in the limit of $t \to 0$?

**26.0** *Lambert function and Michaelis-Menten kinetics.* The Michaelis-Menten kinetics, upon separating the variables of concentration and time and integration, becomes Eq. (6.55). When generating the drug profile, that is, $(t, C(t))$ curve, it can be seen that each point needs the solution of a transcendental equation. The Lambert $W$ function is available in mathematical software such as MATLAB and Maple. The Lambert $W$ function has been studied in the literature to develop a more useful form of Eq. (6.55) [13]. The Michaelis-Menten kinetics in the dimensionless form can be written from Eq. (6.57) as:

$$\frac{du}{d\tau} = -\frac{u}{u + u_M} \qquad (6.118)$$

The integration of Eq. (6.118) can be seen to be:

$$-\tau = u - 1 + \ln(u^{u_M}) \qquad (6.119)$$

The exponentiation of Eq. (6.119) can yield:

$$\frac{u}{u_M} e^{\frac{u}{u_M}} = \frac{1}{u_M} e^{\frac{1-\tau}{u_M}} \tag{6.120}$$

From application of the Lambert $W$ function:

For any
$$Y = Xe^X \tag{6.121}$$
$$X = W(Y) \tag{6.122}$$

So when
$$X = \frac{u}{u_M} ; Y = \frac{e^{\frac{1}{u_M}}}{u_M} e^{-\frac{\tau}{u_M}} \tag{6.123}$$

the solution is:

$$\frac{u}{u_M} = \text{Lambert}\left( \frac{e^{-\frac{\tau}{u_M}}}{u_M e^{-\frac{1}{u_M}}} \right) \tag{6.124}$$

Use MATLAB and sketch the dimensionless concentration $u$ and show the concentration profile as a function of time for a given $u_M$.

**27.0**   *Ethanol in brain tissue.* Pharmacokinetic modeling of ethanol in various tissues plays an important role in the brain's response to ethanol. Ethanol is a naturally produced drug used by humans for thousands of years because of its psychoactive properties. It is beneficial when used in moderation. Excessive use, however, can be devastating. Brain is a high-blood-flow, small-water-volume organ, and ethanol readily crosses the blood-brain barrier. The ethanol is eliminated by an oxidation reaction as follows:

$$C_2H_5OH + 3O_2 \rightarrow 2CO_2 + 3H_2O \tag{6.125}$$

The stoichiometry of the reaction suggests that the rates of reaction are related by:

$$\frac{-r_A}{1} = -\frac{r_{O_2}}{3} \tag{6.126}$$

The kinetics of the reaction can be expected to be of the form:

$$-r_A = kC_A C_{O_2}^3 \tag{6.127}$$

Lumping all the rate effects into the alcohol concentration assume that the rate expression can be given by a fourth-order rate expression such as;

$$-r_A = kC_A^4 \tag{6.128}$$

Let the amount of alcohol ingested by given by Dose. Develop a single-compartment pharmacokinetic model for this fourth-order absorption with elimination of ethanol. Plot the alcohol concentration as a function of time. Does it undergo a maxima? Calculate the amount of ethanol in the urine.

**28.0**  In Problem 27.0, instead of a fourth-order kinetics, assume that ethanol is eliminated by Michaelis-Menten kinetics. Develop a single-compartment pharmacokinetic model for absorption and elimination of ethanol. What can be inferred from the ethanol concentration versus time plot?

**29.0**  Develop a two-compartment pharmacokinetic model for alcohol absorption and elimination, as described in Problem 6.0. How does the concentration of ethanol in the tissue compare with the concentration of ethanol in the blood?

**30.0**  Consider a periodic ingestion of ethanol given by $Dose(2 - \cos(\omega t))$ in Problem 6.0, and construct a single-compartment model for fourth-order absorption and elimination of ethanol. What can be inferred from the concentration versus time plot of ethanol?

**31.0**  Develop a two-compartment pharmacokinetic model for alcohol absorption and elimination as described in Problem 27.0. How does the concentration of ethanol in the tissue compare with the concentration of ethanol in blood?

**32.0**  In Problem 6.0, instead of fourth-order kinetics, assume that ethanol is eliminated by Michaelis-Menten kinetics. Develop a two-compartment pharmacokinetic model for absorption and elimination of ethanol. How does the concentration of ethanol in the tissue compare with the concentration of ethanol in the blood?

**33.0**  Develop a single-compartment model with absorption and elimination. The kinetics of absorption can be described by a reversible first-order reaction, with the equilibrium rate constant given by $K_{eq}$.

**34.0**  Develop a single-compartment model with elimination for intermediate species $R$ formed during absorption. The abortion process kinetics can be represented as a simple set of reactions in series, both first-order. The first-order absorption of the drug process can be described by:

$$n_{anatomy} \rightarrow R_{anatomy} \rightarrow S_{anatomy}$$

$$\frac{dn_{anatomy}}{dt} = -k_{infusion}n_{anatomy}$$

$$\frac{dR_{anatomy}}{dt} = k_{infusion}n_{anatomy} - k_2 R_{anatomy} \qquad (6.129)$$

$$\frac{dS_{anatomy}}{dt} = k_2 R_{anatomy}$$

The solution to Eq. (6.129) can be written as:

$$R_{anatomy} = (f\,Dose)\left(\frac{k_{infusion}}{k_2 - k_{infusion}}\right)(e^{-k_{infusion}t} - e^{-k_2 t}) \qquad (6.130)$$

when $k_2 \neq k_{infusion}$

**35.0**  Develop a single-compartment model for intermediate species $R$ for the special case in Problem 34.0 when the rate constants in the simple reactions in series are equal to each other: $k_2 = k_{infusion}$.

**36.0**  Develop a two-compartment model for intermediate species $R$ for the reaction scheme given in Problem 34.0.

**37.0**  Develop a two-compartment model for intermediate species $R$ for the special case given in Problem 35.0.

**38.0**  Develop a single-compartment model for the drug $n_{anatomy}$ for the scheme of reactions given in Problem 34.0. How is the concentration profile different from Eq. (6.25)?

**39.0**  Develop a single-compartment model for the species $S_{anatomy}$ for the scheme of reactions given in Problem 34.0.

**40.0**  Develop a two-compartment model for the species $S_{anatomy}$ for the scheme of reactions given in Problem 34.0.

**41.0**  Develop a single-compartment model for the species $S_{anatomy}$ for the special case given in Problem 35.0.

**42.0**  Develop a two-compartment model for the species $S_{anatomy}$ for the special case given in Problem 35.0.

**43.0**  Develop a single-compartment model for the special case in Problem 35.0 when the rate constants in the simple reactions in series are equal to each other: $k_2 = k_{infusion}$ for $n_{anatomy}$.

**44.0**  Develop a two-compartment model for the reaction scheme given in Problem 34.0 for $n_{anatomy}$.

**45.0**  Develop a two-compartment model for the special case given in Problem 35.0 for $n_{anatomy}$.

**46.0**  Develop a single-compartment model with elimination for intermediate species $R$ formed during absorption. The absorption process kinetics can be represented as a simple set of reactions in series, with a first-order reaction followed by a zero-order reaction. The first-order absorption of the drug process can be described by:

$$n_{anatomy} \rightarrow R_{anatomy} \rightarrow S_{anatomy}$$

$$\frac{dn_{anatomy}}{dt} = -k_{infusion}\, n_{anatomy} \qquad (6.131)$$

$$\frac{dR_{anatomy}}{dt} = k_{infusion}\, n_{anatomy} - k_2$$

The solution to Eq. (6.131) can be written as:

$$R_{anatomy} = (f\,\text{Dose})\left(1 - e^{-k_{infusion}t} - \frac{k_2 t}{f\,\text{Dose}}\right) \qquad (6.132)$$

**47.0** Develop a two-compartment model for intermediate species $R$ for a series of reactions, with first order followed by zero order, as described in Problem 44.0.

**48.0** Develop a single-compartment model for $n_{anatomy}$ for a series of reactions, with first order followed by zero order, as described in Problem 44.0.

**49.0** Develop a two-compartment model for $n_{anatomy}$ for a series of reactions, with first order followed by zero order, as described in Problem 44.0.

**50.0** Develop a single-compartment model for species $S$ for a series of reactions, with first order followed by zero order, as described in Problem 44.0.

**51.0** Develop a two-compartment model for species $S$ for a series of reactions, with first order followed by zero order, as described in Problem 44.0.

**52.0** Develop a single-compartment model with elimination for intermediate species $R$ formed during absorption. The absorption process kinetics can be represented as a simple set of reactions in series, with a zero-order reaction followed by a first-order reaction. The absorption of the drug process can be described by:

$$n_{anatomy} \rightarrow R_{anatomy} \rightarrow S_{anatomy}$$

$$\frac{dn_{anatomy}}{dt} = -k_{infusion} \tag{6.133}$$

$$\frac{dR_{anatomy}}{dt} = k_{infusion} - k_2 R_{anatomy}$$

The solution to Eq. (6.133) can be written as:

$$n_{anatomy} = f\,\text{Dose}\left(1 - \frac{k_{infusion}\,t}{f\,\text{Dose}}\right)$$

$$R_{anatomy} = \frac{f\,\text{Dose}}{K}(1 - e^{-k_2 t}), t < \frac{f\,\text{Dose}}{k_{infusion}}$$

$$R_{anatomy} = \frac{f\,\text{Dose}}{K} e^{-k_2 t}(e^{Kt} - 1), t > \frac{f\,\text{Dose}}{k_{infusion}} \tag{6.134}$$

$$K = \frac{k_2 f\,\text{Dose}}{k_1}$$

**53.0** Develop a two-compartment model for intermediate species $R$ for a series of reactions, with zero order followed by first order, as described in Problem 50.0.

**54.0** Develop a single-compartment model for $n_{anatomy}$ for a series of reactions, with zero order followed by first order, as described in Problem 50.0.

**55.0** Develop a two-compartment model for $n_{anatomy}$ for a series of reactions, with zero order followed by first order, as described in Problem 50.0.

| # | Dimensionless Time | Dimensionless Concentration of Drug in Plasma |
|---|---|---|
| 1 | 0 | 0 |
| 2 | 1 | 0.16 |
| 3 | 2 | 0.15 |
| 4 | 3 | 0.06 |
| 5 | 4 | 0.03 |
| 6 | 6 | 0.01 |

**TABLE 6.2**    Drug Concentration in Rats

**56.0**   Develop a single-compartment model for species $S$ for a series of reactions, with zero order followed by first order, as described in Problem 50.0.

**57.0**   Develop a two-compartment model for species $S$ for a series of reactions, with zero order followed by first order, as described in Problem 50.0.

**58.0**   A pharmacokinetic study in rats resulted in the following data (Table 6.2):
    Develop a pharmacokinetic model to best describe the data given in Table 6.2.

**59.0**   Develop a single-compartment model for first-order absorption with elimination. Consider a train of pulses as an infusion of the drug.

**60.0**   Develop a single-compartment model for first-order absorption with elimination. Consider a periodic dose of drug infused.

**61.0**   Develop a two-compartment model for first-order absorption with elimination. Consider a periodic dose of drug infused.

**62.0**   Develop a single-compartment model with elimination for species $R$ formed during absorption of drug that reacts in parallel as follows:

$$n_{anatomy} \xrightarrow{k_{infusion}} R$$
$$\downarrow k_2$$
$$S$$

(6.135)

The reaction from $n_{anatomy}$ to $R$ is of the first order, and the reaction from $n_{anatomy}$ to $S$ is of the first order.

**63.0**   Develop a single-compartment model for species $S$ formed during absorption of the drug, as described in Problem 60.0.

**64.0**   Develop a single-compartment model for $n_{anatomy}$ for the scheme of reactions given in Problem 60.0.

**65.0**   Develop a two-compartment model for intermediate species $R$ formed during absorption of the drug, as described in Problem 60.0.

**66.0**   Develop a two-compartment model for intermediate species formed during absorption of the drug, as described in Problem 60.0.

**67.0**   Develop a two-compartment model for $n_{anatomy}$ during absorption of the drug for the parallel scheme of reactions, as shown in Problem 60.0.

# CHAPTER 7

# Tissue Design

## Learning Objectives

- Cell transplantation
- History of tissue design
- Scaffold
- Bioresorption
- Biodegradable and nonbiodegradable materials
- Triple helix structure of collagen
- Electrospinning of nanofibers
- Biomimetic films, membranes, and self-assembly
- Porous polymer supports
- Vascularization
- Response to an implant
- Multicompartment pharmacokinetic models

## 7.1 History

The phrase *tissue engineering* was coined by a pioneer in bioengineering, Y. C. Fung of the University of California at San Diego. The terms *regenerative medicine* and *reparative medicine* are synonymous with tissue design. The first scientific conclave of this emerging field was conducted in 1988 at Lake Granlibakken, California. The scientific literature was seeded with this *de novo* concept. Tissue design emerged as a separate discipline from the field of *biomaterials*. The framework for the field of tissue design arose from the proceedings of this workshop and a review article by Langer and Vacanti [1]. The loss of a tissue or failure of an organ is a frequent, costly, and detrimental problem in human patient care. In tissue design, the principles of biology and engineering can be applied to the development of functional substitutes for damaged tissue, and solutions to tissue creation and repair are provided. Langer and Vacanti defined tissue engineering as "an interdisciplinary field that applies the principles of engineering and life sciences toward the development of biological substitutes that restore, maintain or improve tissue function." The healing process for diseases can be achieved based on the natural healing potential of patients. During tissue regeneration therapy, regeneration of tissues is naturally induced by artificially promoting the potential of cell proliferation and differentiation. The three strategies that are employed in tissue design are:

1. Isolated cells or substitutes

2. Tissue-inducing substances

3. Cells placed on or within matrices

*Vascular grafts* were explored by A. Carrel, who was awarded the Nobel Prize in physiology in 1912. He demonstrated successful technique for the anastomosis of blood vessels and extended these techniques from the transplantation of vessels to the transplantation of entire solid organs. Experimental use of rigid glass and metal tubes for vascular grafts was not successful. Tubes of synthetic fabric were used as arterial prostheses [2]. Thrombogenesis and other problems arising from the interaction between synthetic materials and the blood and perigraft tissues that come in contact with each other have been studied by biomaterials researchers, leading to the development of a range of alternative materials. The resorbable vascular graft was developed in the 1960s, and the fully resorbable graft was introduced in 1978. The creation of biologic vascular structures in their entirety using collagen and cultured vascular cells was developed in 1982 [3].

*Skin grafts* are used to cover severe wounds. These were first attempted using both cadavers and living human beings. The immunologic basis for the rejection of skin allografts began to be investigated in the early 20th century. Skin replacement received increased impetus during World War II. This is because of the tremendous increase in the number of burn victims for whom a skin allograft was not feasible. Billingham and Reynolds [4] demonstrated in animal models that the products of a brief culture of epidermal cells could be applied to a graft bed to reconstitute an epidermis.

Growth factors were identified in the 1960s and 1970s that can be added to culture medium to induce greater proliferation of epidermal cells. In mid-1970s, skin replacement methods were developed, such as the co-culture method and the growth of cultured cells in sheets in a Petri dish. In the co-culture method, human epidermal keratinocytes were cultivated serially from small biopsy samples. Thick multilayered skin is generated to resurface the entire body of a burn victim. Another method was developed in which fibroblasts were used to condense a hydrated collagen lattice to a tissue-like structure potentially suitable for wound healing [5].

*Kidney transplantation* research began in earnest in the early part of the 20th century with renal transplantation efforts in the late 1940s. The first dialysis machine was invented by Kolff in the Netherlands during World War II. For the first time, patients in a Boston hospital in 1948 used a refined dialysis machine. In 1954, the first kidney transplant was performed. The concept of a bioartificial kidney was formulated because of a short supply of organs suitable for transplantation and therapeutic limitations of the dialysis machine. Hollow-fiber bioreactors were developed [6] using renal epithelial cells.

The first *pancreas transplant* was performed by Lillehei in 1966. Prior to this, insulin therapy was used to treat people with diabetes for 70 years. The limitations of insulin therapy were severe, long-term complications for a variety of organ systems. The use of microencapsulated islets as artificial beta cells was proposed in the mid-1960s. A hybrid artificial pancreas consisting of beta cells cultured on synthetic semipermeable hollow fibers was developed [7]. This device was able to restore glucose homeostasis in rats when connected to the circulatory system via a shunt. Studies of implanted microencapsulated islets began in 1980. Investigation of different ways of packaging islet cells to provide effective and durable glycemic control also was initiated in the 1980s.

The first *liver transplant* was carried out by Starzl in 1967. Extracorporeal support to patients suffering from liver failure to replace the hepatic function and nonbiological approaches, such as hemodialysis, hemoperfusion over charcoal or resins or immobilized enzymes, plasmapheresis, and plasma exchange, have been attempted by clinicians. These attempts had limited success because the complex synthetic and metabolic functions of the liver are not completely replaced.

Cell-based therapies and bioartificial systems have the same underlying concept as in the case of pancreatic islet cells. A bioreactor containing a rat hepatoma cell line cultured on the surface of semipermeable hollow fibers within a plastic housing was developed [8]. Transplanted hepatocytes were used in the treatment of drug-induced liver failure in rats in the 1970s. Approaches to the microencapsulation of hepatocytes began in the 1980s.

Metals and alloys have been used to replace damaged bone or to provide support for healing bones. These foreign materials are inert. Some nonbiological materials do not remain inert in the environment of the human anatomy. Reactivity is a function of a number of factors that include implantation site, type of trauma, time of surgery, and precise material in use. In the 1970s, porous glass and hydroxyapatite ceramic were found to be bioactive and to solicit the formation of normal tissue on their surfaces.

The growth and regenerative capabilities of bone have been looked at for decades. Urist [9] proved that a certain material was present in demineralized bone that, when transplanted, can induce growth of additional bone. Subsequent investigators pored over the precise factors that led to trigger bone induction, and bone morpho-genetic proteins (BMPs) were discovered in the 1970s and 1980s as mediatory growth factors. These substances can be isolated from the extracellular matrix of bone. Research work is in progress on isolation, purification, and proliferation of BMPs.

## 7.2  Scaffolds

The 3-D tissue formation is supported by an artificial structure called a scaffold. Cells are implanted into scaffolds. This can be affected *in vivo* or *ex vivo*. Scaffolds enable cell attachment and migration, delivery and retention of cells, diffusion of vital cell nutrients, and modification of cell phase. The cell seeding and diffusion throughout the whole structure of both cells and nutrients requires scaffold specifications such as high porosity and appropriate pore size. Scaffolds need to be absorbed by the surrounding tissue space. This imposes a specification of *biodegradability* on the scaffolds.

The rate of tissue formation should be equal to the rate at which degradation occurs. The implications are that the cells make their own natural matrix structure around themselves. Scaffolds provide structural integrity within the body. When the scaffold breaks down, it leaves the neotissue that will absorb the mechanical load.

Engineering tissues and organs with mammalian cells and a scaffolding material is a recent trend compared to the use of harvested tissues and organs. Biodegradable polymers have been attractive candidates for scaffolding materials because they degrade as the new tissues are formed, eventually leaving nothing foreign to the body. They allow the transplanted cells sufficient time to organize the desired 3-D structure and develop the blood supply. They readily hydrolyze on contact with the body fluids and disappear slowly. Examples of biodegradable polymers are:

- Collagen
- Glycosaminoglycans (GAG)
- Chitosan
- Polylactic acid
- Polyglycolic acid
- Polycaprolactone
- Polyhydroxyalkanoates

*Polylactic acid* (PLA) is biodegradable. PLA has been used either as a D isomer or in DL mixed form to prepare implant materials that are bioresorbable. PLA is biodegradable, although it would take three to five years to be fully resorbed. The *in vivo* degradation of PLA is by an autocatalyzed hydrolytic scission of the ester groups in the polymer chain according to the following reaction:

$$\sim COO \sim + H_2O \rightarrow \sim COOH + \sim OH \qquad (7.1)$$

Research work has been published in the literature on attempts that have been made to increase the carboxylic acid (COOH) functionality, thereby increasing the rate of degradation of PLA. Samples of PLA have been contacted with COOH-containing materials such as oleic acid. No effect on the rate of degradation was reported. The effect of lactic acid monomers in PLA has also been investigated. It was found that the added monomer rapidly leached out of the polymer. Polymer blends containing 15 wt % lactic acid exhibited a total weight loss of about 15 percent within the first week of a 10-week study, and little further loss was found in the remaining weeks. A biodegradable stent was formed from lactide polymers where citric acid or fumaric acid can be included during the polymer processing. Other additives that are used to accelerate stent degradation are t-butyl ester of lauric acid and other esters. It was found [10] that it is possible to control the rate of degradation of lactic acid polymers by includin certain additives that are fully miscible with PLA.

The blending process is simple. Stable polymer blends can be easily thermoformed by injection molding in order to form implantable medical devices that will be both physically strong as well as biodegradable, allowing for proper controls. The additives can be selected from hexanoic acid, octanoic acid, decanoic acid, etc., and isovaleric anhydride, etc. The additive is at a 5 percent composition level in the blend. The mechanical properties of the implant are retained in the early stages of degradation, although molecular weight may decrease remarkably. A critical molecular weight is reached and the implant will cease to possess any mechanical strength and will not have degraded sufficiently for resorption to occur.

$$\% \text{ additive} = 100 MW_{nA} \left( \left( \frac{Ln\left(\frac{MW_{n0}}{MW_{ns}}\right) - tk_1}{tk_2} \right)^2 - \frac{1}{MW_{n0}} \right) \qquad (7.2)$$

The relationship between % additive and the degradation rate constants and molecular weights are given in Eq. (7.2). A graph of the degradation rate of the blend versus the concentration of the COOH groups of the blend is obtained. The slope and intercept of this graph is used to obtain the rate constants $k_1$ and $k_2$ in Eq. (7.2). The degradation rate of the blend is obtained from the slope of a logarithmic graph against time for degradation time in weeks. The terms in Eq. (7.2) are $MW_{n0}$, the polymer initial molecular weight $MW_{ns}$, the number averaged molecular weight of the polymer at the point of loss of strength; and *t*, the duration in weeks where strength retention is required.

Another additive can be included to cause a delay of the onset of the degradation process. This is achieved by the use of substances that may later become acidic. For example, acid anhydrides shall hydrolyze to the corresponding acid. The PLA can be a homopolymer or a copolymer of polylactic acid-co-polyglycolic acid (PLA-co-PGA). The polymer blends may be prepared by solution blending using solvents such as chloroform ($CHCl_3$). These materials may be used in medical devices where only temporary residence is required. For example, sutures, anchors, interference screws, scaffolds, maxilla-facial plates, and fracture fixation plates and rods need biodegradable constituents.

*Collagen* has a triple-helical structure. This was discovered by G. N. Ramachandran [11], a former Ph.D. student of Nobel laureate Sir C. V. Raman. Tissues are made up of cells and extracellular matrix (ECM). The ECM is made up of collagen fibers in large measure. Collagen has high tensile strength. The molecular structure of collagen consists of fibrils that are 300 nm in length and 1.5 nm in diameter. Collagen is a vital protein that is present in mammalian connective tissues such as skin, cartilage, and bone, and it accounts for nearly one-third of all the proteins in the human anatomy. Collagen is an insoluble fibrous protein.

Its x-ray diffraction pattern is diffuse and is difficult to interpret. Ramachandran studied the collagen fiber in kangaroo tail tendon. The shark fin ray collagen that he obtained from the biochemistry lab did not yield good diffraction patterns. A description of the original triple-helical structure of collagen was published in the journal *Nature.* The proposed structure consisted of three separate helical chains with their axes parallel to the fiber axis stacked in a hexagonal ray. The structure was innovative and provided better quantitative agreement with the x-ray data. G. Ambady a graduate student in Ramachandran's laboratory, obtained a collagen x-ray diffraction pattern with the fiber inclined to the incident beam. The data used so far were hitting the fiber perpendicularly. A meridional reflection was revealed that could not be explained by the triple-helical model. The model was revised to yield a structure that had a coiled-coil structure of the triple helix instead of the separate helices.

The structure comprises three left-handed helices intertwining themselves into a second right-handed helix with a common axis in a closely packed formation. The model explained all of the published experimental data on collagen up to that point. As mentioned, this work was published in the journal *Nature* in 1955. Nobel laureate Crick called Ramachandran's proposed structure wrong and gave a modified structure for collagen. However, the Crick structure was not possible because one of the hydrogen bonds gave rise to an apparent situation in which hydrogen atoms came closer than the sum of their radii. Then Crick proposed a structure with a single hydrogen bond. Ramachandran and Chandrasekharan proposed that the second bond could be mediated through water. Water promotes the intratriple-helix stability and the binding of neighboring triple helices together into a matrix, thereby improving the collagen fiber's strength.

Drexel University patented [12] scaffolds used in tissue design comprising collagen or collagen-like peptides. Polymeric fiber matrices can be used for the controlled release of bioactive compounds. These compounds are included in the delivery system, either by suspending the compound particles or dissolving the compound in the polymer solution used to produce the fibers. The scaffold is useful in promoting attachment and growth of chondrocytes and is useful in cartilage repair and replacement.

An electrospinning operation can be used to fabricate organic polymer fibers with micron to nanoscale diameters in linear, 2-D, and 3-D architectures. In the electrospinning process, a high-voltage electric field is generated between an oppositely charged polymer fluid contained in a glass syringe with a capillary tip and a metallic collection screen. As the voltage is increased, the charged polymer solution is attracted to the screen. Once the voltage reaches a critical value, the charge overcomes the surface tension of the suspended polymer cone formed on the capillary tip of the syringe and a jet of ultrafine fibers is produced. As the charged fibers are sprayed, the solvent evaporates and the fibers are accumulated randomly on the surface of the collection screen. This results in a nonwoven mesh of nano- and micron-scale fibers. The fiber diameter and mesh thickness can be controlled by varying the applied voltage, polymer solution concentration, solvent used, and the duration of electrospinning. The fiber matrix properties can also be affected by changing the distance between the needle and the collection plate, the angle of the syringe with respect to the collection plate, and the applied voltage. A steady concentration of bioactive compound is maintained by suitable design of the delivery system.

A delayed release of a bioactive compound also can be affected. The bioactive compound containing a fiber polymer matrix is coated with a layer of nonwoven polymer fiber matrix with no bioactive compound. Different polymers with various degradation times can be used to obtain the desired time delays. The selective bioactive compound promotes cell adhesion and growth, and serves as a scaffold. Even nondegradable polymers that go into solution, such as polyethylenes, polyurethanes, and EVA, can be electrospun into fibers. Biodegradable polymers, such as PLA-co-PGA, PLLA, PGLA, polyglaxanaone and polyphosphazenes can be spun into fibers.

Techniques such as salt leaching, fibrous fabric processing, 3-D printing, foaming, nanofiber self-assembly, textile technologies, solvent casting, gas foaming, emulsification, freeze-drying, thermally induced phase separation, computer-aided designing (CAD)/computer-aided mapping (CAM) technologies, fiber bonding, membrane lamination, melt molding, and *in situ* polymerization can be used to prepare scaffolds. Engineering clinically useful tissues and organs is a challenge. Ideal scaffolds are yet to be developed. The principles of scaffolding are not well understood. Pore size, porosity, and surface area are salient parameters in scaffold design. Pore shape, pore wall morphology, and interconnectivity between pores of the scaffolding materials may become important for cell seeding, migration, growth, mass transport, gene expression, and new tissue generation.

In human anatomy, tissues are organized into 3-D structures as functional organs and organ systems. Each tissue or organ has its specific characteristic architecture, depending on its biological function. Architecture design can provide for appropriate channels for mass transport and spatial cellular organization. Mass transport can include signaling molecules, nutritional supplies, and metabolic waste removal. Spatial cellular organization determines cell-cell and cell-matrix interactions, and is critical to normal tissue and organ function. Matrix materials play a critical role in allowing for appropriate cell distribution and guidance of tissue regeneration in 3-D. In order to develop a scaffold, the architectural design, including cellular distribution, mass transport conditions, and tissue function, is important.

The salt leaching technique enables control of porosity by varying the salt particle size and salt/polymer ratio. The shape of the pores is limited to the cubic salt crystal shape. Textile technologies can enable control of fiber diameter on the order of a magnitude of 15 μm, inter-fiber distance, and porosity.

The carbon nanotube (CNT) is a potential candidate for scaffolds in tissue design. They are biocompatible, resistant to biodegradation, and can be functionalized by biomolecules. However, there are some unresolved toxicological issues regarding the use of CNTs.

CNTs are rolled graphene sheets of atoms about the needle axis. They are 0.7 to 100 nm diameter and a few microns in length. Carbon hexagons are arranged in a concentric manner, with both ends of the tube capped by a pentagon containing Buckminsterfullerene-type structure. They possess excellent electrical, thermal, and toughness properties. Young's modulus of CNT has been estimated at 1 TPa and a yield strength of 120 GPa. S. Ijima verified fullerene in 1991 and observed a multiwalled CNT formed from carbon arc discharge.

Five methods of synthesis of CNTs are discussed [13]. These are:

1. Arc discharge
2. Laser ablation
3. Chemical vapor deposition (CVD)
4. High pressure carbon monoxide (HIPCO) process
5. Surface-mediated growth of vertically aligned tubes

The arc discharge process was developed by NEC in 1992. Two graphite rods are connected to a power supply spaced a few mm apart. At 100 amps carbon vaporizes and forms a hot plasma. The typical yield is 30 to 90 percent. The single-walled nanotube (SWNT) and multi-walled nanotube (MWNT) are short tubes with diameters ranging from 0.6 to 1.4 nm. CNTs can be synthesized in open air, although the product needs to be purified. CVD process was invented in Nagano, Japan. The substrate is placed in an oven, heated to 600°C, and a carbon-bearing gas such as methane is slowly added. As the gas decomposes, it frees up the carbon atoms, which recombine as a nanotube. The yield range is 20 to 100 percent. Long tubes with a diameter ranging from 0.6 to 4 nm were formed. CNT scan be easily scaled up to industrial production levels. The SWNT diameter is controllable. The tubes are usually multiwalled and riddled with defects.

The laser vaporization process was developed by Smalley in 1996. The graphite is blasted with intense laser pulses to generate carbon gas. A prodigious amount of SWNTs are formed, with a yield of up to 70 percent. Long bundles of tubes 5 to 20 $\mu$m with diameters in the range of 1 to 2 nm are formed. The product formation is primarily SWNTs. Good diameter control is possible, and few defects are found in the product. The reaction product is pure, but the process is expensive.

The HIPCO process was also developed by Smalley in 1998. A gaseous catalyst precursor is rapidly mixed with carbon monoxide (CO) in a chamber at high pressure and temperature. The catalyst precursor decomposes, and nanoscale metal particles form the decomposition product. CO reacts on the catalyst surface and forms solid carbon and gaseous carbon dioxide ($CO_2$). The carbon atoms roll up into CNTs. One hundred percent of the product is SWNT, and the process is highly selective. Samsung patented a method for vertically aligning CNTs on a substrate. A CNT support layer is stacked on the substrate, which is filled with pores. A self-assembled monolayer (SAM) is arranged on the surface of the substrate. On one end of each of the CNTs are attached portions of the SAM exposed through the pores formed between the colloid particles present in the support layer. CNTs can be vertically aligned on the substrate with the SAM on it through the help of pores formed between the colloid particles.

CNTs possess interesting physical properties. The thermal conductivity of CNTs is in excess of 2000 W/m/K. They have unique electronic properties. Applications include electromagnetic shielding, electron field emission displays for computers and other high-tech devices, photovoltaics, super capacitors, batteries, fuel cells, computer memory, carbon electrodes, carbon foams, actuators, material for hydrogen storage, and adsorbents.

CNTs can be produced with different morphologies, such as SWNT, DWNT, MWNT, nanoribbon, nanosheet, nanopeapod, linear and branched, conically overlapping bamboolike tubules, branched Y-shaped tubules, nanorope, nanowires, and nanofilm. Processes have been developed to prepare CNTs with a desired morphology. Phase-separated copolymers/stabilized blends of polymers can be pyrolyzed along with sacrificial material to form the desired morphology. The sacrificial material is changed to control the morphology of the product. A self-assembly of block copolymers can lead to 20 different complex phase-separated morphologies. Often, as is the precursor, so is the product. Therefore, an even wider variety of CNT morphologies can be synthesized.

Strong van der Waals forces allow for spontaneous roping of nanostructures, leading to the formation of extended carbon structures. CNTs with a predetermined morphology can be synthesized. For example, this was discussed in a patent by Carnegie Mellon University. Phase-separated copolymers/stabilized blends of polymers can be pyrolyzed to form the carbon tubular morphology. These are precursor materials. One of the comonomers that form the copolymers can be acrylonitrile (AN), for example. Another material is added to the precursor material, called the sacrificial material. The sacrificial material is used to control the morphology, self-assembly, and distribution of the precursor phase. The primary source of carbon in the product is the precursor. The polymer blocks in the copolymers are immiscible at the micro scale. Free energy and entropic considerations can be used to derive the conditions for phase separation.

Lower critical solution temperature (LCST) and upper critical solution temperature (UCST) are also important considerations in the phase separation of polymers. However, they are covalently attached, thus preventing separation at the macro scale. Phase separation is limited to the *nanoscale*. The nanoscale dimensions typical of these structures range from 5 to 100 nm. The precursor phase pyrolyzes to form carbon nanostructures. The sacrificial phase ends with pyrolysis. When the phase-separated copolymer undergoes pyrolysis, it forms two different carbon-based structures, such as a pure carbon phase and a doped carbon phase.

The *topology* of the product depends on the *morphology* of the precursor. Due to the phase separation of the copolymer on the nanoscale, the phase-separated copolymers self-assemble on the molecular level into the phase-separated morphologies. The ABC block copolymer may self-assemble into more than 20 different complex phase-separated morphologies (Fig. 7.1 [12]). Typical morphologies are spherical, cylindrical, and lamellar. Phase-separated domains may also include gyroid morphologies with two interpenetrating continuous phases. The morphologies are dependent on many factors, such as volume ratio of segments, chemical composition of segments, connectivity, Flory-Huggins interaction parameters between segments, and processing conditions.

FIGURE 7.1
Different
morphologies of
precursor material.

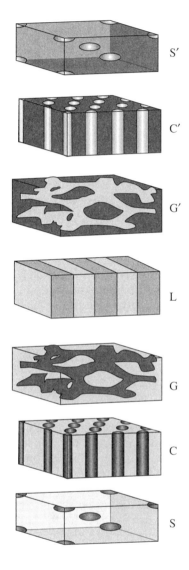

S′

C′

G′

L

G

C

S

Nanoscale morphologies are sensitive to temperature, and they are reversibly formed. Spinning or extrusion of the polymer at the surface prior to carbonization leads to nanowires after pyrolysis. Nanotubes, nanowires, and nanofibers can be formed in this manner. Polyacrylonitrile (PAN) has been used in the industry to form carbon fibers. Cross-linking of microscale phase-separated domains in the polymer blend that forms the precursor can lead to the formation of nanoclusters in the product.

## 7.3   Biomimetic Materials

### 7.3.1   Self-Assembly

Structure-function relations in natural tissues are studied. Then the biomimetic materials [13] are engineered. Biomimetic, according to the dictionary, means a compound that *mimics* a biological material in its structure or function, or a lab protocol to initiate a natural chemical process.

For example, Discher [14] prepared rod-like aggregates referred to as worm micelles. These resemble linear proteins found both inside cells, such as cytoskeleton filament, and outside cells, such as collagen fibers. The research in solvent diluted copolymer systems was initiated in the early 1990s. Discher discovered the formation of cell-mimetic sacs, or vesicles in aqueous solution, using amphiphilic block copolymers.

Molecular assembly in biology uses water a lot. Water constitutes 70 percent of body mass in human anatomy. Protein-folding phenomena may be due to a hydrophobic effect. The binding of molecules to proteins can also be explained in this manner. Nine out of the twenty amino acids that comprise the polypeptide protein molecules are hydrophobic. Cell membranes are made up of lipids that are dual hydrophobic-polar and arranged in a segmented fashion. This sequential arrangement is referred to as amphiphilic. Hydrophobicity can be used to prepare templates for biomimetic nanostructures through self-assembly. Synthetic mimics of cell components and their functions rely on energetics, stability, and fluidity properties.

Block copolymers are segmented into two different monomer units, with sections of the polymer having one or the other monomer-repeating unit. Copolymers with block microstructure have been found to self-assemble and organize into periodic *nanophases,* such as arrays of rods and stacks of lamellar sheets. Hydrophobic-hydrophilic interactions form the driving force for the formation of the structure. The time-average molecular shape of an amphiphile in an aqueous solution in the corresponding forms of cylinder, wedge, cone, etc., will determine the morphology of the membrane formed, such as spherical, rod-like, etc. The average molecular shape is a function of the fraction of the hydrophilic fraction. The solvent effects also have a secondary role.

Copolymers with block microstructure that are amphiphilic assemble into worm micelles and polymer membranes. A polyethylene oxide (PED) and polybutadiene (PBd) copolymer (PEO-PBd) with block architecture is an example of an amphiphilic copolymer. PEO is hydrophilic, and PBd is hydrophobic. Self-assembly of these copolymers in water can lead to the formation of *polymersomes* or *vesicles.* At certain fractions of hydrophilic component, they form rod-like worm micelles. The relatively higher molecular weight results in the aggregate formation.

Lipid vesicles are formed into different morphologies, such as starfish, tube, pear, and string of pearls. Membrane thicknesses of 3 to 4 nm were confirmed using cryotransmission electron microscopy (cryo-TEM). The temperature range of stability of these structures is 273 to 373 K. Flexural Brownian motion and thermal bending modes are important considerations during analysis of stability of the morphologies formed. Membrane bending resistance has been found to increase as the polymer molecular weight increases. The copolymer molecular weight can have a strong effect on vesicle stability and in-plane hydrodynamic properties. Worms less than 10 nm in diameter have been observed using fluorescent labeling.

Diblock copolymer vesicles in aqueous solution have been studied. Protein folding stability is an interesting application of the study. Poly(isocyano-L-alanine)-L-alanine amphiphile, on self-assembly, becomes immunogenic. These are used in biomedical applications. The vesicular shells, upon collapsing are 10 to 100 nm in diameter. They coexist with rod-like filaments as well as chiral super helices. Polymersomes [14] can be formed instantaneously by adding water to lamellar structures of films (Fig. 7.2). Adding chloroform (CHCl$_3$) solutions of a copolymer into water creates vesicles. The chloroform can be removed by dialysis. Cross-linking the PBd can improve the stability of the worms. The applications for these systems are several: the cosmetics and pharmaceutical industries and as an anti-cancer agent, among others.

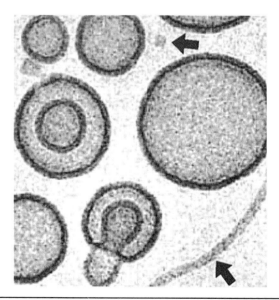

FIGURE **7.2** Cryogenic TEM image of 100-nm polymersomes [14].

Thus, many biological membrane processes can be mimicked by synthetic polymer vesicles. Examples of such processes are protein integration, protein fusion, DNA encapsulation, and DNA compatibility.

The materials in life sciences can be organized into a hierarchical structure. The levels of *structural hierarchy* can be seen in a tendon as follows [15]: (i) 0.5 nm for collagen polypeptide, (ii) 1.5 nm for triple helix, (iii) 3.5 nm for microfibril, (iv) 10 to 20 nm for subfibril, (v) 50 to 500 nm for fibril, (vi) 50 to 300 μm for fascicle, and (vii) 0.1 to 0.5 mm for the tendon. Wood and diarthrodial joints are found to have six levels of structural hierarchy. Thin films can be developed using biomimetics. The sequential adsorption of materials onto the surface observed during biomineralization can be mimicked. In the literature, such film-forming techniques are referred to under different names, such as fuzzy nanoassemblies, polyion multilayers, alternate polyelectrolyte thin films, molecular deposition, bolaform amphile multilayers, polymer self-assembly adsorption, layered composite films, stepwise assembly, and electrostatic self-assembly.

Some polymers have the capability to self-assemble and form complex structures. Some examples of self assembly during thin film formation by the sequential adsorption of materials onto a surface include coil-to-helix formation, formation of a biotin-streptavidin complex, S-layer protein formation on a two-dimensional array, and antigen-antibody interactions. A first-principle-based model of the linear assembly process leading to filaments has been discussed in the literature.

### 7.3.2 Equilibrium Kinetics

This dimerization step is a nucleation process:

$$2A \Leftrightarrow A_2 \tag{7.3}$$

The equilibrium constant for the reversible dimerization reaction can be written as:

$$K_2 = \sigma K \tag{7.4}$$

A linear polymer is formed by adding a monomer to the filament. The recurrent propagation step can be written as:

$$A_{i-1} + A \Leftrightarrow A_i \tag{7.5}$$

The equilibrium constant for the propagation step can be written as:

$$K_i = K \text{ (for } i \geq 3) \tag{7.6}$$

$\sigma$ is the cooperativity parameter. For small values of this parameter, the subsequent propagation steps are thermodynamically favorable, that is, $K \gg 1$. During any self-assembly process, the total monomer + aggregate concentration shall be varied ($C_0$). A certain critical concentration can be derived, $C_{crit}$. The monomer concentration shall remain less than the critical concentration at all times. For values less than the critical concentration, the monomer concentration increases with $C_0$. Above the critical concentration, only the aggregate formation occurs. The dimer formation step is slow during self-assembly. The propagation steps are faster. The kinetics take a sigmoidal growth curve.

An example of a self-assembly process can be found in tropomyosin systems that belong to the KMEF family (keratin-myosin-epidermin-fibrin) category of proteins. The repeat pattern in the amino acids is seven-fold: 1234567. The amino acids in positions 2, 3, 5, 6, and 7 are hydrophilic, and the amino acids in positions 1 and 4 tend to be hydrophobic. This is why the secondary structure of this protein is α-helix. After the formation of the helix conformation, a higher order structure forms on the surface of the helix. The hydrophobic amino acids in positions 1 and 4 form a band on the surface of the helix. The rest of the surface is filled by the hydrophilic amino acids. Two tropomyosin monomers self-assemble into a coiled-coil dimer. The coiled-coil dimer can take on different morphologies, such as filaments from head to tail interactions, muscle fibers, etc.

Self-assembly found in proteins has been attempted to mimicked in synthetic proteins.

The functions in organisms particularly popular are growth and functional adaptation, hierarchical structuring, damage repair and self-healing, capture of light by the eyes, photosensitive erection of plants, wings that enable flight, etc.

One unique property of biomaterials is their capability for self-repair. Investigators have found that sacrificial bonds between molecules break and reform dynamically. For example, during the deformation of wood, bonds were found to undergo reformation and breakage in cycle. This is similar to plastic deformation observed in metal and alloys. *Osteoclasts,* specialized cells in bone, remove material irreversibly, and *osteoblasts* deposit material to form virgin tissue. This cycle allows a continuous structural adaptation to external conditions and removal of damaged material by new tissue. A sensor/actuator system is in place that replaces damaged material. The growth direction of a tree after a landslide is an example in this regard. A fractured or broken tissue is healed naturally. The mechanism involves formation of an intermediate tissue based on the response to inflammation, followed by a scar tissue. Bone tissue is an exception to this empirical observation. It tends to regenerate completely. A great deal of research is underway in self-healing materials and it represents an opportunity for biomimetic materials research.

Biomimetic materials design starts with the observations of structure-function relationships in biomaterials. Systemization of this approach over serendipity is preferred. For example, the cuticle of arthropods was designed to endure infrared (IR) and ultraviolet (UV) irradiation, as well as the demands of sensory movement, transmission, etc. Biomimetic solutions are stored in large databases. These can be retrieved by engineers in search of technical solutions. Biomimetic solutions can be classified according to their functions in the databases. Validation and verification of biomechanisms is an iterative process between life sciences and engineering.

*Hydroxyapatite* [16] and *collagen* bonelike nanocomposite was prepared by Kikuchi et al. [17]. A self-organization mechanism between the hydroxyapetite and collagen surfaces was used in the preparation of nanocomposite. The composite prepared was found to possess good biocompatibility and biointegrative activities. They are equivalent to autogenous bone and perform better compared with other synthetic bone materials. These nanocomposites are poised for use in medical and dental fields in the future. They can reduce the patients' loads, including pain, at the donor sites of autogenous bone after transplantation.

Bone in human anatomy is chiefly composed of hydroxyapetite and collagen. Collagen is a protein abundantly found during the formation of most life the earth, except for insects. Extracellular matrices are constructed, such as tendons, ligaments, skin, and scar tissue, using collagen. Hydroxyapetite is a stable calcium phosphate with a pH of 7.2 to 7.4. It is found in body fluids of vertebrates and has an affinity for organic molecules. It can be used to filter and separate DNA and proteins. Endoskeletons of vertebrates, by evolution, have selected collagen and hydroxyapatite as their constituents. Bone is one of the human organs where turnover occurs by metabolism but the mechanical properties of the bone are intact. The turnover process is triggered by the attachment of osteoclasts to repaired parts of bone.

Hydroxyapetite nanocrystals are dissolved by the release of protons from osteoclasts that are attached to bone, forming the clear zone that distinguishes the resorption from other parts of the bone. Collagen fibrils are decomposed by collagenase and other proteases secreted by the osteoclasts. *Howship's lacunae* are cavities created by osteoclastic bone resorption. Osteoblasts cover the surfaces of the lacunae formed. These osteoblasts form the bone via collagen and subsequent release of calcium and phosphorous. Hydroxyapetite nanocrystals deposit on the c-axis, and a bundle of collagen is formed. This nanostructure plays a salient role in bone metabolism and mechanical properties.

There is a lot of interest in the literature in preparing hydroxyapetite/collagen nanocomposites. Their biocompatibility is tested using implantation techniques. Some of them are self-setting. Hydroxyapetite crystals are grown on collagen fibers using $CaHPO_4$ as precursors of hydroxyapatite. Mimesis of bone nanostructure is required to function as bone in recipient sites. Hydroxyapetite nanocrystals synthesized in the absence and presence of aspartic acid are shown in Fig. 7.3.

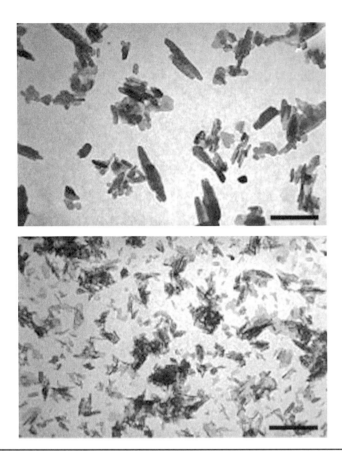

**FIGURE 7.3** Hydroxyapetite nanocrystals in the absence of aspartic acid and the presence of aspartic acid [17].

The detailed description of bone nanostructure formation has not yet been reported. Under healthy conditions, the supersaturated hydroxyapetite and body fluid solution does not deposit on collagen and other organic substances. The presence of $Ca^{2+}$ and $PO_4^{3-}$ ions may have contributed to the calcification of collagen fibrils. The stable formation of hydroxyapetite occurs with an alkaline pH of 8 to 9. Collagen fibrillogenesis occurs at 40°C body temperature. *In vitro* fibrillogenesis of collagen requires physiological saline conditions at a pH of 8 to 9 and 40°C body temperature.

Lengths of hydroxyapetite/collagen fibers grown were 20 μm in length, and those of collagen molecules grown were 300 nm. Electron diffraction pattern of the fibers indicated crescentlike 002 diffraction of hydroxyapetite. The *c*-axes of hydroxyapetite nanocrystals are aligned along the elongation direction of the hydroxyapetite/collagen fibers. The orientation is similar to that of the bone of vertebrates. The fibril length is based upon the degree of self-organization of hydroxyapetite and collagen, as measured by TEM and diffraction patterns. The alkaline pH conditions and body temperature promote calcium ion accumulation and the first phase of hydroxyapetite nucleation on the collagen surface. Collagen fibrillogenesis is promoted by the neutral surface charge of collagen achieved by sodium (Na) and chlorine (Cl) ions in physiological saline. With self-organization, the bending strength of the composite was found to increase. Excess water was removed by consolidation. The pH and temperature were sensitive parameters in determining the bending strength of the resulting nanocomposite.

Hydroxyapetite formation on the Langmuir-Blodgett monolayer indicates a driving force for the surface interaction between hydroxyapetite and collagen, and is an interfacial interaction between these molecules. This can be deduced from the formation of hydroxyapetite nanocrystals on carboxyl-terminated monolayer, but not on an amino-terminated monolayer. The interfacial interaction was studied using Fourier transform infrared (FTIR) spectrometer using the Kramers-Kronig equation for energy shifts of residues at the interface of hydroxyapetite and collagen. Red shifts in the spectra were found, meaning a decrease in bonding energies of C-O bonds. The hydroxyapetite crystal structure consists of two different Ca sites.

*Biocompatibility* of the nanocomposite specimens was studied under transmission electron (TM) and scanning electron (SE) microscopes. Tissue granulation and surface erosion were observed after two to four weeks. Collagen fibers encapsulated the debris of composites. Large cells with round nuclei infiltrated into the regions around the composites. Infiltrating macrophages phagocytize the resulting debris. Composites implanted into the subcutaneous tissue are collapsed from their surfaces. The composites were collapsed continuously and phagocytized for 24 weeks after implantation. Infiltration of macrophages into the nanocomposites occurred in a similar manner as implanted collagen sponges. Lymphocytes were not observed in either the hydroxyapetite/collagen or collagen sponge implantation. This is due to differences in the rejection, mobilization, and activation of granulocytes. The nanocomposites possess good biocompatibility in comparison with collagen sponges. Bone tissue reaction was examined in SD rats to understand this mechanism. Nanocomposite cylinders were implanted into SD rats. They were observed using optical microscopy after 1, 3, 5, 7, 14, and 28 days. The cut sections were stained with hematoxylin, eosin, tartrate-resistant acid phosphatase (TRAP), and alkaline phosphate (ALP) and were studied under the microscope. Progressive resorption of the composites was found.

Good *osteocompatibility* was found after direct bonding between new bone and composites without fibrous connective tissue in the surroundings. Howship's-like resorption was found in stained sections after 14 to 28 days. TRAP activity was raised on day 5 in the surroundings and cracks of composites. ALP activity also showed progress in surroundings of new bone. The substitution process of the composites to new bone occurs as follows: (a) formation of the composite debris by erosion with body fluid, (b) phagocytosis of the debris and composite surface by macrophages, (c) induction of osteoclastic cells on the composite surface and resorption of the composite by an analogous process to that of bone, and (d) induction of osteoblasts to the resorption lacunae created by osteoclastic cells and formation of new bone in the surroundings of the composite.

These steps are similar to the autogenous bone transplantation. Reconstruction of a critical bone defect in the tibias of beagles was examined for possible clinical use. The tibia defect was 20 mm in length and was formed by surgical saw using Ilizarov bone fixator. The nanocomposite resorption and bone growth were observed at each week by soft x-ray photography until the sacrifice date. The interface between the composite and bone was unclear at 10 to 12 weeks after implantation. Nanocomposite resorption and new bone formation in a beagle's tibia is analogous to bone remodeling in rats. The nanocomposite was studied for presence of human bone morphogenetic protein 2 to exploit the large surface area of contained hydroxyapetite nanocrystals with high adsorbability to organic substances. The nanocomposite was found to be a useful carrier of the protein.

### 7.3.3   Thin Films

The iridescence of insects and structural colors of plants are not well understood. The structure and composition of the chromophores are, however. Structural colors are different from pigmented colors. Investigations on optical thin films in biology have been performed for decades. Structural colors can be altered with the application of pressure, swelling or shrinking, or addition of a solvent. Addition of a swelling agent can result in a change of color in a reversible manner in iridescent wing membranes. Thin film optical interference can explain this observation. Light scattering causes the white color seen in insect wings. Structural colors can be studied using electron microscopy, as seen in butterfly and moth scales. These serve as thin film interference filters. Each scale is a flattened stack with two surfaces; the upper lamina contains a grid consisting of raised longitudinal ridges regularly joined by cross-ribs. The ridges and cross-ribs form a series of windows opening into the scale interior. The ridge structure consists of an alternating stack of high- and low-refractive index layers. Each ridge acts as a quarter-wave thin film interference mirror with a phase change upon reflection. The optical thickness, $nt$, of a dielectric stack layer composed of alternating thicknesses $t_a$ and $t_b$ is related by:

$$n_a t_a = n_b t_b \tag{7.7}$$

$$nt = n_a t_a + n_b t_b \tag{7.8}$$

where $nt$ is the optical thickness of the bilayer composed of a high-index and low-index component. The wavelength of maximum reflection is given by:

$$\lambda = 4n_a t_a = 4n_b t_b \qquad (7.9)$$

The wavelength of maximum constructive interference varies from 320 to 348 nm over a wing tilt from 0 to 50°C. Two *Lycaenid* butterflies were studied for the development of iridescence. Two types of internal reflective structures are closely related by development. The diffraction lattice appears to form within the scale cell boundaries through the assistance of a convoluted series of membranes. Membrane cuticle units are produced that are continuous with invaginations of the plasma membrane. Crystallites are formed that grow toward each other by accretion until the adult morphology arises. Thin film interference laminae are formed from the condensation of the network of filaments and tubes secreted outside the boundaries of the cell. Lattice formation occurs through the self-assembly of material into an face-centered cubic (FCC) Bravais lattice structure. The thin film laminae are formed by stretching the lattice. The lead reflectance spectrum from *Lindsea lucida* has a blue-green reflection band at 538 nm. Blue fruits of *Elaeocarpus angustifolius* exhibit a reflection band at 439 nm. A multilayer structure within the epidermis consisting of a parallel network of strands 78 nm thick was detected by electron microscopy. The optical thickness was 109 nm, and the reflectance maximum was 436 nm.

The mechanism of biomineralization in mollusks has been studied by investigating "flat pearls." *In vivo* monitoring is accomplished by placing a glass substrate between the mantle and the inner surface of the mollusk shell. The shell structure contains multiple organic, calcite, and aragonite layers, and the process is sensitive to substrate. The crystal phase during nacre formation is controlled by soluble mollusk proteins. The amino acids contained in the proteins were aspirate, glycine, glutamate, and serine residues. The red abalone shell formed the source of the proteins. The composition of the aragonitic composite was studied using gel electrophoresis denaturisation. These proteins promote the growth of $CaCO_3$ crystals. *Rhombohedral* calcite morphology was found to form in crystals grown in the absence of soluble protein. *Spherulitic* calcite morphology was found to form in crystals grown in the presence of calcitic protein fraction. *Aragonite needles* are formed in the presence of aragonitic fraction. The calcite-to-aragonite transition is caused by the addition of aragonitic polyanionic proteins. When soluble aragonite proteins are depleted, the sequential transition of calcite to aragonite and back to calcite is the result. Atomic force microscopy (AFM) was used to study the mechanism of aragonite tablet growth. Iridescent patches of organic material are formed when organic pearls are demineralized.

Organic sheets with a pore diameter of 5 to 50 nm was found using AFM and scanning ion conductance microscopy.

Crystal CdS synthesis was demonstrated in films made up of polyethylene oxide (PEO). The factors governing the synthesis are strong binding; solubility of the reagents; and an ordered, regular environment to induce nucleation. The crystals formed were found to be uniform in size, phase, and crystallographic orientation. The morphology type of the crystals formed was *rock salt* in nature.

The sequential laying down of inorganic layers forms a critical step in biomineralization in a mollusk shell. A positively charged substrate is dipped into an aqueous solution containing polyelectrolytes of the negative charge. The negatively charged polyelectrolyte adsorbs to the surface. Upon rinsing and drying, the film is dipped into a solution containing a positively charged polyelectrolyte. This process is repeated indefinitely with multiple electrolyte solutions. Multilayer thin film formation is an important aspect to this process. Examples of layer formation in albumin/silica, silica/alumina, treatment of fabrics, and multilayers on metals and mica were shown in the literature.

Clean substrates are needed for an efficient film formation. The glass substrates are acid cleaned using concentrated sulfuric acid ($H_2SO_4$) and hydrogen peroxide ($H_2O_2$). The substrate surface needs to be modified with charged groups some of the time. Silanol groups are sometimes added. Negative charges are imparted to quartz surfaces. Carboxyl groups are added to gold substrates. The substrate used in polymethylformamide (PMF) film is polyethylene terepthalate (PET). Amide linkages are formed between carboxylate groups and some PAH amino groups. A net positive charge is affected to silanized slides by dipping into hydrocholoric acid (1 N HCl).

Polyion multilayer films are characterized by small-angle x-ray reflectance spectroscopy (SAXR). Ultraviolet-visible (UV/Vis) spectroscopy taps into the chromophore present in PMF. Film thicknesses are measured using ellipsometry. The deposition kinetics, as well as film thickness, can be measured using surface plasmon resonance spectroscopy. Material deposition can be studied using IR spectroscopy. Real-time monitoring of rate and amount of monolayer deposition during PMF monolayer formation is allowed for by the use of quartz crystal microbalance (QCM). QCM is a piezoelectric device. Mass charges on the order of nanograms are quantitatively measured. The changes in resonant frequency $f$ of a quartz crystal with the change of mass of material loaded into the crystal is given by the Sauerbrey equation. The mechanism of PMF film formation can be studied by atomic force microscopy (AFM). Hectorite sheets of 25 to 30 nm were imaged using AFM. The charged macromolecules adsorb onto surface defects at short deposition times. These form islands and retain their coil conformation. Homogeneous monolayers composed of flattened polymer chains are found at longer deposition times.

Monolayer formation in PMF can be studied using AFM. Surface coverage is measured as a function of adsorption time and ionic strength. Initial kinetics are diffusion-limited, and at long times, it becomes random sequential adsorption. Surface coverage is sensitive to the ionic strength.

Biopolymers have been prepared, such as polysaccharide-containing PMFs. In chitosan/PSS, film thickness was increased by dipping in a solution with ionic strength ranging from 15 A/bilayer in the absence of salt to 69 A in a molar sodium chloride. Ionic strength was found to be an important parameter in the determination of adsorption kinetics. Saturation condition was achieved at lower molarity. Shielding the chitosan charge by adding salt provided for more conformational flexibility and enabled adsorption at the surface. A film containing alternating layers of DNA and PAH, a polymer/biopolymer hybrid, was prepared. Coil conformation of polypeptides can be detected using circular dichroism. β-sheet conformation formed from self-assembly between two polypeptides can be seen using IR analysis of the film.

Streptavidin-containing films were described in the literature. A precursor was used. The film was irradiated with UV light through a copper mask. The film was then immersed in a solution containing fluorescein isothiocyanate (FITC)–labeled streptavidin. Fluorescence spectroscopy was used to study the protein arrays. Protein-containing PMFs were studied. The molecular weight range of proteins studied was 12,400 to 240,000. A multilayer was prepared consisting of glucose isomerase and the bolaamphiphile in porous trimethylamine polystyrene beads. The carrier pore diameter was 46 nm, and only two layers of enzyme could deposit on the pores. Enzyme activity was comparable to that of soluble and monolayer enzyme preparations. It was studied in films containing up to 40 enzyme layers. In the 40 bilayer film, the average activity per layer decreased by 50 percent of that measured for a 10-layer film. This is apparently because of the inability of the substrate to diffuse deeply into the film.

Sequential adsorption is a low-cost approach to the assembly of thin films. Most polions can be incorporated into a film, including dyes, polymers, proteins, viruses, inorganic nanoparticles, and ceramic plates. Automation of the technique is possible, with a minimal investment in equipment. Films with interesting features can be synthesized using beakers, stopwatch, water, and electrolytes by hand. Complex multilayers are formed using an automated slide stainer. Scale-up of sequential adsorption technique is less expensive compared with the Langmuir-Blodgett technique. Some examples of PMFs were discussed. The properties can be tuned by varying the number of layers or the spacing between the layers. Any substrate where a charge can be placed can be used in the synthesis of these films. Self-healing characteristics are exhibited by the sequential adsorption process. Point defects and dust inclusions have less penetration distance. The ionic strength is a sensitive parameter in varying the bilayer thickness. The sequential adsorption process can be combined with other techniques.

Three-dimensional control of film composition and properties are provided. Spin coating consists of an application of a solution of film material into a rapidly spinning disk. A uniform fluid film forms and solidifies upon evaporation of the solvent. Solvent casting involves drying a polymer solution placed in a well. Oriented thin films are formed when cast in magnetic or electric fields. Films have been formed by direct polymerization into an initiator covalently attached to the substrate. In the Langmuir-Blodgett technique, an amphiphile monolayer is placed on an air-water interface. The temperature in the water bath is controlled. The surface pressure is measured and controlled by a Teflon arm touching the interface containing the monolayer. The monolayer is transferred to the glass slide using a mechanical dipping apparatus. Complex optical films can be prepared this way. Practical application of this technique, however, is precluded by high cost and poor efficiency. The control of molecular structure is made possible using photolithography. This is effected by a combined method of solid-phase peptide synthesis and semiconductor-based photolithography.

The sequential adsorption technique for film formation can be applied in the following ways:

- Light emitting diode (LED)
- Conducting polymer
- Second-order nonlinear optics (NLO)
- Dye-containing optical film
- Polydiacetylene
- Bioreactors
- Molecular recognition by antibody-antigen interaction
- Nonthrombogenic surfaces
- Nanoscale thin film pH electrodes

Starch can be converted to gluconic acid using a sequential adsorption film formation process. The reaction rate achieved is $0.0045 \text{ mol}/\text{m}^2/\text{h}$, compared with $0.017 \text{ mol}/\text{L}/\text{h}$. A $1 \text{ m}^2$ membrane would have about one-third the efficiency of a commercial microbial fermenter. A two-component film has four levels in its structure hierarchy: monolayer, bilayer, and multilayer. A three-component film has a five-level hierarchy.

### 7.3.4 Membranes

A protein scaffold/biomimetic membrane material was developed at Argonne National Laboratory. It is a tool for encapsulating and studying the native behavior and structure of membrane and soluble proteins. The membrane material is a complex fluid made up of a mixture of a lipid, polymer amphiphile, a cosurfactant, and water. It undergoes thermoreversible phase changes and exists as liquid below a certain threshold temperature and as a liquid crystalline gel above that temperature. Dedicated proteins, enzymes, and other biomolecules are mixed and ordered in the liquid state and ordered again by increasing the temperature above room temperature. The orientation of the materials is further increased by the use of magnetic fields. When applied to selected substrate materials, domains can be oriented preferentially. Certain nonionic, amphiphilic triblock copolymers of PEO-PPO-PEO can be employed as an alternative to a more expensive PED that is also architecturally limited. They can be used as lipid conjugates for producing biomimetic nanostructures. The PPO chain length, when approximately similar to the dimensions of the acyl chain region of the lipid bilayer, results in a strongly anchored triblock copolymer.

Medical researchers used biomimetic nanostructures to examine soft tissue cellular wounds such as burns, frostbite, radiation exposure, pressure trauma, electric shock, scrapes and abrasions, heart attack, and stroke. It can be used as a drug screening and development tool in the following applications: nano Band-Aids to augment the healing of cellular wounds, nanocapsules for site-directed delivery of healing agents, ideal polymers for healing soft tissue damage, and nerve regeneration in spinal cord injuries.

Synthetic biological membranes with self-organizing characteristics, such as liquid crystalline gels that change shape and function in response to environmental changes, were developed at the University of Chicago [18]. There is increased interest in the development of "smart materials." The properties of these materials change in response to environmental stimuli such as ionic strength, temperature, and magnetic or electric fields. The basis of molecular machines, chemical valves and switches, sensors, and a wide range of optoelectronic materials is the response of bio-organisms to external stimuli.

A mixture of lipids, a low-molecular-weight polyethylene glycol-derived polymer lipid, and a pentanol surfactant is one such example of a smart material. These gels transform to liquid by heating to an elevated temperature. At higher temperatures, the incorporated proteins and pentanol surfactant rapidly degenerate. The material undergoes phase separation at lower temperatures. It responds to an external stimulus of temperature alone. The material developed at the University of Chicago was in response to a need for materials that are responsive to a variety of external stimuli. The material must also be biocompatible.

The developed material exists as a gel at elevated temperatures and as a liquid at lower temperatures. It is used in drug delivery systems where the body temperature is the high temperature. The developed material is a biocompatible, membrane-mimetic liquid crystalline material.

The biologically active membrane proteins can be encapsulated using the developed material within an organized lipid matrix.

Macroscopic ordering of molecules occurs when a magnetic field, an electric field, or shear is applied to the mixture. The mixture is multipositional, and the material reacts to external stimuli to provide both structural and functional characteristics. It manifests a birefringent phase when subjected to a certain environment, an optically isotropic or transparent phase. An optical cue is given when an intact membrane has formed in response to the application of certain stimuli. The material undergoes a *thermoreversible* phase change. It is comprised of a lipid polymer amphiphile (such as a polymer grafted phospholipid), cosurfactant, and water. When the temperature is increased, the material solidifies.

The material is comprised of 65 to 90 percent water, 3 to 5 percent surfactant, 7 to 27 percent lipid and amphiphilic polymer, and the ratio of polymer to lipid is approximately 4 to 10 mole percent. The mixture undergoing phase change is depicted in Fig. 7.4. The stimulus responsive fluid developed with self-assembling properties switches between two distinct structural states and two distinct functional states in response to several external stimuli.

The material is prepared by a noncovalent self-assembly of a quaternary mixture of a phospholipid; a lipopolymer, a diblock or triblock copolymer or polymer-grafted amphiphile, and a surfactant dispersed in water. The supramolecular, nondenaturing material undergoes a reversible transformation from a liquid crystalline gel to a nonbirefringent fluid upon reduction in temperature. The liquid phase is found to instantaneously organize into a liquid crystalline gel with an increase in temperature. These changes are at the molecular level, but manifest at the macromolecular level. The phase change occurs at a temperature range of 15 to 20°C.

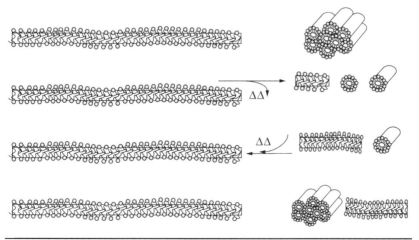

**Figure 7.4**    Gel undergoing phase change.

The gel composition shown in Fig. 7.4 is a mixture of a phospholipid, a polymer amphiphile such as an end-grafted phospholipid or a diblock or a triblock copolymer, and a zwitterionic or cationic cosurfactant dispersed in water. These gels form bilayer membranes, with the hydrophobic ends of the lipid and cosurfactant of each layer oriented inwards. Cavities/spaces in these membranes can accommodate the fluid in which the membranes are immersed. When juxtaposed, the cavities/spaces are differentially organized into planar sheets and channels separated by water-impermeable lipid micelle and membranes. Proteins and other substances generally of a size between 1 and 50 nm may be incorporated in the bilayer membranes or in the aqueous channels. This allows the membranes to be used in packaging or encapsulation for drug delivery applications. Depending on their characteristics, gels can be used as sensors and opto- and microelectronic products.

*Mesoscopic self-assemblage* of the developed fluid is further enhanced when the fluid contacts an appropriate surface. When the gel phase of the mixture interacts with certain surfaces containing OH group, the orientation of lamellar domains of the gel are directed into macroscopic dimensions. This ordering enhancement is because of polar phospholipid head groups contained in the mixture and a similar hydrophilic group on the support substrate. This mechanism enables the mixture to hold target functional groups that are encapsulated by the mixture in a certain orientation.

## 7.4   Design of Bioartificial Organs

The design of bioartificial organs needs a lot of information about the permeability of the capillary wall, tissue space in specific organs, and other transport parameters. Single- and multiple-compartment pharmacokinetic models can be developed. Tracer compounds can be injected, and transient concentration data can be measured in the tissue space and blood vessels in the human anatomy either *in vivo* or *in vitro*. The pharmacokinetic model can be correlated to the concentration of the tracer and a variety of transport information, such as the matrix blood flow rate and the capillary wall and immunoisolation membrane permeability to insulin. With the information available from one tracer, permeabilities of other substances of interest can be predicted using the theory discussed in Chap. 3. This information can be used to *scale up* into bioartificial organs, which will be discussed in Chap. 8.

For example, insulin is injected as a tracer to study the ramifications of an implant device. The initial response to an implant is infiltration of the support of a variety of cell types. The process is similar to that of wound healing and involves overlapping phases [19]:

1. Inflammation
2. Proliferation
3. Maturation

Platelets and neutrophils arrive during the inflammation period, which can take several days. The scaffold for the inflammatory cells is provided by a fibrin network triggered by the activation of the clotting process and release of growth factors such as platelet-derived growth factor (PDGF), epidermal growth factor (EGF), and transforming growth factor (TGF). Foreign material is ingested by phagocytosis, controlled by factors released by neutrophils. Monocytes are attracted and fused into macrophages and transformed into foreign-body giant cells. The site is cleansed, and dead tissue, bacteria, and debris are taken out by the giant cells. Fibroblasts and endothelial cells proliferate, triggered by a variety of growth factors such as PDGF, TGF, EGF, and fibroblast growth factor (FGF) released from the macrophages. A collagen network is formed by extracellular matrix materials released from fibroblasts. Capillary sprouts form due to low oxygen levels and become a vascular bed. Low oxygen levels can regress the vascular supply. The site is remodeled during the maturation phase. The wound is contracted and the collagen matrix is organized. The rate of tissue growth is a salient consideration in tissue design.

Radioactive insulin was injected as a tracer to evaluate a vascularized implant device. The tracer's uptake and elimination data can be evaluated using a multiple-compartment pharmacokinetic model [20]. A schematic of this multiple-compartment pharmacokinetic model is shown in Fig. 7.5. The distribution of insulin within the human anatomy is distributed into four compartments that are assumed to be well mixed. These include: implantation chamber,

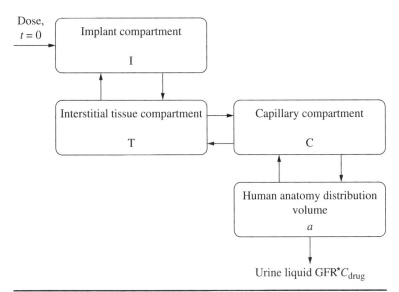

**FIGURE 7.5** Multiple-compartment models for analysis of radioactive insulin injection.

interstitial tissue chamber, blood capillary chamber, and human anatomy distribution volume. The insulin is injected into the implantation chamber at time $t = 0$. Insulin then diffuses across a semipermeable membrane and enters the tissue chamber. Then the insulin goes
into the capillary bed, where the blood is flowing at a rate $Q$. Then the
insulin is carried to the human anatomy distribution, which is shown
as another compartment in Fig. 7.5. The elimination step of insulation
through the kidneys is also included in the model.

The transient concentration of the insulin drug can be obtained
by solving the model developed from mass balances across each compartment. Let the renal clearance or first-order rate constant from
each compartment be given by $k$ in general, and the subscripts denote
the compartment from which the drug originates to the compartment
to which it must reach. The mass balance equations for each compartment can be written as discussed in Chap. 6.

$$\frac{dc_{insulin}^I}{dt} = -k_{IT}C_{insulin}^I + k_{TI}\left(\frac{V_T}{V_I}\right)C_{insulin}^T \tag{7.10}$$

$$\frac{dc_{insulin}^T}{dt} = -(k_{TI} + k_{TB})C_{insulin}^T + k_{IT}\left(\frac{V_I}{V_T}\right)C_{insulin}^I + k_{BT}\left(\frac{V_B}{V_T}\right)C_{insulin}^B \tag{7.11}$$

$$\frac{dc_{insulin}^B}{dt} = -(k_{BT} + k_{BC})C_{insulin}^B + k_{TB}\left(\frac{V_T}{V_B}\right)C_{insulin}^T + k_{aB}\left(\frac{V_T}{V_B}\right)C_{insulin}^a \tag{7.12}$$

$$\frac{dc_{insulin}^a}{dt} = -k_{aB}C_{insulin}^a + k_{Ba}\left(\frac{V_T}{V_a}\right)C_{insulin}^B - (GFR)C_{insulin}^a \tag{7.13}$$

Equations (7.10) to (7.13) are a set of ordinary differential equations
(ODEs) with constant coefficients that are simple and linear. These
can be written in the state form as follows [21]:

$$d/dt\begin{pmatrix} C_{insulin}^I \\ C_{insulin}^T \\ C_{insulin}^B \\ C_{insulin}^a \end{pmatrix} = \begin{pmatrix} -k_{IT} & (k_{TI}V_T)/V_I & 0 & 0 \\ (k_{IT}V_I)/V_T & -(k_{IT}+k_{TB}) & (k_{BT}V_B)/V_T & 0 \\ 0 & (k_{TB}V_I)/V_B & -(k_{BT}+k_{BC}) & (k_{aB}V_a)/V_T \\ 0 & 0 & (k_{Ba}V_I)/V_a & -(k_{aB}+GFR) \end{pmatrix} \begin{pmatrix} C_{insulin}^I \\ C_{insulin}^T \\ C_{insulin}^B \\ C_{insulin}^a \end{pmatrix}$$

$$\tag{7.14}$$

The initial conditions for the system is given by:

$$C_{insulin}^I = \frac{Dose}{V_I}; C_{insulin}^T = 0; C_{insulin}^B = 0; C_{insulin}^a = 0$$

The solution to Eq. (7.14) can be obtained by expanding an arbitrary vector in terms of eigenvectors. Equation (7.14) is in the form of:

$$\frac{dc}{dt} = K^*C \tag{7.15}$$

The solution can be written as:

$$C = Ze^{\lambda t} \tag{7.16}$$

where $Z$ is the unknown matrix of constants and $\lambda$ is the unknown constant(s). Substituting Eq. (7.16) in Eq. (7.15) and realizing that $e^{\lambda t} \neq 0$, it can be seen that:

$$K^*Z = \lambda Z \tag{7.17}$$

Equation (7.16) requires that $Z^t \neq 0$; thus, in Eq. (7.17), $\lambda$ is an eigenvalue of $K$ and $Z$ is an eigenvector that belongs to $\lambda$. Because $K$ is a square matrix ($4 \times 4$ rows and columns) of the order 4, there are four values of $\lambda$ that satisfy the characteristic equation:

$$|K - \lambda_j I| = 0, \; j = 1, 2, 3, \text{ and } 4 \tag{7.18}$$

Corresponding to each eigenvalue is an eigenvector, $Z_j$. There are four solutions of the form presented in Eq. (7.16). Equation (7.15) is homogenous, so there is no need for a particular solution for this system. Each of the solutions correspond to the concentration of insulin in the implant compartment, interstitial tissue compartment, blood capillary compartment, and the human anatomy distribution compartment. The glomerular filtration rate (GFR) was also included in the governing equation of the concentration of insulin in the human anatomy distribution compartment. The initial conditions given below Eq. (7.14) can be invoked to obtain the integration constants.

Equation (7.18) can be expanded to a polynomial equation of the fourth degree in $\lambda$. Laplace's development of the determinant to write the polynomial as:

$$P_n(\lambda) = \lambda^4 + s_1\lambda^3 + s_2\lambda^2 + s_1\lambda + |K| = 0 \tag{7.19}$$

where

$$s_j = (-1)^{4-j}(\Sigma \; principal \; minors \; of \; order \; j \; of \; K) \, j = 1, 2, 3, \text{ and } 4. \tag{7.20}$$

If all the eigenvalues of $C$, that is, $\lambda_j$, have negative real parts, then $C$ would tend to zero at infinite times. If any eigenvalue has a positive real part, then one or more of the insulin compartment concentrations can become *arbitrarily large*. If the eigenvalues are imaginary, the concentration would undergo oscillations. The oscillations would be subcritical damped or growing supercritical, depending on whether the real part of the eigenvalues is negative or positive, respectively.

## Summary

Tissue design evolved as a separate discipline from the field of bio-materials during a scientific conclave in 1988. Langer and Vacanti defined tissue engineering as "an interdisciplinary field that applies the principles of engineering and life sciences toward the development of biological substitutes that restore, maintain, or improve tissue function." The natural healing potential of patients is also tapped into. The three strategies in tissue design are providing substitutes, introducing tissue-inducing substances, and placing cells within matrices. Vascular grafts were introduced for the anastomosis of blood vessels. Resorbable grafts were developed in the 1960s. Skin grafts are used to cover severe wounds. Fibroblasts were used to condense a hydrated collagen lattice to a tissue-like structure suitable for wound healing. The first dialysis machine was introduced during Word War II. The concept of a bioartificial kidney was formulated. The first pancreas transplant was performed in 1966, and the first liver transplant was performed in 1967. Different ways of packing islet cells to provide effective and durable glycemic control were started in the 1980s. Metals and alloys have been used to replace damaged bone to provide support for healing bones. Hydroxyapetite ceramics were found to be bioactive and solicit the formation of normal tissue on their surfaces. Bone induction has been studied, and bone morphogenic proteins (BMPs) were discovered in 1970s and 1980s.

Three-dimensional tissue formation is supported by a structure called a scaffold. Scaffolds need to be biodegradable. The rate of tissue formation should be equal to the rate at which degradation occurs. Biodegradable polymers make attractive candidates for scaffold materials. Examples of biodegradable polymers are collagen, glycosaminoglycans, chitosan, polylactic acid (PLA), polycaprolactone, and polyhydroxyalkanoates. It can take three to five years for PLA to be fully resorbed. Degradation of PLA is via an autocatalyzed hydrolytic scission of the ester groups in the polymer chain. Polymer blends can be used to form implantable devices. When a critical molecular weight is reached, the implant will cease to posses any mechanical strength. The relationship between percent additive and the degradation rate constants and molecular weights are given in Eq. (7.2). Collagen has a triple-helical structure. The extracellular matrix (ECM) is made up of collagen. The molecular structure of collagen consists of fibrils that are 300 nm in length and 1.5 nm in diameter. Ramachandran studied the collagen fiber in kangaroo tail tendon. Electrospinning operation can be used to fabricate organic polymer fibers with micron-to-nanoscale diameters in linear, 2-D, and 3-D architecture. Polyglaxanaone and polyphosphazenes can be spun into fibers. Techniques such as salt-leaching, fibrous fabric processing, 3-D printing, foaming, nanofiber self-assembly, textile technologies, solvent casting, gas foaming, emulsification, freeze-drying, thermally induced phase separation, CAD/CAM technologies, fiber bonding, membrane lamination, melt molding, and *in situ* polymerization can be used to prepare scaffolds. Carbon nanotubes (CNTs) are also potential candidates for scaffolds in tissue design. They are biocompatible, resistant to biodegradation, and can be functionalized by biomolecules. However, some toxicological issues regarding the use of CNTs still need to be resolved.

Lower critical solution temperature (LCST) and upper critical solution temperature (UCST) are also important considerations in the phase separation of polymers. However, they are covalently attached, thus preventing separation at the macroscale. Phase separation is limited to the nanoscale. The ABC block copolymer may self-assemble into more than 20 different complex phase-separated morphologies (Fig. 7.1 [12]). Typical morphologies are spherical, cylindrical, and lamellar. Nanoscale morphologies are sensitive to temperature. They are reversibly formed.

Biomimetic materials are designed to mimic a natural biological material. Copolymers with block microstructure have been found to self-assemble and organize into periodic nanophases. Molecular shape is found to be a function of a fraction of hydrophilic fraction. Polymersomes, or vesicles, can be formed by the self-assembly of PEO-PBd in water. Lipid vesicles are formed into different morphologies such as starfish, tube, pear, and string-of-pearl shapes. Worms that are less than 10 nm in diameter and with a membrane 3 nm thick have been observed. The stability of protein folding can be studied using self-assembly. Many biological membrane processes can be mimicked by synthetic polymer vesicles.

The equilibrium kinetics of self-assembly reactions was discussed. A cooperativity parameter is defined along with the equilibrium rate constant. The example system used to illustrate the mathematical treatment is tropomyosin. Amino acids in position 1 and 4 are hydrophobic and are hydrophilic in positions 2, 3, 5, 6, and 7. Banding on helix structures comes about.

One property of biomaterials worthy of mimicking is capable of self-repair. Biomimetic mechanisms are stored in databases. Hydroxyapetite and collagen were used to prepare bone-like nanocomposite. Howship's lacunae are cavities created by osteoclastic bone resorption. Hydroxyapetite form on Langmuir-Blodgett monolayers. There is interfacial interaction between hydroxyapetite and collagen. The substitution process of composites to new bone occurs in stages similar to autogenous bone transplantation: erosion of body fluid and formation of composite debris, phagocytosis of debris, resorption of composite, and induction of osteoblast to the resorption lacunae. Reconstruction of a critical bone defect in beagles' tibias was examined for possible clinical use.

The iridescence of insects and structural colors of plants are not well understood. Optical thickness of a dielectric stack layer of alternating thickness and a wavelength of maximum constructive interference were quantitated. Two *Lycaenid* butterflies were studied for the development of iridescence. The mechanism of biomineralization in mollusks has been studied by investigating "flat pearls." Rhombohedral calcite morphology, spherulite calcite morphology, and aragonite needles are formed under different conditions. Aragonite tablet growth was studied using AFM. Crystal CdS with rock salt morphology was synthesized in films made up of PEO.

Efficient film formation needs a clean substrate. Polyion multilayer films are characterized by small-angle x-ray spectroscopy (SAXR). Quartz crystal microbalance is used to measure mass charges in nanogram-quantity materials. The mechanism of PMF film formation was studied by using AFM. Polysaccharide-containing PMF biopolymers have been prepared. Adsorption kinetics depends on ionic strength. A polymer/biopolymer hybrid such as DNA and PAH was formed into a film containing alternating layers. Films containing streptavidin, glucose isomerase, etc., were discussed. Assembly of thin films is by sequential adsorption. Three-dimensional control of film composition and properties were discussed.

Protein scaffold/biomimetic membrane material was discussed. Membrane material is a complex fluid made up of a mixture of a lipid, polymer amphiphile, and a cosurfactant. It undergoes thermoreversible phase changes and exists as liquid below a certain threshold temperature and as a liquid crystalline gel above that temperature. Biomimetic nanostructures are used to examine soft tissue cellular wounds and dry sensing and development and nerve regeneration. Smart materials have been developed that undergo a property change in response to environmental stimuli. These materials are used in drug delivery systems. A gel undergoing phase change was shown in Fig. 7.4.

The design of bioartificial organs needs a lot of information about the permeability of the capillary wall, tissue space in specific organs, and other transport parameters. Tracer compounds can be injected, and transient concentration data can be measured in the tissue space and blood vessels in the human anatomy, either *in vivo* or *in vitro*. This information can be used to scale up into bioartificial organs (discussed in the next chapter). Finally, a multicompartment model for the analysis of radioactive insulin injection was discussed.

# References

[1]  R. Langer and J. P. Vacanti, "Tissue engineering," *Science* (1993), 260, 920–926.
[2]  A. B. Voorhees, "How it all began," in P. N. Sawyer and M. J. Kaplitt, *Vascular Grafts*, New York: Appleton-Century-Crofts, 1978, pp. 3–4.

[3] L. Xue and H. P. Greisler, "Blood vessels," in R. P. Lanza, R. Langer, and J. P. Vacanti, *Principles of Tissue Engineering*, New York: Academic Press, 1980, pp. 427–446.

[4] R. E. Billingham and J. Reynolds, "Transplantation studies on sheets of pure epidermal epithelium and on epidermal cell suspensions" (1953), *Br. Journal Plast. Surg.*, 6, 25–36.

[5] E. Bell, H. P. Ehlrich, D. J. Buttle, and T. Nakatsuji, "Living tissue formed in vitro and accepted as skin equivalent tissue of full thickness" (1981), *Science*, 211, 1052–1054.

[6] R. Lewis, "A compelling need" (1995), *The Scientist*, 9, 15.

[7] W. L. Chick, A. A. Like, and V. Lauris, "A hybrid artificial pancreas" (1975), *Trans. Am. Soc. Art. Intern. Organs*, 21, 8–15.

[8] J. A. Allen, T. Hassanein, and S. N. Bhatia, "Advances in bioartificial liver devices" (2001), *Hepatology*, 34(3), 446–455.

[9] M. R. Urist, "Bone formation of autoinduction" (1965), *Science*, 150, 893–899.

[10] J. Rose and S. Hardwick, "Biodegradable polymer systems" (2009), U.S. Patent 7,524,891, Smith & Nephew PLC, York, Great Britain.

[11] G. N. Ramachandran and G. Kartha, "Structure of collagen," *Nature* (1954), 174, 269–270.

[12] A. Fertala and F. Ko, "Collagen or collagen-like peptide containing polymeric matrices" (2004), U.S. Patent 6,753,311, Drexel University, Philadelphia, PA.

[13] K. R. Sharma, *Nanostructuring Operations in Nanoscale Science and Engineering*, New York: McGraw-Hill Professional, 2010.

[14] D. E. Discher and F. Ahmed, "Polymersomes," *Annual Rev. Biomed. Eng.* (2006), 8, 323–341.

[15] T. M. Cooper, "Biomimetic thin films," in *Handbook of Nanostructured Materials and Nanotechnology*, Vol. 5: *Organics, Polymers, and Biological Materials*, H. S. Nalwa (ed.), New York: Academic Press, 2000, 711.

[16] E. Boanini, M. Fini, M. Gazzana, and A. Bigi, "Hydroxyapatite nano-crystals modified with acidic amino acids," *Eur. J. Inorg. Chem.* (2006), 4821–4826.

[17] M. Kikuchi, T. Ikoma, S. Itoh, H. N. Matsumoto, Y. Koyama, K. Takakuda, K. Shinomiya, and J. Tanaka, "Biomimetic synthesis of bone-like nanocomposites using the self-organization mechanism of hydroxyapetite and collage" (2004), *Composites Sci. Techn.*, 64, 819–825.

[18] M. A. Firestone and D. M. Tiede, "Synthetic biological membrane with self-organizing properties," U.S. Patent 6,537,575 (2003), The University of Chicago, Chicago, IL.

[19] F. Arnold and D. C. West, "Angiogenesis in wound healing," *Pharmacol. Therapy* (1991), 52, 407–422.

[20] R. L. Fournier, *Basic Transport Phenomena in Biomedical Engineering*, Lillington, NC: Taylor & Francis 1998.

[21] A. Varma and M. Morbidelli, *Mathematical Methods in Chemical Engineering*, New York: Oxford University Press, 1997.

# Exercises

## Review Questions

**1.0** Where was tissue engineering born?

**2.0** What is the difference between natural healing and tissue repair?

**3.0** What is meant by reparative medicine?

**4.0** What are the three different strategies employed in tissue design?

**5.0** What happened when rigid glass and metal tubes were attempted to be used as vascular grafts?

**6.0** What is a resorbable graft?

**7.0** What are the differences between vascular grafts and skin grafts?

**8.0** What is the immunological basis for skin grafts?

**9.0** Can collagen be used to prepare biologic vascular structures?

**10.0** How did allografts form the impetus for the development of skin grafts during World War II?

**11.0** How did growth factors help?

**12.0** How was wound healing affected using fibroblasts?

**13.0** How is pancreas transplant superior to insulin therapy?

**14.0** Describe the hybrid artificial pancreas made of beta cell cultures on synthetic semipermeable hollow fibers.

**15.0** Was the first liver transplant extracorporeal or *in vivo*?

**16.0** How is a bioreactor helpful in cell-based therapy?

**17.0** Why is hydroxyapetite preferred to replace bone?

**18.0** Why are growth and regenerative capabilities important in tissue design?

**19.0** What are the advantages of scaffolds?

**20.0** Give two examples of biodegradable polymers.

**21.0** How does the mechanical strength of the implant change with molecular weight and other parameters of the blend?

**22.0** How is the collagen triple-helix structure different from protein $\alpha$-helix and DNA double-helix structures?

**23.0** What was the issue in Crick's structure of collagen and Ramachandran's structure of collagen?

**24.0** Describe the electrospinning process to fabricate organic polymer fibers in nanoscale dimensions.

**25.0** How can the salt-leaching technique be used to prepare a scaffold?

**26.0** What does "soot harvesting" mean?

**27.0** What is the degree of control available in the electric arc process to control the diameter of CNTs?

**28.0** What is the degree of control available in the laser ablation process to control the diameter of CNTs?

**29.0**  What is the degree of control available in the CVD process to control the diameter of CNTs?

**30.0**  What is the degree of control available in the HIPCO process for to control the diameter of CNTs?

**31.0**  What is the degree of control available in the surface-mediated process for the vertical alignment of CNTs?

**32.0**  What is the typical length of CNT formed during the electric arc process for the synthesis of CNTs?

**33.0**  What is the typical length of CNT formed during the laser ablation process for the synthesis of CNTs?

**34.0**  What is the typical length of CNT formed during the CVD process for the synthesis of CNTs?

**35.0**  What is the typical length of CNT formed during the HIPCO process for the synthesis of CNTs?

**36.0**  What is the typical length of CNT formed during the surface mediated vertically aligned nanotube process for the synthesis of CNTs?

**37.0**  What are the typical temperature and pressure used during the synthesis of CNTs using the electric arc process?

**38.0**  What are the typical temperature and pressure used during the synthesis of CNTs using the laser ablation process?

**39.0**  What are the typical temperature and pressure used during the synthesis of CNTs using the CVD process?

**40.0**  What are the typical temperature and pressure used during the synthesis of CNTs using the HIPCO process?

**41.0**  What are the typical temperature and pressure used during the synthesis of CNTs using the surface mediated vertically aligned nanotubes?

**42.0**  What is the difference between SWNT, DWNT, and MWNT?

**43.0**  What is the role of the catalyst in the laser ablation process?

**44.0**  What is the time duration of the laser pulse during laser ablation process to prepare CNTs?

**45.0**  What does the pelletization do in the laser ablation process?

**46.0**  Rank the CVD, laser ablation, and electric arc processes to form CNTs.

**47.0**  What does the quartz tube do in the CVD process to form CNTs?

**48.0**  Mention the typical growth rates of CNTs in the CVD process.

**49.0**  Discuss the sequence of steps in the CVD process to prepare CNTs.

**50.0**    Sketch the CVD method, showing clearly the mask, substrate, deposition of solvent-based catalyst, removal of mask region, etc.

**51.0**    What is unique about the "lift-off" process?

**52.0**    What is the role of trench formed and etching in the CVD process?

**53.0**    What is the role of diffusion in the growth mechanism of CNT by CVD?

**54.0**    What is the energy needed for surface reactions and desorption for CNT synthesis?

**55.0**    Comment on the rate-determining step among diffusion, adsorption of species on the surface, surface reactions, desorption, and diffusion of species during the CNT synthesis by CVD.

**56.0**    What are the typical operating conditions during the high pressure carbon monoxide (HIPCO) process developed by Smalley in 1998?

**57.0**    What is the purpose of the annealing zone in the HIPCO process?

**58.0**    Discuss gas-phase nucleation and growth during the HIPCO process.

**59.0**    Discuss the formation of SWNTs by Boudouard reaction.

**60.0**    Distinguish the substrate layer from the material layer during the surface-mediated vertical alignment of CNTs.

**61.0**    Compare CNT's thermal conductivity with that of steel.

**62.0**    Compare CNT's Young's modulus of elasticity with that of steel.

**63.0**    Compare CNT's yield strength with that of steel.

**64.0**    What are the 12 different CNT morphologies?

**65.0**    How many different morphologies does a "phase-separated copolymer blend" exhibit?

**66.0**    Rod-like aggregates referred to as worm micelles were prepared from resemblance with _____.

**67.0**    What is the role of hydrophobic and hydrophilic properties of protein molecules in protein folding?

**68.0**    How are nanophases in copolymers with block microstructures formed?

**69.0**    What is the mechanism of formation of polymersomes from PEO-Pbd system?

**70.0**    What are the four different morphologies that the lipid vesicles can exhibit?

**71.0**    What are some of the important considerations during analysis of stability of the morphologies formed?

**72.0**   Name some applications of vesicles formed from lamellar structures of films.

**73.0**   Discuss the levels of structural hierarchy that can be seen in tendons.

**74.0**   What is the equilibrium reaction that can be used to describe the linear assembly process leading to filaments?

**75.0**   Under what values of the cooperativity parameter is the polymerization propagation step thermodynamically favorable?

**76.0**   What is the connection between aggregate formation and critical monomer concentration?

**77.0**   Explain the helix structure formation in keratin-myosin-epideremis-fibrin category of proteins using the repeat pattern in the amino acids and the hydrophobic and hydrophilic properties?

**78.0**   Who is considered the father of biomechanics? According to him, what is the relation between the shape of an animal's bone and its weight?

**79.0**   Describe the self-repair property of biomaterials.

**80.0**   Discuss the osteoclasts and osteoblasts cycle.

**81.0**   What is the role of self-organization in the preparation of hydroxyapetite-bone nanocomposites?

**82.0**   Explain the properties of biocompatibility and biointegration of synthetic materials.

**83.0**   How are hydroxyapetite nanocrystals dissolved?

**84.0**   What are Howship's lacunae?

**85.0**   Why is dicalcium phosphate used as a precursor to prepare hydroxyapetite?

**86.0**   Explain the formation of bone nanostructure in human anatomy.

**87.0**   Explain *in vitro* fibrillogenesis and the conditions preferred.

**88.0**   What are Langmuir-Blodgett monolayers, and what is their role in hydroxyapetite formation?

**89.0**   Where is the Kramers-Kronig equation used?

**90.0**   Why were Wistar rats and beagle dogs chosen for biocompatibility studies?

**91.0**   What is osteocompatibility?

**92.0**   Discuss the stages involved in the substitution process of composites to new bone structures.

**93.0**   Compare the resorption of synthetic nanocomposite with autogenous bone transplantation.

**94.0** Elaborate on the tibia defect.

**95.0** What is the difference between structural colors and pigmented colors?

**96.0** Discuss the mechanism of iridescence in *Lycaenid* butterflies.

**97.0** How is the mechanism of biomineralization studied using "flat pearls?"

**98.0** What are the three morphologies exhibited by aragonite composite?

**99.0** How is crystal CdS synthesis demonstrated in films made up of PEO?

**100.0** Why is SAXR needed to characterize polyion multilayer films?

**101.0** Where is quartz crystal microbalance used?

**102.0** How is AFM used to study the mechanism of PMF film formation?

**103.0** What happens to chitosan/polysaccharide film thickness when the solution ionic strength is increased?

**104.0** Discuss the pore formation in glucose isomerase and bolaamphiphile multilayers.

**105.0** How is sequential adsorption a low-cost approach to the assembly of thin films?

**106.0** Explain the Langmuir-Blodgett technique for film formation.

**107.0** Name two applications of biomimetic membranes.

**108.0** Explain the niche property of "smart materials."

**109.0** Under what temperatures are the smart materials in gel form and under what temperatures are they in liquid form?

**110.0** What is meant by a thermoreversible phase change?

**111.0** Discuss the gel phase changes shown in Fig. 7.1.

**112.0** What is the size range of proteins that may be incorporated in the bilayer membranes in the aqueous channels?

**113.0** Explain the phenomena of mesoscopic self assemblage.

# CHAPTER 8

# Bioartificial Organ Design

## Learning Objectives

- Need for immunoisolation
- Bioartificial pancreas
- Glycolytic oscillations
- Bioartificial liver
- Bioartificial kidney
- Extracorporeal artificial lung

## 8.1    Immunoisolation

Over time, the need to completely replace some vital organs in the human anatomy has become evident. One of the key technical hurdles in the successful transplantation of a bioartificial organ(s) is *immunoisolation*. The host's immune system may reject the replacement introduced, which is counterproductive. Immunological similarity between cells in *ex vitro* assembly of substitute organs and host cells may be one way of overcoming this.

Immunosuppressive drugs have been developed, but they have to be used with caution for obvious reasons, lest the immune response that was needed be suppressed! In some cases, on account of side effects from such drugs, the cure is worse than the disease. This is not a desirable state.

Cell implantation and proliferation *in vitro* may be an alternate consideration. The immune action consists of two parts: the cell mediatory processes and the humarol processes. B-cells, or B-lymphocytes, and T-lymphocytes form in the bone marrow. B-cells are activated by antigen or foreign materials of which transplanted cells are a subset. Upon activation, they transform into antibodies, which are proteinaceous.

The T-lymphocytes, or T-cells, in a similar fashion stem from the thymus and are present as helper T-cells and killer T-cells. Each B-cell recognizes a specific antigen by using a surface antibody receptor. Antibodies are proteins. These immunoglobins make up 20 percent of all protein present in plasma. Five different classes of antibodies exist: IgM, IgG, IgE, IgD, and IgA. An antibody molecule consists of two light and two heavy polypeptide chains held together by disulfide bridges similar to insulin microstructure. The antibody is Y-shaped. The heavy chain is at the stem and the lighter chains are at the branches of the Y.

T-cells are another category of cells that participate in the cell-mediated immune response. They are characterized by an antigen-specific T-cell receptor. Free antigens activate the T-cells. Activation requires that the antigen be presented by other cells. A foreign material is devoured by an antigen-presenting cell and then broken down. The antigen, along with an MHC class II molecule, is then transported to the cell surface, where the antigen is presented. T-cells that are comprised of CD4 molecules are referred to as helper T-cells.

The immune action is a complex process based on interactions among the T-cells, B-cells, etc. The host's immune response is restricted by immunoisolation. Polymeric membranes can be used to affect immunoisolation. The membrane will prevent the passage of major components of the immune system, such as immune cells, antibodies, and complement. The immune response occurs as a result of antigens emanating from cells that were transplanted. The pore size of the membrane is tailored accordingly. Openings that restrict entry of immune cells will still allow the passage of antibodies and complement. The solute permeability characteristics of the immunoisolation membrane are a salient issue to consider. Several issues need to be considered during the development of an immunoisolation membrane. Toxicity of by-products is a critical issue. Retention of low molecular weight by the membrane is another issue.

The bioartificial organs are made up of live cells and tissue that are immunoisolated by the use of polymeric membranes. These organs are made of a hybrid of synthetic nonliving materials and living cells. A host of diseases, such as diabetes, liver malfunction, and kidney catastrophe; neurological dysfunction such as Parkinson's disease or Alzheimer's disease; control of pain; and delivery of drugs generated by genetically engineered cell lines can be treated using bioartificial organs. The treatment of diabetes involves the secretion of insulin that varies with time to maintain glucose at appropriate allowed levels. Most of the functions of the kidneys and liver are performed by their bioartificial counterparts. Genetic engineered cells have been used to secrete products such as dopamine that tends to be neuroprotective, $\beta$-endorphin that can reduce pain resulting from cancer, and hormones that combat severe combined immunodeficiency disease (SCID).

## 8.2    Bioartificial Pancreas

The bioartificial pancreas can be used to treat insulin-dependent diabetes mellitus (IDDM). IDDM has complications such as blindness, gangrene, heart and kidney dysfunction, stroke, and nonaccidental amputation of limbs, and can lead to a reduction in the quality of life. Over a million people in the United States suffer from IDDM. This is one of the most prevalent causes of death by disease in the United States next to cardiovascular and neoplastic diseases.

Diabetic treatment has been focused largely on achieving metabolic control of blood glucose. Insulin was discovered in 1921 by Banting and Best. Since then, treatment for diabetes has been largely via daily insulin injections. An alternative approach to achieve homeostatic blood glucose levels has been developed that is built on the design of a biofeedback system. With the biofeedback system, insulin is released in response to the rise and fall of glucose concentrations. Three approaches to achieving control of glucose level by the use of insulin are as follows:

1. Computer-aided insulin pump with an implanted glucose sensor

2. Glycosylated-insulin-bound Concanavalin A system in which glycosylated insulin is released in response to blood glucose levels

3. Immunoprotected islets by artificial membrane and development of a hybrid artificial pancreas

Bioartificial pancreases developed on the principle of *microencapsulation* and/or semipermeable membranes have been patented. Microencapsulation of tissue cells such as the islets of Langerhans, which are injected into the human anatomy, has led to interesting results. Microencapsulation provided penetration distances of diffusion of 100 to 200 μm and large surface areas per volume of islet tissue. The small size provided excellent diffusion characteristics for nutrients and oxygen, which improves islet viability.

The microencapsules consists of the islet immersed in a hydrogel material, with another eggshell layer that provides the immunoisolation characteristics and mechanical strength. Different polymer chemistries have been described in the literature to prepare the hydrogel and the immunoprotective layer. One important issue is that the fibrotic capsule formation can limit the diffusion of nutrients and oxygen, resulting in the loss of islet function. Membrane materials have to be carefully selected to minimize the fibrotic reaction.

Soon-Shiong et al. [1], has proposed a microencapsulation approach using poly-L-lysine. They demonstrated their results in large mammals and human patients. Alginates are natural polymers made up of the polysaccharide, mannuronic acid and guluronic acid.

They found that the large mannuronic acid residues in the alginate are the culprit for the fibrotic response. Lymphokines IL-1 and tumor necrosis factor (TNF) are induced and end up promoting and proliferating fibroblasts, which leads to the formation of fibrotic capsules. The fibrotic response was minimized by reducing the alginate's mannuronic acid content and increasing the guluronic acid content. Alginates with higher guluronic contents were found to possess higher mechanical strength. The modified alginate microcapsules were tested in dogs with diabetes. Three dogs received free unencapsulated islets, and the other six dogs received encapsulated islets. The islets were dosed at 20,000 EIN/kg of body mass. Exogenous insulin was stopped four days prior to islet injection, and the plasma glucose levels were reduced to an average of 116 mg/dL in those animals that received encapsulated islets. The rejection of unprotected free islets was found to occur, with hyperglycemia returning in about six days. The animals receiving encapsulated islets exhibited normoglycemia for periods ranging from 63 to 172 days, with a median period of 105 days. Failure of the encapsulated islets was attributed to membrane dysfunction as a result of hydrophilicity of the alginate system.

The treatment of diabetes with peritoneal implants of encapsulated islets has been discussed in the literature. Diabetic models have been developed by Colton et al. [2]. According to Colton et al. [3], the number of islets required to reverse diabetes is up to 5,000 islets/kg. A 70-kg human will require 350,000 islets to maintain glucose levels. The volume of encapsulated islets with a mean diameter of 500 μm would be roughly 18 mL and the surface area would be 2,750 cm$^2$.

One of the salient considerations in the design of a bioartificial pancreas using hybrid materials is to prolong the cell life within the system. Peritoneally implanted membrane encapsulated cells usually have a limited lifespan. This may be due to oxygen deficiency and the inactivation of cells by low-molecular-weight humoral components of the immune system, such as interleukin-1. The membrane, however, will isolate the entrapped islets from the cellular immune system or high-molecular-weight cytokines. Foreign proteins released from cells will accelerate the attack of the cellular immune system upon cell death. Implanted islets should be replaced with fresh islets after certain period. A self-contained miniaturized implant from which the islets are replenished after a certain period was patented by the University of Utah [3]. This device is extravascularly implantable and rechargeable. It consists of a refillable immunoprotective membrane pouch containing an islet-polymer matrix. The polymer is soluble below human anatomy temperatures and insoluble above human anatomy temperatures. They exhibit lower critical solution temperature (LCST) behavior. The LCST exists for some systems. This demarcates the temperature above which the solute is insoluble in the solvent and below which the solute is soluble in the solvent. The pouch also contains means for stimulating the insulin secretion function of β-cells of the islets. Bioactive agents are released that are meant to regulate interactions between the bioartificial pancreas membrane and cellular components in the human anatomy. The membrane isolates the islets from cellular and humoral components in the human immune system.

The volume of the artificial pancreas is minimized by not encapsulating each islet. The islet cells are separated and held within a polymer matrix that is soluble in an aqueous solution below the human anatomical temperature and insoluble in an aqueous solution above the human anatomical temperature: 37°C (LCST behavior). The polymer-islet mixture is contained in a pouch with entry/exit ports. By changing the temperature, the contents of the pouch are replaced. This is affected by the injection of cold saline into the pouch to simulate localized hypothermia. The pouch is made of biocompatible material permeable to insulin and other substances of similar or less molecular weight, such as oxygen, nutrients, and hormones that may pass in and out of the pouch. The pouch also is impermeable to cellular and humoral components of the human anatomical immune system.

The artificial pancreas is comprised of a pouch membrane that requires minimal space while affording optimal implant volume. The implant consists of islets suspended in polymers that exhibit LCST behavior. These polymers exist as a liquid at low temperatures, form into a solid microsphere at human anatomical temperature, and can be sampled or replaced as desired. The suspension may also contain islet-stimulating agents.

Methods have been developed to isolate large quantities of cells from the pancreas of mammals such as pigs. A supply of donor islet tissue is provided in the bioreactor. Immunoisolation is needed for the bioartificial pancreas to operate successfully, as was discussed in the previous sections. IDDM is believed to be caused by an autoimmune process that results in the destruction of insulin-secreting cells found in the islets of Langerhans. Islets of Langerhans form 1 to 2 percent of the mass of the pancreas.

An improved solid support was patented by Seed Capital Investments, Amsterdam, Netherlands [4] for the cultivation of cells. Hollow fibers are provided for the supply and removal of gases such as oxygen and carbon dioxide. Support is provided for improved adhesion between the tissue cells and the support. A bioartificial pancreas can be developed using this support system. The support can be made of gelfoam, polyvinyl fluoride (PVF), polyglycolic acid (PGA), polyvinyl alcohol (PVA), polyglycolic acid/polylactic acid (PGA/PLA), 3-D polyurethane foam, porous silicon rubber foam, etc. The support provided a large surface area, and acceptable porosity formed from a network of fibers. The pore diameter ranges from 10 to 100 μm. The porosity is in the range of 0.6 to 0.95. The fibers are made up of

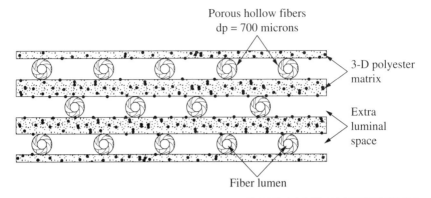

Porous hollow fibers
dp = 700 microns

3-D polyester matrix

Extra luminal space

Fiber lumen

**FIGURE 8.1** Sandwich configuration of bioartificial pancreas.

hydrophobic material, such as silicone, polyethylene, polypropylene, etc. The fibers are evenly distributed throughout the matrix material. The sandwich configuration of the support material is shown in Fig. 8.1. With the sandwich configuration, an improved supply and removal of gases is possible. The fibers also act as baffles and channels for uniform flow and distribution of the liquid medium through the extraluminal space to all parts of the solid support. The fibers provide physical support to the matrix sheets. This becomes more important as the solid support is subjected to high shear, such as during liquid flow. The fibers are at an angle to each other in the sandwich configuration.

## 8.3 Glycolytic Oscillations

Some of the salient considerations in the design of a bioartificial pancreas include the better understanding of the insulin release rate from an islet and its dependence on plasma glucose levels. A step-change in glucose concentration is given to islets that have been isolated from the pancreas of mammals. The islet viability and glucose responsiveness are studied from the F curve. Insulin release has been found to be biphasic.

Nomura [5] used control theory to describe mathematically the insulin release rate during a step-change in glucose concentration. The dynamics of glucose-induced secretion of insulin can be expressed as the sum of the proportional response to the step change and a derivative response to the rate of change in the glucose concentration. Each of them has a first-order lag time. The Laplace transform of the islet insulin release rate can be expressed as follows:

$$r_{isl} = \left( \frac{K_{prop}}{1+s\tau_1} + \frac{T_{der}}{1+s\tau_2} \right) \bar{C}_{glucose} \qquad (8.1)$$

The lag times are $\tau_1$ and $\tau_2$, respectively. The Laplace domain expression in Eq. (8.1) can be inverted to give:

$$r_{isl} = \int_{-\infty}^{t} \frac{K_{prop}C_{glucose}}{\tau_1} e^{-\left(\frac{t-z}{\tau_1}\right)} dz + \int_{-\infty}^{t} \frac{\tau_2}{\tau_1} \frac{dC_{glucose}}{dz} e^{-\left(\frac{t-z}{\tau_2}\right)} dz \qquad (8.2)$$

The lag times can be obtained by using nonlinear regression of islet release-rate experimental data. The ramp function of glucose concentration is written as follows:

$$C_{glucose} = C_{glucose}^0 \qquad \text{for } t < 0 \qquad (8.3)$$

$$C_{glucose} - C_{glucose}^0 = mt \qquad \text{for } 0 \leq t \leq t_0 \qquad (8.4)$$

where $m = \dfrac{C_{glucose}^{ss} - C_{glucose}^0}{t_0}$

$$C_{glucose} = C_{glucose}^{ss} \qquad \text{for } t \geq t_0 \qquad (8.5)$$

Equations (8.3) to (8.5) can be substituted into Eq. (8.2) and integrated to yield:

$$r_{isl} = K_{prop}C_{glucose}^0 e^{-\frac{t}{\tau_1}} + K_{prop}\tau_1 m\left(\frac{t}{\tau_1} - \left(1 - e^{-\frac{t}{\tau_1}}\right)\right)$$

$$+ K_{prop}C_{glucose}^0\left(1 - e^{-\frac{t}{\tau_1}}\right) T_{der}m\left(1 - e^{-\frac{t}{\tau_2}}\right) \qquad \text{for } 0 \leq t \leq t_0 \qquad (8.6)$$

$$r_{isl} = K_{prop}C_{glucose}^0 e^{-\frac{t}{\tau_1}} + K_{prop}\tau_1 m\left(\left(\frac{t_0}{\tau_1} - 1\right)e^{\frac{t-t_0}{\tau_1}} + e^{-\frac{t}{\tau_1}}\right)$$

$$+ K_{prop}C_{glucose}^0\left(e^{-\frac{t-t_0}{\tau_1}} - e^{-\frac{t}{\tau_1}}\right) + K_{prop}C_{glucose}^{ss}\left(1 - e^{-\frac{t-t_0}{\tau_1}}\right)$$

$$+ T_{der}m\left(e^{-\frac{t-t_0}{\tau_2}} - e^{-\frac{t}{\tau_2}}\right) \qquad \text{for } t \geq t_0 \qquad (8.7)$$

Pharmacokinetic models have been developed to describe glucose and insulin metabolism. A model was proposed by Sturis et al. [6] to predict the oscillations of insulin and glucose concentrations with time observed experimentally. Insulin formed in human anatomy has been found to exhibit two kinds of oscillations: a rapid oscillation

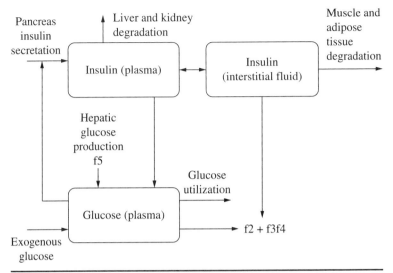

**FIGURE 8.2** Compartment model for glucose and insulin interaction.

with a period of 10 to 15 minutes and small amplitude, and longer or ultradian damped oscillations with a period of 100 to 150 min and larger amplitude. The compartment model proposed by Sturis et al. to describe glucose and insulin interactions is shown in Fig. 8.2.

Four negative feedback loops form their model involving glucose and insulin interactions: insulin formation is triggered when glucose levels reach the more than tolerable limit; an increase in insulin level increases the utilization of glucose and hence, reduces the glucose levels; a rise in glucose level inhibits production of glucose; and increase in glucose levels stimulates its utilization. The glucose and insulin never reach a stable equilibrium. The model includes two time delays that are critical in describing the observed oscillatory dynamics. The suppression of glucose levels by insulin production is captured by one time delay, and the correlation of the biological action of insulin with insulin concentration is captured by another time delay in an interstitial compartment. Six differential equations describe the system. The variables are $C_{glucose}^{plasma}$, concentration of glucose in plasma; $C_{insulin}^{plasma}$, concentration of insulin in the plasma; and $C_{insulin}^{interstitial}$, concentration of insulin in the interstitial fluid. Three additional variables used to describe the insulin and glucose system are the delay between the plasma insulin level and its effect on glucose production $x_1, x_2, x_3$, and time lag, $\tau_{delay}$. The six differential equations can be written as follows:

$$V_{plasma}\frac{dC_{insulin}^{plasma}}{dt} = r_{isl}C_{glucose}^{plasma} - k_E\left(C_{insulin}^{plasma} - C_{insulin}^{interstitial}\right) - \frac{C_{insulin}^{plasma}V_{plasma}}{\tau_{plasma}}$$

(8.8)

$$\frac{dC_{\text{insulin}}^{\text{interstitial}}}{dt} = k_E \left( \frac{C_{\text{insulin}}^{\text{plasma}} - C_{\text{insulin}}^{\text{interstitial}}}{V_{\text{interstitial}}} \right) - \frac{C_{\text{insulin}}^{\text{interstitial}}}{\tau_{\text{interstial}}} \quad (8.9)$$

$$V_{\text{plasmaG}} \frac{dC_{\text{glucose}}^{\text{plasma}}}{dt} = r_{\text{glucose}(in)} - f_2 C_{\text{glucose}}^{\text{plasma}} - f_3 f_4 C_{\text{glucose}}^{\text{plasma}} C_{\text{insulin}}^{\text{interstitial}} + x_3 f_5 \quad (8.10)$$

$$\frac{dx_1}{dt} = 3 \frac{\left( C_{\text{insulin}}^{\text{plasma}} V_{\text{interstitial}} - x_1 \right)}{\tau_{\text{delay}}} \quad (8.11)$$

$$\frac{dx_2}{dt} = 3 \frac{(x_1 - x_2)}{\tau_{\text{delay}}} \quad (8.12)$$

$$\frac{dx_3}{dt} = 3 \frac{(x_2 - x_3)}{\tau_{\text{delay}}} \quad (8.13)$$

The volumes of the insulin plasma compartment, insulin interstitial fluid compartment, and glucose plasma compartments are denoted by $V_{\text{plasma}}$, $V_{\text{interstitial}}$, and $V_{\text{plasmaG}}$, respectively. The $k_E$ is the rate constant that is used to describe the insulin transport rate into the interstitial fluid compartment. The first-order degradation time constants for insulin in a plasma compartment and insulin in interstitial fluid compartments are given by $\tau_{\text{plasma}}$ and $\tau_{\text{interstitial}}$, respectively. Utilization functions are given by $f$. The subscripts 2 and 3 are used to denote the glucose utilization function throughout the glucose plasma compartment. The 4 subscript is used for dependence on interstitial insulin concentration. The glucose inhibition on account of insulin formation is given by subscript 5.

The pharmacokinetic model developed by Sturis et. al. [6] can be combined with an insulin-release model to monitor the glucose control that is achievable using a bioartificial pancreas. Two types of tests are used: intravenous glucose tolerance test (IVGTT) and oral administration of glucose tolerance test (OGTT). The initial conditions for IVGTT or OGTT can be selected based on the fasting levels of the patient.

## 8.4 Bioartificial Kidney

As was discussed in previous sections, several patients with end-stage renal disease can use an artificial kidney. Approximately 800,000 patients worldwide use hemodialysis, at a cost of US $15 billion every year. Current methods of hemodialysis and continuous ambulatory peritoneal dialysis (CAPD) are by no means permanent solutions. Some of the other deficiencies of the use of dialysis machine are that they are expensive, they are large and heavy, and they require detoxification several times a week—200 to 300 liters of dialyzing fluid are needed for each dialysis treatment. The quality of life of users of such devices is poor. The morbidity and mortality associated with the hemodialysis therapy is high.

Inventors of artificial kidneys admit two problems, that is, blood clotting and water removal. Much of the research and development in the field of the artificial kidney have been in the development of novel dialyzing membranes, autosterilizable membranes, reduction in the 200 to 300 liters of dialyzing fluid required, development of blood-compatible polymers for the membranes, etc. The goals in detoxification of impure blood are: removal of urea and uric acid, removal of creatinine and substances with molecular weights between 1,400 and 1,550, and removal of water and phosphate from the blood. Urea is the final product of the decomposition and utilization of proteins in the body. It is eliminated through urine. Uric acid is a product of metabolism found usually in the urine. Creatinine is a waste product found in muscles and animal tissues. Phosphate is a salt of phosphoric acid with an atom or atoms of a metal.

Blood clotting can become an issue during the detoxification process. Blood platelet damage is caused by the requirements for higher hydrostatic pressure for water compared to that of blood for ultrafiltration. The platelet damage causes blood clotting. Calcium adheres to dialyzing membranes so it has to be added to prevent heart damage. Protein tends to adsorb to some polymer walls, as well as dialyzing membranes, causing a reduction in membrane efficiency and an increase in clotting. Other ions also adsorb on the membranes. The smallest known artificial kidney weighs about 70 lb. The pumps, plumbing, and dialysate increase the weight of the entire apparatus.

An artificial kidney that was portable and wearable was patented by Beltz [7]. The basic components of the novel artificial kidney are a blood separator unit, a chemical treatment unit, and a water removal unit. Blood access is provided from the patient via an arteriovenous shunt that is implantable in the patient's arteries and corresponding veins for a period, after which it is removed. The flow of blood is affected by either the hydrostatic pressure difference or a pump with battery and vacuum pressure on the plasma side of the apparatus. The blood plasma separator, which comes in the form of a packet, is replaced every day. It is an elongated outer tube through which passes an inner tube. Whole blood enters one end of the blood plasma separator and is directed into the interior of the inner tube, which is made of a polycarbonate perforated membrane. The hole sizes in the membrane were of the order of 450 nm to 3 microns. The holes are small enough to prevent extrusion of blood elements such as platelets and red and white corpuscles while allowing the passage of plasma. The urease-coated tube coming out of the blood plasma separator and going to the chemical treatment packet is used to break down urea. The breakdown results in ammonia ions and carbon dioxide, with the ammonium ions being picked up by zirconium phosphate. The concentric tubes have a convoluted shape as they pass through the blood plasma separator. The polycythemic blood that exits the inner tube is transmitted back to the patient's body. The plasma that exits the end of the outer tube is next transmitted to the chemical treatment unit. The chemical treatment unit is a removable packet. It is used to relieve the contaminated plasma of uric acid and creatinine by means of activated charcoal. The phosphate ions are removed by zirconium oxide, and the ammonium ions are removed by zirconium phosphate. After being processed through the chemical treatment unit packet, the detoxified plasma passes into the water removal unit packet. The packet has a perforated tube through which passes the detoxified plasma. This tube is a polycarbonate perforated membrane material with holes that have a diameter between 450 nm to 3 $\mu$m and is formed of cellophane. Sephadex desiccant is used. The tube in the water removal packet is convoluted. The water removal packet has a color-coded indicator to show when it is loaded and needs to be replaced. The detoxification period and packet replacement will vary from patient to patient. The advantages of the artificial kidney are as follows:

1. The unit is maintained by the patient, who would change the batteries.
2. The patient may have small amounts of heparin.
3. Blood cell damage will be greatly reduced over that with dialysis.
4. Shock to the body from quick removal of heightened levels of urea, creatinine, uric acid, and the middle molecular substances.

The bioartificial glomerulus and the artificial tubule can be arranged in two different configurations. In the first arrangement, blood flows through the luminal spaces of the hollow fibers that comprise the hemofilter and two tubule sections. A flow control valve placed after the hemofilter regulates the filtration rate of the hemofilter. The hyperosmotic blood then leaves the hemofilter and continues through the proximal and distal tubule modules. The ultrafiltrate generated in the hemofilter flows through the shell side of the tubules. Cells are immunoprotected by the hollow-fiber membrane. Solutes are selectively transported from the shell side to the blood side. Reabsorption of water from the ultrafiltrate is facilitated by the hemofilter.

In the second arrangement, contact of blood is with the artificial membrane in the hemofiltration unit. Blood flows within the hollow fibers. Ultrafiltrate is collected within the hemofilter shell space. Ultrafiltrate enters the shell side. Selective solute transport is affected between the shell side and hollow-fiber luminal space. Shell-side ultrafiltrate exits with the blood from the hemofilter.

## 8.5 Extracorporeal Artificial Lung

Extracorporeal devices are made to work outside the human anatomy and are connected to the patient by an arteriovenous shunt. They may be viewed as artificial organs. They can be used to remove undesired chemicals from the human anatomy or substitute for a damaged or failed organ. Before being allowed to enter these devices, blood is prevented from clotting by the use of heparin. Auxiliary equipment, such as pumps, probes, sensors, and control systems for pressure, temperature, and concentration, can be used to augment the device. These devices do not contain living cells. The ones that contained living cells were discussed in the earlier part of the chapter under different bioartificial organ design.

A hollow-fiber artificial lung used in extracorporeal circulation to remove carbon dioxide from blood and add oxygen to the blood is shown in Fig. 8.3. This was patented by Terumo Corp., Tokyo, Japan [8]. This device uses less blood and has greater mass transfer efficiency of gas transport across the hollow-fiber surface.

Artificial lungs can be classified into two categories: porous variety and membrane variety. The membrane artificial lung, such as stacked membrane type, coil type, or hollow-fiber type, has been found to be superior to the porous-type artificial lung. This is because the blood

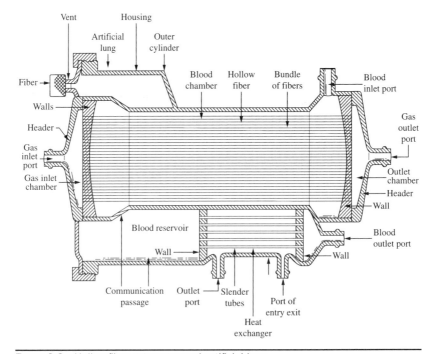

FIGURE 8.3  Hollow-fiber extracorporeal artificial lung.

flowing through the lungs undergoes hemolysis, albumin degenera-
tion, clotting, and affixation with the porous variety. The operating
mechanism of the membrane variety is close to that of the human
lung. The porous artificial lung is used during open heart surgery. In
order to obtain sufficient oxygenation with the membrane-type artificial lung, the blood flow layer must be reduced in thickness, that is, a
narrow blood flow conduit area and large flow resistance. It is not
possible to achieve perfusion of the blood within the artificial lung by
using the head developed between the patient and the lung. Different
blood circuits using a pump, blood reservoir, and heat exchanger
have been attempted to overcome this difficulty. The problem that
arose in such designs was the increase in the internal pressure of the
circuit on the blood feeding side. A proposed solution to this problem
is to the let the blood flow on the outer side of the hollow fibers. This
could not be reduced to practice, however, because of the presence of
air bubbles in the blood in the extracorporeal circuit.

As shown in Fig. 8.3, the mass exchanger consists of a hollow-
fiber bundle accommodated along the axial direction of the housing,
axially extended housing, blood inlet port, blood outlet port, gas
venting port, etc. The hollow fibers are made of a microporous membrane. The fibers are allowed to touch each other. The flow circuit
consists of a blood reservoir, a pump, and a heat exchanger through
which blood is allowed to flow. The ends of the hollow fibers are
retained tightly within the housing via walls. A header is attached to
each end of the housing. All the hollow fibers exit to a gas outlet
chamber. The outer walls of adjacent hollow fibers define channels
through which the blood is allowed to flow. Turbulence can be
expected due to the interconnections of the channels. The area across
which blood flows decreases with an increase in distance from the
blood inlet port. This makes it possible for the flow rate of the blood
flowing axially of the housing within the blood chamber to be made
uniform in relation to the circumferential direction of the hollow-fiber
bundle.

The housing consists of an inner and outer cylinder. The outer
cylinder consists of a rigid material made up of acrylstyrene copolymer, polycarbonate, or polystyrene. The bundle of fibers is housed
within the inner cylinder. Oxygen and other gases supplied from the
gas inlet port pass through the interior of the hollow fibers, while
blood flows under turbulence on the periphery of the hollow fibers.
This allows for gas transport between the blood and the gas stream
within the hollow fibers. In between the inner cylinder and outer cylinder is located a blood reservoir. A blood outlet port and heat
exchanger are provided. The heat exchanger is comprised of a bundle
of slender tubes supported at both ends by respective walls located
within the heat exchanger tank. The ends of the slender tubes open
externally of the blood reservoir on the outer sides of the walls. The
heat transfer medium flows inside the tubes. Cooling and heating

water is allowed to enter via inlet and outlet ports. The heat exchanger tank serves to heat, cool, or maintain the temperature of blood following the gas exchange. The artificial lung is designed to improve the gas exchange performance per unit membrane of the hollow fibers. Perfusion of the blood is allowed due to the pressure head developed between the human anatomy and the artificial lung. The quantity of blood needed to fill the blood circuit in which the artificial lung is used can be estimated.

The contact between the gas stream and blood stream can be in *counter-current, co-current,* and *cross-current* modes of contact. The gas flows through the fibers and blood flows at right angles across the outer surface of the fibers during cross-current contact. For the counter-current mode of contact, the following mathematical model would be applicable. Let the concentration of solute on the blood side and gas side be given by $C_{blood}$ and $C_{gas}$, respectively. The amount of oxygen in blood that is bound to hemoglobin is given by $C'_{blood}$. The interface concentrations of gas for the blood and gas sides of the membrane are given by $C_{bm}$ and $C_{gm}$, respectively. The blood and gas side mass transfer coefficients are given by $k_{blood}$ and $k_{gas}$, respectively. A shell balance on a slice of thickness $\Delta x$ in Fig. 8.3 across the hollow fiber axially located for oxygen solute can be written as:

$$Q_{blood}\frac{d(C_{blood}+C'_{blood})}{dx} = k_{blood}A_{blood}(C_{blood}-C_{bm})$$

$$Q_{gas}\frac{dC_{gas}}{dx} = k_g A_{gas}(C_{gas}-C_{gm})$$

where $A_{blood}$ and $A_{gas}$ are the membrane area per unit length of membrane. Further simplications of these two equations can be made using $m = dC'_{blood}/dC_{blood}$. Furthermore, the concentration of oxygen in the blood and gas can be expressed as partial pressures, as shown in Chap. 5. An overall mass transfer coefficient, $K_o$, can be defined and shown to be:

$$\frac{1}{K_o} = \frac{H}{k_{blood}} + \frac{A_{blood}}{\rho_{STP}A_{mean}P_m} + \frac{RTA_{blood}}{k_{gas}A_{gas}}$$

where $A_{mean}$ is the log mean area of the membrane per unit length, $A_{mean} = (A_{gas}-A_{blood})/k_{gas}A_{gas}$ ; $P_m$ is the permeability of the membrane, and $H$ is the Henry's constant, $p_{O_2} = HC_{blood}$, relating the partial pressure of oxygen to the concentration of oxygen in the blood stream. The density of gas in standard temperature and pressure (STP) conditions is given by $\rho_{STP}$. The overall mass transfer coefficient can be used in the mass balance equations of oxygen in blood and gas streams and the model equations solved for to obtain:

$$p_{O_{2g}} = \left(\frac{Q_{blood}}{Q_{gas}}\right)\left(\frac{RT}{H}\right)(1+m)\left(p_{O_{2b}} - p_{O_{2g}}^{out}\right) + p_{O_{2g}}^{out}$$

$$A_{blood} = \frac{a}{b}\ln\left(\frac{bp_{O_{2b}}^{out} + c}{bp_{O_{2b}}^{in} + c}\right)$$

where $a = \dfrac{Q_{blood}(1+m)}{K_o H}$; $b = \dfrac{Q_{blood}RT}{Q_{gas}H}(1+m) - 1$

$$c = p_{O_{2g}}^{out} - \frac{Q_{blood}RT}{Q_{gas}H}(1+m)p_{O_{2b}}^{in}$$

A similar set of equations can be developed for carbon dioxide ($CO_2$) in the mass exchanger.

## Summary

One of the key technical hurdles in the successful transplantation of bioartificial organ(s) is immunoisolation. The host's immune system may reject the replacement introduced. Immunosuppressive drugs have been developed. The immune action consists of two parts: cell mediatory processes and humarol processes. The role of B-cells and T-cells in the immune process was outlined. The bioartificial organs are made up of live cells and tissue that are immunoisolated by the use of polymeric membranes. These organs are made of a hybrid of synthetic nonliving materials and living cells.

A bioartificial pancreas can be used to treat diabetes mellitus, and is an improved therapy compared with insulin therapy. The pancreas is developed on the principle of microencapsulation. Microencapsulation provided penetration distances of diffusion of 100 to 200 μm and large surface areas per volume of islet tissue. The small size provided excellent diffusion characteristics for nutrients and oxygen, which improves islet viability. A self-contained miniaturized implant from which the islets are replenished after a certain period was patented by the University of Utah [3]. This device is extravascularly implantable and rechargeable. It is comprised of a refillable immunoprotective membrane pouch containing an islet-polymer matrix. The polymer is soluble below human anatomy temperatures and insoluble above human anatomy temperatures. They exhibit LCST behavior. An improved solid support was patented by Seed Capital Investments, Amsterdam, Netherlands [4] for the cultivation of cells. Hollow fibers are provided for the supply and removal of gases such as oxygen and carbon dioxide. The support provides improved adhesion between the tissue cells and the support. A bioartificial pancreas can be developed using this support system. The support can be made of gelfoam, PVF, PGA, PVA, PGA/PLA, 3-D polyurethane foam, porous silicon rubber foam, etc.

Nomura [5] used control theory to describe mathematically the insulin release rate during a step-change in glucose concentration. Pharmacokinetic models have been developed to describe glucose and insulin metabolism. A model was proposed by Sturis et al. [6] to predict the oscillations of insulin and glucose concentrations with time observed experimentally. Insulin formed in the human anatomy has been found to exhibit two kinds of oscillations: a rapid oscillation with a period of 10 to 15 minutes and small amplitude and longer or ultradian damped oscillations with a period of 100 to 150 min and larger amplitude. The compartment model proposed by Sturis et al. to describe glucose and insulin interactions is shown in Fig. 8.2.

Much of the research and development in the field of artificial kidneys have been in the development of novel dialyzing membranes, autosterilizable membranes, reduction in the 200 to 300 liters of dialyzing fluid required, development of blood-compatible polymers for the membranes, etc. An artificial kidney that was portable and wearable was patented by Beltz [7]. The basic components of the novel artificial kidney rare a blood separator unit, a chemical treatment unit, and a water removal unit. Blood access is provided from the patient via an arteriovenous shunt that is implanted in the patient's arteries and corresponding veins for a period, after which it is removed. The flow of blood is affected by either the hydrostatic pressure difference or a pump with battery and vacuum pressure on the plasma side of the apparatus.

Extracorporeal devices are made to work outside the human anatomy and are connected to the patient by an arteriovenous shunt. A hollow-fiber artificial lung used in extracorporeal circulation to remove carbon dioxide from blood and add oxygen to the blood is shown in Fig. 8.3. This was patented by Terumo Corp., Tokyo, Japan [8]. This device uses less blood and has greater mass transfer efficiency of gas transport across the hollow-fiber surface. As shown in Fig. 8.3, the mass exchanger consists of a hollow-fiber bundle accommodated along the axial direction of the housing, axially extended housing, blood inlet port, blood outlet port, gas venting port, etc. The hollow fibers are made of a microporous membrane. The fibers are allowed to touch each other. The flow circuit consists of blood reservoir, a pump, and a heat exchanger through which blood is allowed to flow. The ends of the hollow fibers are retained tightly within the housing via walls. The contact between the gas stream and blood stream can be in *counter-current, co-current,* and *cross-current* modes of contact. The gas flows through the fibers and blood flows at right angles across the outer surface of the fibers during cross-current contact.

# References

[1]  P. Soon-Shiong, E. Feldman, R. Nelson, et al., "Longterm reversal of diabetes by the injection of immunoprotected islets" (1993), *Proc. Natl. Acad. Sci.*, 90, 5843–5847.

[2]  C. K. Colton and E. S. Avgounstiniatos, "Bioengineering in development of the hybrid artificial pancreas" (1991), *J. Biomech. Eng.*, 113, 152–170.

[3]  Y. H. Bae and S. W. Kim, "Implantable and refillable biohybrid artificial pancreas," U.S. Patent 5,262,055 (1993), The University of Utah, Salt Lake City, Utah.

[4]  L. M. Flendrig, "Bioartificial organ containing a matrix having hollow fibers for supplying gaseous oxygen," U.S. Patent 6,372,495 (2002), Seed Capital Investments, Utrecht, Netherlands.

[5]  N. Nomura, M. Schihiri, R. Kawamori, et al., "A mathematical insulin secretion model and its validation in isolated rat pancreatic islets perfusion" (1984), *Comput. Biomed. Res.*, 17, 570–579.

[6]  J. Sturis, C. Knudsen, N. M. O'Meara, et al., "Phase-locking regions in a forced model of slow insulin and glucose oscillations," in J. Belair, L. Glass, U. An der Heiden, and J. Milton (eds.), *Dynamical Disease: Mathematical Analysis of Human Illness*, Woodbury, NY: AIP Press, 1995.

[7]  A. D. Beltz, "Wearable, portable, lightweight artificial kidney," U.S. Patent 5,284,470 (1994), Upland, CA.

[8]  H. Fukusawa and T. Monzen, "Hollow fiber-type artificial lung," U.S. Patent 4,620,965 (1986), Terumo Corp., Tokyo, Japan.

# Exercises

## Problems

**1.0**  Mass exchanger for extracorporeal artificial lung. Consider a mass exchanger similar to the one shown in Fig. 8.3. Calculate the membrane surface area needed to remove carbon dioxide from the blood and supply oxygen to the bloodstream. The membranes are made up of polycarbonate. The length of each fiber is 40 cm, with a wall thickness of 40 microns. The inside diameter of a fiber is 350 microns. The blood flow rate between the fibers is 5,500 mL/min, and the gas flow rate is 5,500 mL/min. The temperature and pressure of operation are 98.6°F and 1 atm, respectively. The partial pressure of oxygen entering the blood is 45 mmHg and the partial pressure of oxygen exiting the blood is 100 mmHg. The partial pressure of entering gas is 700 mmHg. The Henry's law constant of oxygen in blood is 0.74 Hg/μm. Ideal gas can be assumed for the gas side. The diffusion coefficient of oxygen in blood may be taken as 1.7 E-9 $m^2$/sec. The $m = (dC'_{blood}/dC_{blood})$ can be taken to be 26. Is the information given sufficient?

**2.0**  *Oscillations in concentrations of glucose and insulin.* Estimate glucose concentrations during OGTT, oral glucose tolerance test. Fifty grams of glucose are consumed orally. The glucose absorption and elimination rate constants are 0.042 L/min and 0.0083 L/min, respectively. Assume that the body mass is 75 kg. The patient receives a total of 1 million EIN. Assume that the islet insulin secretion rate is given by:

$$r_{isl} = \frac{0.21}{1+\exp\left(-3.33 C^{plasma}_{glucose} + 6.6\right)}$$

**3.0**   The half-thickness of the islet chamber is 50 microns. The void volume within the islet chamber must be at least 50 percent for sufficient oxygen transport. The immunoisolation membrane permeabilities for glucose and insulin are 0.0004 cm/sec for glucose and 8 E-5 cm/sec for insulin. Present the blood insulin and glucose levels as charts with time on the x axis.

# CHAPTER 9

# Bioheat Transport

## Learning Objectives

- First law of thermodynamics and metabolism
- Conservation of energy
- Fourier's law of heat conduction
- Damped wave conduction and relaxation
- Cartesian, cylindrical, and spherical coordinates
- Steady-state and transient temperature profiles
- Metabolic energy production
- Forced and natural convection
- Rayleigh-Bernard instability
- Sweating with evaporation
- Thermal wear design
- Metabolism and regulation of body temperature
- Bioheat transfer equation

The laws of thermodynamics and of heat conduction can be applied to the human anatomical and physiological systems. Two important applications of bioheat transport in medicine are *thermal therapy* and *cryopreservation.* Local destruction of tissue is made to come about by the use of hyperthermic or cryothermic technology. The payoff in studying bioheat transfer toward complete eradication of disease by 2050 is to understand better and find a cure for the disease mechanism of cancer and cardiovascular disease. Statistics on the number of people with cancer and cardiovascular disease in the United States in the 2005 according the American Cancer Society and American Heart Association, respectively, are listed in Table 9.1.

Thermal therapies are accomplished by the use of invasive probes, which either act as a hyperthermic energy source (such as microwave, radiofrequency, high-intensity focused ultrasound, and laser) or as a

| Disease | Number of Patients |
|---|---|
| Prostate and breast cancer | 200,000 |
| Kidney cancer (males) | 20,000 |
| Liver cancer | 20,000 |
| Colorectal cancer | 100,000 |
| Coronary heart disease | 13 million |
| Atrial fibrillation | 200,000 |

TABLE **9.1**   Incidence of Cancer and Cardiovascular Disease in the U.S. Population

cryothermic energy sink (such as Joule-Thomson argon effect and cryogen-circulation probe technologies) [1]. Such probes are used for the treatment of prostate and kidney disease, including cancer and benign prostatic hyperplasia (BPH). Probe-tissue interactions need to be better understood. Nanoscale cellular- and tissue-level events are correlated to the therapeutic outcome. Research is in progress to quantitate temperature, injury, and the mechanisms that relate them to each other. Various types of biopreservation techniques include hypothermic storage (above the freezing point), cryopreservation (below –80°C), and vitrification (freeze-drying).

## 9.1   Five Laws of Thermodynamics and Metabolism

The word *energy* comes from the Greek words *en* meaning "in" and *ergon* meaning "work." The sun is the primary source of energy. Light is converted into chemical energy by the process of *photosynthesis.* This is common in plants, where starch is formed in leaves from carbon dioxide in the presence of sunlight and oxygen is liberated. As in respiration of humans and other species, oxygen is inhaled and carbon dioxide is exhaled, and the composition balance of air in the atmosphere remains the same, invariant with time. Through another process called *chemosynthesis,* bacteria that thrive a mile below the surface of the sea use sunlight for their energy requirements. They use $Fe^{++}$ or $Mn^{++}$ as an energy source. The photon energy present in sunlight can be related to the photon frequency by Planck's law as:

$$E = \frac{hc}{\lambda} = h\nu \tag{9.1}$$

where $h$ is the Planck's constant ($6.63 * 10^{-34}$ Js) and $c$ is the speed of light in a vacuum ($2.998 * 10^8$ m/s). Plants combine trapped energy from sunlight from $CO_2$ and water to make glucose ($C_6H_{12}O_6$), oxygen, and heat. Solar energy is stored in the form of chemical bonds. Red blood cells (RBCs) are derived from stem cells in the bone marrow in accordance with the genetic code stored in DNA and in response to a hormone secreted by the kidneys.

Protein signaling is a salient item in the utilization of the plant's energy by the bioorganism. Molecular oxygen is transported from the lungs in human anatomy and the gills in aquatic species to burn "fuel." Through a series of reactions, glucose, fats, and other organics are oxidized to $CO_2$, water, heat, and other by-products. Energy from digested food is used for locomotion, to regulate body heat, to generate light in some species, to ward off infection from microbes, and for reproduction. A number of biochemical reactions take place that require energy. Protein structures are encoded by the nucleic acids and formed by gene expression. The protein's secondary structure regulates the functions of any organism by signaling the *flow of energy* and information in and between cells. The interplay of energy and information is a theme that is emphasized in the field of biological thermodynamics [2].

Cells use energy to maintain osmotic pressure gradients for the synthesis of proteins. Even at rest, humans have some metabolic functions that require energy: i) autonomic motor activity for respiration, ii) motor activity for contraction of the heart; iii) perfusion of blood and other body fluids; iv) regulation of body temperature; v) phenomena of the mind; vi) digestion of food consumed; and vii) simple anatomy motion. Energy needs to be stored in the human anatomy in the event of increases in metabolic state arising from anatomical responses to environmental stimuli. Adenosine triphosphate (ATP) drives energy-dependent biochemical reactions.

The efficiency with which biological energy transport occurs is the ratio of useful work, defined as the total work less the work done by the system, to the energy input for volume expansion. The energy stored in each mole of ATP that is available to perform useful work is roughly 42 kilojoules (kJ). Aerobic glycolysis of 1 mole of glucose produces roughly 36 moles of ATP. Converting 1 mole of glucose to water and $CO_2$ releases 2.823 megajoules (mJ) of energy. The maximum efficiency during glycolysis is about 53 percent. This is higher than many mechanical devices that operate under a temperature gradient. Energy released from metabolism that is uncoverted to chemical energy or mechanical work is used up to regulate human anatomical temperature. Excess heat generated has to be removed to avoid temperature overshoot in the human anatomy.

There are five laws of thermodynamics. An elaborate treatment of these laws and their applications are beyond the scope of this textbook. These are briefly stated as follows:

**Zeroth law of thermodynamics:** If two systems are in thermal equilibrium with a third system, then the two systems are in thermal equilibrium with each other.

**First law of thermodynamics:** The total quantity of energy is a constant, and when energy is consumed in one form, it appears concurrently in another form. Energy exists in many forms. The mathematical statement of the first law of thermodynamics for closed systems can be written as:

$$Q + W = \Delta U \qquad (9.2)$$

$Q$ is the heat energy needed *for* the system supplied from the surroundings. $W$ is the work done on the system. When the work is done *by* the system, then a negative sign should precede the work contribution to Eq. (9.2). $\Delta U$ is the internal energy change *of* the system. The sign convention used in Eq. (9.2) is recommended by the International Union of Pure and Applied Chemistry. In differential form, Eq. (9.2) may be written as:

$$dQ + dW = dU \qquad (9.3)$$

For an open system, the first law of thermodynamics may be written as:

$$\dot{Q} + \dot{W} - \Delta \dot{m}\left(U + \frac{u^2}{2} + gz\right) = \frac{d(mU)}{dt} \qquad (9.4)$$

where $u$ is the fluid flow rate across the control volume and $z$ is the height of the fluid. The mass flow rate is $\dot{m}$ in (mole/s); heat and work rates are $\dot{Q}$ and $\dot{W}$. In terms of enthalpy, $H = U + PV$, the first law for the open systems can be written as:

$$\dot{Q} + \dot{W} = \Delta \dot{m}\left(H + \frac{u^2}{2} + gz\right) + \frac{d(mU)}{dt} \qquad (9.5)$$

**Second law of thermodynamics:** No process can be effected and no machine can be devised whose sole effect is the complete conversion of heat absorbed to work done by the system. Some heat has to be discarded to the surroundings. In other words, it is impossible to construct a process whose sole effect is the transfer of heat from a low temperature to a higher temperature. Heat flows spontaneously from a higher temperature to a lower temperature and not from a lower temperature to a higher temperature.

The first statement is the Kelvin-Planck statement of the second law of thermodynamics. As a corollary, it is not possible to effect a cyclic process that can convert heat absorbed by a system completely into work done by the system. Mathematically stated, the second law of thermodynamics can be written as:

$$\Delta S_{tot} \geq 0 \qquad (9.6)$$

Thus, each and every process proceeds in such a direction that the total entropy change associated with it is positive. In the limiting case of reversible operation, the entropy change would be zero. It is impossible to effect a process whose entropic change is negative. The inequality given by Eq. (9.6) is also referred to as the Clausius inequality.

## 9.1.1 PMM2: Perpetual Motion Machine of the Second Kind

Some machines and processes are designed in such a fashion that they are infeasible. They claim perpetual motion. They either violate the conservation of energy principle or they disobey the Clausius inequality. The types of designs with sustained, undamped motion that violate the conservation of energy principle are referred to as perpetual motion machines of the first kind (PMM1). The designs with sustained, undamped motion that violate the second law of thermodynamics are referred to as perpetual motion machines of the second kind (PMM2). Although the second law of thermodynamics will be formally introduced later, simply stated, no machine can be devised and no process can be designed whose sole effect is to convert all heat to work. Some heat will have to be discarded to the surroundings. Heat cannot flow from a low temperature to a higher temperature in a spontaneous fashion. Heat can only travel from a hot temperature to a cold temperature in a spontaneous manner.

**Example 9.1** *Water screw perpetual motion machine.* Water from a tray falls and spins a water wheel. This powers a set of gears and pumps, and returns the water to the tray. Can this last forever?

No. Frictional effects will result in reduced water at the water wheel in subsequent cycles. Any other design is a violation of PMM1. The law of conservation of energy is violated.

**Third law of thermodynamics:** The third law of thermodynamics was developed by Nernst and is referred to as the Nernst postulate or Nernst theorem. It states that entropy of pure substances approaches zero when the temperature of the substance is brought to zero Kelvin.

If the entropy of each element in a crystalline state with a perfect structure is zero at the absolute zero of temperature, every substance has a finite positive entropy; however, at the absolute zero of temperature, the entropy may become zero, and does so in the case of crystalline substances with a perfect structure.

**Fourth law of thermodynamics:** The Onsager reciprocal relations are referred to collectively as the fourth law of thermodynamics. The relation between forces and flows for systems not in equilibrium but in a state of local equilibrium are provided in the Onsager relations as follows:

$$J_i = \sum_j L_{ij} F_j \qquad (9.7)$$

$J$ represent the flows, $F$ the forces, and $L$ the phenomenological coefficients. The $i$ and $j$ denote the different flows and forces. Thus, the
concentration difference may be one force, temperature difference
another force, momentum difference another force, etc., and the flows
can be heat transfer, mass transfer, and momentum transfer. Onsager
showed that from analysis of a positive definite matrix, the cross-
coefficients in Eq. (9.7) have to be equal. Thus:

$$L_{ij} = L_{ji} \tag{9.8}$$

**Example 9.2**   *Evaluate the difference.*   $\left(\dfrac{\partial U}{\partial T}\right)_P - \left(\dfrac{\partial U}{\partial T}\right)_V$

Assumption: Fluid is an ideal gas.

$$C_v = \left(\frac{\partial U}{\partial T}\right)_V \tag{9.9}$$

$$\left(\frac{\partial U}{\partial T}\right)_P = \left(\frac{\partial (H - PV)}{\partial T}\right)_P = C_p - P\left(\frac{\partial V}{\partial T}\right)_P \tag{9.10}$$

$$\left(\frac{\partial U}{\partial T}\right)_P - \left(\frac{\partial U}{\partial T}\right)_V = C_p - C_v - PV\beta = R(1 - \beta T) \tag{9.11}$$

where $\beta$ is the compressibility factor.

## 9.1.2   Isobaric Process

Consider an ideal gas expansion in a piston cylinder arrangement from
an initial volume $V_i$ to a final volume $V_f$. The expansion is conducted at
constant pressure. Applying the first law of thermodynamics for $n$
moles of the gas in the cylinder:

$$d(nU) = dQ + dW \tag{9.12}$$

The work done by the system can be written as:

$$dW = -\int Pd(nV) = -nP(V_f - V_i) \tag{9.13}$$

Combining Eqs. (9.12) and (9.13) and writing the internal energy
change in terms of the temperature change of the gas:

$$nC_v(T_f - T_i) = \int dQ - nP(V_f - V_i) \tag{9.14}$$

it can be seen that:

$$dU = \left(\frac{\partial U}{\partial T}\right)_P dT + \left(\frac{\partial U}{\partial P}\right)_T \qquad (9.15)$$

The term $\partial U/\partial P$ is negligible and:

$$\left(\frac{\partial U}{\partial T}\right)_P = \left(\frac{\partial U}{\partial T}\right)_v + \left(\frac{\partial U}{\partial V}\right)_T \left(\frac{\partial V}{\partial T}\right)_P \qquad (9.16)$$

The term $\partial U/\partial V$ is negligible and $dU$ can be written as $C_v dT$.

Assuming that the fluid obeys the ideal gas law, $PV = RT$, for one mole of the gas, Eq. (9.14) becomes:

$$nC_P(T_f - T_i) = Q \qquad (9.17)$$

In other words, during a constant pressure process, the heat supplied for the system from the surroundings should equal the change in enthalpy of the system. For one mole of the ideal gas in the differential form, the first law of thermodynamics for a closed system during a constant pressure process can be written as:

$$dU = dQ - PdV = dQ - d(PV) + VdP \qquad (9.18)$$

or $\qquad dU + d(PV) = dH = dQ \qquad (9.19)$

From Eq. (9.2):

$$dH = TdS + VdP \qquad (9.20)$$

For a constant pressure process, $VdP = 0$ and hence $dH = TdS = dQ$.

### 9.1.3 Isothermal Process

Consider a piston cylinder arrangement with the cylinder filled with one mole of an ideal gas. The piston is pulled, and the volume expands from an initial volume, $V_i$, to a final volume, $V_f$, at constant temperature. In differential form, the first law of thermodynamics for a closed system for one mole of the ideal gas may be written as:

$$dQ - PdV = dU \qquad (9.21)$$

Internal energy refers to the energy internal to a substance. All molecules possess kinetic energy of translation, energy of rotation, and energy of vibration. When heat is added to a closed system at a macroscopic level, the energy of the molecules increases. The internal energy is defined to capture such changes in energy level. The internal energy of a substance includes the potential energy resulting from intermolecular forces. Absolute values of the energy are not as important as the changes in the state of the system. For an isothermal process, $dU = 0$. This is assuming that $U$ is a function of only temperature. Thus:

$$dQ = PdV = TdS \qquad (9.22)$$

For an ideal gas, $PV = RT$ for one mole of gas:

$$Q = \int \frac{RTdV}{V} = RT\ln\left(\frac{V_f}{V_i}\right) = T\Delta S \qquad (9.23)$$

Thus, the heat supplied from the surroundings during an isothermal expansion of an ideal gas can be given by Eq. (9.23). From the relation $dU = TdS - PdV$ for an isothermal process, $dU = 0$ and hence, $TdS = PdV$. For an isothermal process, $dU$ is zero because $U$ is a function of the state of the system only, that is, its temperature in this case and not the path taken to reach it.

### 9.1.4 Adiabatic Process

Consider a piston cylinder arrangement with the cylinder filled with one mole of an ideal gas. The piston is pulled, and the volume expands from an initial volume, $V_i$, to a final volume, $V_f$, adiabatically. In differential form, the first law of thermodynamics for a closed system for one mole of the ideal gas may be written as:

$$dQ - PdV = dU \qquad (9.24)$$

or

$$-PdV = dU \qquad (9.25)$$

For an ideal gas, for one mole of gas, $P = RT/V$ and $dU = C_v dT$. $dU = C_v dT$ is assuming that $U$ is a function of only temperature. Then Eq. (9.25) becomes:

$$-\frac{-dV}{V} = \frac{C_v dT}{RT} \qquad (9.26)$$

Integrating Eq. (9.26) from the initial volume to final volume and initial temperature to final temperature:

$$-\ln\frac{V_f}{V_i} = \frac{C_v}{R}\ln\frac{T_f}{T_i} \qquad (9.27)$$

In the power-potentiated form, Eq. (9.27) becomes:

$$\frac{V_f}{V_i} = \left(\frac{T_i}{T_f}\right)^{C_v/R} \qquad (9.28)$$

Equation (9.28) is valid for an adiabatic expansion of an ideal gas.

**Example 9.3**   *Show that the adiabatic process for an ideal gas is polytropic. Let* $\gamma = C_p/C_v$

For an ideal gas, it was shown earlier that $C_p - C_v = R$, or $(\gamma - 1) = R/C_v$. Applying this to Eq. (9.27):

$$\left(\frac{V_i}{V_f}\right)^\gamma = \left(\frac{V_i T_f}{V_f T_i}\right) = \frac{P_f}{P_i} \tag{9.29}$$

## 9.1.5   Isochoric Process

Consider a piston cylinder arrangement with the cylinder filled with one mole of an ideal gas. The pressure of the closed system is increased from an initial $P_i$ to a final $P_f$ by a constant-volume process. In differential form, the first law of thermodynamics for a closed system for one mole of the ideal gas may be written as:

$$dQ - PdV = dU \tag{9.30}$$

or $\qquad\qquad dQ = dU = TdS - PdV = TdS \tag{9.31}$

**Example 9.4**   *Three-step cycle.* An ideal gas is heated from a temperature $T_0$ to a final temperature $T_f$ at constant pressure $P_0$. Then it is compressed to a pressure $P_1$ at constant temperature. Then it is brought back to its initial state by isothermal expansion. Show the changes in internal energy and enthalpy for each of the three steps. Choose one mole of gas as the basis, and derive the work done in each of the steps. The heat capacity at constant volume $C_v = 3R/R$ and $C_p = 5R/R$.

Step 1–2 in Figure 9.1 is a constant-pressure process. Equation (9.30) may be used:

$$Q = \Delta H = C_p(T_f - T_0) \tag{9.32}$$

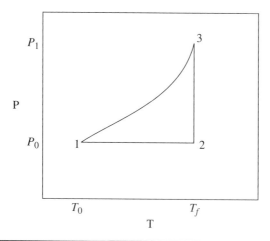

**FIGURE 9.1**   Three-step cycle in a P–T diagram.

Step 2–3 is an isothermal process. Equation (9.30) may be used:

$$Q = RT_f \ln(V_3/V_2) \qquad (9.33)$$

The internal energy change during Step 3–1 may be calculated as follows:

$$0 = \Delta U_{12} + \Delta U_{23} + \Delta U_{31} \qquad (9.34)$$

or

$$\Delta U31 = -Cv(Tf - T0) + 0 \qquad (9.35)$$

Work done by the system: $W = -\int PdV$
From the first law of thermodynamics, $Q + W = \Delta U$.

Thus,

$$Q = \int_{v_3}^{v_1} PdV - C_v(T_f - T_0) \qquad (9.36)$$

**Example 9.5**  *Compressibility factor,* $\beta$. Show that:

$$\left(\frac{\partial \beta}{\partial P}\right)_T = -\left(\frac{\partial \kappa}{\partial T}\right)_P \qquad (9.37)$$

$$\beta = \frac{1}{V}\left(\frac{\partial V}{\partial T}\right)_P \qquad (9.38)$$

$$\kappa = -\frac{1}{V}\left(\frac{\partial V}{\partial P}\right)_T \qquad (9.39)$$

Differentiate Eq. (9.38) with respect to pressure $P$ and Eq. (9.39) with respect to temperature $T$:

$$\left(\frac{\partial \beta}{\partial P}\right) = \frac{\partial^2 V}{V \partial T \partial P} \qquad (9.40)$$

$$\left(\frac{\partial \kappa}{\partial T}\right) = -\frac{\partial^2 V}{V \partial P \partial T} \qquad (9.41)$$

Volume $V$ is a function of pressure $P$ and temperature $T$. $V$ is a continuous differentiable function of two variables, $P$ and $T$. Hence, the order of differentiation should not matter. Thus:

$$\left(\frac{\partial^2 V}{\partial P \partial T}\right) = \left(\frac{\partial^2 V}{\partial T \partial P}\right) \qquad (9.42)$$

Hence, Eqs. (9.40) and (9.41) are equal and Eq. (9.37) is shown.

## 9.1.6 Carnot Cycle

The Carnot cycle is the most efficient of cyclical processes. It consists of two isothermal and two adiabatic steps that alternate. The thermal efficiency of any heat engine can be written as:

$$\eta = 1 - \frac{|Q_h| - |Q_c|}{|Q_h|} \tag{9.43}$$

The heat absorbed from the hot reservoir is given by $Q_h$, the work done is $Q$, and the heat given out to the cold reservoir is $Q_c$. The Carnot engine sets the upper limit on the maximum efficiency achievable. This is because it is operated in a reversible manner. From the Kelvin-Planck statement of the second law of thermodynamics, Eq. (9.43) has to be less than 1 and cannot be equal to or greater than 1. A reversible ideal engine was proposed by Sadi Carnot in 1824. The four steps of the Carnot cycle are shown in Fig. 9.2 in a P–V diagram of the working fluid and in Fig. 9.3 in a T–S diagram of the working fluid.

The Carnot cycle consists of four steps. These are as follows:

A. A reversible isothermal process (1–2) is effected when heat $Q_h$ is taken in by the working fluid from the hot reservoir at the temperature $T_h$.

B. A reversible adiabatic expansion (2–3) of the working fluid, where the temperature of the working fluid changes from $T_h$ to $T_c$.

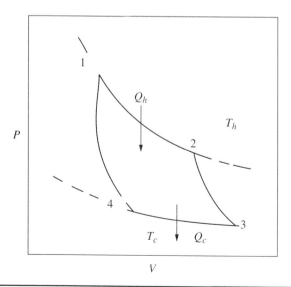

FIGURE 9.2   Four-step Carnot cycle on a P–V diagram of the working fluid.

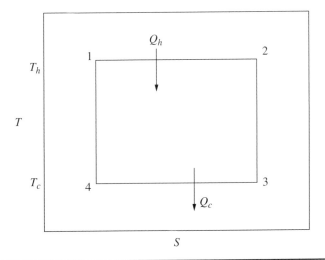

**Figure 9.3** Four-step Carnot cycle on a $T$–$S$ diagram of a working fluid.

C. A reversible isothermal process (3–4) when heat $Q_c$ is discharged from the working fluid into the cold reservoir at the temperature $T_c$.

D. A reversible adiabatic process (4–1) where the working fluid is brought back to the hot reservoir temperature, $T_h$, by compression.

During step 1–2, the first law is applied. The work done during isothermal expansion of an ideal gas was found to be $RT_h \ln(V_2/V_1)$:

$$Q_h - RT_h \ln(V_2/V_1) = \Delta U \qquad (9.44)$$

The internal energy change with volume can be neglected and as step 1–2 is isothermal, $\Delta U = 0$. Hence, Eq. (9.41) becomes:

$$Q_h = RT_h \ln(V_2/V_1) \qquad (9.45)$$

In a similar manner, the first law can be applied to the isothermal step 3–4 and:

$$-Q_c = RT_c \ln(V_4/V_3) \qquad (9.46)$$

Step 2–3 is adiabatic and reversible. Hence:

$$dU = TdS - PdV \qquad (9.47)$$

Neglecting changes of internal energy with volume and assuming that internal energy is a function of only temperature, Eq. 9.47 may be written as:

$$C_v dT = -RT dV/V \tag{9.48}$$

or

$$\frac{dT}{T} = -\frac{R}{C_v}\frac{dV}{V} \tag{9.49}$$

Integrating Eq. (9.49) between the hot reservoir and cold reservoir temperatures:

$$\ln\left(\frac{T_c}{T_h}\right) = \left(\frac{V_3}{V_2}\right)^{-R/C_v} \tag{9.50}$$

In a similar manner, the first law for the adiabatic step 4–1 can be integrated to yield:

$$\ln\left(\frac{T_c}{T_h}\right) = \left(\frac{V_4}{V_1}\right)^{-R/C_v} \tag{9.51}$$

Comparing Eqs. (9.50) and (9.51):

$$\frac{V_3}{V_2} = \frac{V_4}{V_1} \tag{9.52}$$

the thermal efficiency of the cycle can be calculated as:

$$\eta = 1 - \frac{Q_c}{Q_h} = 1 - \frac{RT_c \ln\left(\frac{V_3}{V_4}\right)}{RT_h \ln\left(\frac{V_2}{V_1}\right)} \tag{9.53}$$

Combining Eqs. (9.52) and (9.53):

$$\eta = 1 - \frac{T_h}{T_c} \tag{9.54}$$

Equation (9.54) is the Carnot equation. The ideal gas temperature is in Kelvin scale.

Thus, for a Carnot engine:

$$\left(\frac{-Q_c}{Q_h}\right) = \left(\frac{T_c}{T_h}\right) \tag{9.55}$$

or

$$\left(\frac{Q_h}{T_h}\right) + \left(\frac{Q_c}{T_c}\right) = 0 \tag{9.56}$$

The quantity $Q/T$ can be seen to be the entropy. The entropic change for a reversible cycle is zero. Hence:

$$\oint \frac{dQ_{rev}}{T} = \Delta S = 0 \qquad (9.57)$$

For an irreversible cycle, $\Delta S$ would be positive. For $\Delta S$ for the cycle less than zero, the process is infeasible.

## 9.1.7    Carnot's Theorem

Carnot's theorem may be stated as follows: No machine can be devised and no process can be designed with an efficiency greater than that of the Carnot efficiency.

The proof can be provided by the method of *reduction de abstractum*. Assume that there exists an engine $E$ with an efficiency greater than that of Carnot. Then that engine is operated along with a Carnot refrigerator, as shown in Fig. 9.4.

The heat received from the cold reservoir is $Q_h - Q_h'$. This can be seen to be the heat gained by the hot reservoir at temperature $T_h$ from Fig. 9.4. Thus, the net effect of the engine $E$ and Carnot refrigerator is to take heat from the cold reservoir to the hot reservoir. Per the alternate statement of the second law, no machine can be devised and no process can be designed whose sole effect is to take heat from a cold temperature and discharge it to a hot temperature.

Thus, the engine $E$ cannot have a thermal efficiency greater than that of the Carnot cycle. Hence, by the method of *reduction de abstractum*, the Carnot theorem stands proved. As a corollary, the thermal efficiency of the Carnot engine depends only on the temperatures of the hot and cold reservoir temperatures.

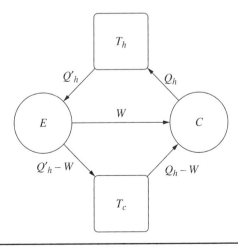

**FIGURE 9.4**  Engine $E$ and Carnot refrigerator $C$.

## 9.1.8 Otto Cycle

Internal combustion engines are used to drive automobiles and airplanes by the thrust generated by a propulsion system. The Otto cycle is shown in Fig. 9.5 on a $PV$ diagram of the working fluid. The intake stroke of the engine begins at step a. The intake valve is opened, and the piston is pulled out of the cylinder through an isobaric process. The fuel is drawn into the cylinder through the open valve along with the oxygen needed for combustion. The compression stroke of the engine begins after the closing of the intake valve. In step bc, work is done, volume decreases, and the pressure rises. Step c marks the beginning of the combustion of fuel/oxygen supplied by the air. The combustion is anh isochoric process. Heat is released from the exothermic reactions. This results in an increase in pressure (step cd). The power stroke of the engine starts at step d. During step de, pressure drops as volume expands and work is done by the gas in the piston-cylinder assembly. The exhaust valve is opened at step e. Heat is expended to the surroundings. Step ef is isochoric. The exhaust stroke of the engine begins at step f. The process can repeat in cycles from there on. Real cycles will be less efficient on account of the heat losses during the compression and power strokes, friction losses, and spontaneous combustion in an isochoric fashion. The area in the enclosure of step bcdef is the work done.

Steps cd and ef are isochoric operations. The first law applied to these steps can lead to:

$$Q_h = C_v(T_d - T_c) \tag{9.58}$$

$$-Q_c = C_v(T_b - T_e) \tag{9.59}$$

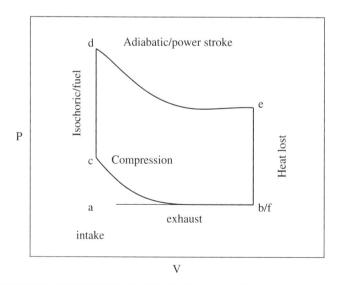

FIGURE 9.5 Six-stroke Otto cycle used in automobiles.

Because steps cd and ef are adiabatic and reversible, the thermal efficiency of an Otto cycle may be calculated as:

$$\eta = 1 - \left( \frac{T_e - T_b}{T_d - T_c} \right) \tag{9.60}$$

Assuming that the fluid obeys the ideal gas law:

$$\eta = 1 - r \left( \frac{P_e - P_b}{P_d - P_c} \right) \tag{9.61}$$

where $r$ is the compression ratio $r = (V_b/V_c) = (V_e/V_d)$.

Realizing that steps bc and de are adiabatic, the polytropic relationships may be used. Thus:

$$P_d V_d^\gamma = P_e V_e^\gamma = P_e V_b^\gamma \tag{9.62}$$

and

$$P_c V_c^\gamma = P_b V_b^\gamma = P_c V_d^\gamma \tag{9.63}$$

Dividing Eq. (9.62) by Eq. (9.63):

$$\left( \frac{P_d}{P_c} \right) = \left( \frac{P_e}{P_b} \right) \left( \frac{V_e^\gamma}{V_b^\gamma} \right) \left( \frac{V_c^\gamma}{V_d^\gamma} \right) = \left( \frac{P_e}{P_c} \right) \tag{9.64}$$

Combining Eq. (9.64) in Eq. (9.61), the thermal efficiency of an Otto cycle can be seen to be:

$$\eta = 1 - \frac{1}{r^{\gamma-1}} \tag{9.65}$$

where $r$ is the compression ratio and $\gamma$ is the polytropic constant for an ideal gas ($C_p/C_v = 5/3$).

or
$$\eta = 1 - \frac{1}{r^{5/3}} \tag{9.66}$$

## 9.2 Conservation of Energy

As discussed in Chap. 1, as in the equation of conservation of momentum, the equation of conservation of energy can be written from a shell balance in the system as follows:

$$\begin{pmatrix} \text{rate of energy} \\ \text{accumulated} \end{pmatrix} = \begin{pmatrix} \text{rate of energy} \\ \text{into the surfaces} \end{pmatrix} - \begin{pmatrix} \text{rate of energy} \\ \text{out of the surfaces} \end{pmatrix}$$

$$- \begin{pmatrix} \text{rate of work done} \\ \text{by the system} \end{pmatrix} + \begin{pmatrix} \text{rate of energy} \\ \text{produced within} \\ \text{the system} \end{pmatrix}$$

$$(9.67)$$

The energy is the sum of the internal energy and kinetic energy of the system. Let $q$ be the heat flux into and out of the surfaces of the shell considered and $T$ the temperature. Equation (9.68) can be written as:

$$\rho C_p = \left( \frac{\partial T}{\partial t} + v.\nabla T \right) = -\nabla q + \dot{Q} - \dot{W} \qquad (9.68)$$

Fourier's law of heat conduction can be written for the heat flux in one dimension as:

$$q_x = -k \frac{\partial T}{\partial x} \qquad (9.69)$$

The generalized Fourier's law of heat conduction can be written as:

$$q_x = -k \frac{\partial T}{\partial x} - \tau_r \frac{\partial q_x}{\partial t} \qquad (9.70)$$

## 9.2.1   Reasons to Seek Generalized Fourier's Law of Heat Conduction

Fourier's law of heat conduction, Fick's law of mass diffusion, Newton's law of viscosity, and Ohm's law of electricity are physical laws that are used to describe transport phenomena of heat, mass, momentum, and electricity, respectively. These phenomenological laws were developed largely from empirical observations at steady states several centuries ago. Although they have been used widely for extended periods, there are a number of applications where massive deviations from theoretical predictions based on these laws have been found. Here are six reasons to seek a generalized Fourier's law of heat conduction:

1. Fourier's law of heat conduction was found to contradict the microscopic theory of reversibility introduced by Onsager [3].

2. Singularities have been found in the description of transient heat conduction using the Fourier parabolic equations. A "blow-up" occurs: a) at short contact times in the expression

for surface flux, for the case of description of transient temperature in a semi-infinite medium subject to constant wall temperature boundary condition; b) surface flux for a finite slab subject to constant wall temperature on either of its edges; c) temperature term in the constant wall flux problem in cylindrical coordinates in a semi-infinite medium solved for by using the Boltzmann transformation, leading to a solution in exponential integral; d) in the short time limit, the parabolic conduction equations for a semi-infinite sphere are solved for by using the similarity transformation.

3. Development of Fourier's law of heat conduction was from observations at steady state and empirical in nature. The observations' use in transient state, such as at a nanoscale level in the time domain, is an extrapolation.

4. Overpredictions of theory to experiment have been found in important industrial processes, such as fluidized bed heat transfer to surfaces, CPU overheating, adsorption, gel acrylamide electrophoresis, restriction mapping, laser heating of semiconductors during the manufacture of semiconductor devices, and drug delivery systems.

5. Landau and Lifshitz examined the solution for transient temperature and noted that for times greater than zero, the temperature is finite at all points in the infinite medium, except at an infinite location. The inference is that the heat pulse has traveled at infinite speed. But light is the speediest of all velocities. Hence, there is a conflict with the light speed barrier stated by Einstein's theory of relativity. The speed of any mobile object, including a thermal wave, ought to be less than the speed of light.

6. Fourier's law breaks down at the nanoscale space level. This is also referred to as the Casimir limit in some quarters. In this regime, the mean free path of the molecules is greater than the dimension of the object under scrutiny.

In order to better describe transient heat conduction events at the nanoscale level, the damped wave conduction and relaxation equation can be used. A comprehensive insight into the characteristics of the analytical solution using the damped wave conduction and relaxation equation was provided. This was originally suggested by Maxwell. The equation can be written as in Eq. (9.70), where $\tau_r$ is the relaxation time (nanoseconds), $q$ is the heat flux $(w/m^2)$, and $k$ is the thermal conductivity of the medium of conduction $(w/m/K)$. This equation was postulated in the mid 20th century by Cattaneo and Vernotte. Reviews have been provided by Joseph and Preziosi. The estimates of the relaxation times are of the order of nanoseconds. Some concerns have been expressed about the generalized Fourier's law of heat conduction violating the second law of thermodynamics. It is going to be shown later in the chapter that investigators who do not use the time and space conditions appropriately end up with solutions that appear in violation of the second law of thermodynamics. But when those conditions are corrected to more physically realistic conditions, well-bounded solutions within the constraints of the second law of thermodynamics can be obtained.

## 9.3 Derivation of Damped Wave Conduction and Relaxation Equation from Free Electron Theory

The damped wave conduction and relaxation equation is derived from the free electron theory. The derivation of Ohm's law of electric conduction is revisited to obtain the damped wave momentum transfer and relaxation equation by analogy. The electrical resistivity of materials differs by 30 orders of magnitude. So a single theory to explain the behavior of all materials may be difficult to develop. In the free electron model, the outermost electrons of the atoms can take part in conduction. They are not bound to the atom, but are free to move through the whole solid. These electrons have been variously called the free electron cloud, the free electron gas, or the Fermi gas. The assumption is that the potential field due to the ion cores is uniform throughout the solid. The free electrons have the same potential energy everywhere in the solid. Due to the electrostatic attraction between a free electron and the ion core, this potential energy will be a finite negative value. Only energy differences are important, and the constant potential can be taken to be zero. Then, the only energy that has to be considered is the kinetic energy. The kinetic energy is substantially lower than that of the bound electrons in an isolated atom, as the field of motion for the free electron is considerably enlarged in the solid as compared to the field around an isolated atom. The free electron theory can be used to better understand electrical conduction.

By Lorenz analogy, the heat conduction can also be predicted in a similar manner. The independent electron assumption was developed by Drude in 1905. The free electron theory assumes that electrons are responsible for all of the conduction. The electrons behave like an ideal gas, occupy negligible volume, undergo collisions, and are perfectly elastic. Electrons are free to move in a constrained flat-bottom well. Electron distribution of energy is a continuum.

The general equation of motion for the drift velocity of the free electron on account of an applied temperature gradient driving force can be given by the following expression from the Drude theory:

$$m\frac{dv_e}{dt} + \frac{mv_e}{\tau} = -\frac{3k_B}{2}\frac{dT}{dx} \qquad (9.71)$$

where $m$ is the mass of the electron, $v_e$ is the drift velocity, $\tau$ is the collision time of the electron with an obstacle, $k_B$ is the Boltzmann constant, and $dT/dx$ is the applied temperature gradient. The drift velocity of the electron is different from the random velocities associated with it. Drift velocity is superimposed on the random motion. It is in a net direction of the superimposed field. This leads to a net flow of charge and the passage of electric current. The electrons encounter obstacles during drift, and the directional motion is lost and reduced to the random motion. The memory gained is lost, and the clock is set back to zero. Collisions occur in the time interval $\tau$. The rate of destruction of momentum by virtue of the collision is given by $mv_e/\tau$. This slows down the electron. The drag force will balance the applied force due to the temperature gradient at steady state to yield the Fourier's law of heat conduction. This can be seen in the following steps:

$$v_e = \frac{-3\tau}{2m} \frac{k_B dT}{dx} \tag{9.72}$$

The heat flux can be defined as:

$$q = n\left(\frac{3}{2}k_B T\right)v_e \tag{9.73}$$

where $n$ is the number of electrons per unit volume and $(3/2\,k_B T)$ is the average energy of the electron from the equipartition energy theorem. Using the Boltzmann relation, the heat flux can also be written as:

$$q = n\frac{1}{2}mv_e^3 \tag{9.74}$$

Multiplying Eq. (9.72) throughout by $n\,(3/2\,k_B T)$ and using Eq. (9.73), Eq. (9.72) becomes:

$$q = \frac{-9nT\tau}{4m}k_B^2 \frac{dT}{dx} = -\frac{k\partial T}{\partial x} \tag{9.75}$$

where the thermal conductivity can be written as:

$$k = \frac{9nT\tau}{4m}k_B^2 \tag{9.76}$$

During transient heat conduction, the acceleration term may become important. Rewriting Eq. (9.72) as:

$$\frac{\tau dv_e}{dt + v_e} = \frac{-3\tau}{2m} \frac{k_B dT}{dx} \tag{9.77}$$

Multiplying Eq. (9.77) throughout by $n\,(3/2\,k_B T)$ and combining it with Eqs. (9.74) through (9.76):

$$n\frac{3}{2}k_B T \frac{\tau dv_e}{dt + q} = \frac{-k\partial T}{\partial x} \tag{9.78}$$

Using the Boltzmann relation ($\frac{1}{2}\,mv_e^2 = 3/2k_BT$), Eq. (9.78) becomes:

$$\frac{1}{2}\frac{n\tau mv_e^2 dv}{dt+q} = \frac{-k\partial T}{dx} \qquad (9.79)$$

Differentiating Eq. (9.74) with respect to $t$:

$$\frac{\partial q}{\partial t} = \frac{3}{2}\frac{nmv_e^2 dv_e}{dt} \qquad (9.80)$$

Combining Eqs. (9.80) and (9.79):

$$\frac{\tau}{3}\frac{\partial q}{\partial t+q} = \frac{-k\partial T}{\partial x} \qquad (9.81)$$

Equation (9.81) is equivalent to the Cattaneo and Vernotte equation given by Eq. (9.70) when $\tau/3 = \tau_r$:

$$\frac{\tau_r \partial q}{\partial t+q} = \frac{-k\partial T}{\partial x} \qquad (9.82)$$

## 9.4 Semi-infinite Cartesian and Infinite Cylindrical and Spherical Mediums

Consider a semi-infinite medium at an initial temperature of $T_0$ (Fig. 9.6). For times greater than 0, the surface at $x = 0$ is maintained at a constant surface temperature at $T = T_s$, $T_s > T_0$. The boundary conditions and initial condition are as follows:

$$t = 0, T = T_0 \qquad (9.83)$$

$$x = 0, T = T_s \qquad (9.84)$$

$$x = \infty, T = T_0 \qquad (9.85)$$

The transient temperature in the semi-infinite medium can be solved for by solving the Fourier parabolic heat conduction equations using the Boltzmann transformation $\eta = x/\sqrt{4\alpha t}$ and shown to be:

$$u = \frac{(T - T_0)}{(T_s - T_0)} = 1 - erf\left(\frac{x}{\sqrt{4\alpha t}}\right) \qquad (9.86)$$

$$T = T_s \qquad\qquad\qquad T = T_0$$

$$x = 0 \qquad\qquad\qquad x = \infty$$

**FIGURE 9.6** Semi-infinite medium with initial temperature at $T_0$.

The heat flux can be written as:

$$q* = \frac{q}{\sqrt{k\rho C_p / \tau_r} (T_s - T_0)} = \frac{1}{\sqrt{\pi\tau}} \exp\left(-\frac{x^2}{4\alpha t}\right) \qquad (9.87)$$

The dimensionless heat flux at the surface is then given by:

$$q_s^* = \frac{1}{\sqrt{\pi\tau}} \qquad (9.88)$$

It can be seen that there is a "blow-up" in Eq. (9.88) as $\tau \to 0$. For applications with substantial industrial importance, such as the heat transfer between fluidized beds to immersed surfaces [4], large deviations have been found between experimental data and mathematical models based upon surface renewal theory. The critical parameter in the mathematical models is the contact time of the packets that are comprised of solid particles at the surface. This contact time is small for gas-solid fluidized beds for certain powder types. Under such circumstances, the microscale time effects may have been significant. These are not accounted for by the parabolic heat conduction models. This is one of the motivations for studying the hyperbolic heat conduction models. It has been shown that the ballistic term in the governing hyperbolic heat conduction equation is the "only" mathematical modification to the parabolic heat conduction equation that can remove the singularity in Eq. (9.86) at short times.

The governing hyperbolic heat conduction equation in one dimension for a semi-infinite medium with constant thermophysical properties $\rho$, $C_p$, $k$, and $\tau_r$, that is, the density, heat capacity, thermal conductivity, and thermal relaxation time, can be obtained by combining the damped wave conduction and relaxation equation with the energy balance equation to yield:

$$\frac{\partial u}{\partial \tau} + \frac{\partial^2 u}{\partial \tau^2} = \frac{\partial^2 u}{\partial X^2} \qquad (9.89)$$

where $\qquad u = \frac{(T - T_0)}{(T_s - T_0)}; X = \frac{x}{\sqrt{4\alpha t}}; \tau = \frac{t}{\tau_r} \qquad (9.90)$

Baumeister and Hamill obtained the Laplace transform of Eq. (9.89) and applied the boundary condition at $x = \infty$, given by Eqs. (9.85) and (9.84), to obtain in the Laplace domain:

$$\bar{u} = \frac{\exp(-X\sqrt{s(s+1)})}{s} \qquad (9.91)$$

They integrated Eq. (9.91) with respect to space to obtain:

$$H(s) = \int \exp\frac{-X\sqrt{s(s+1)}}{s} \, dX = -\frac{1}{s\sqrt{s(s+1)}}\exp\frac{-X\sqrt{s(s+1)}}{s}$$

(9.92)

The inversion of Eq. (9.94) was obtained from the Laplace transform tables and found to be:

$$H(\tau) = \int_0^\tau \exp\left(-\frac{p}{2}\right)I_0 1/2\sqrt{p^2 - X^2}\,dp$$

(9.93)

The dimensionless temperature is obtained by differentiating $H(\tau)$ in Eq. (9.93) with respect to $X$ and for $\tau \geq X$:

$$u = \frac{\partial H}{\partial X} = -X\int_X^\tau \exp\left(-\frac{p}{2}\right)\frac{I_1 1/2\sqrt{p^2 - X^2}}{\sqrt{p^2 - X^2}}\,dp + \exp\left(-\frac{X}{2}\right)$$

(9.94)

Baumeister and Hamill presented their solution in the integral form, as shown in Eq. (9.94). In this study, the integrand is approximated to a Chebyshev polynomial and a useful expression for the dimensionless temperature is obtained. This is used to compare the results to those obtained by relativistic transformation. The dimensionless heat flux can be seen to be:

$$q* = \exp\left(-\frac{\tau}{2}\right)I_0 1/2\sqrt{\tau^2 - X^2}$$

(9.95)

The surface heat flux can be seen to be:

$$q_s^* = \exp\left(-\frac{\tau}{2}\right)I_0\left[\frac{\tau}{2}\right]$$

(9.96)

## 9.4.1 Chebyshev Economization or Telescoping Power Series

In order to further study the dimensionless transient temperature from the hyperbolic damped wave conduction and relaxation equation, the integral expression given by Baumeister and Hamill in Eq. (9.95) can be simplified using a Chebyshev polynomial. Chebyshev polynomial approximations tend to distribute the errors more evenly, with reduced maximum error, by the use of cosine functions. The set of polynomials $T_n(r) = \mathrm{Cos}(n\theta)$ generated from the sequence of cosine functions using the transformation:

$$\theta = \mathrm{Cos}^{-1}(r)$$

(9.97)

is called Chebyshev polynomials (Table 9.2). Coefficients of the Chebyshev polynomials for the integrand in Eq. (9.94) $I_1 1/2\sqrt{p^2 - X^2}/\sqrt{p^2 - X^2}$ can be computed with some effort. The modified Bessel function of the first order and first kind can be expressed as a power series as follows:

$$\frac{I_1 1/2\sqrt{p^2 - X^2}}{\sqrt{p^2 - X^2}} = \sum_{m=0}^{\infty} \frac{(p^2 - X^2)^m}{4^{2k+1}(m!)(m+1)!} = \frac{\psi^m}{4^{2k+1}(m!)(m+1)!} \tag{9.98}$$

where $\psi = p^2 - X^2$.

Each of the $\psi^m$ terms can be replaced with its expansion in terms of the Chebyshev polynomials given in Table 9.2.

The coefficients of like polynomials $T_i(r)$ are collected. When the truncated power series polynomial of the integrand is represented by a Chebyshev polynomial, some of the high-order Chebyshev polynomials can be dropped with negligible truncation error. This is because the upper bound for $T_n(r)$ in the interval $(-1, 1)$ is 1. The truncated series can then be retransformed to a polynomial in $r$ with fewer terms than the original and with modified coefficients. This procedure is referred to as Chebyshev economization, or telescoping a power series.

Prior to expressing Eq. (9.98) in terms of Chebyshev polynomials, the interval $(X, \tau)$ needs to be converted to the interval $(-1, 1)$. So let:

$$r = \frac{2\psi - \tau - X}{\tau - X} \text{ and } \psi = \frac{r(\tau - X) + (\tau + X)}{2} \tag{9.99}$$

Further, let $\quad \xi = (\tau - X) \text{ and } \eta = (\tau + X)$ $\qquad$ (9.100)

$$T_0(r) = 1$$
$$T_1(r) = r$$
$$T_2(r) = 2r^2 - 1$$
$$T_3(r) = 4r^3 - 3r$$
$$T_4(r) = 8r^4 - 8r^2 + 1$$
$$T_5(r) = 16r^5 - 20r^3 + 5r$$
$$T_6(r) = 32r^6 - 48r^4 + 18r^2 - 1$$

**TABLE 9.2** Chebyshev Polynomials

Thus,
$$\psi = \frac{r\xi + \eta}{2} \qquad (9.101)$$

Substituting Eq. (9.101) in Eq. (9.98):

$$\frac{I_1 1/2(p^2 - X^2)}{\sqrt{p^2 - X^2}} = \sum_{m=0}^{\infty} \frac{(r\xi + \eta)^m}{2^k 4^{2k+1} m!(m+1)!} \qquad (9.102)$$

The right-hand side (RHS) of Eq. (9.102) can be written as:

$$RHS \; Eq. (8.22) = \frac{1}{4} + \frac{r\xi + \eta}{256} + \frac{(r\xi + \eta)^2}{49,152} + \cdots \qquad (9.103)$$

A truncation error of $(r\xi + \eta)^3/18,874,368$ is incurred in writing the left-hand side (LHS) of Eq. (9.102) as Eq. (9.103). Replacing the $r$, $r^2$, and $r^3$ terms (see Table 9.3) in Eq. (9.102) in terms of the Chebyshev polynomials given in Table 9.2 and collecting the like Chebyshev coefficients $T_0$, $T_1$, and $T_2$, the RHS of Eq. (9.103) can be written as:

$$T_0(r)\left( \frac{1}{4} + \frac{\eta}{256} + \frac{\eta^2}{49,152} + \frac{\xi^2}{98,304} \right)$$

$$+ T_1(r)\left( \frac{\xi}{256} + \frac{2\eta\xi}{49,152} + \right) \qquad (9.104)$$

$$1 = T_0(r)$$

$$r = T_1(r)$$

$$r^2 = \frac{1}{2}(T_0(r) + T_2(r))$$

$$r^3 = \frac{1}{4}(3T_1(r) + T_3(r))$$

$$r^4 = \frac{1}{8}(3T_0(r) + 4T_2(r) + T_4(r))$$

$$r^5 = \frac{1}{16}(10T_1(r) + 5T_3(r) + T_5(r))$$

$$r^6 = \frac{1}{32}(10T_0(r) + 15T_2(r) + 6T_4(r) + T_6(r))$$

**TABLE 9.3** Powers of $r$ in Terms of the Chebyshev Polynomials

The $T_2(r)$ term can be dropped with an added error of only $\xi^2/98,304$. The order of magnitude of the error incurred is thus $O(\xi^2/98,304)$. Retransformation of the series given by Eq. (9.105) yields:

$$\frac{I_1 1/2\sqrt{p^2-X^2}}{\sqrt{p^2-X^2}} = \frac{1}{4} - \frac{X^2}{128} + \frac{\eta^2}{49,152} + \frac{\xi^2}{98,304} + \frac{(p^2-X^2)}{128} \qquad (9.105)$$

The error involved in writing Eq. (9.105) is $4.1\ 10^{-5}\ \eta\xi$. If Chebyshev polynomial approximation was not used for the integrand and the power series was truncated after the second term, the error would have been $4\ 10^{-3}r^2$. Substituting Eq. (9.105) in Eq. (9.95) and further integrating the expression for dimensionless temperature:

$$u = \exp\left(-\frac{X}{2}\right) + X\exp\left(-\frac{X}{2}\right)\left(\frac{5}{8} + \frac{X}{16} + \frac{\eta^2}{24,576} + \frac{\xi^2}{49.152}\right)$$

$$+ X\exp\left(-\frac{\tau}{2}\right)\left(\frac{3}{8} - \frac{\tau}{16} - \frac{X^2}{64} + \frac{\eta^2}{24,576} + \frac{\xi^2}{49,152}\right) \qquad (9.106)$$

It can be seen that Eq. (9.106) can be expected to yield reliable predictions on the transient temperature close to the wavefront. This is because the error increases as a function of $4.1\ 10^{-5}\ \xi\eta$. Far from the wavefront, that is, close to the surface, the numerical error may become significant.

### 9.4.2   Method of Relativistic Transformation of Coordinates

Sharma developed a relativistic transformation method to solve for the transient temperature by damped wave conduction and relaxation in a semi-infinite medium. A closed-form solution for the transient temperature was obtained. The hyperbolic governing equation [Eq. (9.89)] can be multiplied by $\exp(n\tau)$ and for $n\ \frac{1}{2}$ reduced to Eq. (9.107) below in wave temperature. Thus, the transient temperature was found to be a product of a decaying exponential in time and wave temperature, that is, $u = W\exp(-n\tau)$. This is typical of transient heat conduction applications. Also, the damping term in the hyperbolic PDE, once removed, will lead to an equation of the Klein-Gordon type that can be examined for the wave temperature without being clouded by the damping component. It can be shown that at $n = \frac{1}{2}$, the governing equation for temperature, Eq. (9.89), can be transformed as:

$$\frac{\partial^2 W}{\partial \tau^2} - \frac{W}{4} = \frac{\partial^2 W}{\partial X^2} \qquad (9.107)$$

Equation (9.107) for the wave temperature can be transformed into a Bessel differential equation by the following substitution. Let $\psi = \tau^2 - X^2$.

This substitution variable $\psi$ can be seen to be a spatiotemporal variable. It is symmetric with respect to space and time. It is for the open interval $\tau > X$. Equation (9.107) becomes:

$$4\psi \frac{\partial^2 W}{\partial \psi^2} + 4\frac{\partial W}{\partial \psi} - \frac{W}{4} = 0 \tag{9.108}$$

Equation (9.108) can be seen to be a Bessel differential equation, and the solution can be seen to be:

$$W = c_1 I_0 (1/2\sqrt{\tau^2 - X^2}) + c_2 K_0 (1/2\sqrt{\tau^2 - X^2}) \tag{9.109}$$

It can be seen that at the wavefront, that is, $\psi = 0$, $W$ is finite and, therefore, $c_2 = 0$. Far from the wavefront, close to the surface, the boundary condition can be written as:

$$X = 0, u = 1 \text{ or } W = 1\exp(\tau/2) \tag{9.110}$$

Because $\psi$ is a spatiotemporal variable, the constants of integration $c_1$ can tolerate a function in time up to an exponential relation in time. Applying the boundary condition at the surface, $c_1$ can be eliminated between Eqs. (9.110) and (9.109) to yield in the open interval $\tau > X$:

$$u = \frac{I_0 1/2(\sqrt{\tau^2 - X^2})}{I_0(\tau/2)} \tag{9.111}$$

In the domain $X > \tau$, it can be shown that the solution for the dimensional temperature by a similar approach is:

$$u = \frac{J_0 1/2(\sqrt{X^2 - \tau^2})}{I_0(\tau/2)} \tag{9.112}$$

At the wavefront, $\psi = 0$, Eq. (9.108) can be solved and:

$$\ln(W) = \frac{\psi}{16} \text{ or } W = c_3 \exp\left(\frac{\psi}{16}\right)$$

The temperature at the wavefront is thus $u = c_3\exp(-\tau/2) = c_3\exp(-X/2)$. From the boundary condition at $X = 0$, $c_3 = 1.0$. Thus, at the wavefront:

$$u = \exp\left(\frac{-X}{2}\right) \tag{9.113}$$

From Eq. (9.112), the inertial lag time associated with an interior point in the semi-infinite medium can be calculated by realizing that the first zero of the Bessel function, $J_0(\psi)$, occurs at $\psi = 2.4048$. Thus:

$$2.4048^2 = \frac{x_p^2}{\alpha \tau_r} - \frac{t_{lag}^2}{\tau_r^2}$$

$$t_{lag} = \sqrt{x_p^2 \frac{\tau_r}{\alpha} - 23.132 \tau_r^2} \qquad (9.114)$$

The penetration distance for a given time instant can be developed at the first zero of the Bessel function. Beyond this point, the interior temperatures can be no less than the initial temperature. Thus:

$$X_{pen} = \sqrt{23.132 + \tau^2} \qquad (9.115)$$

The surface heat flux for a semi-infinite medium subject to constant wall temperature solved by the Fourier parabolic heat conduction model and the hyperbolic damped wave conduction and relaxation model are compared with each other using a Microsoft Excel spreadsheet. The parabolic and hyperbolic solutions for surface heat flux are shown side by side in Fig. 9.7. The "blow-up" in the Fourier model can be seen at short times. The hyperbolic model is well bounded at short times and reached an asymotic limit of $q* = 1$ instead of $q* = \infty$. There appears to be a cross-over at $\tau = \frac{1}{2}$. It was found that for $\tau > 3.8$, the prediction of the hyperbolic model is within 10 percent of the parabolic models. It can be seen from Fig. 9.7 that at large times, the predictions of the parabolic and hyperbolic models are the same. For short times, both qualitatively and quantitatively, the predictions of the parabolic and hyperbolic models are substantially different.

It is not clear what happens at $\tau = 1/2$. The hyperbolic governing equation can be transformed using the Boltzmann transformation as follows. Let $\gamma = X/\sqrt{\tau}$. Equation (9.89) becomes:

$$-\left(2\gamma \frac{\partial u}{\partial \gamma} + \frac{\partial^2 u}{\partial \gamma^2}\right) = \frac{1}{\tau}\left(\gamma \frac{\partial u}{\partial \gamma} - \gamma^2 \frac{\partial^2 u}{\partial \gamma^2}\right) \qquad (9.116)$$

For long times, such as $\tau > \frac{1}{2}$, the RHS of Eq. (9.116) can be dropped and the LHS solved for to yield the solution that is identical to that of the Fourier parabolic heat conduction equation, that is:

$$u = 1 - erf\left(\frac{X}{\sqrt{4\tau}}\right) \qquad (9.117)$$

When differentiated and the expression for flux obtained at the surface and $X = 0$, it can be seen that both the parabolic heat conduction

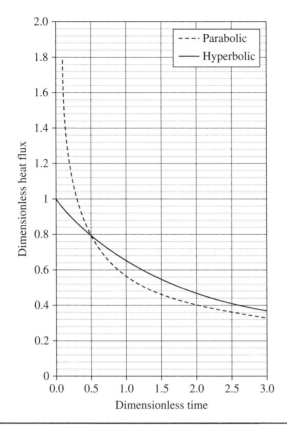

**FIGURE 9.7** Comparison of surface flux from the Fourier parabolic heat conduction and hyperbolic damped wave conduction and relaxation models.

equation and hyperbolic heat conduction equation predict the same fall of heat flux for large times. This is why beyond $\tau > \frac{1}{2}$, the predictions of parabolic and hyperbolic models are close to each other, as seen in Fig. 9.7. For short times, $\tau < \frac{1}{2}$, the microscale time effects become important and when neglected, give rise to a singularity, which also can be seen from Fig. 9.7. So the hyperbolic heat conduction model needs to be used for short-time transient applications.

The temperature solution obtained after the Chebyshev polynomial approximation for the integrand in the Baumeister and Hamill solution and further integration is shown in Fig. 9.8. The condition selected was for a typical $\tau = 5$ hyperbolic solution, and was plotted using a Microsoft Excel spreadsheet. This is shown in Fig. 9.8. The expression for temperature developed by using the method of relativistic transformation for the same condition of $\tau = 5$ is also shown in Fig. 9.8.

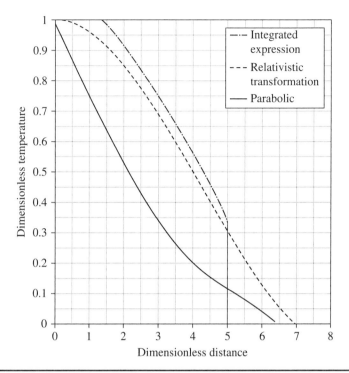

**FIGURE 9.8** Temperature distribution in a semi-infinite medium by damped wave conduction and relaxation $\tau = 8$ and parabolic Fourier heat conduction.

It can be seen that both the Baumeister and Hamill solution and the solution from the relativistic transformation are close to each other, within an average of 12 percent deviation from each other. It can also be seen that close to the surface or far from the wavefront the numerical errors expected from the Chebyshev polynomial approximation are large. For such conditions, the expression developed by the method of relativistic transformation may be used. For conditions close to the wavefront, the further integrated expression developed in this study may be used. The penetration dimensionless distance for $\tau = 5$, beyond which there is no expected heat transfer, is given by Eq. (9.116), and is 6.94 by the method of relativistic transformation.

The Baumeister and Hamill solution is only for $\tau > X$. Both the solutions for transient temperature for the damped wave conduction and relaxation hyperbolic equation from the method of Laplace transforms and Chebyshev economization and the method of relativistic transformation are compared against the prediction for transient temperature by the Fourier parabolic heat conduction model. The transient temperature from the Chebyshev economization was found to be within 25 percent of the error function solution for the parabolic Fourier heat conduction model. The hyperbolic model solutions compare well with the Fourier model solution for transient temperatures close to the wavefront and close to the surface (to within 15 percent of each other). The deviations are at the intermediate values.

### 9.4.3 Method of Relativistic Transformation of Coordinates in an Infinite Cylindrical Medium

Consider a fluid at an initial temperature $T_0$. The surface of the cylinder is maintained at a constant temperature $T_s$ for times greater than zero. The heat propagative velocity is given as the square root of the ratio of the thermal diffusivity and relaxation time: $V_h = \text{sqrt}(\alpha/\tau_r)$. The two time conditions, initial and final, and the two boundary conditions are:

$$t = 0, r > R, T = T_0 \tag{9.118}$$

$$t > 0, r = R, T = T_s \tag{9.119}$$

$$r = \infty, t > 0, T = T_0 \tag{9.120}$$

The governing equation in temperature is obtained by eliminating the second cross-derivative of heat flux with respect to $r$ and $t$ between the non-Fourier damped wave heat conduction and relaxation equation and the energy balance equation in cylindrical coordinates (Fig. 9.9). Considering a cylindrical shell of thickness $\Delta r$:

$$\Delta t(\, 2\pi rL\, q_r - 2\pi(r + \Delta r)L\, q_{r+\Delta r}) = ((\rho C_p)\, 2\pi Lr\Delta r\, \Delta T) \tag{9.121}$$

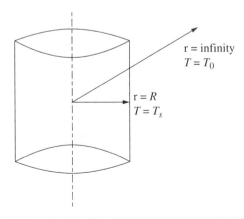

r = infinity
$T = T_0$

r = R
$T = T_s$

**Figure 9.9** Semi-infinite medium in cylindrical coordinates heated from a cylindrical surface.

In the limit of $\Delta r$, $\Delta t$ going to zero, the energy balance equation in cylindrical coordinates becomes:

$$-\frac{\partial(rq_r)}{r\partial r} = \left(\frac{\rho C_p \partial T}{\partial t}\right)$$

(9.122)

The generalized Fourier heat conduction and relaxation equation is:

$$q_r = -k\frac{\partial T}{\partial r} - \tau_r\frac{\partial q_r}{\partial t}$$

(9.123)

Multiplying Eq. (9.123) by $r$ and differentiating with respect to $r$ and then dividing by $r$:

$$\frac{\partial(rq_r)}{r\partial r} = -\frac{k}{r}\frac{\partial}{\partial r}\left(\frac{r\partial T}{\partial r}\right) - \frac{\tau_r}{r}\frac{\partial^2(rq_r)}{\partial t\partial r}$$

(9.124)

Differentiating Eq. (9.125) with respect to $t$:

$$-\frac{1}{r}\frac{\partial^2(rq_r)}{\partial t\partial r} = (\rho C_p)\frac{\partial^2 T}{\partial t^2}$$

(9.125)

Substituting Eqs. (9.125) and (9.124) into Eq. (9.122), the governing equation in temperature is obtained as:

$$(\rho C_p\tau_r)\frac{\partial^2 T}{\partial t^2} + (\rho C_p)\frac{\partial T}{\partial t} = \frac{k}{r}\frac{\partial}{\partial r}\left(r\frac{\partial T}{\partial r}\right)$$

(9.126)

Obtaining the dimensionless variables:

$$u = \frac{(T - T_0)}{(T_s - T_0)}; X = \frac{r}{\sqrt{4\alpha t}}; \tau = \frac{t}{\tau_r}$$

(9.127)

The governing equation in the dimensionless form can be written as:

$$\frac{\partial u}{\partial \tau} + \frac{\partial^2 u}{\partial \tau^2} = \frac{\partial^2 u}{\partial X^2} + \frac{1}{X}\frac{\partial u}{\partial X}$$

(9.128)

The damping term is removed from the governing equation. This is done realizing that the transient temperature decays with time in an exponential fashion. The other reason for this maneuver is to study the wave equation without the damping term. Let $u = w\exp(-\tau/2)$, and the damping component of the equation is removed to yield:

$$\frac{-w}{4} + \frac{\partial^2 w}{\partial \tau^2} = \frac{\partial^2 w}{\partial X^2} + \frac{1}{X}\frac{\partial w}{\partial X}$$

(9.129)

Equation (9.129) can be solved by using the method of relativistic transformation of coordinates. Consider the transformation variable $\eta$ as $\eta = \tau^2 - X^2$ for $\tau > X$. The governing equation becomes a Bessel differential equation for wave temperature:

$$\frac{\partial^2 w}{\partial \eta^2} 4(\tau^2 - X^2) + 6\frac{\partial w}{\partial \eta} - \frac{w}{4} = 0 \qquad (9.130)$$

$$\eta^2 \frac{\partial^2 w}{\partial \eta^2} + \frac{3}{2} - \eta\frac{\partial w}{\partial \eta} - \eta\frac{w}{16} = 0 \qquad (9.131)$$

Comparing Eq. (9.131) with the generalized Bessel equation, the solution is:

$$a = 3/2; \, b = 0; \, c = 0; \, d = -1/16; \, s = \tfrac{1}{2}$$

The order $p$ of the solution is then $p = 2 \, \text{sqrt}(1/16) = \tfrac{1}{2}$:

$$w = c_1 \frac{I_{1/2}\left(\frac{1}{2}\sqrt{\tau^2 - X^2}\right)}{(\tau^2 - X^2)^{1/4}} + c_2 \frac{I_{-1/2}\left(\frac{1}{2}\sqrt{\tau^2 - X^2}\right)}{(\tau^2 - X^2)^{1/4}} \qquad (9.132)$$

$c_2$ can be seen to be zero, as $W$ is finite and not infinitely large at $\eta = 0$. $c_1$ can be eliminated between the boundary condition at the surface and Eq. 9.133. It can be noted that this is a mild function of time, however. The general solution of the PDE consists of $n$ arbitrary functions when the order of the PDE is $n$ compared with $n$ arbitrary constants for ODE. From the boundary condition at $X = X_R$:

$$1 = \exp(-\tau/2) \, c_1 I_{1/2} \, (1/2 \, \text{sqrt}(\tau^2 - X_R^2)/(\tau^2 - X_R^2))^{1/4} \qquad (9.133)$$

$$u = [(\tau^2 - X_R^2)^{1/4}/(\tau^2 - X^2)^{1/4}][I_{1/2} \, (1/2 \, \text{sqrt}(\tau^2 - X^2)/ \\ I_{1/2} \, (1/2 \, \text{sqrt}(\tau^2 - X_R^2))] \qquad (9.134)$$

In terms of elementary functions, Eq. (9.134) can be written as:

$$u = \frac{(\tau^2 - X_R^2)^{1/4}}{(\tau^2 - X^2)^2} \frac{\text{Sinh}\left(\frac{1}{2}\sqrt{\tau^2 - X^2}\right)}{\text{Sinh}\left(\frac{1}{2}\sqrt{\tau^2 - X_R^2}\right)} \qquad (9.135)$$

In the limit of $X_R$ going to zero, the expression becomes:

$$u = \frac{\tau}{\sqrt{\tau^2 - X^2}} \frac{\text{Sinh}\left(\frac{1}{2}\sqrt{\tau^2 - X^2}\right)}{\text{Sinh}\left(\frac{\tau}{2}\right)} \quad \text{for } \tau > X \qquad (9.136)$$

For $X > \tau$:

$$u = \frac{\left(X_R^2 - \tau^2\right)^{1/4}}{(X^2 - \tau^2)^{1/2}} \frac{J_{1/2}\left(\frac{1}{2}\sqrt{X^2 - \tau^2}\right)}{I_{1/2}\left(\frac{1}{2}\sqrt{\tau^2 - X^2}\right)} \qquad (9.137)$$

Equation (9.137) can be written in terms of trigonometric functions as:

$$u = \frac{\left(X_R^2 - \tau^2\right)^{1/4}}{(X^2 - \tau^2)^{1/2}} \frac{\mathrm{Sin}\left(\frac{1}{2}\sqrt{X^2 - \tau^2}\right)}{\mathrm{Sinh}\left(\frac{1}{2}\sqrt{\tau^2 - X^2}\right)} \qquad (9.138)$$

In the limit of $X_R$ going to zero, the expression becomes:

$$u = \frac{\tau}{\sqrt{X^2 - \tau^2}} \frac{\mathrm{Sin}\left(\frac{1}{2}\sqrt{X^2 - \tau^2}\right)}{\mathrm{Sinh}\left(\frac{\tau}{2}\right)} \qquad (9.139)$$

The dimensionless temperature at a point in the medium at $X = 7$, for example, is considered and shown in Fig. 9.10. Three different regimes can be seen. The first regime is that of the thermal lag and consists of no change from the initial temperature. The second regime is when:

$$\tau_{\mathrm{lag}}^2 = X^2 - 4\pi^2 \text{ or } \tau_{\mathrm{lag}} = \mathrm{sqrt}(X_p^2 - 4\pi^2) = 3.09 \qquad \text{when } X_p = 7 \qquad (9.140)$$

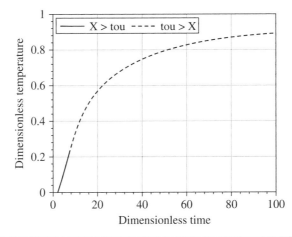

**FIGURE 9.10** Transient temperature at a point $X = 7$ in the infinite medium.

For times greater than the time lag and less than $X_p$, the dimensionless temperature is given by Eq. (9.134). For dimensionless times greater than 7, the dimensionless temperature is given by Eq. (9.137). For distances closer to the surface compared with $2\pi$, the time lag will be zero.

### 9.4.4 Relativistic Transformation of Spherical Coordinates in an Infinite Medium

Consider a fluid at an initial temperature $T_0$. The surface of a solid sphere is maintained at a constant temperature $T_s$ for times greater than zero (Fig. 9.11). The heat propagative velocity is given as the square root of the ratio of the thermal diffusivity and relaxation time: $v_h = \mathrm{sqrt}(\alpha/\tau_r)$.

The two time conditions, initial and final, and the two boundary conditions are:

$$t = 0,\, r > R,\, T = T_0 \tag{9.141}$$

$$t = \infty,\, T = T_s \text{ for all } R \tag{9.142}$$

$$t > 0,\, r = R,\, T = T_s \tag{9.143}$$

$$r = \infty,\, t > 0,\, T = T_0 \tag{9.144}$$

The governing equation in temperature is obtained by eliminating the second cross-derivative of heat flux with respect to $r$ and $t$ between the non-Fourier damped wave heat conduction and relaxation equation and the energy balance equation in spherical coordinates. Considering a shell of thickness $\Delta r$ at a distance $r$ from the center of the solid sphere:

$$\Delta t(4\pi r^2\, q_r - 4\pi(r + \Delta r)^2\, q_{r+\Delta r}) = ((\rho C_p)\, 4\pi r^2 \Delta r\, \Delta T) \tag{9.145}$$

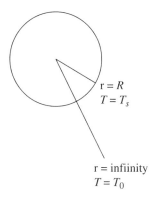

**FIGURE 9.11**
Infinite medium heated from a solid spherical surface.

$r = R$
$T = T_s$

$r = \text{infiinity}$
$T = T_0$

Dividing Eq. (9.145) throughout with $\Delta r \Delta t$, and in the limit of $\Delta r$, $\Delta t$ going to zero, the energy balance equation in cylindrical coordinates becomes:

$$-\frac{\partial(rq_r)}{r\partial r} = \left(\frac{\rho C_p \partial T}{\partial t}\right) \tag{9.146}$$

The generalized Fourier heat conduction and relaxation equation is:

$$q_r = -k\frac{\partial T}{\partial r} - \tau_r\frac{\partial q_r}{\partial t} \tag{9.147}$$

Combining Eqs. (9.146) and (9.147), the governing equation in temperature can be written as:

$$(\rho C_p \tau_r)\frac{\partial^2 T}{\partial t^2} + (\rho C_p)\frac{\partial T}{\partial t} = \frac{k}{r^2}\frac{\partial}{\partial r}\left(r^2\frac{\partial T}{\partial r}\right) \tag{9.148}$$

Obtaining the dimensionless variables:

$$u = \frac{(T - T_0)}{(T_s - T_0)}; X = \frac{r}{\sqrt{4\alpha t}}; \tau = \frac{t}{\tau_r} \tag{9.149}$$

The governing equation in the dimensionless form can be written as:

$$\frac{\partial u}{\partial \tau} + \frac{\partial^2 u}{\partial \tau^2} = \frac{\partial^2 u}{\partial X^2} + \frac{2}{X}\frac{\partial u}{\partial X} \tag{9.150}$$

The damping term is removed from the governing equation. This is done realizing that the transient temperature decays with time in an exponential fashion. The other reason for this maneuver is to study the wave equation without the damping term. Let $u = w \exp(-\tau/2)$, and the damping component of the equation is removed to yield:

$$\frac{-w}{4} + \frac{\partial^2 w}{\partial \tau^2} = \frac{\partial^2 w}{\partial X^2} + \frac{2}{X}\frac{\partial w}{\partial X} \tag{9.151}$$

Equation (9.151) can be solved by using the method of relativistic transformation of coordinates. Consider the transformation variable $\eta$ as $\eta = \tau^2 - X^2$ for $\tau > X$. The governing equation becomes a Bessel differential equation for wave temperature:

$$\frac{\partial^2 w}{\partial \eta^2} 4(\tau^2 - X^2) + 8\frac{\partial w}{\partial \eta} - \frac{w}{4} = 0 \tag{9.152}$$

$$\eta^2\frac{\partial^2 w}{\partial \eta^2} + 2\eta\frac{\partial w}{\partial \eta} - \eta\frac{w}{16} = 0 \tag{9.153}$$

Comparing Eq. (9.153) with the generalized Bessel equation, the solution is:

$$a = 2; b = 0; c = 0; d = -1/16; s = \frac{1}{2}$$

The order $p$ of the solution is then $p = 1$; $sqrt(d/s) = \frac{1}{2}$:

$$w = c_1 \frac{I_1\left(\frac{1}{2}\sqrt{\tau^2 - X^2}\right)}{(\tau^2 - X^2)^{1/4}} + c_2 \frac{I_{-1}\left(\frac{1}{2}\sqrt{\tau^2 - X^2}\right)}{(\tau^2 - X^2)^{1/4}} \tag{9.154}$$

$c_2$ can be seen to be zero, as $W$ is finite and not infinitely large at $\eta = 0$. $c_1$ can be eliminated between the boundary condition at the surface and Eq. (9.154). It can be noted that this is a mild function of time, however. The general solution of the PDE consists of $n$ arbitrary functions when the order of the PDE is $n$ compared with $n$ arbitrary constants for the ODE. From the boundary condition at $X = X_R$:

$$1 = \exp(-\tau/2)\, c_1 I_{1/2}\, (1/2\, sqrt(\tau^2 - X_R^2)/(\tau^2 - X_R^2)^{1/4}) \tag{9.155}$$

$$u = \left(\frac{\tau^2 - X_R^2}{\tau^2 - X^2}\right)^{1/2} \frac{I_1\left(\frac{1}{2}\sqrt{\tau^2 - X^2}\right)}{I_1\left(\frac{1}{2}\sqrt{\tau^2 - X_R^2}\right)} \tag{9.156}$$

This is applicable for $\tau > X$.

For $X > \tau$, the solution can be written as:

$$u = \left(\frac{\tau^2 - X_R^2}{\tau^2 - X^2}\right)^{1/4} \frac{J_1\left(\frac{1}{2}\sqrt{X^2 - \tau^2}\right)}{J_1\left(\frac{1}{2}\sqrt{X_R^2 - \tau^2}\right)} \tag{9.157}$$

Equation (9.157) can be written for $X > \tau$. For $X = \tau$, the solution at the wavefront is the result. This can be obtained by solving Eq. (9.152) at $\eta = 0$. In the limit of $X_R$ going to zero:
for $\tau > X$.

$$u = \left(\frac{\tau}{\sqrt{X^2 - \tau^2}}\right) \frac{I_1\left(\frac{1}{2}\sqrt{\tau^2 - X^2}\right)}{I_1\left(\frac{\tau}{2}\right)} \tag{9.158}$$

For $X > \tau$:

$$u = \left(\frac{\tau}{\sqrt{X^2 - \tau^2}}\right) \frac{J_1\left(\frac{1}{2}\sqrt{X^2 - \tau^2}\right)}{I_1\left(\frac{\tau}{2}\right)} \tag{9.159}$$

Seventeen terms were taken in the series expansion of the modified Bessel composite function of the first kind and first order and the Bessel composite function of the first kind and first order, respectively, and the results plotted in Fig. 9.12 for a given $X_p = 9$ using a Microsoft Excel spreadsheet on a Pentium IV desktop microcomputer. Three regimes can be identified. The first regime is that of the thermal lag and consists of no change from the initial temperature. The second regime is when:

$$\tau_{lag}^2 = X^2 - (7.6634)^\wedge 2 \text{ or } \tau_{lag} = sqrt(X_p^2 - 7.6634^2)$$
$$= 4.72 \text{ when } X_p = 9 \tag{9.160}$$

The first zero of $J_1(x)$ occurs at $x = 3.8317$. The 7.6634 is twice the first root of the Bessel function of the first order and first kind. For times greater than the time lag and less than $X_p$, the dimensionless temperature is given by Eq. (9.160). For dimensionless times greater than 9, the dimensionless temperature is given by Eq. (9.159). For distances closer to the surface compared with 7.6634 $sqrt(\alpha\tau_r)$, the thermal lag time will be zero. The ballistic term manifests as a thermal lag at a given point in the medium.

The parabolic Fourier model and hyperbolic model for transient heat flux at the surface for the problem of transient heat conduction in a semi-infinite medium subject to constant surface temperature boundary condition was found to be within 10 percent of each other for times $t > 2\tau_r$ (Fig. 9.7). This checks out with the Boltzmann transformation—the hyperbolic governing equation reverts to the parabolic at long times. At short times, there is a "blow-up" in the parabolic model. In the hyperbolic model there is no singularity. This has significant implications in several industrial applications, such as fluidized bed heat transfer, CPU overheating, gel acrylamide electrophoresis, etc.

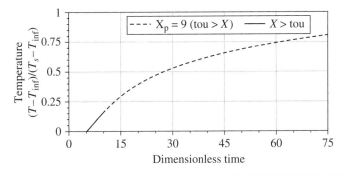

**Figure 9.12**  Transient temperature at a point $X = 9$ in the infinite spherical medium.

The solution developed by Baumeister and Hamill by the method of Laplace transforms was further integrated into a useful expression. A Chebyshev polynomial approximation was used to approximate the integrand with a modified Bessel composite function of space and time of the first kind and first order. The error involved in Chebyshev economization was $4.1 \ 10^{-5} \ \eta\xi$. The useful expression for transient temperature was shown in Fig. 9.8 for a typical time of $\tau = 5$. The dimensionless temperature as a function of dimensionless distance also is shown in Fig. 9.8. The predictions from Baumeister and Hamill and the solution by the method of relativistic transformation are within 12 percent of each other on the average. Close to the wave-front, the error in the Chebyshev economization is expected to be small and verified accordingly. Close to the surface, the numerical error involved in the Chebyshev economization can be expected to be significant. This can be seen in Fig. 9.8 close to the surface. The method of relativistic transformation yields bounded solutions without any singularities. The transformation variable $\psi$ is symmetric with respect to space and time. It transforms the PDE that governs the wave temperature into a Bessel differential equation. Three regimes are identified in the solution: an inertial zero-transfer regime, a regime characterized by Bessel composite function of the zeroth order and first kind in space and time, and a third regime characterized by modified Bessel composite function of the zeroth order and first kind in space and time.

Earlier attempts by other investigators to obtain an analytical solution for the damped wave conduction and relaxation equation in an infinite cylindrical medium were made by using the method of Laplace transformation. Singularities were found in the results for a step-change in temperature at the surface. In this study, the method of relativistic transformation is used in order to obtain an analytical solution to infinite cylindrical coordinates for the case of a step-change in boundary temperature. The transformation $\eta = \tau^2 - X^2$ was found to transform the governing equation in wave temperature into a Bessel differential equation in one variable, that is, the transformation variable. This was done for the case of an infinite spherical medium as well. The governing equation for wave temperature from the governing equation for transient temperature can be obtained either by multiplying the transient temperature equation with $\exp(\tau/2)$ or removing the damping component from the governing equation by a $u = w \exp(-\tau/2)$ substitution. The analytical solution for an infinite cylinder was characterized by a modified Bessel composite function in space and time of the first kind and half-order in the open interval of $\tau > X$. This is when the wave speed ($\sim r/t$) is smaller than the diffusion speed $\sqrt{\alpha/\tau_r}$. For values of times less than the dimensionless distance $X$, the solution is characterized by a Bessel composite function in space and time of the first kind and half order. This is when the wave speed is greater than the diffusion speed. The inertial time-lagging regime marked the third regime of transfer. For the infinite sphere, the solutions were characterized by a modified Bessel composite function in space and time of the first kind and first order, and by a Bessel composite function in space and time of the first kind and first order for the open intervals of $\tau > X$ and $X > \tau$. The initial condition can be verified in the asymptotic limits of zero time. The transformation variable is symmetric with respect to space and time. No singularities were found in the analytical solutions for semi-infinite slab, infinite cylinder, and infinite sphere.

## 9.5    Finite Slab and Taitel Paradox

Taitel [9] considered a finite slab (Fig. 9.13) with two boundaries of width 2a heated from both sides. Both the sides are maintained at a constant temperature $T_s$ for times $t > 0$. At initial time $t = 0$, the temperature at all points in the slab is $T_0$. The governing equation is given by Eq. (9.161). The four conditions used by Taitel—two in space and two in time—that are needed to completely describe a hyperbolic PDE that is second order with respect to space and with respect to time are:

$$\frac{\partial u}{\partial \tau} + \frac{\partial^2 u}{\partial \tau^2} = \frac{\partial^2 u}{\partial X^2} \tag{9.161}$$

where $u = T - T_s / T_0 - T_s ; \tau = t/\tau_r ; X = x/\sqrt{\alpha \tau_r}$

$$t = 0, -a < x < +a, T = T_0, \text{ or } u = 1 \tag{9.162}$$

$$t > 0, x = \pm a, T = T_s, u = 0 \tag{9.163}$$

$$t = 0, \partial u/\partial \tau = 0 \tag{9.164}$$

Taitel solved for Eq. (9.161), and for the conditions stated previously, obtained the analytical solution for damped wave conduction and

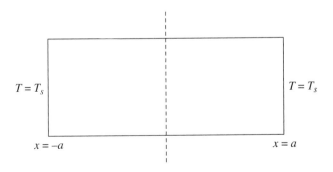

**FIGURE 9.13**    Finite slab with two boundaries heated from both sides.

relaxation in a finite slab. The solution obtained by Taitel for the centerline temperature of the finite slab follows. They considered a constant wall temperature, and the initial time conditions included a $\partial T / \partial t = 0$ term in addition to the initial temperature condition. The solution they presented is as follows:

$$u = \sum_0^\infty b_n \exp(-\tau/2) \exp(-\tau/2 \text{sqrt}(1 - 4(2n+1)^2 \pi^2 \alpha \tau_r / a^2))$$
$$+ \sum_0^\infty c_n \exp(-\tau/2) \exp(+\tau/2 \text{sqrt}(1 - 4(2n+1)^2 \pi^2 \alpha \tau_r / a^2)) \qquad (9.165)$$

Multiplying both sides of Eq. (9.166) by $\exp(\tau/2)$:

$$u\exp(\tau/2) = W = \sum_0^\infty b_n \exp(-\tau/2 \text{sqrt}(1 - 4(2n+1)^2 \pi^2 \alpha \tau_r / a^2))$$
$$+ \sum_0^\infty c_n \exp(+\tau/2 \text{sqrt}(1 - 4(2n+1)^2 \pi^2 \alpha \tau_r / a^2)) \qquad (9.166)$$

At infinite times, the LHS of Eq. (9.166) is zero times $\infty$ and is zero. The RHS does not vanish. For certain values of space and time, Taitel found that the analytical solution predicted values of temperature above the surface temperature. This is referred to in the literature as the temperature overshoot paradox. The temperature overshoot may be as a result of the growing exponential term in Eq. (9.165).

## 9.5.1 Final Condition in Time for a Finite Slab

Consider the finite slab shown in Fig. 9.13 subject to the following four conditions—two in space and two in time—that are required to complete a problem in a hyperbolic PDE that is second order with respect to space and second order with respect to time:

$$t = 0, -a < x < +a, T = T_0, \text{ or } u = 1 \qquad (9.167)$$
$$t > 0, x = \pm a, T = T_s, u = 0 \qquad (9.168)$$
$$t = \infty, u = 0 \qquad (9.169)$$

Equation (9.167) is the final condition in time. Equation (9.161) is now solved for as follows.

Multiplying throughout Eq. (9.161) by $\exp(n\tau)$:

$$\frac{\partial^2 (u e^{n\tau})}{\partial X^2} = e^{n\tau} \frac{\partial u}{\partial \tau} + e^{n\tau} \frac{\partial^2 u}{\partial \tau^2} \qquad (9.170)$$

Let $w = u e^{n\tau}$. Then:

$$\frac{\partial w}{\partial \tau} = e^{n\tau} \frac{\partial u}{\partial \tau} + n e^{n\tau} u = nw + e^{n\tau} \frac{\partial u}{\partial \tau}$$
$$\frac{\partial^2 w}{\partial \tau^2} = n \frac{\partial w}{\partial \tau} + n e^{n\tau} \frac{\partial u}{\partial \tau} + e^{n\tau} \frac{\partial^2 u}{\partial \tau^2} \qquad (9.171)$$

Combining Eqs. (9.171) and (9.170):

$$\frac{\partial^2 w}{\partial X^2} = \frac{\partial w}{\partial \tau} - nw + \frac{\partial^2 w}{\partial \tau^2} - 2n\frac{\partial w}{\partial \tau} + n^2 w \qquad (9.172)$$

For $n = \frac{1}{2}$, Eq. (9.172) becomes:

$$\frac{\partial^2 w}{\partial X^2} = \frac{\partial^2 w}{\partial \tau^2} - \frac{w}{4} \qquad (9.173)$$

The $w$ in Eq. (9.173) is the wave temperature. Equation (9.173) can be solved by the method of separation of variables:

Let $\qquad\qquad u = V(\tau)\,\phi(X) \qquad\qquad (9.174)$

Equation (9.173) becomes:

$$\phi''(X)/\,\phi(X) = (V'(\tau) + V''(\tau))/V(\tau) = -\lambda_n^2 \qquad (9.175)$$

$$\phi(X) = c_1\text{Sin}(\lambda_n X) + c_2\text{Cos}(\lambda_n X) \qquad (9.176)$$

From the boundary conditions:

$$\text{At } X = 0,\ \partial\phi/\partial X = 0, \qquad \text{so, } c_1 = 0 \qquad (9.177)$$

$$\phi(X) = c_1\text{Cos}(\lambda_n X) \qquad (9.178)$$

$$0 = c_1\text{Cos}(\lambda_n X_a) \qquad (9.179)$$

$$(2n - 1)\pi/2 = \lambda_n X_a \qquad (9.180)$$

$$\lambda_n = (2n - 1)\pi\,\text{sqrt}(\alpha\,\tau_r)/2a,\ n = 1,2,3\ldots \qquad (9.181)$$

The time domain solution would be:

$$
\begin{aligned}
V = \exp(-\tau/2)\,(c_3\exp(\text{sqrt}(1/4 -\lambda_n^2)\,\tau\,) \\
+ c_4\exp(-\text{sqrt}(1/4 - \lambda_n^2)\tau))
\end{aligned}
\qquad (9.182)
$$

or $\qquad$
$$
\begin{aligned}
V\exp(\tau/2) = (c_3\exp(\text{sqrt}(1/4 - \lambda_n^2)\,\tau\,) \\
+ c_4\exp(-\text{sqrt}(1/4 - \lambda_n^2)\tau))
\end{aligned}
\qquad (9.183)
$$

From the final condition $u = 0$ at infinite time, so is $V\phi\exp(\tau/2) = W$, the wave temperature at infinite time. Although $0*\infty$ is of the indeterminate form, the compound function $V\exp(\tau/2)$ can be transformed into 0/0 the form and shown to become zero in the limit of infinite time. The wave temperature is that portion of the solution that remains after dividing the damping component either from the solution or the governing equation. For any nonzero $\phi$, it can be seen that at infinite time, the LHS of Eq. (9.183) is a product of zero and infinity and a function of $x$, and is zero. Hence, the RHS of Eq. (9.183) is also zero, and in Eq. (9.182), $c_3$ needs to be set to zero. Thus:

$$u = \sum_{1}^{\infty} c_n\exp(-\tau/2)\exp(-\text{sqrt}(1/4 - \lambda_n^2)\,\tau)\,\text{Cos}(\lambda_n X) \qquad (9.184)$$

where $\lambda_n$ is described by Eq. (9.181) and $c_n$ can be shown using the orthogonality property to be $4(-1)^{n+1}/(2n-1)\pi$. It can be seen that Eq. (9.184) is bifurcated. As the value of the thickness of the slab changes, the characteristic nature of the solution changes from monotonic exponential decay to subcritical damped oscillatory. For $a < \pi$ sqrt $(\alpha \tau_r)$, even for $n = 1$, $\lambda_n > \frac{1}{2}$. This is when the argument within the square root sign in the exponentiated time domain expression becomes negative and the result becomes imaginary. Using De Moivre's theorem and taking a real part for the small width of the slab:

$$u = \sum_{1}^{\infty} c_n \exp(-\tau/2) \, \text{Cos}(\text{sqrt}(\lambda_n^2 - 1/4) \, \tau) \, \text{Cos}(\lambda_n X) \qquad (9.185)$$

Equations (9.184) and (9.185) can be seen to be well bounded. These become zero at long times. This would be the time taken to reach steady state. Thus, for $a \geq \pi$ sqrt$(\alpha \, \tau_r)$:

$$u = \sum_{1}^{\infty} c_n \exp(-\tau/2) \, \exp(-\text{sqrt}(1/4 - \lambda_n^2) \, \tau) \, \text{Cos}(\lambda_n X) \qquad (9.186)$$

where $c_n = 4(-1)^{n+1}/(2n-1)\pi$ and $\lambda_n = (2n-1)\pi$ sqrt$(\alpha \, \tau_r)/2a$

The centerline temperature for a particular example is shown in Fig. 9.14. Eight terms in the infinite series given in Eq. (9.186) were taken and the values calculated on a 1.9-GHz Pentium IV desktop personal computer. The number of terms was decided on the incremental change or improvement obtained by doubling the number of terms. The number of terms was arrived at a 4 percent change in the dimensionless temperature.

The Taitel paradox is obviated by examining the final steady-state condition and expressing the state in mathematical terms. The $W$ term, which is the dimensionless temperature upon removal of the damping term, needs to go to zero at infinite time. This resulted in our solution, which is different from previous reports and is well bounded. The use of the *final* condition is what is needed for this

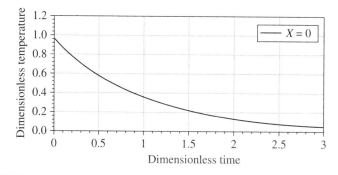

**FIGURE 9.14** Centerline temperature in a finite slab at constant wall temperature (large a) ($a = 0.86$ m, $\alpha = 10^{-5}$ m$^2$/s, $\tau_r = 15$ s).

problem to be used extensively in engineering analysis without being branded as violating the second law of thermodynamics. The conditions that were the touted violations of the second law are not physically realistic. A bifurcated solution results. For a small width of the slab, $a < \pi \sqrt{\alpha \tau_r}$, the transient temperature is subcritical damped oscillatory. The centerline temperature is shown in Fig. 9.15.

$$u = \sum_{0}^{\infty} c_n \exp(-\tau/2)\, \mathrm{Cos}(\mathrm{sqrt}(\lambda_n^2 - 1/4)\, \tau)\, \mathrm{Cos}(\lambda_n X) \qquad (9.187)$$

The subcritical damped oscillations in the centerline temperature at various values of large relaxation times are shown in Fig. 9.16. The

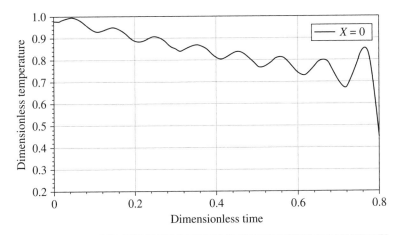

FIGURE **9.15**   Centerline temperature in a finite slab at CWT for small a (a = 0.001 m, $\alpha = 10^{-5}$ m$^2$/s; $\tau_r = 15$ s).

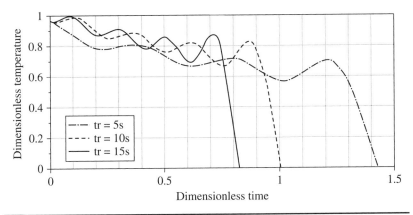

FIGURE **9.16**   Centerline temperature in a finite slab at CWT for large relaxation times (a = 0.001 m, $\alpha = 10^{-5}$ m$^2$/s).

relaxation time value that is greater than the subcritical damped oscillations can be given by:

$$\tau_r > \frac{a^2}{\pi^2 \alpha} \qquad (9.188)$$

## 9.6 Finite Sphere Subject to Constant Wall Temperature

Consider a sphere at initial temperature $T_0$. The surface of the sphere is maintained at a constant temperature $T_s$ for times greater than zero. The heat propagative velocity is given as the square root of the ratio of thermal diffusivity and relaxation time, $V_h = \sqrt{\alpha / \tau_r}$. The initial, final, and boundary conditions are:

$$t = 0, 0 \le r < R, T = T_0 \qquad (9.189)$$

$$t = \infty, 0 \le r < R, T = T_s \qquad (9.190)$$

$$t > 0, r = 0, \partial T / \partial r = 0 \qquad (9.191)$$

$$t > 0, r = R, T = T_s \qquad (9.192)$$

The governing equation can be obtained by eliminating $q_r$ between the generalized Fourier's law of heat conduction and the equation from the energy balance of in – out = accumulation. This is achieved by differentiating the constitutive equation with respect to $r$ and the energy equation with respect to $t$ and eliminating the second cross-derivative of $q$ with respect to $r$ and time. Thus:

$$\tau_r \frac{\partial^2 T}{\partial t^2} + \frac{\partial T}{\partial t} = \alpha \frac{\partial^2 T}{\partial r^2} + \frac{2}{r} \frac{\partial T}{\partial r} \qquad (9.193)$$

In the dimensionless form, Eq. (9.193) can be written in cylindrical coordinates as:

$$\frac{\partial u}{\partial \tau} + \frac{\partial^2 u}{\partial \tau^2} = \frac{\partial^2 u}{\partial X^2} + \frac{2}{X} \frac{\partial u}{\partial X} \qquad (9.194)$$

The solution is obtained by the method of separation of variables. First, the damping term is removed by the substitution $u = e^{-\tau/2} w$. With this substitution, Eq. (9.194) becomes:

$$-\frac{w}{4} + \frac{\partial^2 w}{\partial \tau^2} = \frac{\partial^2 w}{\partial X^2} + \frac{2}{X} \frac{\partial w}{\partial X} \qquad (9.195)$$

The method of separation of variables can be used to obtain the solution of Eq. (9.195):

Let $$w = V(\tau) \, \phi \, (X) \qquad (9.196)$$

Plugging Eq. (9.196) into Eq. (9.195) and separating the variables that are a function of $X$ only and $\tau$ only, the following two ODEs, one in space and another in time, are obtained:

$$\frac{d^2\phi}{dX^2} + \frac{2}{X}\frac{d\phi}{dX} + \lambda^2\phi = 0 \qquad (9.197)$$

$$\frac{d^2V}{d\tau^2} = \left(\frac{1}{4} - \lambda^2\right) = 0 \qquad (9.198)$$

The solution for Eq. (9.197) is the Bessel function of half order and first kind:

$$\phi = c_1 J_{1/2}(\lambda X) + c_2 J_{-1/2}(\lambda X) \qquad (9.199)$$

It can be seen that $c_2 = 0$ as the concentration is finite at $X = 0$. Now from the boundary condition (BC) at the surface:

$$\phi = c_1 J_{1/2}\left(\frac{\lambda R}{\sqrt{\alpha\tau_r}}\right) + c_2 J_{-1/2}\left(\frac{\lambda R}{\sqrt{\alpha\tau_r}}\right) \qquad (9.200)$$

$$\frac{\lambda_n R}{\sqrt{\alpha\tau_r}} = (n-1)\pi \text{ for } n = 2, 3.4\ldots \qquad (9.201)$$

The solution for Eq. (9.198) is the sum of two exponentials in time, one that decays with time and another that grows exponentially with time:

$$V = c_3 \exp\left(\tau\sqrt{0.25 - \lambda_n^2}\right) + c_4 \exp\left(-\tau\sqrt{0.25 - \lambda_n^2}\right) \qquad (9.202)$$

The term containing the positive exponential power exponent will drop out because with increasing time, the system may be assumed to reach steady state and the points within the sphere will always have temperature values less than that at the boundary. From the final condition in time, that is, at steady state:

$$w = ue^{\tau/2} \qquad (9.203)$$

Thus, $w$ will have to be zero at infinite time. Therefore, $c_3$ in Eq. (9.202) is found to be zero. The term containing the positive exponential power exponent will drop out because with increasing time, the system may be assumed to reach steady state and the points within the sphere will always have temperature values less than at the boundary:

Thus:
$$V = c_4 \exp\left(-\tau\sqrt{0.25 - \lambda_n^2}\right) \qquad (9.204)$$

or
$$u = \sum_{0}^{\infty} c_n J_{1/2}(\lambda_n X) \exp\left(\frac{-\tau}{2} - \tau\sqrt{0.25 - \lambda_n^2}\right) \qquad (9.205)$$

The $c_n$ can be solved for from the initial condition by using the principle of orthogonality for Bessel functions. At time zero, the LHS and RHS are multiplied by $J_{1/2}(\lambda_m X)$. Integration between the limits of 0 and $R$ is performed. When $n$ is not $m$, the integral is zero from the principle of orthogonality. Thus, when $n = m$:

$$c_n = \frac{-\int_0^R J_{1/2}(\lambda_n X)}{\int_0^R J_{1/2}^2(c\lambda_n X)} \qquad (9.206)$$

It can be noted from Eq. (9.205) that when:

$$1/4 < \lambda_n^2 \qquad (9.207)$$

the solution will be periodic with respect to the time domain. This can be obtained by using De Moivre's theorem and obtaining the real part to $\exp(-i\tau\sqrt{\lambda_n^2 - 0.25})$. Thus, for materials with relaxation times greater than a certain limiting value, the solution for temperature will exhibit subcritical damped oscillations. Therefore:

$$\tau_r > \frac{R^2}{12.57\alpha} \qquad (9.208)$$

And a bifurcated solution is obtained. From Eq. (9.207), it also can be seen that all terms in the infinite series will be periodic, that is, even for $n = 2$ when Eq. (9.208) is valid:

$$u = \sum_{0}^{\infty} c_n J_{1/2}(\lambda_n X) \cos\left(\tau\sqrt{\lambda_n^2 - 0.25}\right) \qquad (9.209)$$

Thus, the transient temperature profile in a sphere is obtained for a step-change in temperature at the surface of the sphere using the modified Fourier's heat conduction law. For materials with relaxation times greater than $R^2/12.57\alpha$, subcritical damped oscillations can be seen in the transient temperature profile. The exact solution for a transient temperature profile using finite-speed heat conduction is derived by the method of separation of variables. It is a bifurcated solution. For certain values of $\lambda$, the time portion of the solution is cosinous and damped, and for others it is an infinite series of Bessel functions of the first kind and half order and decaying exponential in time. Also, it can be shown that for terms in the infinite series with $n$ greater than 2, the contribution to the solution will be periodic for small $R$. The exact solution is bifurcated.

## 9.7  Finite Cylinder Subject to Constant Wall Temperature

Consider a cylinder at initial temperature $T_0$. The surface of the sphere is maintained at a constant temperature $T_s$ for times greater than zero. The heat propagative velocity is given as the square root of the ratio of thermal diffusivity and relaxation time, $V_h = \sqrt{\alpha / \tau_r}$. The initial, final, and boundary conditions are the same as given for the sphere. The governing equation can be obtained by eliminating $q_r$ between the generalized Fourier's law of heat conduction and the equation from energy balance of in − out = accumulation. This is achieved by differentiating the constitutive equation with respect to $r$ and the energy equation with respect to $t$ and eliminating the second cross-derivative of $q$ with respect to $r$ and time. Thus:

$$\tau_r \frac{\partial^2 T}{\partial t^2} + \frac{\partial T}{\partial t} = \alpha \frac{\partial^2 T}{\partial r^2} + \frac{\alpha}{r}\frac{\partial T}{\partial r} \tag{9.210}$$

The governing equation in the dimensionless form is then:

$$\frac{\partial u}{\partial \tau} + \frac{\partial^2 u}{\partial \tau^2} = \frac{\partial^2 u}{\partial X^2} + \frac{1}{X}\frac{\partial u}{\partial X} \tag{9.211}$$

The solution is obtained by the method of separation of variables. First the damping term is removed by the substitution $u = e^{-\tau/2}w$:

$$-\frac{w}{4} + \frac{\partial^2 w}{\partial \tau^2} = \frac{\partial^2 w}{\partial X^2} + \frac{1}{X}\frac{\partial w}{\partial X} \tag{9.212}$$

The method of separation of variables can be used to obtain the solution of Eq. (9.212):

Let 
$$w = V(\tau)\,\phi\,(X) \tag{9.213}$$

Plugging Eq. (9.213) into Eq. (9.212) and separating the variables that are a function of $X$ only and $\tau$ only, the following two ODEs, one in space and another in time, are obtained:

$$\frac{d^2\phi}{dX^2} + \frac{1}{X}\frac{d\phi}{dX} + \lambda^2\phi = 0 \tag{9.214}$$

$$\frac{d^2V}{d\tau^2} = \left(\frac{1}{4} - \lambda^2\right) = 0 \tag{9.215}$$

The solution to Eq. (9.214) can be seen to a Bessel function of the zeroth order and first kind and Bessel function of the zeroth order and second kind:

$$\phi = c_1 J_0\left(\frac{\lambda R}{\sqrt{\alpha \tau_r}}\right) + c_2 Y_0\left(\frac{\lambda R}{\sqrt{\alpha \tau_r}}\right) \tag{9.216}$$

It can be seen that $c_2 = 0$ as the temperature is finite at $X = 0$. Now from the BC at the surface:

$$\frac{\lambda_n R}{\sqrt{\alpha \tau_r}} = 2.4048 + (n-1)\pi \text{ for } n = 2,3.4... \tag{9.217}$$

The solution for Eq. (9.215) is the sum of two exponentials in time, one that decays with time and another that grows exponentially with time:

$$V = c_3 \exp\left(\tau\sqrt{0.25 - \lambda_n^2}\right) + c_4 \exp\left(-\tau\sqrt{0.25 - \lambda_n^2}\right) \tag{9.218}$$

The term containing the positive exponential power exponent will drop out because with increasing time, the system may be assumed to reach steady state and the points within the sphere will always have temperature values less than at the boundary. From the final condition in time, that is, at steady state:

$$w = ue^{\tau/2} \tag{9.219}$$

Thus, $w$ will have to be zero at infinite time. Therefore, $c_3$ in Eq. (9.218) is found to be zero. The term containing the positive exponential power exponent will drop out because with increasing time, the system may be assumed to reach steady state and the points within the sphere will always have temperature values less than at the boundary.

Thus: 
$$V = c_4 \exp\left(-\tau\sqrt{0.25 - \lambda_n^2}\right) \tag{9.220}$$

or 
$$u = \sum_0^\infty c_n J_0(\lambda_n X) \exp\left(\frac{-\tau}{2} - \tau\sqrt{0.25 - \lambda_n^2}\right) \tag{9.221}$$

The $c_n$ can be solved for from the initial condition by using the principle of orthogonality for Bessel functions. At time zero, the LHS and RHS are multiplied by $J_0(\lambda_m X)$. Integration between the limits of zero and $R$ is performed. When $n$ is not $m$, the integral is zero from the principle of orthogonality. Thus, when $n = m$:

$$c_n = \frac{-\int_0^R J_0(\lambda_n X)}{\int_0^R J_0^2(c\lambda_n X)} \tag{9.222}$$

It can be noted from Eq. (9.221) that when:

$$1/4 < \lambda_n^2 \qquad (9.223)$$

the solution will be periodic with respect to the time domain. This can be obtained by using De Moivre's theorem and obtaining the real part to $\exp\left(-i\tau\sqrt{\lambda_n^2 - 0.25}\right)$. Thus, for materials with relaxation times greater than a certain limiting value, the solution for temperature will exhibit subcritical damped oscillations, and:

$$\tau_r > \frac{R^2}{9.62\alpha} \qquad (9.224)$$

Thus, a bifurcated solution is obtained. Also from Eq. (9.221) it can be seen that all terms in the infinite series will be periodic, that is, even *for n = 2*, when Eq. (9.224) is valid:

$$u = \sum_0^\infty c_n J_0(\lambda_n X)\cos\left(\tau\sqrt{\lambda_n^2 - 0.25}\right) \qquad (9.225)$$

The transient temperature profile in a cylinder is obtained for a step-change in temperature at the surface of the cylinder using the modified Fourier's heat conduction law. For materials with relaxation times greater than $(R^2/9.62\alpha)$ where $R$ is the radius of the cylinder, subcritical damped oscillations can be seen in the transient temperature profile. The exact solution for a finite cylinder subject to constant wall temperature using finite speed heat conduction is derived by the method of separation of variables. It is a bifurcated solution. For certain values of lambda, the time portion of the solution is cosinous and damped, and for others it is an infinite series of Bessel functions of the first kind and half order and decaying exponential in time. Also, it can be shown that for terms in the infinite series with $n$ greater than 2, the contribution to the solution will be periodic for small $R$.

The temperature overshoot found in the analytical solution of Taitel for the case of a finite slab subject to constant wall temperature was a cause for alarm as a possible violation of the second law of thermodynamics. In this study, the final condition in time is posed as one of the two space conditions and two time conditions needed in order to fully describe a second-order hyperbolic partial differential equation in two variables. In addition to the initial time condition, the constraint from the steady-state attainment is translated to a fourth time condition. The wave dimensionless temperature has to become zero at steady state, and the wave temperature itself has to attain equilibrium. When this condition is applied, a growing exponential in time vanishes and a well-bounded solution results for a finite sphere and a finite cylinder. Taitel used a condition at time zero that the time derivative of the temperature will be zero. This means any initial temperature distribution. It turns out this cannot be a physically realistic fourth condition. The fourth condition in this study comes from what can be expected at steady state. The time derivative of temperature at zero time may have to be calculated from the model solution. In terms of degrees of freedom in time conditions, the constraint from steady state has to take precedence. The method of separation of variables was used to obtain the analytical solution. The solutions were found to be bifurcated for all three cases of finite slab, finite sphere, and finite cylinder.

When the relaxation time of the material under consideration becomes large, the temperature can be expected to undergo oscillations in the time domain. These oscillations were found to be subcritical damped oscillatory. For a finite sphere, when the relaxation times are greater than $R^2/(12.57\alpha)$, the solution becomes subcritical damped oscillatory from monotonic exponential decay in time and is given as an infinite Bessel series solution of the half order and first kind. For a finite cylinder, when the relaxation times are greater than $R^2/(9.62\alpha)$, the solution becomes subcritical damped oscillatory from monotonic exponential decay in time and is given as an infinite Bessel series solution of the zeroth order and first kind. For a finite slab, when the relaxation times are greater than $a^2/\pi\alpha$, the solution becomes subcritical damped oscillatory from monotonic exponential decay in time and is given by an infinite Fourier series solution.

The expressions for heat flux can be obtained from the energy balance equation and the convergence of the infinite series confirmed at the surface. Thus, the singularities found in the solution to the Fourier parabolic equations for the same geometry are now absent in the solution to the damped wave conduction and relaxation hyperbolic equations. The main conclusions from the study are:

1. The use of the final condition in time leads to bounded solutions.

2. The temperature overshoot problem can be attributed to use of a physically unrealistic time condition.

3. An analytical solution obtained for finite sphere, finite cylinder, and finite slab is found to be bifurcated.

4. For materials with large values of relaxation times, such as given in Eq. (9.187) for a finite slab, Eq. (9.208) for the case of a finite cylinder, and Eq. (9.224) for the case of finite sphere subcritical damped oscillations in temperature can be found.

## 9.8 Thermophysical Properties

The thermophysical properties of the biological tissues and other materials are provided in Table 9.2 [11]. The role of fat as an insulator under the skin in the human anatomy can be evaluated using the thermophysical properties provided in Table 9.4. Consider the

| S.No. | Substance | T (K) | Thermal Conductivity k (W/m/K) | Thermal Diffusivity α (m²/s) | Heat Capacity $C_p$ (J/Kg/K) | Relaxation Time τ, sec | Mass Density ρ (kg/m³) | Pressure (N/m²) |
|---|---|---|---|---|---|---|---|---|
| 1 | Air | 300 | 0.025 | 2.11 E-5 | 1006 | 2.457E-10 | 1.177 | 101330 |
| 2 | Water | 300 | 0.609 | 1.5 E-07 | 4183 | 1.438E-09 | 996 | 101330 |
| 3 | Bone | 298 | 0.44 | 1.5E-07 | 1440 | 3.015E-09 | 1920 | 101330 |
| 4 | Blood | 298 | 0.642 | 1.7E-07 | 3889 | 1.629E-09 | 937 | 101330 |
| 5 | Tooth enamel | 310 | 0.92 | 4.2E-07 | 750 | 1.217E-08 | 2900 | 101330 |
| 6 | Ice | 273 | 2.22 | 1.1E-06 | 2050 | 1.068E-08 | 917.6 | 101330 |
| 7 | Ethanol | 300 | 783.5 | 4.1E-04 | 2454 | 3.150E-06 | 784 | 101330 |
| 8 | Copper | 300 | 401 | 1.2E-04 | 385 | 1.027E-05 | 8930 | 101330 |
| 9 | Gold | 298 | 318 | 1.3E-04 | 129 | 2.432E-05 | 19,300 | 101330 |
| 10 | Gold | 298 | 318 | 1.3E-04 | 129 | 4.6 ms | 19,300 | 533.32 |
| 11 | Titanium | 273 | 22.4 | 9.4E-06 | 523 | 4.227E-07 | 4540 | 101330 |
| 12 | Skin | 310 | 0.442 | 1.2E-07 | 3471 | 1.25669E-09 | 1070 | 101330 |
| 13 | Fat | 298 | 0.21 | 6.9E-08 | 3258 | 6.367E-10 | 937 | 101330 |

TABLE 9.4  Thermophysical Properties of Biological Properties and Other Materials

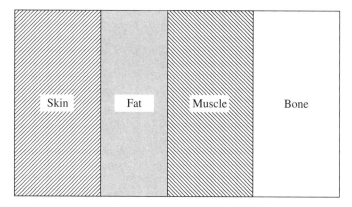

**FIGURE 9.17** Steady heat conduction through skin, fat, muscle, and bone layers.

thicknesses of skin, fat, muscle, and bone to be 2.5 mm, 10 mm, 20 mm, and 7.5 mm, respectively (Figure 9.17). The thermal conductivity of the muscle and bone is 0.5 W/m/K and 0.6 W/m/K, respectively. The effect of the layer of fat on the heat flux from the human anatomy can be evaluated as as follows;

The governing equation for steady-state temperature in the composite assembly of skin, fat, muscle, and bone can be written as follows:

$$\frac{d^2T}{dz^2} = 0 \qquad (9.226)$$

The temperature profile can be seen to be linear with respect to the space coordinate. The heat flux can be seen to be a constant through the composite assembly. The effective thermal conductivity of the composite assembly can be written as:

$$\frac{L}{k_{eff}} = \frac{L_{skin}}{k_{skin}} + \frac{L_{fat}}{k_{fat}} + \frac{L_{muscle}}{k_{muscle}} + \frac{L_{bone}}{k_{bone}} \qquad (9.227)$$

where $k_{skin}$, $k_{fat}$, $k_{muscle}$, and $k_{bone}$ are the thermal conductivities of the skin, fat, muscle, and bone, respectively. Examples of insulators of heat used to cover the human anatomy are fur, hair, and sweat. The effective thermal conductivity of hair on the human skull can be calculated from the idealized model, where the hair is reduced to a composite of cylindrical fibers aligned parallel to the axis and parallel to the flow of air. Let the thermal conductivity of fiber and air be taken as $k_{fiber}$ and $k_{air}$, respectively. The effective thermal conductivity of a composite assembly of hair and air is shown to be [10]:

$$\frac{k_{eff,zz}}{k_{air}} = 1 + \left(\frac{k_{fiber} - k_{air}}{k_{air}}\right)\phi \qquad (9.228)$$

$$\frac{k_{eff,xx}}{k_{air}} = 1 + \frac{2\phi}{\left(\left(\frac{k_{fiber} + k_{air}}{k_{fiber} - k_{air}}\right) - \phi + \left(\frac{k_{fiber} - k_{air}}{k_{fiber} + k_{air}}\right)(0.30584\phi^4 + 0.013363\phi^8 + \dots)\right)}$$

$$(9.229)$$

where $\phi$ is the volume fraction of the fibers. The preceding expressions tend to capture the role of fat and hair on the heat insulation process.

## 9.9    Warm/Cool Sensations and Thermal Wear

The effect of damped wave conduction and relaxation on warm/cool feeling of the human skin was studied by Sharma [12]. In this study a two-layer mathematical model was developed to study the transient heat conduction of the human skin and thermal fabric layer during use to protect the human body from cold weather outdoors.

A schematic of the relevant aspects from a cross-section of human skin near a finger pad is shown in Fig. 9.18. Many kinds of receptors in human skin are known to transmit information about the surroundings to the central nervous system. The role of these receptors in generating sensations caused by stimuli from the surroundings is analyzed. The response in the receptors is physicochemical in nature.

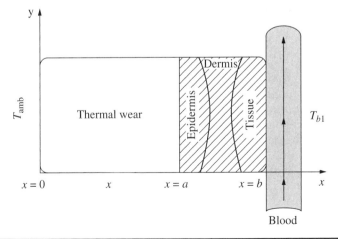

**Figure 9.18**   Transient heat conduction in human skin and thermal wear in winter.

The correlation between mechanical stimuli and sensation of touch from a neurophysiological standpoint was studied experimentally by earlier investigators. They discuss the relation between the surface roughness of fabric and the sense of touch. "Krause's end bulb" is the receptor attributed with the sensation of coolness, and "Ruffini's ending" is the receptor that is responsible for the sensation of warmth. The transient heat conduction in the neighborhood of these receptors as the outside temperature plummets to low levels, which is typical of winter in the northern part of United States, needs to be modeled. The skin layer and the thermal layer assembly are approximated, as shown in Fig. 9.1. The blood flow in the vessels results in a constant temperature environment at $x = b$ at $T = T_{bl}$, where $T_{bl}$ is the blood temperature. The origin is taken at the interface between the winter surroundings and the outer surface of the thermal wear used to protect the skin from the winter weather. The thickness of the thermal fabric is $a$, and the interface of the thermal wear and human skin occurs at $x = a$. Let the ambient temperature be $T_{amb}$. In winter, this can be expected to be much lower than the blood temperature, that is, $T_{amb} \ll T_{bl}$. The temperature difference between the blood vessels and the cold surroundings drives the problem. Let the dimensionless temperature time and penetration distance be defined with respect to the human skin parameters as follows:

$$u = \frac{T - T_{amb}}{T_{bl} - T_{amb}} \; ; \; \tau = \frac{t}{\tau_{rs}} \; ; \; X = \frac{x}{\sqrt{\alpha_s \tau_{rs}}} \qquad (9.230)$$

## 9.9.1 Steady State

At steady state, the governing equation for temperature in the skin layer can be written as follows. A heat source in the skin layer generates heat to provide the warm feeling that comes in winter. Let this heat source be dependent on the temperature difference with the ambient temperature. The energy balance on a thin slice with thickness $\Delta x$ and cross-sectional area $A$ can be written as:

$$-\frac{\partial q_x}{\partial x} + U'''(T - T_{amb}) = 0 \qquad (9.231)$$

here $U'''$ is the temperature-difference-dependent heat source. Writing the Fourier's law of heat conduction for the heat flux, $q_x$, in the dimensionless form, Eq. (9.231) may be written as:

$$\frac{\partial^2 u}{\partial X^2} = -U * u \qquad (9.232)$$

where $U* = U''' / S ; S = \rho_s C_{ps} / \tau_{rs}$

The solution for Eq. (9.233) can be written as:

$$u = A \, Sin(\sqrt{U} * X) + B \, Cos(\sqrt{U} * X) \qquad (9.233)$$

The boundary conditions at the interface of the skin tissue layer and the blood vessels can be written as:

$$\text{At } X = X_b = \frac{b}{\sqrt{\alpha_s \tau_{rs}}} \, , u = 1 \text{ since } T = T_{bl} \qquad (9.234)$$

Further, it can be assumed that there is no heat transfer from the skin layer to the blood vessels or vice versa. Thus:

$$\text{at } X = X_b = \frac{b}{\sqrt{\alpha_s \tau_{rs}}} \, , \frac{\partial u}{\partial X} = 0 \qquad (9.235)$$

Applying these boundary conditions to the integration constants, $A$ and $B$ can be solved for and found to be $Sin(\sqrt{U}*X_b)$ and $Cos(\sqrt{U}*X_a)$, respectively, and Eq. (9.234) can be seen to be:

$$u = Cos\left[\sqrt{U} *(X_b - X)\right] \qquad (9.236)$$

In a similar fashion, the governing equation for the temperature in the thermal wear layer, that is, for $0 \le x \le a$, or in terms of dimensionless distance, $0 \le X \le X_a$ can be written in the dimensionless form at steady state as follows:

$$\frac{\partial^2 u}{\partial X^2} = 0 \qquad (9.237)$$

The solution to Eq. (8) can be seen to be:

$$u = CX + D \qquad (9.238)$$

The boundary conditions for the temperature in the thermal wear layer can be written as:

$$X = 0, u = 0 \qquad (9.239)$$

$$X = X_a, u \text{ (thermal wear)} = u \text{ (skin)} = Cos(X_b - X_a) \qquad (9.240)$$

Thus, $C$ and $D$ in Eq. (9.238) can be solved for using the boundary conditions given by Eqs. (9.239) and (9.240). Thus:

$$u = \frac{xCos\left[\sqrt{U} *(X_b - X_a)\right]}{a} \qquad (9.241)$$

## 9.9.2 Transient State In Human Skin Layer

The energy balance equation in the skin layer during transient state can be written as:

$$-\frac{\partial q}{\partial x} + U'''(T - T_{bl}) = (\rho_s C_{ps})\frac{\partial T}{\partial t} \tag{9.242}$$

Combining Eq. (9.242) with the damped wave conduction and relaxation equation given by:

$$q = -k_s \frac{\partial T}{\partial x} - \tau_{rs}\frac{\partial q}{\partial t} \tag{9.243}$$

and making the terms dimensionless, the following governing equation can be written for the skin layer undergoing transient heat conduction:

$$\frac{\partial^2 u}{\partial \tau^2} + \frac{\partial u}{\partial \tau}(1 - U*) = \frac{\partial^2 u}{\partial X^2} + uU* \tag{9.244}$$

where $u$, $\tau$, $U*$, and $X$ are defined by Eq. (9.230). It can be seen that the boundary conditions are not homogeneous. Hence, the solution is assumed to take the form:

$$u = u^t + u^s \tag{9.245}$$

where $u^t$ is the transient temperature and $u^s$ = steady-state temperature. The solution for $u^s$ was solved for in the case 1 above (Fig. 9.1) and given by Eq. (9.241). The rest of the problem is obtaining the solution of the transient temperature subject to the following time and space conditions:

$$\frac{\partial^2 u^t}{\partial \tau^2} + \frac{\partial u^t}{\partial \tau}(1 - U*) = \frac{\partial^2 u^t}{\partial X^2} + u^t U* \tag{9.246}$$

The initial and final time and space conditions can be written as:

$$\text{at } \tau = 0, u^t = 1 \tag{9.247}$$

$$\text{at } \tau = \infty, u^t = 0 \tag{9.248}$$

$$\text{at } X = X_b, u^t = 0 \tag{9.249}$$

$$\text{at } X = X_a = u^t = u - u \tag{9.250}$$

Assuming that equilibrium is established rapidly at the interface, $u - u^s$ in the RHS of Eq. (9.250) can be taken as zero. Hence, the boundary condition at $X = X_a$, $u^t = 0$.

The transient temperature, $u^t$, can be expected to have an exponential decaying component, or the damping component in Eq. (9.246) can be removed by the following substitution: $u^t = w\exp(-n\tau)$.

For $n = (1 - U*)/2$, Eq. (9.246) is transformed into a governing equation of the wave temperature $w$:

$$\frac{\partial^2 w}{\partial \tau^2} - \frac{w(1+U*)^2}{4} = \frac{\partial^2 w}{\partial X^2} \tag{9.251}$$

Equation (9.251) can be solved for by the method of separation of variables. Let $w = V(\tau)\phi(X)$. Then the terms in Eq. (9.251) can be separated into two equations, one in space and one in the time domain:

$$\frac{\phi''}{\phi} = -\lambda_n^2 = \frac{V''}{V} - \frac{(1+U*)^2}{4} \tag{9.252}$$

The general solution in the space domain for the second-order ODE with constant coefficients can be seen to be:

$$\phi = c_1 \text{Sin}(\lambda_n X) + c_2 \text{Cos}(\lambda_n X) \tag{9.253}$$

The boundary conditions given by Eqs. (9.249) and (9.250) are applied to Eq. (9.254). Thus:

$$c_2 = -c_1 \text{Tan}(\lambda_n X_b) \tag{9.254}$$

$$\lambda_n = \frac{2n\pi}{(X_b - X_a)}, \; n = 1,2,3,\ldots\ldots \tag{9.255}$$

The time domain portion of the solution for the second-order ODE with constant coefficients can be written as:

$$V = c_3 \exp\left(+\tau\sqrt{\frac{(1+U*)^2}{4} - \lambda_n^2}\right)$$

$$+ c_4 \exp\left(-\tau\sqrt{\frac{(1+U*)^2}{4} - \lambda_n^2}\right) \tag{9.256}$$

From the final condition given by Eq. (9.248) at $\tau = \infty$, $w = u^t \exp$ $(+n\tau) = 0*\infty = 0$. Hence, $c_3$ can be seen to be zero. The general solution for the transient temperature can be written:

$$u = \sum_{n=1}^{\infty} c_n e^{-\tau\frac{(1-U*)}{2}} e^{-\tau\sqrt{\frac{(1+U*)^2}{4} - \lambda_n^2}} (\text{Tan}(\lambda_n X_b)(\text{Sin}(\lambda_n X)$$

$$+ \text{Cos}(\lambda_n X)) \tag{9.257}$$

$\lambda_n$ is given by Eq. (9.255). $c_n$ can be solved for from the initial condition given by Eq. (9.247) and using the principle of orthogonality. $c_n$ is found to be:

$$c_n = \frac{2(1-(-1)^n)(X_b - X_a)}{n^2\pi^2 Tan(\lambda_n X_b)} \tag{9.258}$$

### 9.9.3 Transient State In Thermal Fabric Layer

In a similar fashion, the transient temperature in the thermal wear layer can be calculated as follows. The governing equation for transient temperature in the thermal fabric layer can be written as:

$$\beta\frac{\partial^2 u}{\partial X^2} = \frac{\partial u}{\partial \tau} + \gamma\frac{\partial^2 u}{\partial \tau^2} \tag{9.259}$$

where $\beta = \alpha_f/\alpha_s$; $\gamma = \tau_{fr}/\tau_{sr}$ and $u$, $\tau$, and $X$ are the same as defined in Eq. (9.230).

It can be seen that the boundary conditions are not homogeneous. Hence, the solution is assumed to take the form:

$$u = u^t + u^s \tag{9.260}$$

where $u^t$ is the transient temperature and $u^s$ = steady-state temperature. The solution for $u^s$ was solved for in the section on steady-state previously and given by Eq. (9.241). The rest of the problem is obtaining the solution of the transient temperature subject to the following time and space conditions:

$$\gamma\frac{\partial^2 u^t}{\partial \tau^2} + \frac{\partial u^t}{\partial \tau} = \beta\frac{\partial^2 u^t}{\partial X^2} \tag{9.261}$$

The initial and final time and space conditions can be written as:

$$\text{at } \tau = 0, u^t = 1 \tag{9.262}$$

$$\text{at } \tau = \infty, u^t = 0 \tag{9.263}$$

$$\text{at } X = 0, u^t = 0 \tag{9.264}$$

$$\text{at } X = X_a = u^t = u - u^s \tag{9.265}$$

Assuming that equilibrium is established rapidly at the interface, $u - u^s$ in the RHS of Eq. (9.265) can be taken as zero. Hence, the boundary condition at $X = X_a$, $u^t = 0$.

The transient temperature, $u^t$, can be expected to have an exponential decaying component, or the damping component in Eq. (9.261) can be removed by the following substitution: $u^t = w\exp(-n\tau)$. For

$n = 1/2\gamma$, Eq. (9.261) is transformed into a governing equation of the wave temperature $w$:

$$\gamma \frac{\partial^2 w}{\partial \tau^2} - \frac{w}{4\gamma} = \beta \frac{\partial^2 w}{\partial X^2} \tag{9.266}$$

Equation (9.266) can be solved for by the method of separation of variables. Let $w = \theta(\tau)g(X)$. Then the terms in Eq. (9.266) can be separated into two equations, one in space and another in the time domain:

$$\beta \frac{g''}{g} = -\lambda_m^2 = \gamma \frac{\theta''}{\theta} - \frac{1}{4\gamma} \tag{9.267}$$

The general solution in the space domain for the second-order ODE with constant coefficients can be seen to be:

$$g = c_1 Sin(\lambda_m X) + c_2 Cos(\lambda_m X) \tag{9.268}$$

The boundary conditions given by Eqs. (9.264) and (9.265) are applied to Eq. (9.268). Thus:

$$c_2 = 0 \tag{9.269}$$

$$\lambda_m = \frac{m\pi}{X_a}, m = 1,2,3,\ldots\ldots \tag{9.270}$$

The time domain portion of the solution for the second-order ODE with constant coefficients can be written as:

$$\theta = c_5 \exp\left( +\tau \sqrt{\frac{1}{4\gamma^2} - \frac{\lambda_m^2}{\gamma}} \right) + c_6 \exp\left( -\tau \sqrt{\frac{1}{4\gamma^2} - \frac{\lambda_m^2}{\gamma}} \right) \tag{9.271}$$

From the final condition given by Eq. (9.265) at $\tau = \infty$, $w = u^i \exp(+n\tau) = 0*\infty = 0$. Hence, $c_5$ can be seen to be zero. The general solution for the transient temperature can be written:

$$u = \sum_{n=1}^{\infty} d_n e^{\frac{-\tau}{2\gamma}} e^{-\tau \sqrt{\frac{1}{4\gamma^2} - \frac{\lambda_m^2}{\gamma}}} Sin(\lambda_m X) \tag{9.272}$$

$\lambda_m$ is given by Eq. (9.270). $d_n$ can be solved for from the initial condition given by Eq. (9.262) and using the principle of orthogonality. $d_n$ is found to be:

$$d_n = \frac{2(1-(-1)^n)}{m^2\pi^2}, n = 1,2,3,\ldots. \tag{9.273}$$

The solutions to the transient heat conduction in the human skin layer and the thermal fabric layer, including the damped wave conduction and relaxation effects, were derived using the method of separation of variables. The use of the final condition in time leads to well-bounded, physically realistic solutions within the bounds of Clausius inequality. The transient temperature in the two layers at steady state is shown in Fig. 9.19. The nature of the temperature profile is cosinous in the human skin layer and linear in the thermal fabric layer. The parameters the profile is derived for are $X_a = 3$, $X_b = 5$, and $U* = 2.0$. The heat flux at steady state for the human skin layer and the thermal fabric layer is shown in Figure 9.20 for the parameters

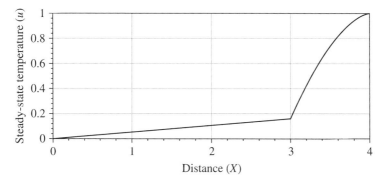

**FIGURE 9.19** Steady-state temperature in human skin layer and thermal fabric layer for $X_a = 3.0$, $X_b = 4.0$, and $U* = 2.0$.

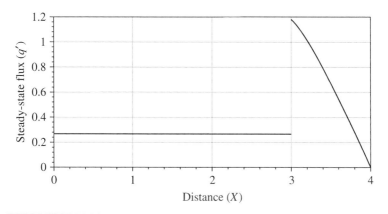

**FIGURE 9.20** Steady-state temperature in human skin layer and thermal fabric layer for $X_a = 3.0$, $X_b = 4$, $k_f/k_s = 5$, and $U* = 2.0$.

$X_a = 3$, $X_b = 5$, $k_f/k_s = 5.0$, and $U* = 2.0$. It can be seen from Fig. 9.20 that the interface heat flux undergoes a maxima. For the heat flux in dimensionless form with respect to the skin's thermophysical properties, to be continuous, the ratio of the thermal conductivities of the thermal fabric layer and skin layer have to be a certain value and cannot be specified independently. This can be deduced from model solutions. The maximum heat flux at the interface of the two layers may be related to the heat flux reported by Yoneda and Kawabata [12].

It can be seen from the model solutions that the transient temperature will undergo subcritical damped oscillations under certain conditions. These conditions are for large relaxation times of the skin and $f$ for the thermal fabric layer.

Thus, for large relaxation time values of the skin, that is, $\tau_{rs} > (1 + U*)^2 (b - a)^2 / 16\pi^2 \alpha_s$, it can be seen that the temperature in the skin layer can be expected to exhibit oscillations. The nature of the oscillations will depend on the strength of the heat source, which is a function of the temperature difference with the ambient cold winter temperature. For heat source $U* > 1$, oscillations that grow with time may be expected. For heat source $U* < 1$, subcritical damped oscillations can be expected. The solution for these materials is then given by:

$$u = \sum_{n=1}^{\infty} c_n e^{-\tau \frac{(1-U*)}{2}} \mathrm{Cos}\left( \tau \sqrt{\lambda_n^2 - \frac{(1+U*)^2}{4}} \right)$$

$$(\mathrm{Tan}(\lambda_n X_b)(\mathrm{Sin}(\lambda_n X) + \mathrm{Cos}(\lambda_n X)) \tag{9.274}$$

where $\lambda_n$ and $c_n$ are given by Eqs. (9.270) and (9.258), respectively.

Eight terms in the infinite series in Eq. (9.274) were plotted in a Microsoft Excel spreadsheet, and the dimensionless temperature is plotted as a function of dimensionless time in Fig. 9.21. It can be seen that for human skin materials with large relaxation times, as discussed previously, the dimensionless temperature exhibits growing oscillations at a heat source of $U* = 2.0$. The general solution is given by Eq. (9.274).

In a similar fashion, under certain conditions, the temperature in the thermal fabric layer can be expected to undergo subcritical damped oscillations. Thus, for:

$$\tau_{rf} > \frac{a^2}{4\pi^2 \alpha_s} \tag{9.275}$$

the transient temperature in the thermal fabric layer can expect to undergo subcritical damped oscillations. Under these conditions, the transient temperature in the thermal fabric layer is given by:

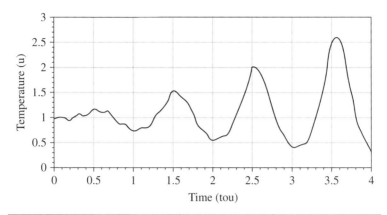

**FIGURE 9.21** Transient temperature at $X = 3.1$ in human skin layer for large relaxation time values $U^* = 2.0$.

$$u = \sum_{n=1}^{\infty} d_n e^{\frac{-\tau}{2\gamma}} \mathrm{Cos}\left(\tau\sqrt{\frac{\lambda_n^2}{\gamma} - \frac{1}{4\lambda^2}}\right) \mathrm{Sin}(\lambda_m X) \qquad (9.276)$$

where $d_n$ and $\lambda_m$ is given by Eqs. (9.273) and (9.270), respectively.

Six terms in the infinite series given by Eq. (9.276) were taken, and the dimensionless temperature was plotted against dimensionless time in the thermal layer for fabric materials with large relaxation times, as shown in Fig. 9.22. The $\gamma$ chosen for the study was 0.15, and the temperature at $X = 1.1$ was obtained for $X_a = 3.0$. It can be seen that the temperature undergoes subcritical damped oscillations.

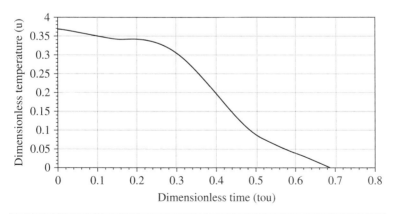

**FIGURE 9.22** Transient temperature in the thermal fabric layer for large relaxation times at $X = 1.1$.

## 9.10    Regulation of Human Anatomical Temperature

The body temperature in humans is held in dynamic balance by the generation of heat through metabolic activities within the human anatomy and by transfer of heat outside the human anatomy to the surrounding environment. The heat gain, heat storage, and heat transfer mechanisms coexist in human anatomy. A number of research studies have been undertaken to investigate this dynamic balance through the use of laser irradiation of tissue. As seen from the discussions in the previous sections, the thermophysical properties of the blood, skin, fat, and bone are different from each other. The modes of heat transfer can be molecular heat conduction, heat convection, heat radiation, and a fourth mode of heat transfer called damped wave conduction. Metabolism includes all the chemical reactions taking place within the human anatomy. Energy is liberated from chemical reactions that are exothermic. This is used to sustain life and to perform the various functions, basic and chosen. The work is done by the human anatomy. In plants, however, the energy is supplied by sunshine, which is converted to chemical energy by photosynthesis. The minimal rate of metabolism needed to sustain life is referred to as the basic rate of metabolism. This rate is obtained while the patient is awake and resting and is at a stressless state. Digestive activities should cease. The external hot weather does not cause any heat exchange or thermoregulation. There is enough energy generated at this state for the heart to pump blood throughout the human anatomy, retain normal electrical activity in the nervous system, and generate calories of energy. The basic rate of metabolism can be measured using the rate at which oxygen is consumed and the energy generated from the metabolism of oxygen. Some work done by the human anatomy is allowed. The energy needed for metabolic activity is obtained from chemical reactions that are coupled, resulting in a net decrease in free energy. The basic rate of metabolism in an average patient is roughly 75 watts. The major organs, such as brain, skeletal muscle, liver, heart, gastrointestinal tract, kidneys, lungs, etc., participate in the base metabolism. The muscles in the human skeleton require less energy at the resting state compared with the state of exercise. When the patient is asleep, the metabolic rate falls below the basic rate of metabolism. The metabolic rate of all other activities, such as walking, sitting, mating, eating, cooking, growing, etc., are higher than the basic rate of metabolism. The active rate of metabolism can exceed the basic rate by a factor of 10 to 20 during strenuous exercise.

The basic rate of metabolism in *Homo sapiens* varies with the body mass $m$ as $m^{0.75}$. The relationship between the basic rate of metabolism and human anatomical parameters can be expressed in terms of the surface to volume ratio of the patient as $(S/V)^{-1.5}$. As discussed in the earlier section, the thermodynamic efficiency can be written as:

$$\eta = \frac{|W|}{|W_{max}|} = \frac{|W|}{\Delta G} \qquad (9.277)$$

The thermodynamic efficiency can be seen from Eq. (9.277) to be a ratio of the work done by the human anatomy to the maximum work capable of the same human anatomy. Neglecting the mass transferred into the human anatomy, the maximum work capacity of the human can be said to be equal to the free energy change of the system $\Delta G$. During any real process, according to the second law of thermodynamics, as discussed earlier section in this chapter, the entropy change will always be positive [Eq. (9.6)]. It can be realized that:

$$\Delta G = \Delta H - T\Delta S \qquad (9.278)$$

It can be seen that $|W_{max}| > \Delta H$ for real processes. When the process is reversed, the entropic change is zero. Thus, the maximum thermodynamic efficiency achievable can be seen to be:

$$\eta_{max} = 1 - \frac{T_2}{T_1} \qquad (9.279)$$

Within the human anatomy, mechanisms are in place that will take effect to cool the anatomy when the average temperature reaches 35°C. The average temperature within the human anatomy is usually 37°C. When the skin temperature drops below 22.5°C, cellular mechanisms will take effect that will result in the generation of heat. The core human anatomical temperature is maintained within a narrow range by use of insulation and heat production.

Two mechanisms that can cause cooling within the human anatomy are vasodilation and evaporative cooling affected by sweat. After strenuous exercise, on account of vasodilation, the skin exterior appears a bit reddish. The blood near the skin surface is cooled and flows back to the veins and arteries, thereby affecting energy transfer. The human anatomy reduces heat loss in the temperature range of 24 to 32°C by reducing blood flow to the dermis. Below 24°C, the vasoconstriction mechanism is not sufficient and heat production is through shivering or physical activity. There appears a set point in thermoregulation. This regulatory process is a bit more complicated than a first-order feed-forward control process. Transient receptor potential (TRP) ion channels are sensitive to hot and cold temperatures. TRP channels are activated upon a control action from the hypothalamus and stimulate the nerves. Nerve signals and hormone signals result in vasodilation/vasoconstriction, or blood flow regulation and changes in metabolism and heat generation.

### 9.10.1 Bioheat Transfer Equation

The heat generated within the human anatomy on account of the several metabolic reactions and the heat transfer to the surroundings can be described using the bioheat transfer equation. This was first introduced by Pennes [15]. Combining the energy balance equation and Fourier's law of heat conduction, the bioheat transfer equation can be written as:

$$\rho C_p \frac{\partial T}{\partial t} = k\left(\frac{\partial^2 T}{\partial x^2} + \frac{\partial^2 T}{\partial y^2} + \frac{\partial^2 T}{\partial z^2}\right) + U_{met} + U_{blood} \qquad (9.280)$$

where $U_{met}$ is the heat generated per unit volume on account of metabolism and $U_{blood}$ is the heat removed per unit volume on account of blood flow. Expressions for heat removed per unit volume by blood flow can be written as:

$$U_{blood} = Q_{blood} C_{pb}(T_b - T) \qquad (9.281)$$

where $Q_{blood}$ is the volumetric rate of blood flow and $C_{pb}$ and $T_b$ are the heat capacity of the blood and temperature of the artery, respectively.

Some investigators have found that the assumption that the venous blood temperature and tissue temperature are equal may not be valid. They attempted to provide separate energy balances for arterial and venous blood and examined the distribution of capillaries. Experimental data vindicate the separate energy balances for blood and tissue. An expression for effective thermal conductivity for blood and tissue was suggested by Charny [14]:

$$\frac{k_{eff}}{k_{tissue}} = 1 + \frac{\rho_n}{\pi} \cosh^{-1}\left[\frac{d}{d_a}\right]\left(\frac{k_{bl}\pi d_a P e_{heat}}{2 k_{tissue}}\right)^2 \qquad (9.282)$$

where $d_a$ is the vessel diameter, $d$ is the spacing between blood vessels, $Pe_{heat}$ is the thermal Peclet number, and $\rho_n$ is the number density of capillaries. The Peclet number is the ratio of convection to conduction modes of heat transfer.

### 9.10.2 Damped Wave Conduction and Relaxation Effects

The damped wave conduction effects may become important in the time frame associated with heat transfer between tissue and blood. They are not considered in Eq. (9.266). Here is an attempt to account for the damped wave conduction and relaxation effects in bioheat transfer.

Consider a rod of length l, with one end maintained at temperature $T_s$. The other end is at the zero temperature (0 K). This is the lowest temperature achievable according to the third law of thermodynamics. The entropy is zero at 0 K. At time $t = 0$, the rod is at 0 K. For times

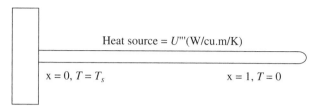

**FIGURE 9.23** Temperature-dependent heat source in a rod.

greater than zero, the temperature-dependent heat source is allowed to heat the rod. It is of interest to study the temperature distribution in the rod using the non-Fourier damped wave conduction and relaxation equation.

A temperature-dependent heat source with the strength U''' w/ m³/K is present in the rod (Figure 9.23) The energy balance on a thin section with thickness $\Delta x$ is considered at a distance $x$ from the origin for an incremental time $\Delta t$. Thus, in one dimension:

$$(q\,A\,|_x - q\,A\,|_{x+\Delta x} + U'''\,TA\,\Delta x)\,\Delta t = A\,\Delta x\,(\rho C_p)\,\Delta T \qquad (9.283)$$

Dividing throughout the equation with respect to $x$ and $t$ and taking the limits as $\Delta x, \Delta t$ goes to zero, and at a constant cross-sectional area, the energy balance equation becomes:

$$-\frac{\partial q}{\partial x} + U'''T = \rho C_p \frac{\partial T}{\partial t} \qquad (9.284)$$

The non-Fourier damped wave heat conduction and relaxation equation can be written as:

$$q = -k\frac{\partial T}{\partial x} - \tau_r \frac{\partial q}{\partial t} \qquad (9.285)$$

The governing equation for the temperature can be obtained by eliminating the heat flux between the energy balance equation and the constitutive law for heat conduction. Thus, differentiating the energy balance equation with respect to time and the constitutive equation with respect to $x$ and eliminating the second cross-derivative of flux with respect to time and space yields:

$$U'''T + k\frac{\partial^2 T}{\partial x^2} = (\rho C_p - U'''\tau_r)\frac{\partial T}{\partial t} \qquad (9.286)$$

Using the dimensionless variables:

$$u = \frac{T}{T_s}\,; \tau = \frac{t}{\tau_r}\,; X = \frac{x}{\sqrt{\alpha \tau_r}} \qquad (9.287)$$

The governing equation in temperature becomes:

$$U'''u + S\frac{\partial^2 u}{\partial X^2} = S\frac{\partial^2 u}{\partial \tau^2} + (S - U''')\frac{\partial u}{\partial \tau} \qquad (9.288)$$

Let $U* = U'''/S$, then the dimensionless governing equation can be written as:

$$U*u + \frac{\partial^2 u}{\partial X^2} = \frac{\partial^2 u}{\partial \tau^2} + (1 - U*)\frac{\partial u}{\partial \tau} \qquad (9.289)$$

where $S = (\rho C_p/\tau_r)$ is the storage coefficient. It has units of w/m$^3$/K. It can be seen from the governing equation that when $S = U'''$, the damped wave conduction and relaxation equation simplifies to a wave equation. The equation reverts to the governing equation seen for the finite slab at constant wall temperature when $U''' = 0$. Thus, when $U'''/S = 1$:

$$U*u + \frac{\partial^2 u}{\partial X^2} = \frac{\partial^2 u}{\partial \tau^2} \qquad (9.290)$$

Let
$$\eta = X^2 - \tau^2 \qquad (9.291)$$

For $X > \tau$, the governing equation will transform to:

$$U*u + 4\eta\frac{\partial^2 u}{\partial \eta^2} + 4\frac{\partial u}{\partial \eta} = 0 \qquad (9.292)$$

$$\eta^2\frac{\partial^2 u}{\partial \eta^2} + \eta\frac{\partial u}{\partial \eta} + \frac{\eta U*u}{4} = 0 \qquad (9.293)$$

Comparing Eq. 9.293 with the generalized Bessel equation; $a = 1$; $b = 0$; $s = \frac{1}{2}$; $d = U*/4$; $c = 0$. $p = 2$. sqrt(0) = 0; sqrt($d$)/$s$ = $U*^{1/2}$.

$$u = c_1 J_0\left(\sqrt{U*(X^2 - \tau^2)}\right) + c_2 Y_0\left(\sqrt{U*(X^2 - \tau^2)}\right) \qquad (9.294)$$

$c_2$ can be seen to be zero, as $u$ is finite at zero $\eta$.

Thus:
$$u = c_1 J_0\left(\sqrt{U*(X^2 - \tau^2)}\right) \qquad (9.295)$$

This function exhibits damped wave behavior. This is valid until the first zero.

At the first zero of the Bessel function:

$$5.7831 = U*(X^2 - \tau^2) \qquad (9.296)$$

$$X \geq \sqrt{\frac{5.7831}{U*} + \tau^2} \qquad (9.297)$$

The temperature $u$ will be zero. For short times, a good portion of the rod will not have any temperature, even at an infinite heat source strength. This is a clear manifestation of the finite speed propagation of the heat. For $\tau > X$:

$$u = c_1 I_0 \sqrt{U * (\tau^2 - X^2)} \tag{9.298}$$

From the boundary condition at $X = 0$:

$$1 = c_1 I_0 \left( \sqrt{U * \tau^2} \right) \tag{9.299}$$

Eliminating $c_1$ between the two equations, an approximate solution for $u$ can be written as:

$$u = \frac{I_0 \left( \sqrt{U * (\tau^2 - X^2)} \right)}{I_0 \left( \sqrt{U * \tau^2} \right)} \tag{9.300}$$

The general solution for the temperature in the rod with a temperature-dependent heat source can be obtained as follows. Let the solution be expressed as a sum of steady-state and transient-state components of the dimensionless temperature.

Let

$$u = u^{ss} + u^{\tau} \tag{9.301}$$

Then Eq. (9.289) can be written as:

$$U * u^{ss} + \frac{\partial^2 u^{ss}}{\partial X^2} = \frac{\partial^2 u^{\tau}}{\partial \tau^2} + (1 - U*) \frac{\partial u^{\tau}}{\partial \tau} - U * u^{\tau} + \frac{\partial^2 u^{\tau}}{\partial X^2} \tag{9.302}$$

The steady-state component will obey the equation:

$$U * u^{ss} + \frac{\partial^2 u^{ss}}{\partial X^2} = 0 \tag{9.303}$$

with the boundary conditions:

$$X = 0, \ u^{ss} = 1 \tag{9.304}$$
$$X = X_1, \ u^{ss} = 0 \tag{9.305}$$

The solution to the second-order ODE will then be:

$$u^{ss} = c' \text{Sin}(U*^{1/2}X) + c'' \text{Cos}(U*^{1/2}X) \tag{9.306}$$

From the boundary condition given in Eq. (9.304), $c''$ can be seen to be 1. From the boundary condition given in Eq. (9.305):

$$c' = -\text{Cot}(U*^{1/2}X_1) \tag{9.307}$$

The steady-state solution to the temperature is given by:

$$u^{ss} = \text{Cos}(U*^{1/2}X) - \text{Cot}(U*^{1/2}X_l)\,\text{Sin}(U*^{1/2}X) \tag{9.308}$$

$$u^{ss} = \text{Cos}\left(\sqrt{U*X_l^2}\right)\left(\frac{\text{Cos}\left(\sqrt{U*X^2}\right)}{\text{Cos}\left(\sqrt{U*X_l^2}\right)} - \frac{\text{Sin}\left(\sqrt{U*X^2}\right)}{\text{Sin}\left(\sqrt{U*X_l^2}\right)}\right) \tag{9.309}$$

### 9.10.3 Critical Point of Null Heat Transfer

It can be seen that at steady state, the temperature is periodic with respect to position. This is an interesting result. It can also be noted that the mathematical expression given in Eq. (2.310) can take on negative values. A negative temperature cannot exist, as according to the third law of thermodynamics, the lowest temperature attainable is 0 K. At 0 K, the entropy of any system would be zero. The interpretation of the model solution in terms of the wave conduction and relaxation is that after a certain location in the rod, the temperature will be zero. This can be referred to as the critical point of zero heat transfer. This is shown in Fig. 9.24. This was generated using Microsoft Excel on a 1.9-GHz Pentium IV personal computer. In Fig. 9.24, the heat source $U* = U'''/S$ is 0.5, the length of the rod is 10 cm, the thermal diffusivity is $10^{-5}$ m$^2$/s, and the relaxation time, $\tau_r$, is 15 seconds. For $X \geq$ 3.75, the temperature comes to the end temperature of ) K imposed on the right end of the rod. Beyond this region there is no heat transfer.

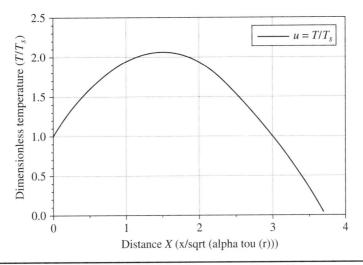

**FIGURE 9.24** Dimensionless temperature along a rod heated by a temperature-dependent heat source.

It can also be noted that the temperature within the rod exceeds the surface temperature. In this case, this is caused by the temperature-dependent heat source. The maximum in temperature occurs at $X = 1.5$. Further, since the temperature is higher within the rod, the heat flux direction will the reverse of what it was to begin with neat the zero time. Thus, the heat flow will be from the maximum location at $X = 1.5$ toward the $X = 0$ location, as well as from the $X = 1.5$ location to the $X = X_1$ location. To begin with, the problem was one where the initial temperature was at 0 K. The surface temperature at $X = 0$ was $T_s$. At short times, the heat flux would be from the $X = 0$ location toward the right side of the surface. This has been reversed by the time the system reaches steady state. The heat source term is contributes the energy.

The heat flux expression at steady state can be written as:

$$q* = -\frac{\partial u^{ss}}{\partial X} = \sqrt{U} * Cos\left(\sqrt{U} * X_I^2\right)\left(\frac{Sin\left(\sqrt{U} * X^2\right)}{Cos\left(\sqrt{U} * X_I^2\right)} - \frac{Cos\left(\sqrt{U} * X^2\right)}{Sin\left(\sqrt{U} * X_I^2\right)}\right)$$

(9.310)

where $q* = q/T_s\sqrt{k\rho C_p/\tau_r}$

It can be seen from Fig, 9.25 that several things happen when the steady-state heat flux is plotted as a function of the distance in the

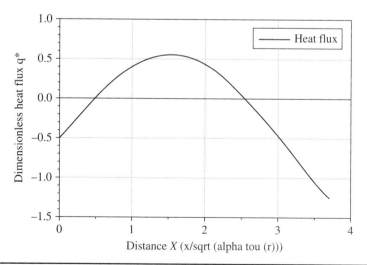

**FIGURE 9.25** Dimensionless heat flux along a rod heated by a temperature-dependent heat source.

rod, according to the solution given in Eq. (9.310). The following distinctions can be recognized from this illustration. There exists a location of maximum heat flux. In Fig. 9.25, for the assumed values of the length of the rod, the relaxation time, the ratio of heat source strength with the storage coefficient, and the location where the maximum heat flux occurs are at the dimensionless distance $X = 1.5$. There is a critical location in the rod beyond which there is no heat transfer. This is found to occur at $X = 3.75$ in Figs. 9.24 and 9.25. There are two locations in the rod where the heat flux changes direction. The cross-over locations occur at $X = 0.5$ and $X = 2.55$. These are locations of minimum heat flux. The transient portion of the solution will then be:

$$\frac{\partial^2 u^{\tau}}{\partial \tau^2} + (1 - U*)\frac{\partial u^{\tau}}{\partial \tau} = U * u^{\tau} + \frac{\partial^2 u^{\tau}}{\partial X^2} \tag{9.311}$$

The boundary conditions are:

$$X = 0, \ u^{\tau} = 0 \tag{9.312}$$

$$X = X_l, \ u^{\tau} = 0 \tag{9.313}$$

Equation (9.311) can be solved for by the method of separation of variables. First the damping term is removed by the substitution $u^{\tau} = W\exp(-\tau/2)$. Equation (9.311) becomes:

$$\left(1 - U*\right)\left(\frac{\partial W}{\partial \tau} - nW\right) + \left(\frac{\partial^2 W}{\partial \tau^2} + n^2 W - 2n\frac{\partial W}{\partial \tau}\right) = U * W + \frac{\partial^2 W}{\partial X^2} \tag{9.314}$$

$$\frac{\partial^2 W}{\partial \tau^2} + \frac{\partial W}{\partial \tau}(1 - U * -2n) + W(-n(1 - U*) + n^2 - U*) = \frac{\partial^2 W}{\partial X^2} \tag{9.315}$$

Letting $n = (1 - U*)/2$, Eq. (9.311) becomes:

$$\frac{\partial^2 W}{\partial \tau^2} - \frac{W}{4} = \frac{\partial^2 W}{\partial X^2} \tag{9.316}$$

Equation (9.316) can be solved by the method of separation of variables.

Let                         $w = V(\tau)\phi(X)$ \tag{9.317}

Equation (9.316) becomes:

$$V''/V - 1/4(1 + U*)^2 = \phi''/\phi = -\lambda_n^2 \tag{9.318}$$

The solution for the second-order ODEs can be written as follows:

$$\phi = c_1 \text{Sin}(\lambda_n X) + c_2 \text{Cos}(\lambda_n X) \qquad (9.319)$$

From the boundary condition given in Eq. (9.312), $c_2 = 0$. From the boundary condition given in Eq. (9.313):

$$c_1 \text{Sin}(\lambda_n X_1) = 0 \qquad (9.320)$$

$$\frac{\lambda_n l}{\sqrt{\alpha \tau_r}} = n\pi \qquad n = 0,1,2,3\ldots\ldots \qquad (9.321)$$

The time portion of the solution can be written as:

$$V = c_1 \exp\left(\frac{\tau}{2}\sqrt{(1+U*)^2 - 4\lambda_n^2}\right) + c_2 \exp\left(-\frac{\tau}{2}\sqrt{(1+U*)^2 - 4\lambda_n^2}\right)$$

$$(9.322)$$

The transient portion of the solution will decay out to leave the steady-state portion of the solution. The zero temperature at $x = l$ does the job of removing heat as it is generated in the rod. At infinite time $w = u^\tau \exp(\tau/2) = 0$ times infinity $= 0$. Thus, $w = V\phi$ at infinite time $= 0$. Therefore, $V = 0$ at steady state. Hence, the constant $c_1$ is zero in Eq. (9.322):

$$V = c_2 \exp\left(-\frac{\tau}{2}\sqrt{(1+U*)^2 - 4\lambda_n^2}\right) \qquad (9.323)$$

The general solution can be written as:

$$u^\tau = \sum_0^\infty c_n \exp\left(-\frac{\tau(1-U*)}{2}\right)\exp\left(-\frac{\tau}{2}\sqrt{(1+U*)^2 - 4\lambda_n^2}\right)\text{Sin}(\lambda_n X)$$

$$(9.324)$$

where $\lambda_n$ is given by Eq. (9.321). The $c_n$ can be solved for from the initial condition and is found to be:

$$c_n = \frac{2(1-(-1)^n)}{n\pi} \qquad (9.325)$$

The general solution for dimensionless temperature can be written as follows:

$$u = \text{Cos}\left(\sqrt{U * X_l^2}\right)\left(\frac{\text{Cos}\left(\sqrt{U * X^2}\right)}{\text{Cos}\left(\sqrt{U * X_l^2}\right)} - \frac{\text{Sin}\left(\sqrt{U * X^2}\right)}{\text{Sin}\left(\sqrt{U * X_l^2}\right)}\right)$$

$$+ \sum_0^\infty c_n \exp\left(-\frac{\tau(1-U*)}{2}\right)\exp\left(-\frac{\tau}{2}\sqrt{(1+U*)^2 - 4\lambda_n^2}\right)\text{Sin}(\lambda_n X)$$

$$(9.326)$$

where $\lambda_n$ is given by Eq. (9.321). It can be seen from the general solution that even for $n = 1$ in the infinite series:

$$\lambda_n > (1+U*)/2 \text{ or when } 1 < 2\pi\text{sqrt}(\alpha\tau_r)/(1 + U*) \qquad (9.327)$$

the temperature will undergo subcritical damped oscillations. This is for the cases when $U* < 1$. It can be seen that for $U* > 1$, the "damping term" will begin to grow in amplitude with time and cause a runaway reaction.

## Summary

Two important applications of bioheat transport in medicine are *thermal therapy* and *cryopreservation*. The word *energy* comes from the Greek words *en* meaning "in" and *ergon* meaning "work." The interplay of energy and information is a theme that is emphasized in the field of biological thermodynamics [2]. The efficiency at which biological energy transport occurs is the ratio of useful work, defined as the total work less the work done by the system, to the energy input for volume expansion. The zeroth law of thermodynamics, the first law of thermodynamics, the second law of thermodynamics, the third law of thermodynamics, and the fourth law of thermodynamics were reviewed. The perpetual motion machine of the second kind (PMM2) was discussed, including how it violates the second law of thermodynamics. Isobaric, isothermal, isentropic, and isochoric processes were analyzed. The three-step cycle, Carnot cycle, Otto cycle, and efficiencies achieved were reviewed. The law of conservation of energy was written.

Nanoscale effects in the time domain are important in a number of applications. Fourier's law of heat conduction, Fick's law of mass diffusion, Newton's law of viscosity, and Ohm's law of electricity are derived from empirical observations at steady state. There are six reasons to seek a generalized Fourier's law of heat conduction: i) The microscopic theory of reversibility of Onsager is violated; ii) Singularities were found in a number of important industrial applications of the transient representation of temperature, concentration, and velocity; iii) The development of Fourier's law was from observations at steady state; iv) An overprediction of theory to experiment has been found in a number of industrial applications; v) Landau and Lifshitz observed the contradiction of the infinite speed of propagation of heat with Einstein's light speed barrier; and vi) Fourier's law breaks down at the Casimir limit. The generalized Fourier's law of heat conduction is given by Eq. (9.983) and was postulated independently by Cattaneo and Vernotte.

Consider a semi-infinite medium at an initial temperature of $T_0$ subject to a constant surface temperature boundary condition for times greater than zero. The hyperbolic PDE that forms the governing equation of heat conduction is solved for by a new method called relativistic transformation of coordinates. The hyperbolic PDE is multiplied by $e^{\tau/2}$ and transformed into another PDE in wave temperature. This PDE is converted to an ODE by the transformation variable that is spatiotemporal and symmetric. The resulting ODE is seen to be a generalized Bessel differential equation. The solution with this approach is within 12 percent of the exact solution obtained by Baumeister and Hamill using the method of Laplace transforms. There are no singularities in the solution, but there are three regimes: an inertial regime, a regime characterized by a Bessel composite function of the zeroth order and first kind, and a regime characterized by a modified Bessel composite function of the zeroth order and first kind.

Expressions for penetration length and inertial lag time are developed. The comparison between the solution from the method of relativistic transformation of coordinates and the method of Laplace transforms was made by use of Chebyshev polynomial approximation and numerical integration. The dimensionless temperature as a function of dimensionless distance for the parabolic and hyperbolic models are shown in Fig. 9.8. The hyperbolic models were solved for by using the methods of relativistic transformation and method of Laplace transforms.

In a similar manner, the exact solution to the hyperbolic PDE is solved for by the method of relativistic transformation of coordinates for the infinite cylindrical and infinite spherical media.

When heating a finite slab, the Taitel paradox problem is revisited. Taitel found that when the hyperbolic PDE was solved for, the interior temperature in the slab was found to exceed the wall temperature of the slab. This is in violation of the second law of thermodynamics. By using the final condition in time at steady state, the wave temperature was found to be become zero at steady state. When mathematically posed as the fourth condition for the second-order PDE, this condition leads to well-bounded solutions within the bounds of the second law of thermodynamics. For systems with large relaxation times, that is, $\tau_r > a^2/\pi^2\alpha$, subcritical damped oscillations can be seen in the temperature. This is shown in Fig. 9.16. In a similar manner, the transient temperature for a finite sphere and finite cylinder are derived.

The heat generated within the human anatomy on account of the several metabolic reactions and the heat transfer to the surroundings can be described using the bioheat transfer equation. This was first introduced by Pennes [15]. The issues in body regulation of temperature were discussed. The thermophysical properties of biological properties and other materials were discussed. The bioheat transfer equation may be modified by the damped wave conduction and relaxation equation in order to account for the finite speed of propagation of heat.

# References

[1]   J. C. Bischof, "Micro and nanoscale phenomenon in bioheat transfer"(2006), *Heat Mass Transfer*, 42, 955–966.

[2]   D. T. Haynie, *Biological Thermodynamics*, 2d ed., Cambridge, UK: Cambridge University Press, 2008.

[3]   L. Onsager, "Reciprocal relations in reversible processes" (1931), *Phys. Rev.,* 37, 405–426.

[4]   L. Landau and E. M. Lifshitz, *Fluid Mechanics*, Oxford, UK: Pergamon, 1987.

[5]   K. R. Sharma, *Damped Wave Conduction and Relaxation*, Amsterdam, Netherlands: Elsevier.

[6]   D. D. Joseph and L. Preziosi, "Heat waves" (1989), *Reviews of Modern Physics,* 61(1), 41–73.

[7]   K. J. Baumeister and T. D. Hamill, "Hyperbolic heat conduction equation—a solution for the semi-infinite medium" (1971), *ASME J of Heat Transfer*, 93(1), 126–128.

[8]   K. R. Sharma, "Manifestation of acceleration during transient heat conduction" (2006), *Journal of Thermophysics and Heat Transfer*, 20(4), 799–808.

[9]   Y. Taitel, "On the parabolic, hyperbolic, and discrete formulation of the heat conduction equation" (1972), *Int. Journal of Heat and Mass Transfer*, 15(2), 369–371.

[10]   R. B. Bird, W. Stewart, and E. Lightfoot, *Transport Phenomena*, 2d ed., New York: John Wiley & Sons, 2002.

[11]   G. A. Truskey, F. Yuan, and D. F. Katz, *Transport Phenomena in Biological Systems*, 2d ed., Upper Saddle River, NJ: Pearson Prentice Hall, 2009.

[12]   K. R. Sharma, "Transient damped wave conduction and relaxation in human skin layer and thermal wear during winter," Jacksonville, FL, *ASME Summer Heat Transfer Conference*, August 10–14, 2008.

[13]   H. H. Pennes "Analysis of tissue and arterial blood temperature in the resting human forearm" (1998), *Journal Appl. Physiol.,* 85, 35–41.

[14]   C. K. Channy, *Mathematical Models of Bioheat Transfer*, Y. I. Cho (ed.), New York: Academic Press, 1992, pp. 157–358.

# Exercises

## Problems

**1.0**   *Maxwell's demon.* Consider two containers filled with gas at the same temperature T. When a molecule with a higher-than-average velocity in one container moves toward the wall separating the two containers, a gate-keeper demon opens the partition, grabs the molecule, and allows the molecule to reach the second container. On account of this, the average velocity of the remaining molecules in the first container would be lower and hence, the first container's temperature would have lowered from T. The molecules in the second container will have an average velocity higher than the initial velocity, and on account of which the temperature of the second container is expected to rise. Heat has transferred from container A to container B. Is this a violation of the second law of thermodynamics?

**2.0**   *Refrigerator and turbine.* A gentleman tried to do something with the heat discarded by a Carnot refrigerator. He wanted to use the heat as a hot temperature reservoir to do work and generate electricity. Can the electricity generated be used to power the refrigerator? Will this cycle last forever? If not, is this a PMM2 or PMM1?

**3.0**   *Waste heat recovery.* There are a number of discussions about waste heat recovery from steam power plants in the literature. What will happen to the

Carnot efficiency during waste heat recovery? Is this within the laws of thermodynamics?

**4.0** *Zeroth law of thermodynamics.* If two systems are in thermal equilibrium with a third system, then those two are in thermal equilibrium with each other. Prove this and substantiate with examples.

**5.0** *Novel fuel from boiling wood chips.* A lady raised hot water using a gas stove. Then she took the hot water and cooked wood chips into a novel fuel. This was a light-colored gas that emanated from the cooking utensils. She wanted to collect this gas and power up the stove. Can this cycle last forever? Is it a PMM1 or PMM2?

**6.0** *Seebeck effect.* Two bodies with hot and cold temperatures are brought in contact with other. Electricity is generated due to the Seebeck effect. Can a graduate student use this electricity and power a heater that can raise the temperature of a body to a hot temperature? Can this cycle last forever? Can he pull some electrical energy to power up his CD player? Why not? Is this a PMM1 or PMM2?

**7.0** *When the pressure of the ideal gas is cut in half, what happens to the* velocity of the gas molecules?

**8.0** *Brownian ratchet.* A gear referred to as ratchet allows for rotation in one direction, and a pawl prevents rotation in the other direction. The ratchet is connected to a paddle wheel immersed in a bath at temperature $T_A$. The molecules undergo Brownian motion. The molecular collisions with the paddle wheel result in a torque on the ratchet. Continuous motion of the ratchet may be expected. Work can be extracted with no heat gradient. Is this a PMM1 or PMM2?

**9.0** *Bhaskara's wheel.* Bhaskara (1114–1185) was a 12th-century mathematician and astronomer. He headed up the astronomical observatory at Ujjain. Several moving weights are attached to a wheel (Fig. 9.26). The weights fall to a position further from the center of the wheel after half a rotation. Since weights further from the center apply a greater torque, the wheel may be expected to rotate forever. Moving weights may be hammers on pivoted arms, rolling balls, mercury in tubes, etc. Is this a PMM1 or PMM2?

**10.0** *Self-flowing flask.* Robert Boyle suggested that the siphon action may be used to fill a flask by itself. Is this possible? Why not?

**11.0** *Orffyreus wheel.* In 1712, Bessler demonstrated a self-moving wheel that was later capable of lifting weights once set in motion. In 1717, he constructed a wheel 3.7 m in diameter and 14 in. thick. After two weeks, officials found the wheel moving at 2 RPM. Where does the energy for the motion come from?

**12.0** Distinguish between the wave and Fourier regimes.

**13.0** Examine the problem of heating an infinite medium with constant thermal diffusivity from a cylindrical surface with a radius R. Assume a dimensionless

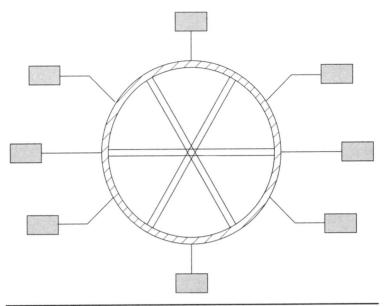

**FIGURE 9.26**    Bhaskara's unbalanced wheel.

heat flux at the wall as 1. Obtain the transient temperature using the parabolic Fourier equation. Is there a singularity in the solution expression?

**14.0**    Examine the problem of heating an infinite medium with constant thermal diffusivity from a spherical surface with a radius $R$. Assume a dimensionless heat flux at the wall as 1. Obtain the transient temperature using the parabolic Fourier equation. Is there a singularity in the solution expression?

**15.0**    Would Nernst's observation of thermal inertial and oscillatory discharge be a seventh reason for seeking a generalized Fourier's law of heat conduction?

**16.0**    Can the generalized Fourier's law of heat conduction be derived from kinetic theory of gases? If so, what is the physical significance of the ballistic term?

**17.0**    Can the generalized Fourier's law of heat conduction be derived from the Stokes-Einstein expression for diffusion coefficients? If so, what is the physical significance of the ballistic term?

**18.0**    Can the generalized Fourier's law of heat conduction be derived from the free electron theory? If so, what is the physical significance of the ballistic term?

**19.0**    What is the time taken to reach the steady state in the problem of heating a finite slab when solved for by the hyperbolic PDE? How is this different from the solution from the parabolic PDE?

**20.0**    How would the solution to Exercise 19.0 change when the boundary condition of the finite slab is changed from constant wall temperature to convective boundary condition?

**21.0**   Can the method of relativistic transformation of coordinates be used to solve for the transient temperature in a semi-infinite medium subject to constant wall temperature in Cartesian coordinates in three dimensions? If so, what would the transformation variable be?

**22.0**   *Can the method of relativistic transformation of coordinates be used to solve* for the transient temperature in an infinite medium subject to constant wall temperature in cylindrical coordinates in three dimensions? If so, what would the transformation variable be?

**23.0**   *Can the method of relativistic transformation of coordinates be used to solve* for the transient temperature in an infinite medium subject to constant wall temperature in spherical coordinates in three dimensions? If so, what would the transformation variable be?

**24.0**   What are the expressions for penetration distance and inertial lag times for the transient temperature in a semi-infinite medium subject to constant wall temperature in one-dimensional Cartesian coordinates?

**25.0**   What are the expressions for penetration distance and inertial lag times for the transient temperature in an infinite medium subject to constant wall temperature in one-dimensional cylindrical coordinates?

**26.0**   What are the expressions for penetration distance and inertial lag times for the transient temperature in an infinite medium subject to constant wall temperature in one-dimensional spherical coordinates?

**27.0**   What are the expressions for penetration distance and inertial lag times for the transient temperature in a semi-infinite medium subject to constant wall temperature in three-dimensional Cartesian coordinates?

**28.0**   What are the expressions for penetration distance and inertial lag times for the transient temperature in an infinite medium subject to constant wall temperature in three-dimensional cylindrical coordinates?

**29.0**   What are the expressions for penetration distance and inertial lag times for the transient temperature in an infinite medium subject to constant wall temperature in three-dimensional spherical coordinates?

**30.0**   What is the time taken to reach the steady state in the problem of heating a finite cylinder when solved for by the hyperbolic PDE? How is this different from the solution from the parabolic PDE?

**31.0**   What would the solution to Exercise 30.0 be should the boundary condition of the finite cylinder be changed from constant temperature to convective boundary condition?

**32.0**   What is the time taken to reach the steady state in the problem of heating a finite sphere when solved for by the hyperbolic PDE? How is this different from the solution from the parabolic PDE?

**33.0**   What would the solution to Exercise 32.0 be should the boundary condition of the finite sphere be changed from constant temperature to convective boundary condition?

**34.0**   In the analysis of transient temperature using the generalized Fourier's law of heat conduction in one-dimensional Cartesian coordinates in a semi-infinite medium, say that the temperature in the interior point $p$ is given as a function of time. Obtain the general solution for the transient temperature. What is the temperature at $X = 0$?

**35.0**   *Obtain the transient temperature in a right circular cone of infinite height* with a constant apex temperature for times greater than 0. What is the effect of the change in area with distance from the apex of the cone?

**36.0**   Consider a finite slab subject to the convective boundary condition. Using a space-averaged expression for temperature, obtain the governing equation for transient temperature for the slab using the generalized Fourier's law of heat conduction equation. The heat transfer coefficient is periodic in time and expressed as:

$$h = h_0 + h_A \text{Cos}(\omega t)$$

**37.0**   *Derive the transient temperature for the entire slab. Comment on the nature of the solution for materials with large relaxation times. Discuss the attenuation* and phase lag.

**38.0**   Consider the earth's crust heated by the sun. The initial temperature of the earth is at $T_0$ imposed by a periodic temperature at the crust by $T_0 + T_s\text{Cos}(\omega t)$.

**39.0**   By the method of Laplace transforms obtain the transient temperature in an infinite cylindrical medium subject to constant wall temperature for times greater than zero.

**40.0**   By the method of Laplace transforms obtain the transient temperature in an infinite spherical medium subject to constant wall temperature for times greater than zero.

**41.0**   How does the solution by the method of relativistic transformation compare with the solution obtained in Exercise 40.0?

**42.0**   How does the solution by the method of relativistic transformation compare with the solution in Exercise 39.0?

**43.0**   At what values of relaxation times of the materials above which subcritical damped oscillations in temperature would be expected for a finite cylinder subject to constant wall temperature?

**44.0**   At what values of relaxation times of the materials above which subcritical damped oscillations in temperature would be expected for a finite sphere subject to constant wall temperature?

**45.0**   Repeat Exercise 43.0 for the convective boundary condition.

**46.0**   Repeat Exercise 44.0 for the convective boundary condition.

**47.0**   What happens to the convex temperature profile obtained from the Fourier equation when a hyperbolic PDE is used? What is the physical significance of the change from concave to convex curvature in the transient temperature?

**48.0**   What happens at the wavefront in a semi-infinite medium in one-dimensional Cartesian coordinates subject to a constant wall temperature boundary condition?

**49.0**   What happens at the wavefront in a semi-infinite medium in one-dimensional Cartesian coordinates subject to a constant wall flux boundary condition?

**50.0**   What happens at the wavefront in a semi-infinite medium in three-dimensional Cartesian coordinates subject to a constant wall temperature boundary condition?

**51.0**   What happens at the wavefront in an infinite medium in three-dimensional cylindrical coordinates subject to a constant wall temperature boundary condition?

**52.0**   *What happens at the wavefront in an infinite medium in three-dimensional spherical* coordinates subject to a constant wall temperature boundary condition?

# APPENDIX A

# Generalized Bessel Differential Equation

The linear second-order differential equation shown in Eq. (A.1) is referred to as Bessel's equation, and the solutions are termed "Bessel" functions. The solutions for this equation are available in Watson [1]. Certain types of differential equations are amenable to a solution expressed as a power series. Such a series is said to converge if it approaches a finite value as $n$ approaches infinity. The simplest test for convergence is the ratio test within the interval of convergence. The method of Frobenius is a convenient method to obtain a power series solution to a linear, homogeneous, second-order differential equation with variable convergent coefficients:

$$x^2 \frac{d^2 y}{dx^2} + x \frac{dy}{dx} + (x^2 - p^2)y = 0 \qquad (A.1)$$

Equation (A.1), when expressed in the standard form, can be written as:

$$\frac{d^2 y}{dx^2} + \frac{1}{x} \frac{dy}{dx} + \frac{x^2 - p^2}{x^2} y = 0 \qquad (A.2)$$

The second-order, homogeneous, general ordinary differential equation can be expressed in the standard form as shown in Varma and Morbidelli [2]:

$$R(x) \frac{d^2 y}{dx^2} + \frac{P(x)}{x} \frac{dy}{dx} + \frac{V(x)}{x^2} y = 0 \qquad (A.3)$$

Comparing Eq. (A.2) with the standard form:

$$R(x) = 1; P(x) = 1; V(x) = x^2 - p^2 \qquad \text{(A.4)}$$

The functions $R(x)$, $P(x)$, and $V(x)$ need be expanded as a power series and the coefficients of the power series calculated as:

$$R_0 = 1; R_1 = R_2 = R_3 \ldots\ldots R_n = 0 \qquad \text{(A.5)}$$

$$P_0 = 1; P_1 = P_2 = P_3 \ldots\ldots P_n = 0 \qquad \text{(A.6)}$$

$$V_0 = -p^2; V_2 = 1; \qquad \text{(A.7)}$$

$$V_1 = V_3 = V_4 = \ldots\ldots = V_n \qquad \text{(A.8)}$$

The solution to Eq. (A.3) by the method of Frobenius (Mickley, Sherwood and Reed [3]) states that there is at least one solution of the following form:

$$y = x^s \sum_0^\infty A_n x^n \qquad \text{(A.9)}$$

Substituting Eq. (A.9) into Eq. (A.3), the indicial equation obtained can be written as:

$$s^2 + (P_0 - 1)s + V_0 = 0 \text{ or } s^2 - p^2 = 0 \qquad \text{(A.10)}$$

$$s_1 = p; s_2 = -p \qquad \text{(A.11)}$$

The recurrence relation for $A_n$ in Eq. (A.9) can be seen to be:

$$A_n = \frac{-\displaystyle\sum_{k=1}^n g_k(s+n)A_{n-k}}{f(s+n)} \qquad \text{(A.12)}$$

The two solutions for the two roots in Eq. (A.11) are:

$$y_1(x) = A_0 x^p \left[ 1 + \sum_{k=1}^\infty \frac{(-1)^k x^{2k}}{(1+p)(2-p)\ldots(k-p)2^{2k}k!} \right] \qquad \text{(A.13)}$$

$$y_2(x) = B_0 x^{-p} \left[ 1 + \sum_{k=1}^\infty \frac{(-1)^k x^{2k}}{(1+p)(2-p)\ldots(k-p)2^{2k}k!} \right] \qquad \text{(A.14)}$$

Equations (A.13) and (A.14) can be expressed in a more useful form by making use of the gamma function. The gamma function can be defined in the Euler form as:

$$\Gamma(p) = \int_0^\infty e^{-x} x^{p-1} dx$$

$$p > 0 \qquad \text{(A.15)}$$

Some mentionable properties of the gamma function are:

$$\Gamma(p+1) = p\Gamma(p) = , p > 0$$
$$\Gamma(p + k) = (p+ k - 1)(p + k - 2)...(p+1)(p)\Gamma(p)$$

If $p$ is a positive integer:

$$\Gamma(n + 1) = n! \tag{A.16}$$

The gamma function generalizes the use of a factorial to noninteger positive values of $p$. Thus:

$$\Gamma(9/2) = (7/2)(5/2)(3/2)(1/2)\ \Gamma(1/2)$$

$$\Gamma\left(\frac{1}{2}\right) = \int_0^\infty e^{-p} p^{-\frac{1}{2}} dp = \sqrt{\pi} \tag{A.17}$$

The definition can be extended to negative noninteger values, but not for zero and negative numbers. For large values of the argument, the Stirling approximation may be used:

$$\Gamma(p) \sim \sqrt{\frac{2\pi}{p}} \left(\frac{p}{e}\right)^p \tag{A.18}$$

With the use of the gamma function, Eq (A.13) becomes:

$$y_1(x) = A_0 \left[ \sum_{k=0}^\infty \frac{(-1)^k \left(\dfrac{x}{2}\right)^{2k+p}}{k!\Gamma(k + p + 1)} \right] \tag{A.19}$$

Using the notation for the Bessel function of the first kind and $p$th order:

$$J_p(x) = \left[ \sum_{k=0}^\infty \frac{(-1)^k \left(\dfrac{x}{2}\right)^{2k+p}}{k!(k + p)!} \right] \tag{A.20}$$

Equation (A.19) becomes:

$$y_1(x) = C_1 J_p(x) \tag{A.21}$$

In a similar vein, Eq. (A.14), when $p$ is neither zero nor a positive integer, can be written as:

$$y_2(x) = C_2 J_{-p}(x) \tag{A.22}$$

where

$$J_p(x) = \left[ \sum_{k=0}^{\infty} \frac{(-1)^k \left(\dfrac{x}{2}\right)^{2k-p}}{k!(k-p)!} \right] \tag{A.23}$$

When $p$ is zero or a positive integer, it can be shown that:

$$y_2(x) = C_2 \, Y_p(x) \tag{A.24}$$

where $Y_p(x)$ is the Bessel function of the second kind and $p$th order. The Weber form can be written as:

$$Y_p(x) = \frac{2}{\pi} \left[ \left( \gamma + ln\left(\frac{x}{2}\right) J_p(x) + \frac{1}{2} \sum_{k=0}^{\infty} (-1)^{k+1} \big( \phi(k) + \phi(k+p) \big) \frac{\left(\dfrac{x}{2}\right)^{2k+p}}{k!(k+p)!} \right) \right] \tag{A.25}$$

where $\gamma$ is Euler's constant:

$$\gamma = 0.577215\ldots$$

In this case, the roots of the indicial equation are both equal to zero, and the second linearly independent form can be written as a Bessel function multiplied with a logarithmic function and a second infinite power series. Thus, the complete solution of the Bessel equation when $p$ is a positive integer or zero can be written as:

$$y = C_1 J_p(x) + C_2 \, Y_p(x) \tag{A.26}$$

When $p$ is neither an integer nor zero:

$$y = C_1 J_p(x) + C_2 J_{-p}(x) \tag{A.27}$$

The linear second-order ordinary differential equation given in Eq. (A.28) can be transformed into the Bessel equation given in Eq. (A.1) by a substitution: $z = ix$

$$x^2 \frac{d^2 y}{dx^2} + x \frac{dy}{dx} - (x^2 + p^2)y = 0 \tag{A.28}$$

$$z^2 \frac{d^2 y}{dz^2} + z \frac{dy}{dz} + (z^2 - p^2)y = 0 \tag{A.29}$$

When $p$ is a positive integer or zero, the solution can be written as:

$$y = C_1 J_p(ix) + C_2 Y_p(ix)$$

or

$$y = C_1 I_p(x) + C_2 K_p(x) \tag{A.30}$$

When $p$ is neither an integer nor zero:

$$y = C_1 J_p(ix) + C_2 J_{-p}(ix) \tag{A.31}$$

or

$$y = C_1 I_p(x) + C_2 I_{-p}(x) \tag{A.32}$$

$I_p(x)$ is referred to as the modified Bessel function of the first kind and $p$th order, and is defined by the expression:

$$I_p(x) = \left[ \sum_{k=0}^{\infty} \frac{\left(\dfrac{x}{2}\right)^{2k+p}}{k!(k+p)!} \right] \tag{A.33}$$

$K_p(x)$ is referred to as the modified Bessel function of the second kind and $p$th order, and is defined by the expression:

$$K_p(x) = \pi/2 \; i^{p+1} \left( J_p(ix) + i Y_p(ix) \right) \tag{A.34}$$

The generalized form of Bessel's equation can be written as:

$$x^2 \frac{d^2 y}{dx^2} + x(a + 2bx^r)\frac{dy}{dx} + (c + dx^{2s} - b(1 - a - r)x^r + b^2 x^{2r})y = 0 \tag{A.35}$$

Equation (A.35) can be reduced to Eq. (A.1) after suitable transformations (Mickley, Sherwood, and Reed [3]). The generalized solution for Eq. (A.35) may be written as:

$$Y = x^{\frac{1-a}{2}} e^{-\frac{bx^r}{r}} \left[ c_1 Z_p \left( \frac{|d|^{1/2}}{s} x^s \right) + c_2 Z_{-p} \left( \frac{|d|^{1/2}}{s} x^s \right) \right] \tag{A.36}$$

where

$$p = \frac{1}{s} \sqrt{\left( \frac{1-a}{2} \right)^2 - c} \tag{A.37}$$

$p$ is the order of the Bessel equation. The different forms the Bessel solution assumes depend on the nature of sqrt($|d|$)/s and $p$, as

| S.No | $\|d\|^{1/2}/s$ | $p$ | $Z_p$ | $Z_{-p}$ |
|------|-----------------|-----|-------|----------|
| 1. | Real | Neither zero nor integer | $J_p$ | $J_{-p}$ |
| 2. | Real | Either zero or integer | $J_p$ | $Y_p$ |
| 3. | Imaginary | Neither zero nor integer | $I_p$ | $I_{-p}$ |
| 4. | Imaginary | Either zero or integer | $I_p$ | $K_p$ |

**TABLE A.1**  Forms of Bessel Solution

given in Table A.1. For small values of $x$, the following approximations can be made for the Bessel functions:

$$J_p(x) \sim \frac{1}{2^p p!} x^p \tag{A.38}$$

$$J_{-p}(x) \sim \frac{2^p}{(-p)!} x^{-p} \tag{A.39}$$

$$Y_p(x) \sim \frac{2^p(p-1)!}{\pi} x^{-p}, \qquad p \neq 0 \tag{A.40}$$

$$Y_0(x) \sim \frac{2}{\pi} ln(x) \tag{A.41}$$

$$I_p(x) \sim \frac{1}{2^p p!} x^p \tag{A.42}$$

$$I_{-p}(x) \sim \frac{2^p}{(-p)!} x^{-p} \tag{A.43}$$

$$K_n(x) \sim 2^{n-1}(n-1)! \, x^{-n} \tag{A.44}$$

$$K_0(x) \sim -ln(x) \tag{A.45}$$

For large values, the general character may be obtained by the following substitution:

$$y = x^{-1/2} u \tag{A.46}$$

Equation (A.1) then becomes:

$$x^2 \frac{d^2y}{dx^2} = \frac{3y}{4} - 2x^{\frac{1}{2}} \frac{dy}{dx} + x^{\frac{3}{2}} \frac{d^2u}{dx^2} = 0 \tag{A.47}$$

$$x \frac{dy}{dx} = -\frac{y}{2} + x^{\frac{1}{2}} \frac{du}{dx} \tag{A.48}$$

$$\frac{d^2u}{dx^2} + u\left(1 - \frac{1}{x^2}\left(p^2 - \frac{1}{4}\right)\right) = 0 \qquad (A.49)$$

For large values of $x$, it can be shown that:

$$J_p(x) \sim \sqrt{\frac{2}{\pi x}} Cos\left(x - \frac{\pi p}{2} - \frac{\pi}{4}\right) \qquad (A.50)$$

$$Y_p(x) \sim \sqrt{\frac{2}{\pi x}} Sin\left(x - \frac{\pi p}{2} - \frac{\pi}{4}\right) \qquad (A.51)$$

In a similar vein, the modified Bessel function can be approximated as:

$$I_p(x) \sim \frac{e^x}{\sqrt{2\pi x}} \qquad (A.52)$$

$$K_p(x) \sim \sqrt{\frac{\pi}{2x}} e^{-x} \qquad (A.53)$$

The first zero of the Bessel function of the first kind occurs for the zeroth order $J_0(x)$ at 2.4048, for the first order $J_1(x)$ at 3.8317, for the second order $J_2(x)$ at 5.1356, for the third order $J_3(x)$ at 6.3802, and for the fourth order $J_4(x)$ at 7.5883. The zeros of the Bessel function of the second kind occur for the zeroth order $Y_0(x)$ at 0.8936, for the first order $Y_1(x)$ at 2.1971, for the second order $Y_2(x)$ at 3.3842, for the third order $Y_3(x)$ at 4.5270, and for the fourth order $Y_4(x)$ at 5.6451. Both $J_p(x)$ and $Y_n(x)$ oscillate like damped sinusoidal functions and approach zero as $x$ tends to infinity. The amplitude of the oscillations about zero decreases as $x$ increases, and the distance between successive zeros of both functions decreases toward a limit of $\pi$ as $x$ increases. The zeros of $J_{p+1}(x)$ separate the zeros of $J_p(x)$. $I_p(x)$, in contrast, increases continuously with $x$, and $K_n$ decreases continuously. Bessel functions of order equal to half an odd integer can be represented in terms of the elementary functions:

$$J_{1/2}(x) = \sqrt{\frac{2}{\pi x}} Sin(x) \qquad (A.54)$$

$$J_{-1/2}(x) = \sqrt{\frac{2}{\pi x}} Cos(x) \qquad (A.55)$$

$$I_{1/2}(x) = \sqrt{\frac{2}{\pi x}} \mathrm{Sin} h(x) \tag{A.56}$$

$$I_{-1/2}(x) = \sqrt{\frac{2}{\pi x}} \mathrm{Cos} h(x) \tag{A.57}$$

The recurrence relations among Bessel functions can be given by:

$$J_{n+1/2}(x) = \frac{2n-1}{x} J_{n-1/2}(x) - J_{n-3/2}(x) \tag{A.58}$$

$$I_{n+1/2}(x) = -\frac{2n-1}{x} I_{n-1/2}(x) + I_{n-3/2}(x) \tag{A.59}$$

The following relations can be proved using Eqs. (A.23) and (A.33):

$$\frac{d}{dx}\left(x^p Z_p(\alpha x)\right) = \alpha x^p Z_{p-1}(\alpha x), \, Z = J, Y, I \tag{A.60}$$

$$\frac{d}{dx}\left(x^p Z_p(\alpha x)\right) = -\alpha x^p K_{p-1}(\alpha x) \tag{A.61}$$

$$\frac{d}{dx}\left(x^{-p} Z_p(\alpha x)\right) = -\alpha x^p Z_{p+1}(\alpha x), \, Z = J, Y, K \tag{A.62}$$

$$\frac{d}{dx}\left(x^{-p} I_p(\alpha x)\right) = \alpha x^{-p} I_{p+1}(\alpha x), \tag{A.63}$$

$$\frac{d}{dx}\left(Z_p(\alpha x)\right) = \alpha Z_{p-1}(\alpha x) - p/x Z_p(\alpha x), \, Z = J, Y, I \tag{A.64}$$

$$\frac{d}{dx}\left(K_p(\alpha x)\right) = -\alpha x^p K_{p-1}(\alpha x) - p/x \, K_p(\alpha x) \tag{A.65}$$

$$\frac{d}{dx}\left(Z_p(\alpha x)\right) = -\alpha Z_{p+1}(\alpha x) + p/x Z_p(\alpha x), \, Z = J, Y, K \tag{A.66}$$

$$\frac{d}{dx}\left(I_p(\alpha x)\right) = \alpha x^p I_{p+1}(\alpha x) + p/x \, I_p(\alpha x) \tag{A.67}$$

$$2\frac{d}{dx} I_p(\alpha x) = \alpha(I_{p-1}(\alpha x) + I_{p+1}(\alpha x)) \tag{A.68}$$

$$2\frac{d}{dx} K_n(\alpha x) = -\alpha(K_{n-1}(\alpha x) + K_{n+1}(\alpha x)) \tag{A.69}$$

$$Z_p(\alpha x) = \frac{\alpha x}{2p} \left(Z_{p+1}(\alpha x) + Z_{p-1}(\alpha x)\right), \ Z = J, Y \qquad \text{(A.70)}$$

$$I_p(\alpha x) = -\frac{\alpha x}{2p} \left(I_{p+1}(\alpha x) - I_{p-1}(\alpha x)\right) \qquad \text{(A.71)}$$

$$K_n(\alpha x) = \frac{\alpha x}{2p} \left(K_{n+1}(\alpha x) - K_{n-1}(\alpha x)\right) \qquad \text{(A.72)}$$

When $n$ is zero or an integer:

$$J_{-n}(\alpha x) = (-1)^n J_n(\alpha x) \qquad \text{(A.73)}$$

$$I_{-n}(\alpha x) = I_n(\alpha x) \qquad \text{(A.74)}$$

$$K_{-n}(\alpha x) = K_n(\alpha x) \qquad \text{(A.75)}$$

The Bessel function $J_n(x)$ and $I_n(x)$ for various orders are plotted in Figs. A.1 and A.2.

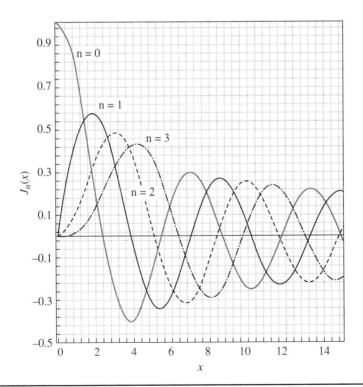

**FIGURE A.1** Bessel function of the first kind and $p$th order ($p = 0,1,2,3,4,...$).

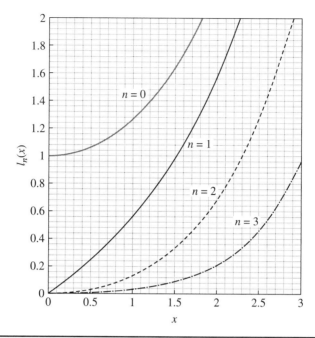

**FIGURE A.2** Modified Bessel function of the first kind (order $p = 0,1,2,3,...$).

## References

[1] G. N. Watson, *A Treatise on the Theory of Bessel Functions*, 2nd ed., Cambridge, UK: Cambridge University Press, 1966.

[2] A. Varma and M. Morbidelli, *Mathematical Methods in Chemical Engineering*, Oxford, UK: Oxford University Press, 1997.

[3] H. S. Mickley, T. S. Sherwood, and C. E. Reed, *Applied Mathematics in Chemical Engineering*, New York: McGraw-Hill, 1957.

# Inverse of Laplace Transforms

| S.No. | Transform | Function f(t) |
|-------|-----------|---------------|
| 1. | $\dfrac{\Gamma(k)}{s^k}$, $k > 0$ | $t^{k-1}$ |
| 2. | $\dfrac{1}{s^{n+\frac{1}{2}}}$, $n = 0,1,2,3....$ | $\dfrac{2^n t^{n-\frac{1}{2}}}{(1.3.5....(2n-1)\sqrt{\pi}}$ |
| 3. | $\dfrac{1}{(s-a)^n}$, $n = 1,2,3.....$ | $\dfrac{e^{at}t^{n-1}}{(n-1)!}$ |
| 4. | $\dfrac{1}{s^2+a^2}$ | $\dfrac{Sin(at)}{a}$ |
| 5. | $\dfrac{s}{s^2+a^2}$ | $Cos(at)$ |
| 6. | $\dfrac{1}{s^2-a^2}$ | $\dfrac{Sinh(t)}{a}$ |
| 7. | $\dfrac{s}{s^2-a^2}$ | $\dfrac{Cosh(t)}{a}$ |
| 8. | $s/(s^4+4a^4)$ | $Sin(at)Sinh(at)/(2a^2)$ |
| 9. | $1/s\,(s-1)^n/s^n$ | $Exp(t)/n!\ d^n/dt^n(t^n exp(-t))$ Laquerre polynomial of degree $n$ |
| 10. | $(s-a)^{1/2}-(s-b)^{1/2}$ | $\frac{1}{2}(exp(bt)-exp(at))/(\pi t^3)$ |
| 11. | $1/(s^{1/2}+a)$ | $(\pi t)^{-1}-aexp(a^2 t)erfc(at^{1/2})$ |
| 12. | $1/(s+a)/(s+b)^{1/2}$ | $1(b-a)^{1/2} exp(-at)$ $erf(t(b-a))^{1/2}$ |
| 13. | $(1-s)^n/s^{n+1/2}$ | $(\pi t)^{-1/2}\,n!/(2n)!\ H_{2n}(t)$ Hermite polynomial $H_n(t) = exp(-t^2)d^n/dt^n\,exp(-t^2)$ |

*(Continued)*

| S.No. | Transform | Function f(t) |
|---|---|---|
| 14. | $(s + a)^{-1/2}(s + b)^{-1/2}$ | $\exp(-(a + b)t/2)I_0((a - b)t/2)$ |
| 15. | $(s^2 + a^2)^{-1/2}$ | $J_0(at)$ |
| 16. | $(s^2 - a^2)^{-k}, k > 0$ | $(\pi)^{1/2}/\Gamma(k)\,(t/2a)^{k-1/2}\,I_{k-1/2}(at)$ |
| 17. | $\exp(-sk)/s$ | $S_k(t) = 0, 0 < t < k = 1$, $t > k$, Heaviside unit step function |
| 18. | $\exp(-sk)/s^j, j > 0$ | $0, 0 < t < k$ $(t - k)^{j-1}/\Gamma(j), t > k$ |
| 19. | $(1 - \exp(-sk))/s$ | 1 when $0 < t < k$ 0 when $t > k$ |
| 20. | $1/2s + \coth(sk/2)/2s$ | $S(k, t) = n$, when $(n - 1)$ $k < t < nk$ $n = 1,2,3...$ |
| 21. | $1/[s\,(\exp(sk) - a)]$ | 0, when $0 < t < k$ $1 + a + a^2 +... + a^{n-1}$, when $nk < t < (n + 1)k$, $n = 1,2,3...$ |
| 22. | $1/s\,\tanh(sk)$ | $M(2k, t) = (-1)^{n-1}$ when $2k(n - 1) < t < 2kn$ |
| 23. | $1/[s(1 + \exp(-sk))]$ | $1/2M(k, t) + \tfrac{1}{2} =$ $(1 - (-1)^n)/2$ when $(n - 1)k < t < nk$ |
| 24. | $1/s^2\,\tanh(sk)$ | $H(2k, t) = t$ when $0 < t < 2k =$ $4k - t$ when $2k < t < 4k$ |
| 25. | $1/(s\mathrm{Sinh}(sk))$ | $2S(2k, t + k) - 2 = 2(n - 1)$ when $(2n - 3)k < t$ $< (2n - 1)k, t > 0$ |
| 26. | $1/(s\mathrm{Cosh}(sk))$ | $M(2k, t + 3k) + 1 = 1 + (-1)^n$ when $(2n - 3)k < t$ $< (2n - 1)k, t > 0$ |
| 27. | $\mathrm{Coth}(sk)/s$ | $2S(2k, t) - 1 = 2n - 1$ when $2k(n - 1) < t < 2kn$ |
| 28. | $k/(s^2 + k^2)\,\mathrm{Coth}(\pi s)/2k$ | $\|\sin(kt)\|$ |
| 29. | $1/[(s^2 + 1)(1 - \exp(-\pi s)]$ | $\mathrm{Sin}\,t$ when $(2n = 2)\pi < t$ $< (2n - 1)\pi$ 0 when $(2n - 1)\pi < t < 2n\pi$ |
| 30. | $1/s\,\exp(-k/s)$ | $J_0(2(kt)^{1/2})$ |
| 31. | $\exp(-k/s)/s^{1/2}$ | $\mathrm{Cos}(2(kt)^{1/2})/(\pi t)^{1/2}$ |
| 32. | $\exp(k/s)/s^{1/2}$ | $\mathrm{Cosh}(2(kt)^{1/2})/(\pi t)^{1/2}$ |

*(Continued)*

| S.No. | Transform | Function f(t) |
|---|---|---|
| 33. | $\exp(-k/s)/s^{3/2}$ | $\operatorname{Sin}(2(kt)^{1/2})/(\pi k)^{1/2}$ |
| 34. | $\exp(k/s)/s^{3/2}$ | $\operatorname{Sinh}(2(kt)^{1/2})/(\pi k)^{1/2}$ |
| 35. | $1/s^j \exp(-k/s), j > 0$ | $(t/k)^{(j-1)/2} J_{j-1}(2(kt)^{1/2})$ |
| 36. | $1/s^j \exp(k/s), j > 0$ | $(t/k)^{(j-1)/2} I_{j-1}(2(kt)^{1/2})$ |
| 37. | $\exp(-k(s)^{1/2}), k > 0$ | $k/[2(\pi t^3)^{1/2}] \exp(-k^2/4t)$ |
| 38. | $1/s \exp(-ks^{1/2}), k > 0$ | $\operatorname{erfc}(k/2t^{1/2})$ |
| 39. | $1/s^{1/2} \exp(-ks^{1/2}), k > 0$ | $1/(\pi t)^{1/2}\exp(-k^2/4t)$ |
| 40. | $1/s^{3/2} \exp(-ks^{1/2}), k > 0$ | $2(t/\pi)^{1/2} \exp(-k^2/4t) - \operatorname{kerfc}(k/2t^{1/2})$ |
| 41. | $a\exp(-s^{1/2}k)/(s(a + s^{1/2})),$ $k > 0$ | $(-\exp(ak)\exp(a^2t)\operatorname{erfc}(at^{1/2} + k/2t^{1/2}) + \operatorname{erfc}(k/2t^{1/2})$ |
| 42. | $\exp(-s^{1/2}k)/s^{1/2}/(a + s^{1/2})$ | $\operatorname{Exp}(ak)\exp(a^2t)\operatorname{erfc}(at^{1/2} + k/2t^{1/2})$ |
| 43. | $\exp(-k(s(s + a))^{1/2})/$ $(s(s+1))^{1/2}$ | $0, 0 < t < k \exp(-at/2)I_0$ $(a/2 (t^2 - k^2)^{1/2})$ |
| 44. | $\exp(-k(s^2 + a^2)^{1/2})/$ $(s^2 + a^2)^{1/2}$ | $0$, when $0 < t < k$ $J_0 (a(t^2 - k^2)^{1/2}$, when $t > k$ |
| 45. | $\exp(-k(s^2 - a^2)^{1/2}/(s^2 - a^2)^{1/2}$ | $0$, when $0 < t < k$ $I_0 (a(t^2 - k^2)^{1/2}$, when $t > k$ |
| 46. | $\exp(-k(s^2 + a^2)^{1/2} - s)/$ $(s^2 + a^2)^{1/2}$ | $J_0(a(t^2 + 2kt)^{1/2})$ |
| 47. | $\exp(-sk) - \exp(-k(s^2 + a^2)^{1/2}$ | $0$, when $0 < t < k$ $ak/(t^2 - k^2) J_1(a(t^2 - k^2))^{1/2}$ |
| 48. | $\exp(-k(s^2 - a^2)) - \exp(-sk)$ | $0$, when $0 < t < k$ $ak/(t^2 - k^2)^{1/2} I_1(a(t^2 - k^2)^{1/2}),$ $t > k$ |
| 49. | $a^j\exp(-k(s^2 + a^2)^{1/2})/$ $(s^2 + a^2)^{1/2}/(s^2 + a^2)^{1/2} + s)^j,$ $j > -1$ | $0$, when $0 < t < k$ $[(t - k)/(t + k)]^{1/2j}$ $J_j(a(t^2 - k^2)^{1/2}), t > k$ |
| 50. | $1/s \ln s$ | $\lambda - \operatorname{Int}, \lambda = -.5772$ |
| 51. | $1/s^k \ln s$ | $t^{k-1} (\lambda/\Gamma(k)^2 - \operatorname{Int}/\Gamma(k))$ |
| 52. | $\ln s/(s - a)$ | $\exp(at) (\ln a - E_i (-at))$ |
| 53. | $\ln s/(s^2 + 1)$ | $\operatorname{Cos} t \operatorname{Si}(t) - \operatorname{sin} t \operatorname{Ci}(t)$ |
| 54. | $s\ln s/(s^2 + 1)$ | $-\operatorname{sin} t \operatorname{Si}(t) - \operatorname{cos} t \operatorname{Ci}(t)$ |
| 55. | $1/s \ln(1 + sk)$ | $-Ei (-t/k)$ |
| 56. | $\ln[(s - a)/(s - b)]$ | $1/t (\exp(bt) - \exp(at))$ |
| 57. | $1/s \ln(1 + k^2s^2)$ | $-2Ci(t/k)$ |

*(Continued)*

| S.No. | Transform | Function f(t) |
|---|---|---|
| 58. | $1/s \ln(s^2 + a^2)$, $a > 0$ | $2\ln a - 2\,Ci(at)$ |
| 59. | $1/s^2 \ln(s^2 + a^2)$, $a > 0$ | $2/a(at\ln a + \operatorname{Sin}at - atCi(at))$ |
| 60. | $\ln(s^2 + a^2)/s^2$ | $2/t(1 - \operatorname{Cos}(at))$ |
| 61. | $\ln(s^2 - a^2)/s^2$ | $2/t(1 - \operatorname{Cos}h(at))$ |
| 62. | $\tan^{-1}(k/s)$ | $1/t \operatorname{Sin}kt$ |
| 63. | $1/s \tan^{-1}k/s$ | $Si(kt)$ |
| 64. | $\exp(s^2k^2)\operatorname{erfc}(sk)$, $k > 0$ | $1/(k\pi^{1/2}) \exp(-t^2/4k^2)$ |
| 65. | $1/s \exp(s^2k^2)\operatorname{erfc}(sk)$, $k > 0$ | $\operatorname{erf}(t/2k)$ |
| 66. | $\exp(sk) \operatorname{erfc}(sk)^{1/2}$, $k > 0$ | $k^{1/2}/[\pi(t)^{1/2}(t+ k)]$ |
| 67. | $1/s^{1/2} \operatorname{erfc}(sk)^{1/2}$ | $0$, $0 < t < k$ $(\pi t)^{-1/2}$, $t > k$ |
| 68. | $1/s^{1/2} \exp(sk)$ $\operatorname{erfc}(sk)^{1/2}$, k > 0 | $(\pi(t + k))^{-1/2}$ |
| 69. | $\operatorname{erf}(k/s^{1/2})$ | $1/\pi t \operatorname{Sin}(kt^{1/2})$ |
| 70. | $1/s^{1/2} \exp(k^2/s) \operatorname{erfc}(k/s^{1/2})$ | $1/(\pi t)^{1/2} \exp(-2kt^{1/2})$ |
| 71. | $K_0(sk)$ | $0$, $0 < t < k$ $(t^2 - k^2)^{-1/2}$, when $t > k$ |
| 72. | $K_0(ks^{1/2})$ | $1/2t \exp(-k^2/4t)$ |
| 73. | $1/s \exp(sk) K_1(sk)$ | $1/k (t(t + 2k))^{1/2}$ |
| 74. | $1/s^{1/2} K_1(ks^{1/2})$ | $1/k \exp(-k^2/4t)$ |
| 75. | $1/s^{1/2} \exp(k/s) K_0(k/s)$ | $2/(\pi t)^{1/2} K_0(2kt)^{1/2}$ |
| 76. | $\pi \exp(-sk) I_0(sk)$ | $(t (2k - t))^{-1/2}$, $0 < t < 2k$ $0$, $t > 2k$ |
| 78. | $-(\gamma + \ln s)/s$, $\gamma =$ Euler's constant $= 0.5772156$ | $\ln t$ |
| 79. | $1/as^2 \tanh(as/2)$ | Triangular wave function |
| 80. | $1/s \tanh(as/2)$ | Square wave function |
| 81. | $\pi a/(a^2s^2 + \pi^2) \coth(as/2)$ | Rectified sine wave function |
| 82. | $\pi a/[(a^2s^2 + \pi^2)(1 - \exp(-as)]$ | Half-rectified sine wave function |
| 83. | $1/as^2 - \exp(-as)/$ $s(1 - \exp(-as))$ | Saw tooth wave function |
| 84. | $\operatorname{Sin}hsx/(s\operatorname{Sin}hsa)$ | $x/a + 2/\pi\sum_1^\infty (-1)^n/n$ $\operatorname{Sin}(n\pi x)/a \operatorname{Cos}(n\pi t/a)$ |
| 85. | $\operatorname{Sin}hsx/(s\operatorname{Cos}hsa)$ | $4/\pi\sum_1^\infty (-1)^n/(2n - 1)$ $\operatorname{Sin}(2n - 1\pi x)/2a$ $\operatorname{Sin}(2n - 1)\pi t/2a)$ |

(*Continued*)

| S.No. | Transform | Function f(t) |
|-------|-----------|---------------|
| 86. | Cos$hsx$/$s$Sin$hsa$ | $t/a + 2/\pi \sum_1^\infty (-1)^n/n$ Cos$(n\pi x)/a$ Sin$(n\pi t/a)$ |
| 87. | Cos$hsx$/$s$Cos$hsa$ | $1 + 4/\pi \sum_1^\infty (-1)^n/(2n-1)$ Cos$(2n-1)\pi x)/2a$ Cos$(2n-1)\pi t/2a$ |
| 88. | Sin$hsx$/$s^2$ Cos$hsa$ | $x + 8a/\pi^2 \sum_1^\infty (-1)^n/$ $(2n-1)^2$Sin$(2n-1)\pi x)/2a$ Cos$(2n-1)\pi t/2a$ |
| 89. | Cos$hsx$/$s^2$Sin$hsa$ | $t^2/2a + 2a/\pi^2 \sum_1^\infty (-1)^n/(n^2)$ Cos$(n\pi x)/a$ $(1 - $Cos$(n\pi t/a)$ |
| 90 | Sin$hxs^{1/2}$/Sin$has^{1/2}$ | $2\pi/a^2 \sum_1^\infty (-1)^n n$ exp$(-n^2\pi^2 t)/$ $a^2$ Sin$(n\pi x/a)$ |
| 91. | Cos$hxs^{1/2}$/$s^{1/2}$Sin$has^{1/2}$ | $1/a + 2/a \sum_1^\infty (-1)^n n$ exp$(-n^2\pi^2 t)/a^2$ Cos$(n\pi x/a)$ |

# Index